Advances in World Aquaculture, Volume 3
Managing Editor, Paul A. Sandifer

Aquaculture and Water Quality

Edited by

David E. Brune
and
Joseph R. Tomasso

Clemson University
and the
South Carolina Agricultural Experiment Station
Clemson, South Carolina USA 29634

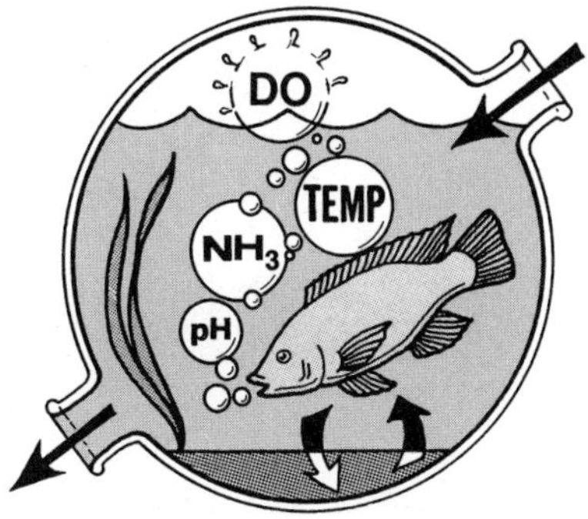

Published by

THE WORLD AQUACULTURE SOCIETY

The World Aquaculture Society
143 J.M. Parker Coliseum
Louisiana State University
Baton Rouge, LA 70803

Library of Congress Catalog Number: 90-050540

ISBN: 0-9624529-1-2

PREFACE

Aquacultural production is a significant and rapidly-growing segment of total worldwide food production. In order for aquaculture to continue to grow and provide both staple and specialty foods to a growing population, new and improved production technology must continue to become available. The optimal management of water quality in aquaculture systems is one aspect of production technology that is becoming critical as many user groups compete for limited supplies of public water. This volume is the proceedings of a symposium organized with the goal of bringing specialists in water quality and aquaculture together to present and discuss ideas relating to the optimal use of water in aquaculture.

The "First International Symposium on Water Quality and Aquaculture" was held in Los Angeles, California, at the Westin Bonaventure Hotel on February 14-15, 1989. This site and time were chosen to encourage the participation of individuals from seven other groups which were simultaneously meeting in the same location as a part of the "Aquaculture 89" conference.

The editors wish to express their appreciation to several groups and individuals for their support in this endeavor. Funds to assist the authors in traveling to the meeting were provided by the South Carolina Sea Grant Consortium and member groups of the South Carolina Aquaculture, Fisheries and Wildlife Cooperative (South Carolina Wildlife and Marine Resources Department, Clemson University/South Carolina Agricultural Experiment Station). All of these groups have been conducting cooperative research activities within the State for several years. The symposium steering committee consisted of John Manzi (chairman), Rick DeVoe, Arnie Eversole, Phil Kirk and Billy McTeer.

TABLE OF CONTENTS

SIMULATION AND MANIPULATION OF WATER QUALITY IN POND CULTURE

IMPACTS OF AQUACULTURE PRODUCTION ON WATER QUALITY AND RESOURCES

CONTRIBUTORS

Anat Alkon
Fish and Aquaculture Research Station
Dor, Hof HaCarmel 30-820
Israel

M. C. M. Beveridge
Institute of Aquaculture
University of Stirling
Stirling, Scotland FK9 4LA
UK

Claude E. Boyd
Department of Fisheries and Allied Aquacultures
Alabama Agricultural Experiment Station
Auburn University, Alabama 36849
USA

David E. Brune
Department of Agricultural and Biological Engineering
Clemson University
Clemson, South Carolina 29634
USA

James D. Bryan
Department of Wildlife and Fisheries Sciences
Texas A&M University System
College Station, Texas 77843-2258
USA

R. M. Clarke
Institute of Aquaculture
University of Stirling
Stirling, Scotland FK9 4LA
UK

John A. Colman
U.S. Geological Survey
10 Causeway St.
Room 926
Boston, Massachusettes 02222-1040
USA

John Colt
Suite 365
14150 NE 20th Street
Bellevue, Washington 98007
USA

Charles R. Gibbs
2904 Bent Drive
Loveland, Colorado 80538
USA

John Hochheimer
Agriculture Engineering Department
University of Maryland
College Park, Maryland 20742
USA

Andrew R. Jacobson
U.S. Geological Survey
102 E. Main Street
Urbana, Illinois 71801
USA

G. Ellen Kaiser
Agriculture Engineering Department
University of Maryland
College Park, Maryland 20742
USA

Malik Laher
Fish and Aquaculture Research Station
Dor, Hof HaCarmel 30-820
Israel

Hiram W. Li
Oregon Cooperative Fishery Research Unit
U.S. Fish and Wildlife Service
Oregon State University
Corvallis, Oregon 97331
USA

Thomas M. Losordo
Departments of Zoology and Biological
and Agricultural Engineering
North Carolina State University
Campus Box 7617
Raleigh, North Carolina 27695-7617
USA

John F. Martin
College of Veterinary Medicine
Mississippi State University
Mississippi State, Mississippi 39762
USA

Ronald D. Mayo
c/o James M. Montgomery
Consulting Engineers
2375 130th Avenue N.E.
Bellevue, Washington 98005
USA

William H. Neill
Department of Wildlife and Fisheries Sciences
Texas A&M University System
College Station, Texas 77843-2258
USA

Christina Orwicz
Suite 365
14150 NE 20th St,
Bellvue, Washington 98007
USA

M. J. Phillips
Institute of Aquaculture
University of Stirling
Stirling, Scotland FK9 4LA
UK

Raul H. Piedrahita
Department of Agricultural Engineering/
Aquaculture and Fisheries Program
University of California-Davis
Davis, California 95616
USA

David Randall
Department of Zoology
University of British Columbia
6270 University Boulevard
Vancouver, B.C. V6T 2A9
Canada

Rosemarie C. Russo
Environmental Research Laboratory
U.S. Environmental Protection Agency
Athens, Georgia 30613
USA

Carl B. Schreck
Oregon Cooperative Fishery Research Unit
U.S. Fish and Wildlife Service
Oregon State University
Corvallis, Oregon 97331
USA

Gerald L. Schroeder
Fish and Aquaculture Research Station
Dor, Hof HaCarmel 30-820
Israel

Daniel Wilkins Smith
BCM Engineers
One Plymouth Meeting
Pltmouth Meeting, Pennsylvania 19462
USA

Robert R. Stickney
School of Fisheries
University of Washington
Seattle, Washington 98195
USA

Robert V. Thurston
Fisheries Bioassay Laboratory
Montana State University
Bozeman, Montana 59717
USA

Joseph R. Tomasso, Jr.
Department of Aquaculture, Fisheries and Wildlife
Clemson University
Clemson, South Carolina 29634
USA

Craig S. Tucker
Mississippi Agricultural and Forestry
Experiment Station
Delta Branch Experiment Station
Stoneville, Mississippi 38776
USA

Donald P. Weston
Horn Point Environmental Laboratory
University of Maryland
P.O. Box 775
Cambridge, Maryland 21613
USA

Fred Wheaton
Agriculture Engineering Department
University of Maryland
College Park, Maryland 20742
USA

AQUACULTURAL WATER QUALITY: THE EMERGENCE OF AN APPLIED DISCIPLINE

J. R. Tomasso and D.E. Brune

ABSTRACT

The importance of aquaculture production in providing high protein foods and enhancing economies worldwide is increasing. A more complete understanding of the interrelationships between aquatic productivity and water quality is essential to ensure the continued growth of the aquaculture industry. The development of this knowledge base will require the talents of a diverse group of engineers and scientists.

In an effort to facilitate such a cooperative effort, this Symposium on Water Quality and Aquaculture was designed to bring together a group of researchers capable of addressing an issue of immediate importance to aquaculture. These presentations provide insight into: a) the biological interaction of aquatic organisms with important water quality parameters, b) the physical requirements for, and limiting constraints to, providing optimal water quality for aquatic organisms, and c) the possible environmental impacts of creating or manipulating aquatic environments for the purpose of sustained aquacultural production.

INTRODUCTION

Aquaculture of various kinds has been practiced throughout the world for centuries. During most of this time, aquatic organisms have been raised for consumption either by the culturist or were sold in local markets. Only in the past few decades has an aquaculture industry developed, with large farms producing products that can be channeled into modern processing and marketing systems. Accurate worldwide aquaculture production statistics are difficult to obtain; however, the United Nations Food and Agriculture Organization (FAO) has been developing a reporting system since 1976.

Recently, FAO (as summarized by Nash 1988) reported that total world aquaculture production in 1985 was 10,587,300 tons. Finfish represented 44.5% of the 1985 production while crustaceans, mollusks and seaweeds represented 2.5%, 26.5% and 26.2%, respectively. China was the leading country with respect to aquacultural production accounting for nearly half of the 1985 total. Ponds and tanks were the most popular types of production systems with nearly 41% of total production coming from these systems. Some 186 "species" were reportedly cultured throughout the world in 1985, although inaccurate identification of species in many cases made it difficult to develop an accurate total. Carps (25 species) were the most widely cultured group in 1985 (2.5 million tons). Worldwide aquaculture production in 1985 was about 3.7% greater than in 1983. The continued need for high protein foods in underdeveloped countries and the trend toward healthier diets (meats containing less fats) in developed countries provide a strong basis for continued growth in aquaculture production at or above the present rate.

Modern aquaculture production systems, like other agricultural production systems, are composites of contributions from many diverse technical specialties within the fields of biological science and engineering. Within the biological arena, we find the specialties of nutrition, fish health, environmental physiology, genetics, reproductive biology, aquatic ecology, and product quality assurance. Engineering specialties include system construction (ponds, cages, etc.), water delivery systems, aeration systems, water filtration, water quality dynamics, and design of various specialized equipment.

The study of water quality as it relates to aquaculture production is of paramount importance if aquaculturists are to successfully and efficiently produce a commodity. Water quality researchers focus on three general questions. First, what are the environmental requirements of a particular species? Second, how can these requirements be met and maintained in an economically viable manner? Third, what is the environmental impact of large scale aquaculture operations? To begin answering these questions, researchers have drawn from almost all of the biology and engineering specialties mentioned above.

EXPECTATIONS AND LIMITATIONS

Because of the similarities shared between conventional agriculture systems and aquaculture production systems, many individuals have assumed that aquaculture production is capable of immediate scale-up parallelling the success of conventional animal or plant agriculture. An often heard statement is that aquaculture researchers will "do with aquatic organisms what the poultry people did with the chicken when they took it out of the yard and put it into a cage." However, there are a number of important differences between terrestrial and aquatic farming that must be understood and dealt with before this prediction can be realized.

In 1945 Samuel Brody published a summary and analysis of an impressive body of knowledge on the growth and metabolism of domestic livestock. Brody's quantitative approach, perhaps more than any other single factor, laid the groundwork for the agricultural scientists and engineers to later design and develop modern animal production systems.

The development of efficient terrestrial animal production systems has depended in large part on a precise understanding of the environmental requirements of the species being reared. In contrast the existing data base for aquatic organisms is meager. Furthermore, existing data are based on essentially "wild" stocks of aquatic organisms. Brody's work was based on animals which through centuries of selection and hybridization were well suited for agricultural production.

The nature of a water environment versus that of a gaseous environment also must be considered when attempting to apply knowledge of terrestrial farming to that of aquaculture. Water is one of the best natural solvents on earth. Hence, aquatic animals are exposed to more dissolved substances (both advantageous and toxic) than terrestrial organisms. Less oxygen is present in a given volume of water than in an equal volume of air requiring a greater rate of ventilation for aquatic organisms. This higher rate of ventilation coupled with more dissolved substances in the water provides the basis for a greater potential exposure of the organisms to toxic substances.

There are a number of major limitations which the aquaculturist faces when attempting to manipulate or modify the surroundings of aquatic organisms. Probably the most severe is the simple fact that water is heavy. The prohibitive economics of pumping water require that aquaculture systems are designed for maximum efficiency and maximum utilization of locally available elevation head or other locally available energy sources to achieve desired culture conditions.

The movement of nutrients and toxic metabolites through water is extremely slow when compared to an air environment. The aquacultural engineer will need to carefully apply the concepts of fluid mechanics to ensure high productivity. Through innovative designs, many such basic challenges may result in certain advantages; for example, assuming that systems can be developed to provide uniform water distribution, aquatic animal confinement systems have the advantage of full utilization of a three dimensional space such as proposed by Van Olst, et al. (1979) for high density lobster culture.

The low "buffer" capacity of high density aquatic animal systems poses additional limitations. For example, the low solubility of oxygen in water necessitates the continuous input of oxygen to high density fish maintenance systems, as well as the availability of a back-up system in the event of failure of the primary system.

Any individual attempting to design systems for culture of either marine or freshwater organisms will need to fully appreciate and incorporate local site and species specific characteristics into the design. The failure to adequately match local water availability or quality to species requirements has led to the failure of many culture schemes.

WATER QUALITY IMPACTS ON AQUATIC ORGANISMS

To maximize aquacultural production, it will be necessary to develop a quantitative understanding of the relationship between the organism being cultured and the system in which it is being maintained. As examples, nutritionists must learn what, how much, and when to feed a species during different periods in the life of the animal and during the different seasons; geneticists must discover and/or develop strains of organisms that are optimal for culturing; and health specialists must develop prevention and treatment regimens to limit losses due to disease.

A good example of the difficulty in determining optimal conditions for organisms may be found in the study of the toxicity of nitrite to fishes. Nitrite is an intermediate of the bacterial oxidation of ammonia to nitrate. Early studies established causal relationships between environmental nitrite and mortality, reduced growth and methemoglobinemia. However, as more information was gathered, it became apparent that these relationships within a species were confounded by other water quality characteristics, nutrition and genetics (see reviews in Russo, 1985, Lewis and Morris, 1986). Multispecies testing demonstrated a very large range of species susceptible to nitrite poisoning. While we have learned a great deal in the past decade about nitrite toxicity to fishes, we still lack a comprehensive, quantitative model of its toxicity to a species under culture conditions.

Obviously, the successful culture of any aquatic organism must begin with a comprehensive understanding of the specific requirements to sustain life. This was our starting point in the symposium.

CULTURE SYSTEMS

If the environmental requirements of a particular organism are known, then the task confronting the culturist is to continuously maintain the aquatic environment within the required limits. Economic constraints dictate that these systems be operated at the limits of design capacity. Four major approaches to this challenge have been attempted (Table 1).

Static Pond Culture

The first approach to be developed was static pond culture. In this system, capital and operating investment is kept at a minimum and biomass carrying capacity is accordingly low. The cultured animals may be free to

Table 1. The nature and environmental impact considerations of the major types of aquaculture systems.

Nature of System	Environmental Impact Considerations
STATIC PONDS	
Pond designed specifically for aquaculture. Oxygenation and nitrification primarily occur by natural lentic processes. Animals may be caged or free ranging.	Large volumes of water are required for initially filling and later maintaining water levels. Effluent may have an impact on receiving waters.
RECIRCULATING SYSTEMS	
Water flows through a fish holding system (pond, raceway, or tank) into a filter, and then back into the holding system.	Limited impact.
NET-PEN CULTURE IN PUBLIC WATERS	
Animals are cultured in net-pens located in large bodies of water (lakes, bays, fjords). Water quality is maintained by natural processes.	Primary impact considerations are the effects of fecal material and uneaten feed on local benthos and water quality. Aesthetics may be a consideration in highly populated areas.
SYSTEMS USING DIVERTED LOTIC WATER	
Animals are cultured in ponds, raceways or tanks provided with a constant supply of water diverted from a nearby lotic system. Effluent is returned to lotic system.	Requires constant diversion of large quantities of surface water. Quality of effluent is usually less than that of receiving waters.

move within the pond or confined in cages and net pens. Acceptable water quality is maintained in static pond systems by complex physio-chemical and biotic interactions which resemble those of natural lentic ecosystems. The major challenge in stillwater pond culture comes in the form of a fluctuating environment. The nature of lentic systems is such that water quality characteristics, such as dissolved oxygen and pH, will fluctuate over a 24 hour period. The extremes of these fluctuations are biologically based, but, to a large degree, are ultimately dependent on energy input in the form of wind action on the surface of the water and sunlight. Consequently, the maximum carrying capacity is limited by the worst case in the range of physical and biological cycling taking place within the system. The pressures from competitive commodity markets require that these systems are operated near their limits. The technical aspects of predicting where the limits lie are extremely complex. Successful high density pond culture requires the ability to make detailed predictions concerning changes in pond water quality variables (pH, DO, etc.) in response to minor changes in pond inputs (feeding rate, sunlight, wind speed). A variety of researchers have invested considerable effort in an attempt to expand this capability. The current status of these efforts is discussed in the section on simulation and manipulation of the aquatic environment.

Water Reuse

A second approach to the challenge of providing sustained life support for aquatic organisms is that of recirculating or water reuse culture. This requires high capital outlay and high operating costs. The productivity per unit of contained volume must therefore be maintained at correspondingly high levels. The major advantage of this system lies in the ability to maintain a constant optimal environment for the organism of interest. Technically, it is not difficult to develop a system to completely meet the environmental needs of a particular aquatic organism. The challenge is to build and operate an economically viable system. As technology improves and as land and water resources become less available for aquaculture, or as demand for specialty products grows, water reuse may become economically feasible on a large scale. A variety of workers have focused research on development of recirculating systems for aquaculture. The state of the art of this aspect of aquaculture was addressed by looking specifically at two of the more critical water quality requirements in recirculating systems, namely, gas exchange and nitrogen control.

Net-Pen Culture in Public Waters and Culture Systems Using Diverted Lotic Waters

A rapidly growing culture system in many parts of the world is net-pen culture in large bodies of public waters. These waters may be freshwater as in the case of lakes or saltwater as in the case of bays, sounds and fjords. Net-pen culture systems utilize natural processes for the most part to provide high quality water for the culture animals. Occasionally, mechanical circulators and aerators are also used. Systems which use diverted lotic water typically pass the diverted water through ponds, raceways or tanks where the cultured animals are held and then immediately discharge the water back into the lotic system.

Both of these systems may be described as "water borrowing" systems in that both rely on public natural systems to provide high quality water on a continuous basis for the culture animals. Hence, the major water quality concerns related to these systems are not usually production oriented but, rather, oriented toward real and perceived environmental impacts.

ENVIRONMENTAL IMPACT

Assuming that the previously discussed limitations can be met and an optimal environment can be continuously provided for the organisms of interest, one final over-riding challenge remains. Aquaculture must be conducted within the constraints of acceptable environmental impact. This constraint presents a number of complicated facets.

The resolution of this issue will require a prioritization of conflicting demands on surface and groundwater, as well as, decisions as to the relative economic value of aquatic production as opposed to the value of the embodied water required for such production. These questions must be dealt with by society through its elected officials. We, as scientists and engineers, can only strive to provide a quantitative understanding of the effects of a given level of aquatic production upon water quality. This final issue is addressed in Section Four - Impacts of Aquaculture Production.

A COMMON GOAL

In 1987 catfish production in the state of Mississippi generated approximately 225 million dollars of farm income. This type of success establishes the economic incentive required to focus aquaculture research. Much of the technology to support recirculating aquaculture has come from research applied to domestic and industrial wastewater treatment. However, only limited effort has been expended in adapting such technologies to the low levels of contaminants present in aquaculture systems or in developing low cost, modular water treatment units needed for aquaculture systems. In addition, environmental scientists have spent much effort in defining lethal limits of exposure of aquatic organisms to various toxic substances, and in understanding in a general way, changes in ecosystem diversity as a result of a particular impact. This type of information is of limited value with respect to establishing optimal criteria for culturing animals. Similarly, biologists have invested considerable effort in understanding single factor dietary and environmental impacts on aquatic organisms, but have only begun to focus efforts on understanding the more complex multi-factor interactions of diet, disease and environment, this latter type of work is essential to ensuring successful long term culture of an aquatic organism.

It has been said that nothing stimulates success like success. The potential of aquaculture is clearly established. What remains is for the diverse group of aquaculture researchers to expand the goal-oriented type of research that formed the basis of Brody's work on domestic livestock to aquatic organisms.

The determination of optimal environmental criteria of aquaculture species and the design of efficient and economical systems to meet these criteria, within the constraints of an acceptable level of environmental impact, are the goals of researchers working in the field of aquacultural water quality. Specifically, these researchers should strive to build a knowledge base to allow for the development of a number of quantitative models which predict (1) the optimal combination of environmental conditions for individual culture species, (3) the performance of individual culture system components, and (3) the response of the environment at different levels of controlled aquatic cultivation.

LITERATURE CITED

Brody, S. 1945. Bioenergetics and Growth. Reinhold Publishing Corporation, New York, USA.

Lewis, W.M. and D.P. Morris. 1986. Toxicity of Nitrite to Fish: A Review. Transactions of the American Fisheries Society 115: 153-195.

Nash, C.E. 1988. A Global Overview of Aquaculture Production. Journal of the World Aquaculture Society 19: 51-58.

Russo, R.C. 1985. Ammonia, nitrite, nitrate. Pages 455- 474 *in* G.M. Rand and S.R. Petrocelli, editors, Fundamentals of Aquatic Toxicology. Hemisphere Publishing Company, Washington, USA.

Van Olst, J.C., J.M. Carlberg and J.T.Hughes. 1980. Aquaculture. Pages 333-384 *in* J.S. Cobb and B.F. Phillips (editors) The Biology and Management of Lobsters (Volume II). Academic Press, New York, USA.

PERFORMANCE CAPACITY OF FISH: STRESS AND WATER QUALITY

Carl B. Schreck and Hiram W. Li

ABSTRACT

The capacity of aquatic organisms to perform necessary life functions (e.g.; grow, resist disease, reproduce, tolerate thermal extremes) is ultimately prescribed by their genetic constitution. As such, each individual can be thought of as possessing a potential performance capacity, and the nature of the variation in individual performance capacities describes the potential performance capacity of each population or stock. The ambient environment delineates the expression of the organisms' genotypes and hence establishes the actual, realized performance capacity for each trait of interest. Because fish are in such intimate association with their environment, water quality factors are of extreme importance in setting the limits of performance traits of concern to aquaculturists. In essence, any suboptimal environmental factor that is toxicological, causes a stress response, or otherwise results in an energetic load will reduce the net capacity of the cultured organism to express its genetically established potential of performing activities important to fish culturists.

INTRODUCTION

Fish being reared under an aquaculture program are expected to meet certain production criteria, many of which are associated with survival, growth, reproduction, and flesh quality. Of course, hatchery goals differ widely, depending on species being reared, intensity of the fish husbandry, and particular market for the final product. It is important to recognize that aquacultural systems are ecosystems; it is simple to become too narrowly focused on technical matters that rely on the paradigm of terrestrial agriculture. Populations or communities in artificial rearing conditions are under selective forces and evolve. Selection may be intentional, as in selective breeding programs, or unintentional, due to either inadvertent spawning practices or the fish farm environment (Tave 1986). The difference between cultured populations and those in the wild is that man is the dominant force in the hatchery.

Man operates aquacultural systems to maximize some production criterion and generally does this by imposing some degree of stability on the system (e.g.; provides food, controls water quality, controls diseases, etc.). Natural systems can frequently be much more variable. In essence, the adaptive strategy that man most often forces on cultured populations is the Maximin Strategy as discussed by Lewontin (1961) for Game Theory; that is, maximizing fitness over a minimum number of environmental states. This is accomplished by expenditure of large amounts of energy by man to provide a steady state system that removes some of the costs of homeostasis by the animal (Knights 1985), thereby hopefully maximizing the capacity of the fish to perform necessary life functions for aquacultural objectives.

The capacity of fish to perform necessary life functions and meet production goals is ultimately prescribed by their genotypes; however, the environment provides both the conditions necessary for expression of the genotype (Williams 1963) and delimits genotypic expression. In other words, the variability in the phenotype of a population is equal to the variability between the genotypes of the individuals in the population, plus the variability in the environment, plus some interactive effect between the genotypes and the environment (Falconer 1960; Kapuscinski and Jacobson 1987). Based on this genetic concept, the scope for activity concept of Fry (1947), and the niche concept of Hutchinson (1957), a geometric metaphor has been proposed that describes the performance capacity of fish (Schreck 1981). Basically, the genotype determines the maximum ability of an individual or a population to perform all of the individual activities required (potential performance capacity), and the environment determines how much of that potential, if any, will actually be expressed (realized performance capacity). The potential performance capacity can vary with ontogeny, whereas the realized performance capacity is also influenced by the animal's prior history. Our goals in this paper are to present a philosophical framework for this symposium, premised on the performance capacity paradigm (Figure 1) and to suggest that any fish cultural condition such as water quality when less than optimal will reduce performance.

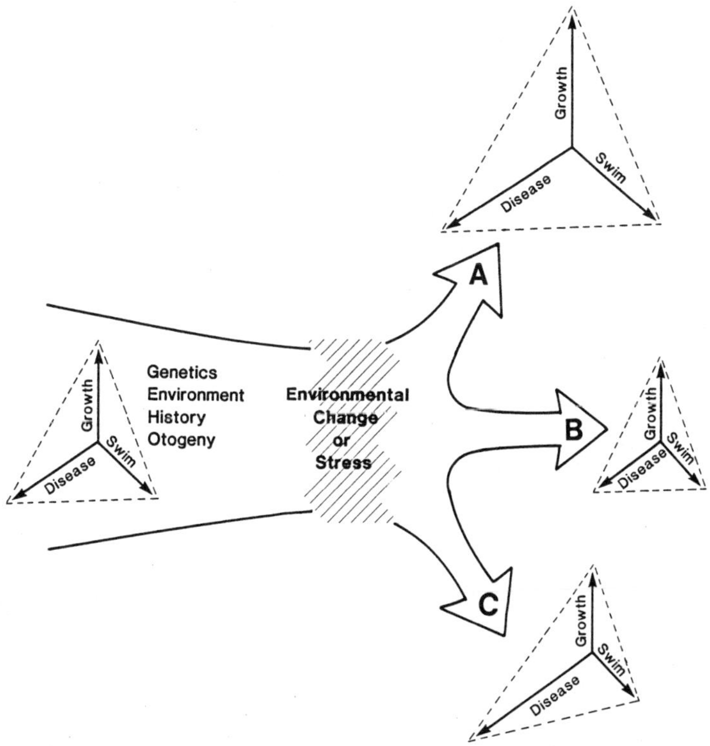

FIGURE 1. - Three performance phenotypes (disease resistance, growth and swimming ability among many possibilities) of an individual or stock, depicted as vectors on a trajectory through time with magnitudes set by the genotype, environmental and rearing history, development, and present environment, can serve to illustrate the realized performance capacity concept. If the individual or stock encounters an environmental change or stressful situation, then each performance vector may respond by increasing or decreasing in magnitude, creating a new realized performance capacity, depending on the individual's or stock's original trajectory and the nature, magnitude and duration of the new environmental conditions. If, for example, environmental conditions become more optimal, then all realized performances may become larger as illustrated in "A". On the other hand, if environmental conditions change for the worse, then all individual performances may become impaired to some degree as in "B". It is also possible that an environmental change could differentially affect individual performances, as in "C" where growth was unaffected, disease resistance was heightened, and swimming ability was depressed.

ARGUMENT

Potential Performance Capacity

From an aquacultural perspective, performance traits such as growth, disease resistance, reproductive timing, fecundity, flesh quality, coloration, and the ability to cope with stresses are determined by the interaction of the genetic make-up of the stock and their rearing conditions. It appears that the genetic component in determining these traits in fish is about 50% on the average, although such estimates of heritability have extremely large errors associated with them (Tave 1986). This implies that the environment is responsible for determining these phenotypes as is genetics. Therefore, fish husbandry conditions and practices are approximately equally as important as genetics in maximizing the realized performance capacity of the fish.

Fish either adapt to hatchery conditions, or the hatchery conditions or practices must be adapted to the needs of the fish. One consequence of natural and artificial selection is that traits other than those responding directly to selective forces are selected for or against as well because of genetic linkage (Tave 1986). For example, it was found that when coho salmon, *Oncorhynchus kisutch*, were selectively bred to improve their resistance to a bacterial pathogen, they were less tolerant of other oceanic conditions they encountered following release from the hatchery (McIntyre and Johnson 1977; Suzumoto et al. 1977). It is important to bear in mind that certain individual potential performances covary with other specific performance traits during evolution because of genetic linkage. Aulstad et al. (1972) discussed the genetic and environmental sources of variation in traits such as length and weight in rainbow trout *O. mykiss*. With regard to water quality, Hirshfield et al. (1980) reviewed genetic differences in the physiological tolerances between populations of pupfish, *Cyprinodon nevadensis*, while Rahel (1981) discussed the response by fish to selection for resistance to a toxicant. As hatchery populations evolve by whatever selective pressure, their potential performance capacities change, but it is difficult to predict the entire scope or nature of this change because we do not as yet understand the relationships and linkages of each individual performance vector to one another.

Realized Performance Capacity

It is well known that water quality is an important consideration in aquaculture. It is also well documented that each water quality factor not only affects fish directly but also interacts with other water quality factors, resulting in a combined effect on performance. One can conceptualize the importance of water quality to fish performance if one considers that performance is energy demanding (Warren 1971), and these energetic principles are directly applicable to fish farming (Knights 1985). The energetic cost of each performance trait is dependent on the quality of the conditions under which it is to be performed. If water quality is good, then it can be viewed as simply being permissive; that is, it does not become a limiting factor and does not to any significant degree constrain the magnitude of genetic expression. Under good water quality conditions the relationship between phenotypic expression (i.e., the realized capacity to perform) and the water quality variable are usually not linear; this can be readily visualized by considering temperature and its effects according to the Q10 phenomenon. The energy costs necessary to perform may follow similar patterns. Alternatively, when water quality conditions are poor and particularly when they are life threatening, the animal must invest a very substantial amount of energy in persisting or simply resisting death; hence, there is usually not much energy to spare for conducting other aquaculturally important performances. Further, under deteriorated water quality conditions physiological responses may or may not be graded (i.e., continuous), depending on the severity of the conditions. Rather, they may have a response threshold; for example, at temperature extremes enzyme activity may be a presence or absence phenomenon, not a matter of degree (Ready 1967).

Water quality and other hatchery factors network to produce the nature of the environment in which the genotype must be expressed. Warren (1971) conceptualized the interaction of water quality factors as part of the bioenergetic equation, and Vernberg (1978) provided an overview of the effects of multiple environmental factors in aquatic systems. This can be visualized by considering how the scope for growth can be affected by the need to channel energy into other physiological processes (Priede 1985). The capacity to conduct physiological functions changes ontogenetically, and this may in part be explainable by the energetic costs associated with development. For example, undergoing metamorphosis or parr-smolt transformation are energetic loads (Schreck 1982), and reproductive maturation is a particularly costly developmental process (Diana and MacKay 1979; Bell 1980; Hirshfield 1980; Guderley et al. 1986). Hence, during major developmental episodes there will likely be less energy available to parcel out to other performances, be they growth or tolerance of deteriorated water quality.

There may also be shifts in environmental optima for carrying on various physiological activities. For example, growth (Averett 1969), swimming speed (Griffiths and Alderdice 1972) in coho salmon and metabolism in bluegill, *Lepomis machrochirus* (Wohlschlag and Juliano 1959) all had different temperature optima at different seasons. Additionally, energy allocation for various activities are often in conflict; for example, fishes with low metabolic scope can either swim or digest food (Priede 1985).

Other hatchery practices, such as those involving handling or physical disturbance of the fish, are energy loads that reduce the net amount of energy available for the fish to channel into activities desirable by the aquaculturist (Schreck 1982). Stressful situations to which the fish are subjected can thus be viewed as reducing further the realized performance capacity of the fish (Schreck 1981); however, the magnitude of this reduction is dependent on the quality of all other environmental factors as well. The nature of the performance capacity also differs over time following an acute stressful experience. For example, the ability to generate an immune response and the capacity to resist a bacterial pathogen were found to oscillate below and above resting values several times in chinook salmon, *O. tshawytscha*, before true recovery was achieved (Maule et al. 1989).

In summary, because fish are in such intimate association with their environment, water quality factors are of extreme importance in setting the limits of realization of performance traits of interest to aquaculturists. In essence, any environmental factor that is less than optimal, that is, toxicological, causing a stress response, or otherwise resulting in an energetic load will result in a lowered ability of the cultured organism to express its genetic potential at performing activities of concern to aquaculturists.

ACKNOWLEDGEMENT

We thank B. E. Coblentz for his words of wisdom regarding this manuscript and T. B. Kahuna for inspiration.

LITERATURE CITED

Aulstad, D., T. Gjedrem, and H. Skjervold. 1972. Genetic and environmental sources of variation in length and weight of rainbow trout (*Salmo gairdneri*). Journal of the Fisheries Research Board of Canada 219:237-241.

Averett, R. C. 1969. Influence of temperature on energy and material utilization by juvenile coho salmon. Ph.D. thesis. Oregon State University, Corvallis, USA.

Diana, J. S., and W. C. MacKay. 1979. Timing and magnitude of energy deposition and loss in the body, liver, and gonads of northern pike (*Esox lucius*). Journal of the Fisheries Research Board of Canada 36:481-487.

Bell, G. 1980. The costs of reproduction and their consequences. American Naturalist 116:45-76.

Falconer, D. S. 1960. Introduction to quantitive genetics. The Ronald Press Company, New York, New York, USA.

Fry, F. E. J. 1947. Effects of the environment on animal activity. University of Toronto Studies. Biology Series No. 55. Publications of the Ontario Fisheries Research Laboratory No. 68. Toronto University: Toronto Press, Toronto, Ontario, Canada.

Griffiths, J. S., and D. F. Alderdice. 1972. Effects of acclimation and acute temperature experience on the swimming speed of juvenile coho salmon. Journal of the Fisheries Research Board of Canada 29:251-264.

Guderley, H., P. Blier, and L. Richard. 1986. Metabolic changes during the reproductive migration of two sympatric coregonines, *Coregonus artedii* and *Coregonus clupeaformis*. Canadian Journal of Fisheries and Aquatic Sciences 43:1859-1865.

Hirshfield, M. F. 1980. Experimental analysis of reproductive effort and cost in the Japanese medaka. Ecology 61:282-292.

Hirshfield, M. F., C. R. Feldmeth, and D. L. Soltz. 1980. Genetic differences in physiological tolerances of Amargosa pupfish (*Cyprinodon nevadensis*) populations. Science 207:999-1001.

Hutchinson, G. E. 1957. Concluding remarks. Cold Spring Harbor Symposium on Quantitative Biology 22:414-427.

Kapuscinski, A. R., and L. D. Jacobson. 1987. Genetic guidelines for fisheries management. Minnesota Sea Grant, Sea Grant Research Report No. 17. University of Minnesota, Deluth, Minnesota, USA.

Knights, B. 1985. Energetics and fish farming. Pages 309-340 in P. Tytler and P. Calow, editors. Fish energetics new perspectives. Johns Hopkins University Press, Baltimore, Maryland, USA.

Lewontin, R. C. 1961. Evolution and the theory of games. Journal of Theoretical Biology 1:382-403.

Maule, A. G., R. A. Tripp, S. L. Kaattari, and C. B. Schreck. 1989. Stress alters immune function and disease resistance in chinook salmon (*Oncorhynchus tshawytscha*). Journal of Endocrinology 120:135-142.

McIntyre, J. D., and A. K. Johnson. 1977. Relative yield of two transferrin phenotypes in coho salmon. Progressive Fish-Culturist 39:175-178.

Priede, I. G. 1985. Metabolic scope in fishes. Pages 33-64 in P. Tytler and P. Calow, editors. Fish energetics new perspectives. Johns Hopkins University Press, Baltimore, Maryland, USA.

Rahel, F. J. 1981. Selection for zinc tolerance in fish: results from laboratory and wild populations. Transactions of the American Fisheries Society 110:19-28.

Read, K. R. H. 1967. Thermostability of proteins in poikilotherms. Pages 93-106 in C. L. Prosser, editor. Molecular mechanisms of temperature adaptation. American Association for the Advancement of Science, Publication No. 84, Washington, D.C., USA.

Schreck, C. B. 1981. Stress and compensation in teleostean fishes: response to social and physical factors. Pages 295-321 in A.D. Pickering, editor. Stress and fish. Academic Press, New York, New York, USA.

Schreck, C. B. 1982. Stress and rearing of salmonids. Aquaculture 28:241-249.

Suzumoto, B. K., C. B. Schreck, and J. D. McIntyre. 1977. Relative resistance of three transferrin genotypes of coho salmon (*Oncorhynchus kisutch*) and their hematological responses to bacterial kidney disease. Journal of the Fisheries Research Board of Canada 34:1-8.

Tave, D. 1986. Genetics for fish hatchery managers. AVI Publishing Company, Westport, Connecticut, USA.

Vernberg, F. J. 1978. Multiple-factor and synergistic stresses in aquatic systems. Pages 726-747 in J. H. Thorp and J. W. Gibbons, editors. Energy and environmental stress in aquatic systems. Technical Information Center, United States Department of Energy, Springfield, Virginia, USA.

Warren, C. E. 1971. Biology and water pollution control. W. B. Saunders Company, Philadelphia, Pennsylvania, USA.

Williams, R. J. 1963. Biochemical individuality. John Wiley and Sons, Inc., New York, New York, USA.

Wohlschlag, D. E., and R. O. Juliano. 1959. Seasonal changes in bluegill metabolism. Limnology and Oceanography 4:195-209.

RESPONSES OF FISH TO TEMPERATURE AND OXYGEN, AND RESPONSE INTEGRATION THROUGH METABOLIC SCOPE

William H. Neill and James D. Bryan

ABSTRACT

Temperature and dissolved oxygen are of paramount importance in aquaculture, because these two environmental factors dominate the physiology and behavior of aquatic organisms. In the context of F. E. J. Fry's categorization of environment, temperature is the principal controlling factor; whereas, oxygen acts primarily as a limiting factor. Temperature sets the pace of metabolism by controlling molecular dynamics and biochemical reaction rates. The biochemical reactions controlled by temperature are subject to limitation by oxygen, which is an essential reactant of aerobic metabolism. Aquatic organisms have effective behavioral responses to temperature and oxygen as directive factors, whereby local environment in heterogeneous habitats is optimized.

Temperature control and oxygen limitation of physiological work interact through their joint effects on metabolic scope. Metabolic scope is proposed as an integrative measure of environmental quality for fish production. A useful derivative of metabolic scope, the marginal metabolic scope, can be estimated via routine respirometry.

INTRODUCTION

F.E.J. Fry (1947, 1971) envisioned a physiological classification of environment in which temperature and oxygen are the archetypal factors "controlling" and "limiting" the metabolism of fish. Metabolism can be considered the physiological engine that powers such activities as swimming, growth, and reproduction. The engine uses oxygen (together with nutrients from food) and produces wastes such as carbon dioxide and ammonia. Temperature controls the pace, or metabolic rate, of the engine.

Restrictions in the supply of oxygen (or nutrients) limit the engine's capacity to perform metabolic work. Either a deficiency of oxygen itself or the presence of dissolved substances that interfere with oxygen uptake, transport, or use (e.g., carbon dioxide, ammonia, detergents, metals, some organic biocides) can restrict oxygen supply to the engine. In water with non-optimal levels of certain dissolved solids (in aggregate, "salinity"), the engine must do extra work to maintain proper internal concentrations of water and ions (ion-osmoregulation); this metabolic loading reduces the engine's capacity for other work, such as growth.

Fry (1947, 1971) also emphasized that both temperature and oxygen can serve as important "directive" factors that influence fish movement and distribution within the habitat. Typically, fish exploit spatial heterogeneities of temperature and oxygen (which exist even in well-stirred tanks and raceways) to achieve a certain degree of independence from the distribution of these variables in the habitat. Such behavioral enviroregulation not only results in uneven spatial distribution of fish but also feeds back to influence the values of temperature and oxygen that control and limit metabolism.

It is from these perspectives that we will consider the roles of temperature and oxygen in fish biology. Our purpose is not an encyclopedic review of literature or compilation of data. Rather, we will elaborate on the ideas introduced above and explore their implications for aquaculture. Most of what we have to say conforms with the information to be found in any relevant textbook (e.g., Bond 1979, Smith 1982, Moyle and Cech 1988) and in the more technical reference works and review/synthesis articles cited at the ends of the next two sections. Other citation of literature will be mostly limited to that required for properly crediting important ideas and for documenting our analysis of problematic issues.

OXYGEN AS A LIMITING FACTOR AND THE CONCEPT OF METABOLIC SCOPE

Oxidation of food materials or stored energy reserves requires oxygen that generally must be acquired from the water via diffusion at the gills. Respiratory and circulatory systems of fish have evolved to function effectively over a range of dissolved oxygen concentration, from above air saturation down to some concentration at which oxygen-demanding activities become constrained. This limiting oxygen concentration (LOC) is increased by anything that increases total oxygen demand: e.g., faster swimming, increased food processing, elevated temperature, and imposition of environmental stressors such as ammonia and non- optimum levels of salinity.

Consider a typical fish living in water with gradually declining oxygen concentration. Given no other cause for changing total activity, the fish maintains a relatively constant rate of oxygen uptake at all oxygen concentrations above LOC. Such oxygen- independence typically is achieved by increasing gill ventilation (water flow) and perfusion (blood flow) as environmental oxygen declines. At LOC, these regulatory measures are exhausted and, as dissolved oxygen declines still farther, respiration becomes oxygen-dependent. If the fish had been engaged in instantaneously non-obligatory activities (swimming, digesting food, growing), those activities must now be curtailed. When the fish has been reduced to activities obligatory for continued existence (maintenance activities supported by "standard metabolism"), the fish must engage in anaerobic metabolism or die. Because most fish have very limited capacity for anaerobic metabolism, LOC for a fish metabolizing at the standard rate is, in effect, the incipient lethal oxygen concentration. To fish physiologists who strive to estimate LOC only for inactive, fasted fish, LOC is understood in this last, restricted sense. But, to those of us who are interested in the limits oxygen imposes on fish performance and production, LOC is a continuum (Fig. 1). For fish that are forced to be maximally active (and, so, have "active metabolism"), the continuum can extend from the standard LOC to an active LOC beyond air saturation. Fry (1947) called these endpoints the "level of no excess activity" and the "incipient limiting level," respectively. Others have identified various intermediate points, frequently calling them "critical" oxygen concentrations (or tensions). Perhaps the commonest of these is the critical oxygen concentration for routine metabolism, which is the metabolism of an undisturbed, fasted fish engaged in voluntary swimming activity.

Values of LOC for a variety of fish species and activity levels have been compiled by Davis (1975) and Hughes (1981). At some risk of overgeneralization, it may be said that standard LOC ranges from 1 to 3 ppm and that for "normal" levels of activity LOC ranges upwards from 3 ppm, with a mean near 5 ppm.

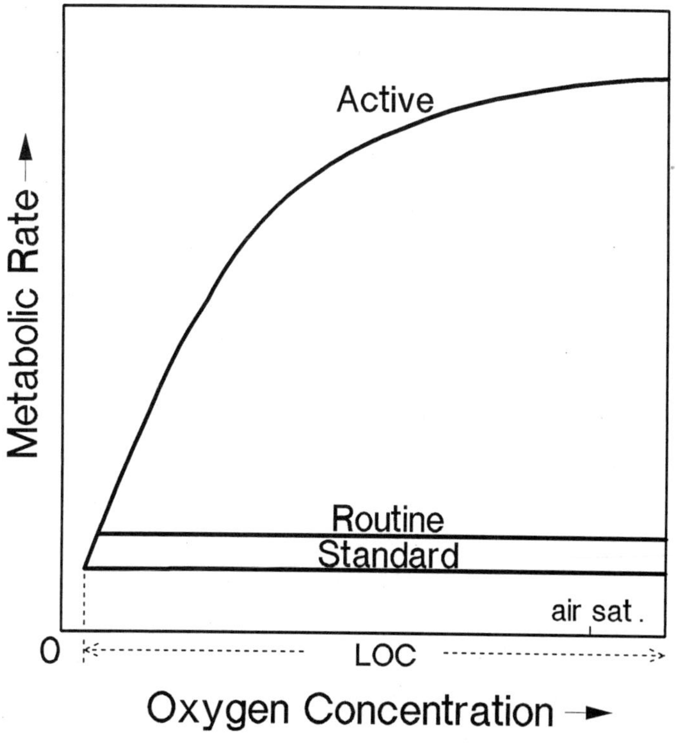

Figure 1. Active, routine, and standard metabolic rates of an idealized fish as functions of dissolved oxygen concentration. The difference between active and standard rates is the metabolic scope (as traditionally defined). LOC = limiting oxygen concentration.

The difference between active and standard metabolism is defined as metabolic scope (Fry 1947). It increases monotonically with oxygen concentration, from zero at standard LOC, to a value 5-15 times standard metabolic rate at the active LOC corresponding with air saturation. Metabolic scope estimates the maximum rate at which oxygen can be supplied in support of swimming, growth, and other activities. Because metabolic scope provides a mechanistic basis for estimating the joint effects of environment on the capacity for fish performance, we will return to this important concept in subsequent sections.

Other sources of information about oxygen transport and metabolism include publications by Fry (1957), Randall (1970a,b; 1982), Beamish (1978), Jones and Randall (1978), Brett and Groves (1979), Langille et al. (1983), Priede (1985), and Heath (1987).

TEMPERATURE AS A CONTROLLING FACTOR AND ITS INTERACTION WITH OXYGEN

Except for tunas and lamnid sharks, fishes are poikilotherms. That is, heat moves so readily between the water and the fish's body that its internal tissues are typically at nearly the same temperature as the water—except when water temperature is changing rapidly, on time-scales of seconds or minutes. Although thermal lag occasioned by such rapid changes may play an important role in the fish's ability to detect and orient to temperature gradients (Neill 1979), we may conclude that rates of metabolic work inside the fish generally are set by time-averaged temperature of water surrounding the fish.

Temperature sets the pace of metabolism by controlling molecular dynamics (diffusivity, solubility, fluidity) and biochemical reaction rates. The resulting patterns of metabolism and metabolic scope are generalized in Figure 2. A resting, fasted fish's metabolic demand for oxygen (standard metabolism) generally increases exponentially with temperature, doubling or trebling for each 10 C rise (i.e., $2 < Q_{10} < 3$). However, the physiological capacity to supply this oxygen rises sigmoidally to a maximum at some high temperature (and then may decline precipitously as temperature increases farther). Thus, metabolic scope reaches a maximum at a species-specific optimum temperature, which is somewhat lower than the temperature permitting maximum supply. Under otherwise favorable conditions, the optimum temperatures for many "cold-water" and "warm-water" fishes are 14-18 C and 24-30 C, respectively. Deficiencies of dissolved oxygen limit active metabolism progressively as temperature rises, and thus lower the optimum temperature (Fig. 3). Other environmental stressors can both elevate obligatory demand for oxygen and reduce physiological capacity to supply it; these stressors can reduce metabolic scope at all temperatures and shift (in general, reduce) the optimum temperature for maximizing the surplus metabolic capacity that, in combination with adequate nutrition, leads to growth.

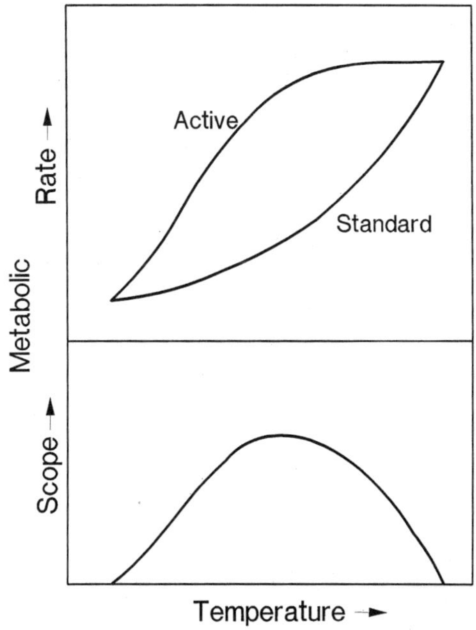

Figure 2. Controlling effect of temperature on metabolic rates and scope of an idealized fish.

Beyond lower and upper extremes of temperature, fish not only can't grow, they can't live—or they can live only for a limited time. The specific relationship depends on the fish species (and perhaps size/age), past thermal experience and general condition of the individual, and levels of non-thermal factors such as salinity. At very high and low temperatures, fish may enter thermal shock and die almost instantly, with the body rigid and opercles flared. But, in general, the logarithm of survival (= resistance) time upon abrupt transfer to lethally warm or cold water is linearly related to temperature (Fig. 4, left panel): If a fish can resist heat death at 32 C for 1,000 min and at 34 C for 100 min, it will die at 36 C in only 10 min. However, the linearity holds only down to the upper incipient lethal temperature, this being the highest temperature at which just 50 % of the individuals in a sample of fish survive indefinitely; there also exists an analogously defined lower incipient lethal temperature. Between the two incipient lethal temperatures lies the zone of thermal tolerance; beyond them lies the zone of thermal resistance.

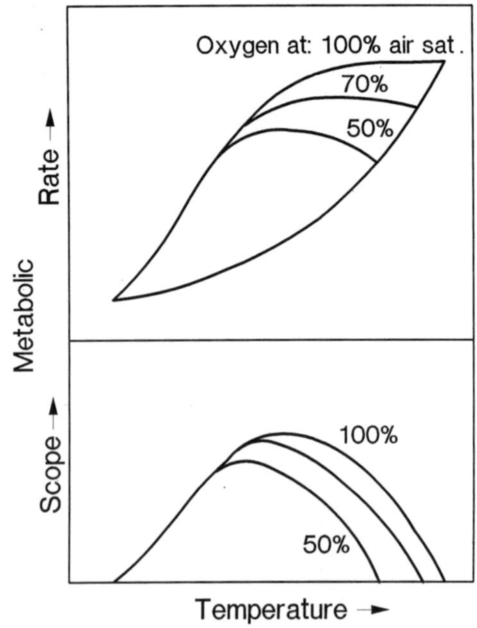

Figure 3. Limiting effect of oxygen on active metabolism of an idealized fish and consequent reduction in its metabolic scope, especially at high temperature.

Most of the concepts and associated terminology mentioned in the previous paragraph originated in two seminal papers by Fry and his students (Fry et al. 1942, 1946). In the second of these papers, Fry et al. (1946) also described a lethal-dose model for predicting the consequences of exposure to time-varying lethal temperature. The idea is that resistance-time data from step-change experiments can be used to calculate a thermal dose for any temperature-time series by summing incremental doses of the form, t_a/t_r, where t_a and t_r are exposure time and resistance time, respectively, for each temperature in the series; death is predicted if and when the cumulative dose reaches unity. Although this approach was presented again in two of Fry's well-read review papers (Fry 1947, 1971), it seems not to have attracted the attention we think it deserves (but see Kilgour and McCauley 1986). Perhaps, the apparent neglect is attributable to a lack of the submodels for acclimation (see below) and dose-decay that will be required to make the

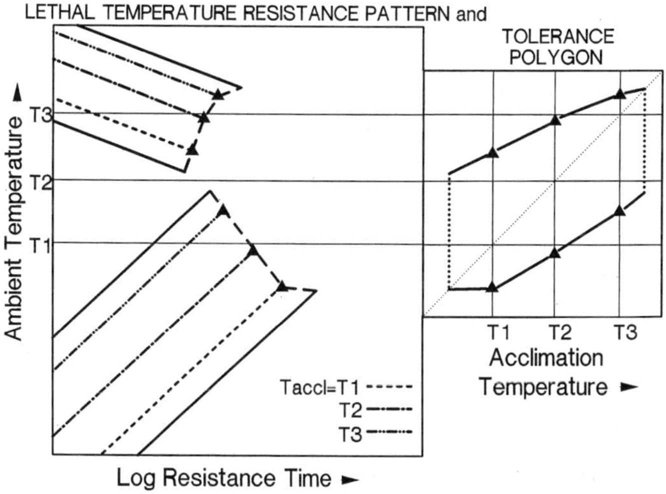

Figure 4. Lethal temperature resistance pattern amd corresponding temperature-tolerance polygon for an idealized fish. Triangles indicate incipient lethal temperatures for three values of acclimation temperature (Taccl).

lethal-dose model useful in the gradually-changing thermal regimes of greatest interest to aquaculturists and fisheries managers.

Other sources of information about the thermal biology of fish include reviews by Fry (1967), Fry and Hochachka (1970), Magnuson et al. (1979), and Elliot (1981).

ADAPTATIONS OF FISH TO ENVIRONMENTAL VARIATION: ENVIROREGULATION AND ACCLIMATION

In the previous section, we saw that fish have characteristic physiological responses to important components of the physico-chemical environment, and that the response to one component typically depends on levels of other components, giving rise to complex physiological interactions. Having established this, it is important to emphasize that the environment varies from time to time and from place to place, and that fish cope with such variation both by behavioral and physiological means.

Fish, like other organisms, have evolved an array of behaviors that allow them to exploit spatial variation of environment; behaviors that tend to optimize environment we call "habitat selection" or "enviroregulation." Thus, fish move toward the surface when dissolved oxygen is deficient and into deeper water when a passing cold front drops water temperature in the shallows. Such responses are obvious. What may be less obvious is that fish continuously enviroregulate, taking advantage of environmental gradients and patchiness that may be inconspicuous to the casual observer. The benefit to the fish is that it keeps itself in those parts of the habitat that are environmentally best. The disadvantage to the fish farmer is that the fish aggregate at high densities in only a fraction of the habitat, leaving the remainder under-used. From this, it follows that aquaculturists should strive to maintain all of the habitat (pond, raceway, etc.) at environmental values near those behaviorally selected by, and thus presumably optimum for, the species being cultured.

In experimental gradients, fish avoid both high and low temperatures but only low oxygen. Temperature-avoidance limits typically span 3-5 C; in this interval is the preferred temperature. Preferred temperatures are species/size-specific and depend on past thermal experience. Limits for low-oxygen avoidance also vary with species and probably with size; whether these limits are well-correlated with routine LOC, as reason would suggest, cannot be judged from available data. In at least one laboratory study (Bryan et al. 1984), fish in oxygen-deficient water markedly reduced their preferred temperature; this enviroregulatory behavior had the effect of decreasing routine LOC to the oxygen concentration that existed in the temperature-gradient tank.

Variations in habitat quality over time cannot be exploited through enviroregulatory behavior. In dealing with temporal variation, fish must rely on physiological mechanisms of adaptation that are termed "acclimation" (responses to change in environmental factors individually) or "acclimatization" (responses to change in the entire complex of environment). In essence, exposure to a change in the level of some environmental factor results in physiological adjustments that enable the fish to better perform at the new level of the factor and shift the fish's avoidance and lethal limits in the direction of the change. For example, transfer of a 20-C acclimated fish to 30 C eventually might result in increased metabolic scope at 30 C, an increase in the preferred temperature, greater heat tolerance, and lowered cold tolerance.

Fish make acclimatory adjustments to temperature, dissolved oxygen, salinity, ammonia, and many other physico-chemical factors. But these adjustments take time—from several hours to several weeks. The rates of acclimatory processes seem to have evolved to keep pace with normal rates of seasonal (not diel) change in environment. Temperature acclimation occurs at a rate that increases (exponentially?) with ambient temperature; so, upward acclimation is faster than downward acclimation over the same range of temperature. For example, if acclimation from 20 to 30 C is 90 % complete in 2 days, the reverse process could take 2 weeks. Acclimation to low oxygen may require from one week to one month.

What are the mechanisms involved in acclimation? With regard to temperature, the fundamental changes seem to be at the level of enzyme and membrane biophysics. Acclimation to low oxygen involves both facilitation of oxygen uptake (increased hemoglobin concentration and hematocrit) and restraint in oxygen use (decreased metabolism).

The effect of thermal acclimation on lethal-temperature resistance and thermal tolerance is generalized in Figure 4. This idealized fish presents the classical resistance pattern, in which 1) the logarithm of resistance time is a linear function of ambient temperature—so long as ambient temperature is more extreme than the incipient lethal temperature; 2) the more extreme the lethal temperature, the faster the fish dies; and, 3) upward thermal acclimation results in increased resistance time at upper lethal temperatures, decreased resistance time at lower lethal temperatures, and increased incipient lethal temperatures (downward acclimation has opposite effects). The extreme resistance lines in Figure 4 (left panel) are for fish acclimated to the ultimate upper and lower incipient lethal temperatures; these are the highest and lowest temperatures, respectively, to which the fish can be acclimated. Ultimate upper incipient lethal temperatures range from near 20 C for some salmonids to about 35 C for many fish species of the temperate zone. Only tropical and subtropical species have ultimate lower incipient lethal temperatures that are well above the freezing point of water. Median resistance time at the upper incipient lethal temperature tends to increase with acclimation temperature, to a maximum of about 10,000 min (one week) for fish acclimated to a temperature just below their ultimate upper incipient lethal temperature; the pattern for cold resistance is just the opposite, except median resistance time at the incipient lethals tends to be greater.

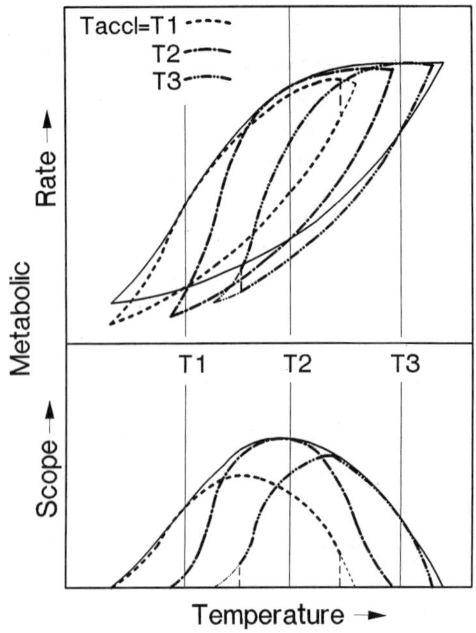

Figure 5. Effect of thermal acclimation on metabolic rates and scope of an idealized fish as functions of ambient temperature. Responses are indicated for three values of acclimation temperature (Taccl).

A plot of the incipient lethal temperatures versus acclimation temperature gives Fry's (1947) temperature-tolerance polygon (Fig. 4, right panel). Within it lies the zone of tolerance; outside lies the zone of resistance. The "area" of the polygon provides a quantitative measure of a fish's position on the stenotherm-eurytherm continuum. Brett (1956) applied this concept to calculate thermal tolerance for a variety of fishes; the values ranged from near 500 C^2 for the Pacific salmons to 1,220 C^2 for goldfish.

The effect of thermal acclimation on the relationship between metabolism and temperature is generalized in Figure 5. The metabolism of our idealized fish undergoes acclimatory adjustments that partially compensate for temperature variation. For any ambient temperature, metabolic scope is maximum when the fish is acclimated to that temperature. However, metabolic scope for fish acclimated to extreme temperatures (T1 and T3) reaches its maximum at an ambient temperature between the acclimation temperature and the temperature that is the steady-state optimum. In this example, the steady-state optimum, where scope reaches its global maximum, is at T2. The ambient-temperature extremes at which scope goes to zero also shift with acclimation temperature. Note the truncation of metabolic scope caused by lethal-temperature shock in T1-acclimated fish exposed to extremely high temperatures and in T3- acclimated fish exposed to extremely low temperatures. One might expect a considerable degree of correspondence between the temperatures for zero scope and the incipient lethal temperatures; indeed, we have made sure this is the case for our hypothetical fish (cf. Figs. 4 and 5), although data are insufficient to test this hypothesis.

Figure 5 implies that the immediate consequences of temperature change can be either good or bad, from the perspective of fish performance. A fish subjected to an abrupt change in temperature can be expected to experience an immediate change in metabolic scope. If the temperature change is such that scope increases, the fish thrives. But if the temperature change causes a decline in metabolic scope, one would expect reductions in general vigor (e.g., increased susceptibility to disease) and feed consumption. If the temperature change is survivable (if scope does not decline to zero), the fish ultimately will acclimate to the new temperature, and its metabolic scope will increase—from whatever transient value obtained just after the temperature change, to the new and higher steady-state value. Whether the new steady-state value is higher or lower than that before the temperature change depends on the specific relationship between steady-state scope and temperature (in Fig. 5, the scope "envelope," which is the same as the scope curve graphed in Fig. 2).

In Figure 6, we have overlaid the thermal-tolerance polygon from Figure 4 with a plot of the ambient temperatures giving maximum scope in

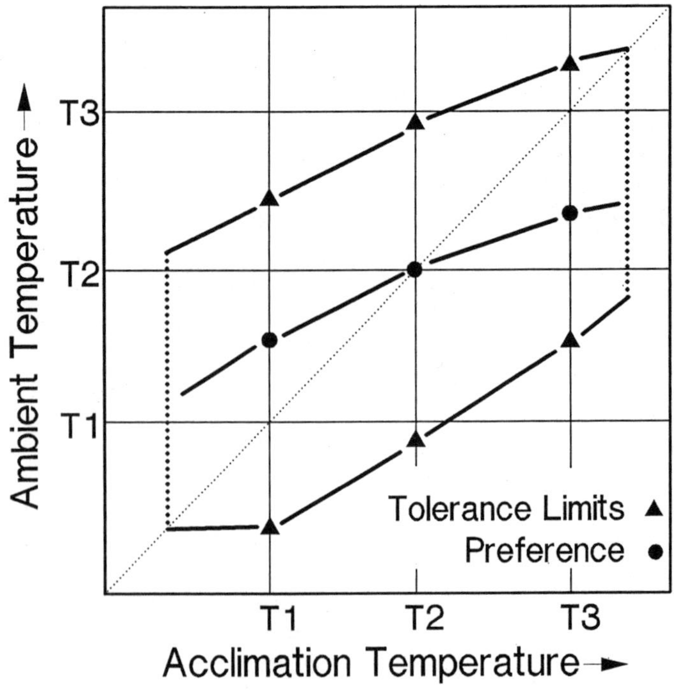

Figure 6. Temperature preference in relation to thermal tolerance for an idealized fish.

Figure 5; the latter points are labeled "Preference." The implication is that fish prefer temperatures that afford maximum metabolic scope. Metabolic scope reaches its global maximum when acclimation and ambient temperatures are both at T2. If our hypothetical fish is acclimated to either T1 or T3 and then put into a temperature gradient, it initially will select the temperature giving maximum scope; this temperature lies between its acclimation temperature and T2. But exposure to the new temperature will cause acclimation toward T2 and a corresponding change in preference. In time (several hours to several days), the fish arrives at T2 and becomes acclimated to it. The preferred temperature equal to acclimation temperature Fry (1947) called the "final preferendum." Coutant (1977) and Jobling (1981) have compiled final temperature preferenda for a variety of fishes; the values range from near 10

C for some salmonids to about 31 C for carp and some centrarchids. In general, final temperature preferenda correspond very closely with optimum temperatures for growth (Jobling 1981)—just as one would expect if fish prefer temperatures that maximize metabolic scope.

MARGINAL METABOLIC SCOPE: A PHYSIOLOGICAL INDEX OF ENVIRONMENTAL QUALITY

The foregoing presentation makes evident our conviction that metabolic scope provides a rational basis for predicting the overall suitability of environment for fish production, where "environment" is meant in the multivariate sense and includes changes through time. But before such predictions can be made, one needs estimates of metabolic scope itself—over the ranges of fish size and environments of interest. Unfortunately, the literature provides values only for a few fish species, and those only for restricted sets of fish size and environment.

The paucity of published values for metabolic scope indicates the difficulty in making prerequisite measurements. The problem is two-fold: 1) Measurement of active metabolism requires that the fish be made to swim at maximum sustainable speeds, which demands the use of elaborate and costly swim-chamber respirometers; and, 2) measurement of standard metabolism requires that the fish be completely inactive, a state that is foreign to most fish. Standard metabolism sometimes is estimated as the minimum oxygen-uptake rate of fasted, routinely active fish. An alternative approach to estimation of standard metabolism involves extrapolation to zero-speed of the regression of oxygen-uptake rate on swimming speed from active respirometry; but, the reliability of such extrapolation is uncertain (e.g., Smit 1965), owing to the tendency for many kinds of fish to change their mode of swimming, from caudal "sculling" to pectoral "paddling," as speed drops below one body length per second. Recently, even the assumption that active metabolism measures maximum metabolism has come into question (Priede 1985), because some fish have higher metabolic rates following a large meal than they do when swimming at maximum sustainable speed.

We believe the practical difficulties in estimating metabolic scope can be lessened considerably—if one accepts some adjustment in the concept. In the following paragraph, we suggest that maximum metabolism can be estimated via routine respirometry. If so, neither standard nor active metabolic rates need be measured to estimate a metabolic scope that is defined as the difference between maximum (aerobic) and routine rates, rather than that between active and standard rates. This redefinition of metabolic scope should meet with no objection from aquaculturists, who are more interested in the growth of routinely active, well-fed fish than in how fast hungry fish can swim.

We propose that maximum metabolism as a function of dissolved-oxygen concentration can be modeled from the results of routine respirometry experiments, given sufficient understanding of the biophysics and physiology involved in oxygen transport. Conservation of mass requires that

$$VO2 = Vg \cdot (O2wi - O2wo),$$

where VO2 is oxygen-uptake rate ($mg \cdot g^{-1} \cdot h^{-1}$), Vg is ventilation rate ($L \cdot g^{-1} \cdot h^{-1}$), and O2wi and O2wo are the oxygen concentrations ($mg \cdot L^{-1}$) of incurrent and excurrent water, respectively. Oxygen utilization (U) is defined as the fraction of dissolved oxygen removed from the ventilating water:

$$U = (O2wi - O2wo)/O2wi.$$

Therefore,

$$VO2 = Vg \cdot U \cdot O2wi.$$

Fundamental to our scheme is the assumption that gill ventilation is maximum at O2wi = LOC, and that the maximum is the same at all values of LOC for a given fish in a given environment. This view is consistent with that of Fry (1957, p. 40), who wrote that "...the major factor in producing the phenomenon of respiratory dependence is the limitation imposed by the maximum amount of water which can be presented to the gills by the ventilatory apparatus." Subsequent literature (see reviews by Randall 1970b and 1982, Heath 1987) does not contradict our assumption (even if there seem to be no data that directly validate it). With the assumption in place,

$$VO2_{LOC} = Vg_{max} \cdot U_{LOC} \cdot LOC;$$

and, in particular, when LOC is the limiting oxygen concentration for routine metabolism, LOCr,

$$VO2_{LOCr} = Vg_{max} \cdot U_{LOCr} \cdot LOCr.$$

This implies that routine metabolic rate and LOCr could be used to estimate Vg_{max}—if only U_{LOCr} were known:

$$Vg_{max} = (VO2_{LOCr}/LOCr)/U_{LOCr}.$$

Of course, U_{LOCr} typically is not measured in routine respirometry. But the relationship still is useful, because from it comes the conclusion that maximum metabolic rate at any O2wi is simply

$$VO2_{max} = (VO2_{LOCr}/LOCr) \cdot (U_{O2wi}/U_{LOCr}) \cdot O2wi.$$

Therefore, *if* oxygen utilization under conditions of maximum uptake at every LOC were the same as that at LOCr, maximum metabolic rate at any O2wi could be predicted as the product of O2wi and the ratio $VO2_{LOCr}/LOCr$, and maximum metabolism would increase linearly with O2wi. Because the typical relationship between active metabolism and dissolved oxygen concentration departs from linearity as the concentration approaches air saturation (Fig. 1), the ratio of utilizations generally must differ from unity. For brook trout (*Salvelinus fontinalis*) at 15 C (Fry 1971, Fig. 24), the required utilization ratios are, approximately, 1.4 at 5 and 8 ppm and 1.2 at 10 ppm (with LOCr being approximately 3.5 ppm).

In the utilization ratio's trend with LOC must be hidden all the biophysics and physiology that goes on inside the fish. Among factors probably involved are changes in gill perfusion-rate (cardiac output), the extent of hemoglobin saturation in the gill-efferent blood, and the partial pressure of oxygen in venous blood returning to the gills. These changes, in turn, must depend on the blood's oxygen-capacity and the affinity of its hemoglobin for oxygen—both of which are responsive to environmental changes. Measured utilizations range from 0.1 to 0.9, with "normal" values being near 0.5. With increasing ventilation rate, utilization tends to decrease; it may (e.g., Gerald and Cech 1970) or may not (e.g., Campagna and Cech 1981) decrease with hypoxia.

Forgetting (for a moment) all the uncertainties about utilization, a model for metabolic scope, MS, follows directly from our redefinition:

$$MS = VO2_{max} - VO2_{LOCr}$$

$$= (VO2_{LOCr}/LOCr) \cdot (U_{O2wi}/U_{LOCr}) \cdot O2wi - VO2_{LOCr}.$$

But, is this model of metabolic scope really useful, without more complete understanding of utilization's trend with LOC? We suggest that the relative quality of environment can be estimated as a marginal metabolic scope, MMS, which we define as MS at the value of O2wi just 1 $mg \cdot L^{-1}$ above LOCr, assuming that the utilization ratio remains near unity over this small interval (but making no assumption about the actual value of utilization):

$$MMS = (VO2_{LOCr}/LOCr) \cdot (LOCr + 1) - VO2_{LOCr}$$

$$= (VO2_{LOCr}/LOCr) \cdot (1\ mg \cdot L^{-1}).$$

So, marginal metabolic scope is simply the ratio of routine metabolic rate to the limiting oxygen concentration for that rate, multiplied by 1 $mg \cdot L^{-1}$. MMS is numerically equivalent to the estimated derivative of MS with respect to LOC, at the value LOCr; $VO2_{LOCr}/LOCr$ estimates that derivative. Multiplication of $VO2_{LOCr}/LOCr$ by 1 $mg \cdot L^{-1}$ serves only to make the units of MMS the same as those of MS. Figure 7 presents the geometric interpretation of these relationships.

Estimation of MMS requires only data from routine respirometry, which is relatively simple and can be automated (Springer and Neill 1988). Therefore, generation of MMS values for a large variety of species, sizes, and environments is feasible.

Papers by Herreid (1980) and Withers and Hillman (1988) helped lead us to the concept of marginal metabolic scope.

SAMPLE APPLICATIONS OF MARGINAL METABOLIC SCOPE

Our argument is that metabolic scope (MS) measures the overall

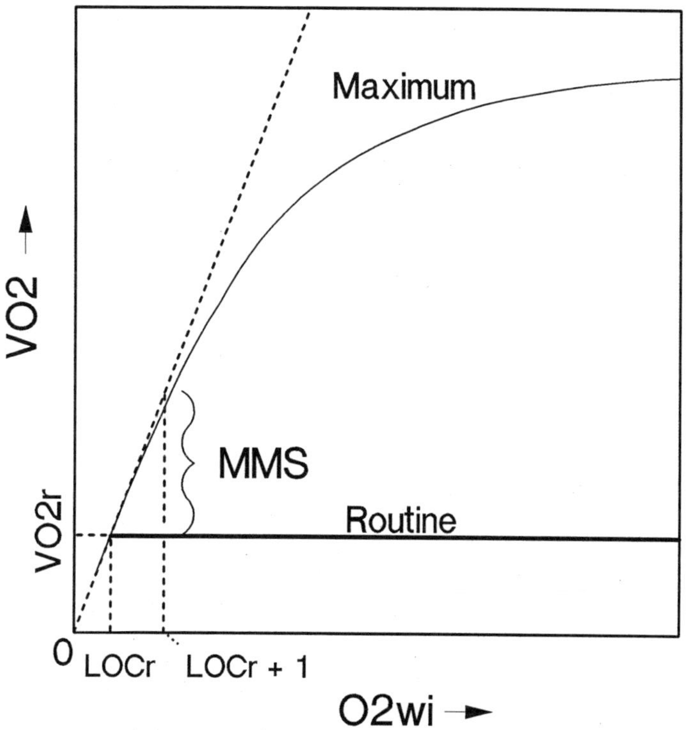

Figure 7. Geometric interpretation of the marginal metabolic scope (MMS) concept. See text for definition of labels.

suitability of a particular environment for production of a particular fish, and that marginal metabolic scope (MMS) provides a convenient index of that suitability. In this section, we draw upon data from the literature to present some examples that demonstrate the practicality of the MMS concept and support our claim of its utility as an indicator of environmental quality.

Ultsch et al. (1978) measured VO2r and LOCr for six species of darter (*Etheostoma*) in an attempt to understand their niche relationships; tests were done both winter and summer at 10 and 20 C, respectively. Data from these tests presented an unanticipated pattern (Fig. 8A): Whereas the expected elevation of metabolic rate with temperature was evident, the relationship between LOCr and temperature was, if anything, inverse. We interpret these

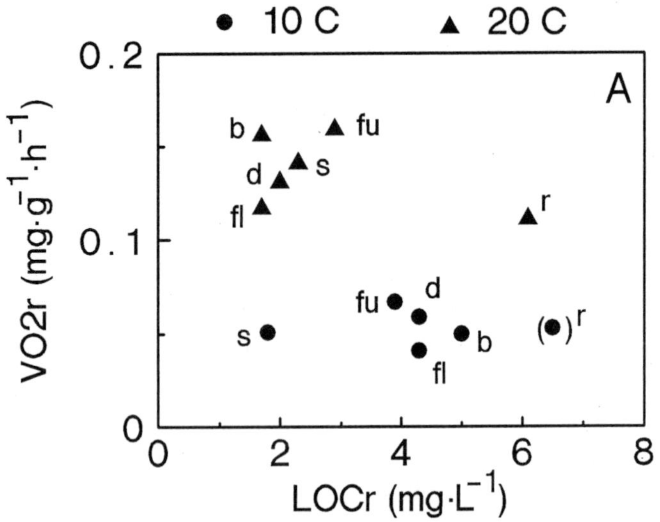

Figure 8A. Routine metabolic rate (VO2r) versus limiting oxygen concentration (LOCr) at 10 and 20 C for six species of darter--*Etheostoma rufilineatum* (r), *E. flabellare* (fl), *E. duryi* (d), *E. squamiceps* (s), *E. boschungi* (b), and *E. fusiforme* (fu). Data are mean values presented by Ultsch et al. (1978), except that we estimated LOCr for the point in parentheses from the author's Fig. 1.

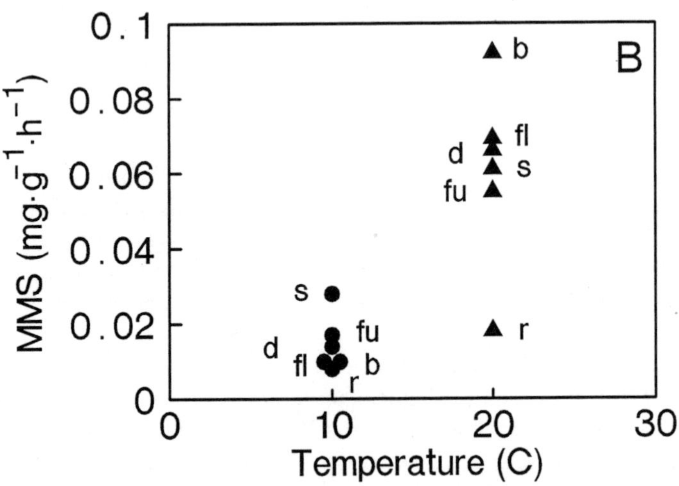

Figure 8B. Marginal metabolic scope (MMS) versus temperature, computed from data presented in Fig. 8A.

data as indicating acclimatory increase in MMS with temperature (and perhaps photoperiod) and predict that tests at various temperatures for darters acclimated to a single temperature (and photoperiod) would yield increasing LOCr with temperature. A plot of MMS versus temperature (Fig. 8B) suggests that thermal niches of these darters are disparate in summer, with *E. boschungi* (a slow-water form) and *E. rufilineatum* (a fast-water form) being, respectively, most and least adapted to high temperature.

A less problematic application of MMS to resolution of thermal optima is afforded by data from Ott et al. (1980). They measured VO2r and LOCr for rainbow trout (*Salmo gairdneri*) and carp (*Cyprinus carpio*) at a series of temperatures to which the fish were acclimated. MMS of carp increased monotonically with temperature, over the range 5 to 32 C; whereas, MMS of rainbow trout was higher at 15 C than at either 10 or 20 C (Fig. 9). The optimum temperatures for growth of rainbow trout and carp are about 17 and 31 C, respectively (Jobling 1981).

Ultsch et al. (1980) investigated the effect of water acidification on VO2r and LOCr of rainbow trout and carp at 15 C. Whereas the effect of

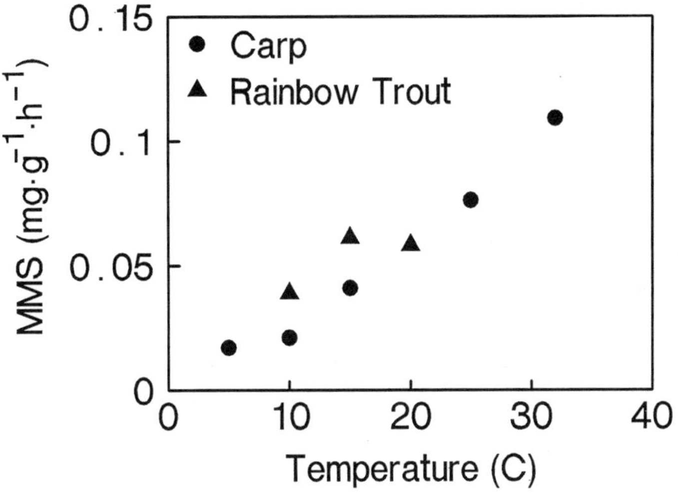

Figure 9. Marginal metabolic scope (MMS) versus temperature for carp and rainbow trout. Data are mean values from Ott et al. (1980).

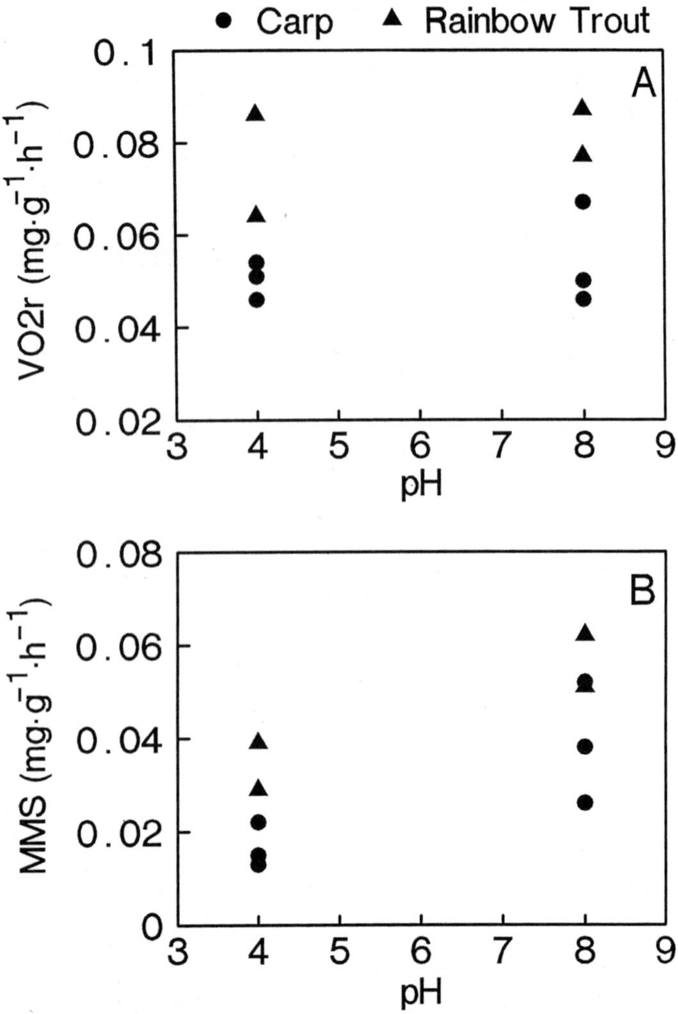

Figures 10A (top) and 10B (bottom). Routine metabolic rate (VO2r) and marginal metabolic scope (MMS) versus pH for carp and rainbow trout at 15 C. Data are individual-fish values from Ultsch et al. (1980).

acidification on neither species' VO2r was significant (Fig. 10A), lowered pH caused a marked decrease of MMS in both species (Fig. 10B).

Procarione (1986) estimated ultimate lower incipient lethal temperatures for juvenile red drum (*Sciaenops ocellatus*) at a variety of salinity levels. Subsequently, Schlechte (1989) measured VO2r and LOCr for another group of juvenile red drum at the same levels of salinity, but all at 24 C. MMS indicated greater capacity for performance by this euryhaline fish at salinities between 5 and 20 ppt than in either freshwater or seawater (Fig. 11A). This result is consistent with the trend in cold tolerance with respect to salinity (Fig. 11B,C).

CONCLUSION

Oxygen limits and temperature controls the metabolism of fish. Together, these two environmental variables are dominant determinants of every fish's environmental niche. That niche is actively sought out and maintained through the processes of behavioral enviroregulation and physiological acclimation.

Metabolic scope is an integrative measure of environmental niche. Direct measurement of metabolic scope is problematic. In this paper, we suggest an approach for estimating metabolic scope via routine respirometry; unfortunately, this approach requires knowledge of cardiovascular physiology that we do not have. So, we offer a "poor man's" substitute: the marginal metabolic scope.

We conclude that marginal metabolic scope has substantial promise as an indicator of environmental quality for fish production. However, only metabolic scope itself will serve as a quantitative measure. This is because aquaculturists (and fisheries managers and conservationists) ultimately will want to measure environmental quality in units of fish-production potential, and that potential depends on the way environment dictates oxygen supply and demand for metabolism, given the *actual* concentration of oxygen in the water. Metabolic scope is our best hope for translating the complexities of environment into potential fish. To realize that hope, we will need to learn considerably more about fish physiology.

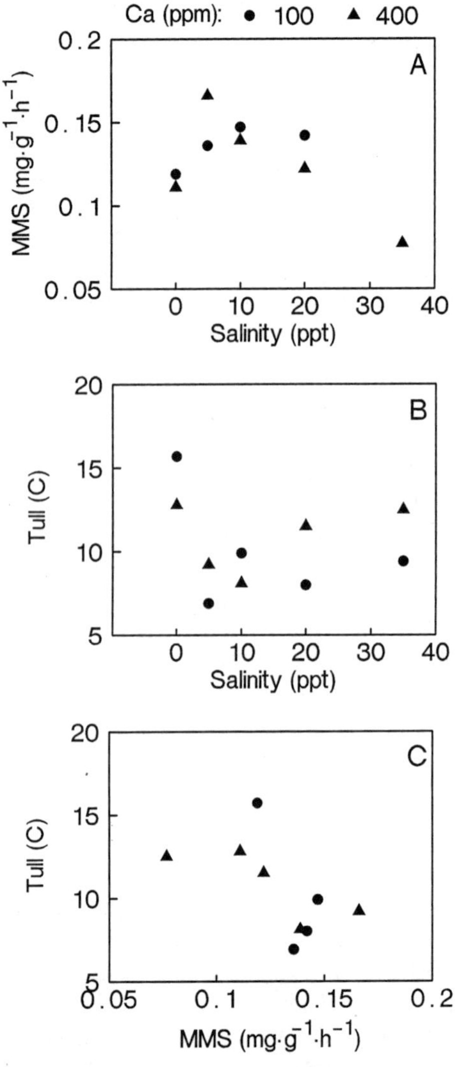

Figure 11. A. Marginal metabolic scope (MMS) versus salinity for juvenile red drum at 24 C and two levels of water hardness. Data are mean values from Schlechte (1988). B. Ultimate lower (incipient) lethal temperature (Tull) versus salinity for juvenile red drum at the same hardness levels as in panel A. Data are mean values from Procarione (1986). C. Tull from panel B versus MMS from panel A, coded by hardness. The correlation between Tull and MMS is -0.59 (P ~ 0.1).

ACKNOWLEDGEMENTS

We thank Tim Springer, Joe Tomasso, and David Brune for critically reviewing the manuscript. The research was funded in part by the Texas Agricultural Experiment Station (TAES) through project H-6295, and this paper is TAES Contribution TA 25163.

LITERATURE CITED

Beamish, F.W.H. 1978. Swimming capacity. Pages 101-187 *in* W.S. Hoar and D.J. Randall, editors. Fish physiology, Volume VII (Locomotion). Academic Press, New York, New York, USA.

Bond, C.E. 1979. Biology of fishes. W.B. Saunders Company, Philadelphia, Pennsylvania, USA.

Brett, J.R. 1956. Some principles in the thermal requirements of fishes. The Quarterly Review of Biology 31:75-87.

Brett, J.R., and T.D.D. Groves. 1979. Physiological energetics. Pages 279-352 *in* W.S. Hoar, D.J. Randall, and J.R. Brett, editors. Fish physiology, Volume VIII (Bioenergetics and growth). Academic Press, New York, New York, USA.

Bryan, J. D., L.G. Hill, and W.H. Neill. 1984. Interdependence of acute temperature preference and respiration in the plains minnow. Transactions of the American Fisheries Society 113:557-562.

Campagna, C.G., and J.J. Cech, Jr. 1981. Gill ventilation and respiratory efficiency of Sacramento blackfish, *Orthodon microlepidotus* Ayres, in hypoxic environments. Journal of Fish Biology 19:581-591.

Coutant, C.C. 1977. Compilation of temperature preference data. Journal of the Fisheries Research Board of Canada 34:739-745.

Davis, J.C. 1975. Minimal dissolved oxygen requirements of aquatic life with emphasis on Canadian species: a review. Journal of the Fisheries Research Board of Canada 32:2295-2332.

Elliott, J.M. 1981. Some aspects of thermal stress on freshwater teleosts. Pages 209-245 *in* A.D. Pickering, editor. Stress and fish. Academic Press, New York, New York, USA.

Fry, F.E.J. 1947. Effects of the environment on animal activity. University of Toronto Studies, Biological Series 55:1-62.

Fry, F.E.J. 1957. The aquatic respiration of fish. Pages 1-63 *in* M.E. Brown, editor. The physiology of fishes, Volume I (Metabolism). Academic Press, New York, New York, USA.

Fry, F.E.J. 1967. Response of vertebrate poikilotherms to temperature. Pages 375-409 *in* A.H. Rose, editor. Thermobiology. Academic Press, New York, New York, USA.

Fry, F.E.J. 1971. The effect of environmental factors on the physiology of fish. Pages 1-98 *in* W.S. Hoar and D.J. Randall, editors. Fish physiology, Volume VI (Environmental relations and behavior). Academic Press, New York, New York, USA.

Fry, F.E.J., J.R. Brett, and G.H. Clawson. 1942. Lethal limits of temperature for young goldfish. Revue Canadienne de Biologie 1:50-56.

Fry, F.E.J., J.S. Hart, and K.F. Walker. 1946. Lethal temperature relations for a sample of young speckled trout, *Salvelinus fontinalis*. University of Toronto Studies, Biological Series 54:1-35.

Fry, F.E.J., and P.W. Hochachka. 1970. Fish. Pages 79-134 *in* G.C. Whittow, editor. Comparative physiology of thermoregulation, Volume I (Invertebrates and nonmammalian vertebrates). Academic Press, New York, New York, USA.

Gerald, J.W., and J.J. Cech, Jr. 1970. Respiratory responses of juvenile catfish (*Ictalurus punctatus*) to hypoxic conditions. Physiological Zoology 43:47-54.

Heath, A.G. 1987. Water pollution and fish physiology. CRC Press, Boca Raton, Florida, USA.

Herreid, C.F., II. 1980. Hypoxia in invertebrates. Comparative Biochemistry and Physiology 67A:311-320.

Hughes, G.M. 1981. Effects of low oxygen and pollution on the respiratory systems of fish. Pages 121-146 *in* A.D. Pickering, editor. Stress and fish. Academic Press, New York, New York, USA.

Jobling, M. 1981. Temperature tolerance and the final perferendum—rapid methods for the assessment of optimum growth temperatures. Journal of Fish Biology 19:439- 455.

Jones, D.R., and D.J. Randall. 1978. The respiratory and circulatory systems during exercise. Pages 425-501 *in* W.S. Hoar and D.J. Randall, editors. Fish physiology, Volume VII (Locomotion). Academic Press, New York, New York, USA.

Kilgour, D.M., and R.W. McCauley. 1986. Reconciling the two methods of measuring upper lethal temperatures in fishes. Environmental Biology of Fishes 17:281-290.

Langille, B. L., E. D. Stevens, and A. Anantaraman. 1983. Cardiovascular and respiratory flow dynamics. Pages 92-139 *in* P.W. Webb and D. Weihs, editors. Fish biomechanics. Praeger Publishers, New York, New York, USA.

Magnuson, J.J., L.B. Crowder, and P.A. Medvick. 1979. Temperature as an ecological resource. American Zoologist 19:331-343.

Moyle, P.B., and J.J. Cech, Jr. 1988. Fishes: An introduction to ichthyology. Second edition. Prentice Hall, Englewood Cliffs, New Jersey, USA.

Neill, W.H. 1979. Mechanisms of fish distribution in heterothermal environments. American Zoologist 19:305-317.

Ott, M.E., N. Heisler, and G.R. Ultsch. 1980. A re-evaluation of the relationship between temperature and the critical oxygen tension in freshwater fishes. Comparative Biochemistry and Physiology 67A:337-340.

Priede, I.G. 1985. Metabolic scope in fishes. Pages 33-64 *in* P. Tytler and P. Calow, editors. Fish energetics: New perspectives. The Johns Hopkins University Press, Baltimore, Maryland, USA.

Procarione, L.S. 1986. Ultimate lower lethal temperature of red drum *Sciaenops ocellatus* as a function of water hardness and salinity. Master's thesis. Texas A&M University, College Station, Texas, USA. 45 pages.

Randall, D.J. 1970a. The circulatory system. Pages 133-172 *in* W.S. Hoar and D.J. Randall, editors. Fish physiology, Volume IV (The nervous system, circulation, and respiration). Academic Press, New York, New York, USA.

Randall, D.J. 1970b. Gas exchange in fish. Pages 253-292 *in* W.S. Hoar and D.J. Randall, editors. Fish physiology, Volume IV (The nervous system, circulation, and respiration). Academic Press, New York, New York, USA.

Randall, D.J. 1982. The control of respiration and circulation in fish during exercise and hypoxia. Journal of Experimental Biology 100:275-288.

Schlechte, J.W. 1989. Routine metabolism and critical oxygen concentration for juvenile red drum *Sciaenops ocellatus* as functions of water hardness and salinity. Master's thesis. Texas A&M University, College Station, Texas, USA. 60 pages.

Smit, H. 1965. Some experiments on the oxygen consumption of goldfish (*Carassius auratus* L.) in relation to swimming speed. Canadian Journal of Zoology 43:623-633.

Smith, L.S. 1982. Introduction to fish physiology. T.F.H. Publications, Neptune, New Jersey, USA.

Springer, T.A., and W.H. Neill. 1988. Automated determination of critical oxygen concentration for routinely active fish. Environmental Biology of Fishes 23:233-240.

Ultsch, G.R., H. Boschung, and M.J. Ross. 1978. Metabolism, critical oxygen tension, and habitat selection in darters (*Etheostoma*). Ecology 59:99-107.

Ultsch, G.R., M.E. Ott, and N. Heisler. 1980. Standard metabolic rate, critical oxygen tension, and aerobic scope for spontaneous activity of trout (*Salmo gairdneri*) and carp (*Cyprinus carpio*) in acidified water. Comparative Biochemistry and Physiology 67A:329-335.

Withers, P.C., and S.S. Hillman. 1988. A steady-state model of maximal oxygen and carbon dioxide transport in anuran amphibians. Journal of Applied Physiology 64:860-868.

TOXICITY OF AMMONIA, NITRITE, AND NITRATE TO FISHES

Rosemarie C. Russo and Robert V. Thurston

ABSTRACT

Ammonia and nitrite are highly toxic to fishes, with ammonia occurring in surface waters more commonly than nitrite. Nitrate is a related compound but is not significantly toxic to fishes. The acute toxicity of ammonia to aquatic organisms is affected by water pH, dissolved oxygen, temperature, concentration fluctuations, degree of salinity, presence of other chemicals, and prior acclimation. The acute toxicity of nitrite is known to be affected by water pH, and the presence of chloride and calcium. More research is needed on the effects of these and other variables on the acute toxicity of both ammonia and nitrite, as well as on the chronic effects of both of these toxins.

INTRODUCTION

Ammonia, although toxic to fishes, is common to many aquatic systems because it is a naturally occurring product of metabolic degradation. In addition, it is frequently added to aquatic systems from industrial, municipal, and agricultural sources.

Nitrite, although also toxic to fishes, usually does not occur in natural aquatic systems at concentrations considered to be toxic. In aquaculture systems, however, where ammonia concentrations may be high, nitrite has a much greater likelihood of reaching toxic concentrations.

Nitrate, although generally present in all but the most oligotrophic surface waters, is relatively nontoxic to fishes. Where it is present in high concentrations, the problem is usually one of contributing to eutrophication rather than to toxicity.

These three compounds are interrelated through the process of nitrification, the biological oxidation of ammonia to nitrate. Under aerobic conditions ammonia (NH_3) is readily oxidized to nitrite (NO_2^-) by *Nitrosomonas* bacteria:

$$2NH_3 + 3O_2 \xrightarrow{\textit{Nitrosomonas}} 2NO_2^- + 2H^+ + 2H_2O$$

Nitrite, in turn, is oxidized to nitrate (NO_3^-) by Nitrobacter bacteria:

$$2NO_2^- + O_2 \xrightarrow{\textit{Nitrobacter}} 2NO_3^-$$

In well-oxygenated aquatic systems, the conversion of ammonia to nitrite is the rate-limiting step in the total process, and the conversion of nitrite to nitrate occurs fairly rapidly; nitrite concentrations in most natural systems, therefore, are usually low.

AMMONIA

Ammonia assumes two chemical forms in aqueous solution, shown by the equilibrium equation

$$NH_3 + nH_2O = NH_3{\cdot}nH_20 = NH_4^+ + OH^- + (n\text{-}1)H_2O.$$

These species are the un-ionized form (NH_3), hydrogen-bonded to at least three ($n > 3$) water molecules (Butler 1964), and the ionized form (NH_4^+). The toxicity of ammonia is generally attributed to NH_3, which is an extremely soluble gas. Total ammonia is the sum of NH_3 and NH_4^+, and it is total ammonia that is most commonly measured; the concentration of NH_3 is then calculated. Tables of NH_3 concentration, as a function of total ammonia concentration, over a range of pH values from 5 to 12 and range of temperatures from 0 to 40 C are available (Emerson et al. 1975; Thurston et al. 1979).

The relative concentrations of ionized and un-ionized ammonia in a given solution are a function of the pH, temperature, and ionic strength of that solution. As pH increases, the concentration of NH_3 increases, while that of NH_4^+ decreases. For example, a pH increase from 7.0 to 8.0 in the temperature range 0 to 30 C results in a nearly 10-fold increase in the concentration of NH_3. Temperature increase also favors the NH_3 species, but to a lesser extent; a temperature increase of 5 degrees between 0 and 30 C at pH 7.0 results in an NH_3 concentration increase of 40-50%. An increase in the ionic strength of a solution, at low concentrations, favors the NH_4^+ species. In natural waters with low to moderate amounts of dissolved solids (200-1000 mg/l), this effect will slightly reduce the concentration of NH_3, and the magnitude of this effect will vary with the composition of the water (Thurston et al. 1979).

Short-term exposure of fishes to high concentrations of ammonia causes increased gill ventilation, hyperexcitability, loss of equilibrium, convulsions, and then death (Smart 1978; Thurston et al. 1981c). These effects are most likely the result of a direct effect of ammonia on the central nervous system. Chronic exposure of fishes to lesser concentrations of ammonia include tissue damage, decrease in reproductive capacity (numbers of eggs produced, egg viability, delay in spawning), decrease in growth, and increase in susceptibility to disease (Thurston et al. 1984, 1986). Chronic exposure may also cause progressive deterioration of other physiological functions, any one of which may be the ultimate cause of death (Randall and Wright 1987).

Acute Toxicity

There is some variation in the susceptibility of fishes to the acutely toxic effects of ammonia. Representative 96-hour acute toxicity values (96-hour median lethal concentrations, or LC50 values) are listed in Table 1. From available data, salmonids appear to be among the most sensitive to acute ammonia exposure, and centrarchids, catfish, and some minnows the most tolerant. In addition to differences in susceptibility among fish species, there can also be differences in susceptibility at different life stages of a given species (Thurston and Russo 1983). Moreover, several factors can affect the toxicity of ammonia, including environmental pH, dissolved oxygen, temperature, salinity and ionic composition, previous acclimation and intermittency of exposure, and presence of other toxicants.

Table 1. Representative acute toxicity values for un-ionized ammonia. (Values reported are ranges.)

Species	96-hour LC50 (mg/l NH3)	Reference
Pink salmon (*Oncorhynchus gorbuscha*)	0.08-0.1	Rice & Bailey (1980)
Mountain whitefish (*Prosopium williamsoni*)	0.14-0.47	Thurston & Meyn (1984)
Brown trout (*Salmo trutta*)	0.50-0.70	Thurston & Meyn (1984)
Rainbow trout (*Oncorhynchus mykiss*)	0.16-1.1	Broderius & Smith (1979) DeGraeve et al. (1980) Calamari et al. (1981) Thurston & Russo (1983)
Largemouth bass (*Micropterus salmoides*)	0.9-1.4	Roseboom & Richey (1977)
Smallmouth bass (*Micropterus dolomieui*)	0.69-1.8	Broderius et al. (1985)
Common carp (*Cyprinus carpio*)	2.2	Hasan & Macintosh (1986a)
Red shiner (*Notropis lutrensis*)	2.8-3.2	Hazel et al. (1979)
Fathead minnow (*Pimephales promelas*)	0.75-3.4	Thurston et al. (1983)
Channel catfish (*Ictalurus punctatus*)	0.50-3.8	Colt & Tchobanoglous(1976) Roseboom & Richey (1977) Arthur et al. (1987)
Bluegill (*Lepomis macrochirus*)	0.55-3.0	Emery & Welch (1969) Roseboom & Richey (1977)

Effect of pH

Chipman (1934), Wuhrmann et al. (1947), and Wuhrmann and Woker (1948) observed that the toxicity of ammonia solutions was greater at higher pH values, and because the percentage of NH_3 in total ammonia increases as solution pH increases, these authors concluded that NH_3 was the toxic form of ammonia in the water. They also concluded that NH_4^+ was non-toxic or appreciably less toxic. More recently, several researchers have reported that NH_3 is more toxic at low pH, separate from the effect of pH on the NH_3/NH_4^+ equilibrium in the water. This greater toxic effect has been demonstrated with prawn larvae and Daphnia, as well as with fishes (rainbow trout *Oncorhynchus mykiss*, fathead minnow *Pimephales promelas*, coho salmon *Onchorhynchus kisutch*, chinook salmon *O. tshawytscha*, channel catfish *Ictalurus punctatus*, and green sunfish *Lepomis cyanellus*) (Tabata 1962; Sousa et al. 1974; Robinson-Wilson and Seim 1975; Armstrong et al. 1978; Hillaby and Randall 1979; Tomasso et al. 1980a; Thurston et al. 1981c; McCormick et al. 1984; Broderius et al. 1985; Sheehan and Lewis 1986). Results of one such study (Thurston et al. 1981c) showed that as pH decreased there was a decrease in the 96-hour LC50 value for NH_3 for rainbow trout (Fig. 1) and fathead minnows, indicating that fishes were more susceptible to NH_3 at low pH. In longer exposures, Broderius et al. (1985) also reported a marked effect of pH on the toxicity of NH_3 in a study on smallmouth bass (*Micropterus dolomieui*) from late embryo through 32 days of exposure. The 32-day "no-observed- effect" concentrations radically decreased from 0.61 to 0.044 mg/l NH_3 as pH was lowered from 8.68 to 6.60. In a preliminary study, we have observed that the toxicity of NH_3 also increased at pH values above 9 (Fig. 2) (Russo et al. 1988).

Effect of Dissolved Oxygen

Several researchers, working with a variety of fish species, have reported that the toxicity of ammonia increases when dissolved oxygen concentrations decrease (Wuhrmann 1952; Wuhrmann and Woker 1953; Downing and Merkens 1955; Merkens and Downing 1957; Alabaster et al. 1979; Thurston et al. 1981b). Ammonia discharges are frequently associated with reduced oxygen concentrations in the receiving water, so the effect of dissolved oxygen on ammonia toxicity can be important.

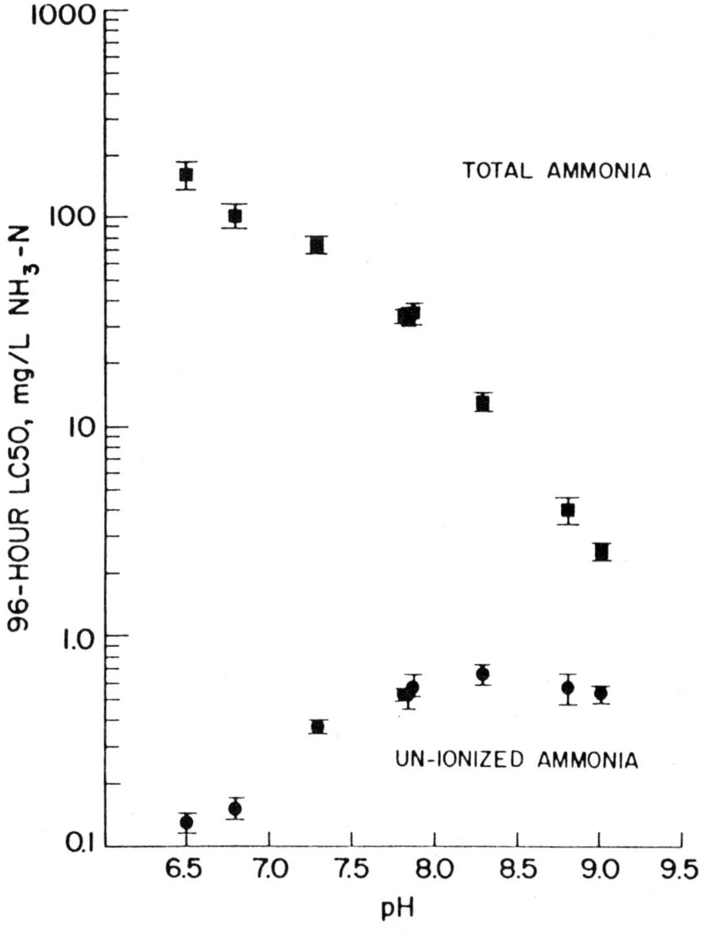

Figure 1. Acute toxicity of ammonia (96-hour LC50 values) to rainbow trout at various environmental pH values. Error bars are 95% confidence intervals. (Reprinted with permission from Thurston et al. 1981c; copyright 1981, American Chemical Society)

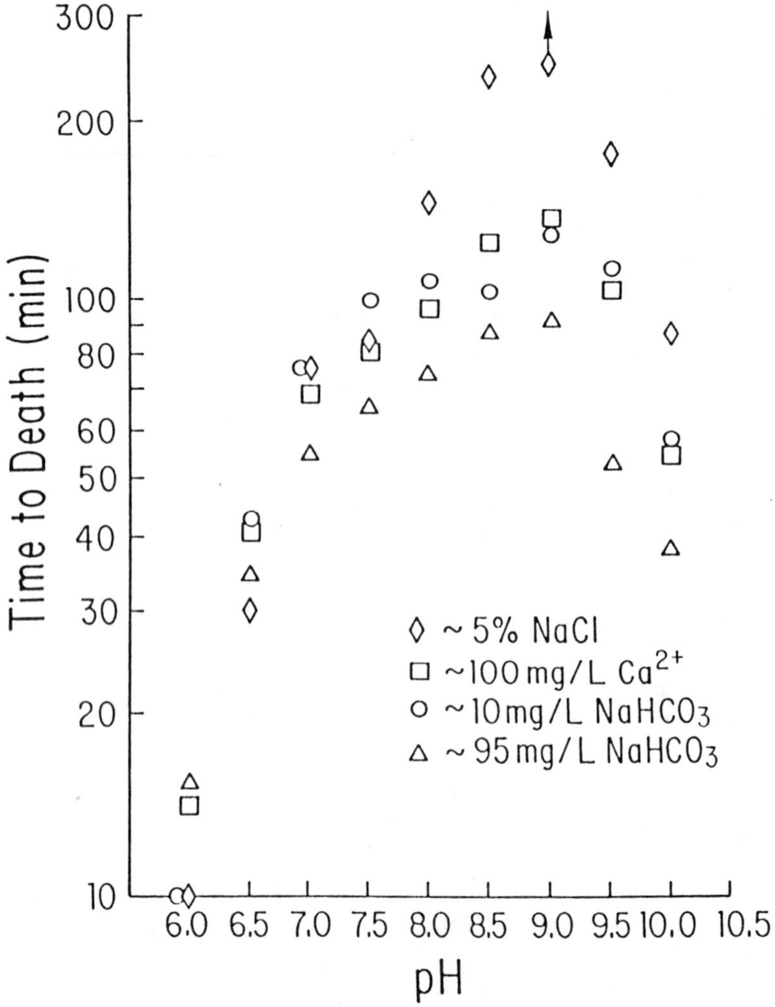

Figure 2. Effect of pH and different water chemistry variables on the acute toxicity of un-ionized ammonia to coho salmon alevins. (Reprinted with permission from Russo et al. 1988; American Fisheries Society)

Effect of Temperature

Many researchers have reported that ammonia is less toxic to fishes at temperatures near the higher end of their normal environmental range than near the lower end, but not all studies support this. Colt and Tchobanoglous (1976) reported a decrease in acute toxicity of ammonia to channel catfish with an increase in temperature over the range 22-30 C, and Roseboom and Richey (1977) reported that bluegill (*Lepomis macrochirus*), largemouth bass (*Micropterus salmoides*), and channel catfish were more susceptible to ammonia toxicity at 22 than at 28-30 C. Arthur et al. (1987) conducted 96-hour ammonia toxicity tests on white sucker (*Catostomus commersoni*), walleye (*Stizostedion vitreum vitreum*), rainbow trout, fathead minnow, and channel catfish at temperatures ranging from 3.4 to 26 C. Except for channel catfish, none of the tests showed a progressive increase in LC50 with increasing water temperature, and no clear relationship was found between ammonia toxicity and temperature; the 96-hour LC50 values for channel catfish were 0.50, 0.98, and 1.29 mg/liter NH_3 at 3.5, 14.6, and 19.6 C.

Fluctuating Concentrations and Acclimation

Fish are frequently subjected to fluctuating concentrations of ammonia as a consequence of diurnal changes. Also, episodic "slugs" of ammonia, as a result of accidental spills or intentional discharges into rivers and lakes, are not uncommon. Thurston et al. (1981a) studied the acutely toxic effects of fluctuating concentrations of ammonia on rainbow trout and cutthroat trout (*Salmo clarki*), and reported that fish were able to withstand short-term excursions slightly above acutely toxic concentrations without any apparent long-term adverse effects, provided the high ammonia concentrations were followed by periods of low ammonia concentration. However, they also reported that fish were better able to withstand steady concentrations of ammonia, over 96 hours, than they were fluctuating concentrations having mean values comparable to those of the steady concentration for the same length of time. Increased tolerance to acute toxicity from ammonia as a consequence of prior exposure to low ammonia concentrations has been reported by Vamos (1963), Malacea (1968), Schulze-Wiehenbrauck (1976), Redner and Stickney (1979), Thurston et al. (1981a), and Alabaster et al. (1983).

Soderberg and co-workers (1983), however, have reported that tissue damage in rainbow trout reared for 4 months in earthen ponds was correlated with ammonia concentration extremes, as opposed to ammonia averages.

Mixtures of Ammonia and Other Chemicals

Information is limited on the increase or decrease of the toxic effects of ammonia as a result of exposure in combination with other chemicals, but an increase in ionic strength appears to reduce ammonia toxicity, separate from the effect of ionic activity on the NH_3/NH_4^+ equilibrium. Increased calcium concentrations in the water have been reported to decrease ammonia toxicity to channel catfish (Tomasso et al. 1980a), and an increase in water salinity has been reported to reduce the toxicity of ammonia to Atlantic salmon (*Salmo salar*) (Alabaster et al. 1979). A decrease in the toxicity of ammonia to juvenile coho salmon with the addition of sodium chloride to the test solution has also been observed in our laboratories (Fig. 2).

Wuhrmann and Woker (1948) and Broderius and Smith (1979) studied combinations of ammonia and hydrogen cyanide and reported synergistic effects of the two toxicants. The toxicity of ammonia and phenol combined was reported to approximate the toxicity of either ammonia or phenol when tested separately (Herbert 1962). Research has also been conducted on the effect of ammonia in combination with copper (Herbert and Vandyke 1964), zinc (Ministry of Technology, U.K. 1962), and combinations of these (Brown et al. 1969); the toxicity of ammonia in combination with these other toxicants appears to be simply additive.

Chronic Toxicity

The longest ammonia exposure study reported in the literature is a life-cycle test on rainbow trout, conducted over a 5-year period (Thurston et al. 1984). Fish were tested at five exposure concentrations over the range 0.01-0.07 mg/l NH_3, parental fish were exposed for 11 months, the first filial generation (F1) for 4 years, and the second filial generation (F2) for 5 months. The parental fish spawned of their own volition, and F1 fish were manually spawned at 4 years of age. Blood ammonia concentrations in F1 fish were positively correlated with ammonia concentrations in the test waters, and histopathological lesions were common in parental and F1 fish at 0.02 mg/l NH_3 and higher. No other major effects were observed. Viable eggs were produced by both generations at all ammonia concentrations tested, and there was no significant correlation between ammonia concentration and numbers of egg lots spawned, total numbers of eggs produced, numbers of viable eggs, growth of progeny, or mortality of parents or progeny in any of the generations tested.

Effects of ammonia on fish reproduction and survival were observed, however, in two life-cycle tests on fathead minnows (Thurston et al. 1986). Fish were tested at five exposure concentrations for 1 year over the range 0.07-0.96 mg/l NH_3. Growth, survival, and egg production of parental fish were all affected at 0.96 mg/l NH_3; no effects were observed on growth or survival of parental fish at 0.44 mg/l, or on egg production or viability at 0.37 mg/l. Growth and survival of F1 larvae were not affected at 0.36 mg/l NH_3 (the highest concentration at which they were tested), and egg hatching success was not affected at 0.19 mg/l, but was at 0.37 mg/l. Brain lesions were common in parental fish at all stages of development at exposure concentrations of 0.21 mg/l NH_3 and higher, but not at 0.11 mg/l. In summary, the chronic effects threshold concentration for these two tests, based on survival, growth and reproductive success, was estimated to be 0.27 mg/l NH_3; based on histological damage it was estimated to be only half that, 0.15 mg/l NH_3.

Other studies at sublethal concentrations of ammonia, lasting from 1 week to 3 months, have been reported for both salmonid and non-salmonid species (Department of the Environment, U.K. 1972; Robinette 1976; Schulze-Wiehenbrauck 1976; Burkhalter and Kaya 1977; Thurston et al. 1978; Rice and Bailey 1980). At exposure concentrations as low as 0.002-0.15 mg/l NH_3, fish showed reduced food uptake and assimilation, accompanied by growth inhibition. In the range 0.04-0.4 mg/l NH_3, effects reported for a variety of fish species, under a wide range of water test conditions, included leucopenia and diminished numbers of red blood cells, inflammation and degeneration of gills and kidneys, and lowered resistance to disease (Reichenbach-Klinke 1967; Flis 1968; Smart 1976; Thurston et al. 1978; Peters et al. 1984; Soderberg 1985; Dabrowska and Wlasow 1986; Lang et al. 1987).

Carline et al. (1987) tested the effects of domestic wastewater on the survival, growth, swimming performance, and gill tissue of brown trout (*Salmo trutta*) for 12 months at concentrations from 0.005 to 0.066 mg/l NH_3. There were no significant effects reported for survival, growth, or swimming performance, but degree of damage to the gills was directly related to effluent concentrations. Mitchell and Cech (1983) have implicated chlorine as confounding ammonia-caused gill damage. Burrows (1964) tested rainbow trout for 6 weeks at concentrations as low as 0.003 mg/l NH_3 (recalculated from the author's original data) and reported extensive gill damage, but Doust and Ferguson (1984) tested that same species at concentrations up to 0.4 mg/l NH_3 for 90 days and did not observe "characteristic" gill damage. Meade and Herman (1986) reported that lake trout (*Salvelinus namaycush*), reared for 8 weeks in a water re-use system, showed gill and kidney damage of the kind reported to be caused by ammonia, and yet the ammonia concentrations to which these fish were exposed averaged 0.001 mg/l NH_3 or less within each 2-week period, and the highest recorded value was 0.003 mg/l. Their conclusion was that "parameters other than ammonia concentrations are significant factors in branchial irritation" of fish under culture.

In an extensive review article, Meade (1985) states as part of his summary: "The accumulating evidence indicates that gill hyperplasia, reported as characteristic of ammonia poisoning, is probably not caused by un-ionized ammonia. ... All end products of metabolism probably have not been identified, and certainly their interactions in water of various qualities and in various fish culture systems have not been determined."

NITRITE

Although nitrite is usually present in natural water systems only in trace amounts, the oxidation process that converts it to nitrate in wastewater treatment facilities and in aquaculture water reuse systems is sometimes impeded, causing potentially toxic amounts of nitrite to accumulate. Factors that affect the nitrite conversion process include pH, temperature, dissolved oxygen, number of nitrifying bacteria, and presence of inhibiting compounds (e.g. nitrous acid and NH_3).

In aqueous nitrite solution the following equilibrium is established:

$$NO_2^- + H^+ = HNO_2$$

This equilibrium is affected by pH; as the H^+ concentration increases, the concentration of ionized nitrite (NO_2^-) will decrease, shifting the equilibrium toward an increase in the concentration of un-ionized nitrous acid (HNO_2). It has been demonstrated that the nitrification process is inhibited by the presence of either HNO_2 or NH_3 (Anthonisen et al. 1976). Nitrous acid inhibits both nitrobacters and nitrosomonads at concentrations between 0.22 and 2.8 mg/l. Un-ionized ammonia inhibits nitrobacters at concentrations appreciably lower than those that inhibit nitrosomonads (0.1-1.0 mg/l NH_3 as opposed to 10-150 mg/l NH_3). Both situations impede the conversion of nitrite to nitrate, causing nitrite to accumulate.

Organic compounds can also inhibit the nitrification process. Hockenbury and Grady (1977) found that dodecyclamine, aniline, and *n*-methylaniline at concentrations less than 1 mg/l caused 50% inhibition of ammonia oxidation by *Nitrosomonas*, and *p*-nitrobenzaldehyde, *p*-nitroaniline, and n-methylaniline at 100 mg/l inhibited nitrite oxidation by *Nitrobacter*. Loss of nitrification flora, resulting from use of antibiotics, has been mentioned by Patrick et al. (1979).

Acute Toxicity

Data from acute toxicity tests for a variety of fishes reveal a very wide range of results, not just between different families of fishes, but within families as well (Table 2). The breadth of this range of sensitivity is probably attributable both to differences in water chemistry for the different bioassay tests, and to physiological differences among species. Salmonids appear to be the most susceptible group of fishes.

McCoy (1972) tested 13 fish species and observed a wide variation in susceptibility. At 5 mg/l NO_2-N, mortality occurred in less than 3 hours in logperch (*Percina caprodes*) and in 3-5 hours in brook stickleback (*Culaea inconstans*). In contrast, 100% of common carp (*Cyprinus carpio*) survived

Table 2. Representative acute toxicity values for nitrite. (Values reported are ranges.)

Species	48- or 96-hour LC50 (mg/l NO2-N)	Reference
Rainbow trout (*Oncorhynchus mykiss*)	0.19-0.39	Russo et al. (1974) Brown & McLeay (1975)
Cutthroat trout (*Salmo clarki*)	0.48-0.56	Thurston et al. (1978)
Chinook salmon (*Oncorhynchus tshawytscha*)	0.88	Westin (1974)
Common carp (*Cyprinus carpio*)	2.6	Hasan & Macintosh (1986b)
Channel catfish (*Ictalurus punctatus*)	7.1-13	Konikoff (1975) Colt & Tchobanoglous (1976) Palachek & Tomasso (1984a)
Labyrinth catfish (*Clarias lazera*)	28-32	Hilmy et al. (1987a)
Fathead minnow (*Pimephales promelas*)	2.3-3.0 43	Russo & Thurston (1977) Palachek & Tomasso (1984b)
Goldfish (*Carassius auratus*)	52	Tomasso (1986)
Bluegill (*Lepomis macrochirus*)	80	Tomasso (1986)
Largemouth bass (*Micropterus salmoides*)	140	Palachek & Tomasso (1984a)
Smallmouth bass (*Micropterus dolomieui*)	160	Tomasso (1986)
Green sunfish (*Lepomis cyanellus*)	160	Tomasso (1986)
Guadalupe bass (*Micropterus treculi*)	190	Tomasso & Carmichael (1986)

40 mg/l NO_2-N for 48 hours, and white sucker survived 100 mg/l for 48 hours. Within a species, there is some indication from data on rainbow trout and coho salmon that younger fish are somewhat less susceptible to nitrite than older fish (Russo et al. 1974; Smith and Williams 1974; Perrone and Meade 1977). However, Russo (1980) reported that a review of toxicity tests on rainbow trout over the size range 2- 387 g showed no relationship between fish size and susceptibility to nitrite.

A major cause of nitrite toxicity is the oxidation of blood hemoglobin iron to its ferric state, forming methemoglobin (Bodansky 1951; Kiese 1974) which is not capable of combining reversibly with oxygen. High levels of methemoglobin in fish blood result in hypoxia and death. The presence of high levels of methemoglobin is visually apparent in that the blood becomes brown colored.

Under conditions without stress from environmental nitrite, the total hemoglobin present as methemoglobin in rainbow trout generally falls in the range 1-4%, although a value as high as 17% has been reported (Shterman 1970; Cameron 1971; Smith and Russo 1975; Brown and McLeay 1975). Increased environmental nitrite concentrations, even as low as 0.015 mg/l NO_2-N, produce elevated methemoglobin levels in fish blood (Smith and Williams 1974; Smith and Russo 1975; Brown and McLeay 1975; Crawford and Allen 1977; Perrone and Meade 1977; Bortz 1977; Wedemeyer and Yasutake 1978; Huey and Beitinger 1982; Scarano et al. 1984; Hilmey et al. 1987a).

Thurston et al. (1978) conducted nitrite toxicity tests on cutthroat trout fry and found LC50 values at 36 days to be only slightly lower than the 96-hour values. Wedemeyer and Yasutake (1978) exposed steelhead trout to low NO_2-N concentrations (0.015- 0.060 mg/l) over a 6-month period and reported that growth and the ability of the fish to adapt to seawater were not impaired; varying degrees of gill hyperplasia and lamellar separation were observed early in the test, but after 28 weeks these abnormalities were no longer observed. Krous et al. (1982) acclimated rainbow trout to isotonic seawater for 24 and 48 hours and then transferred these fish to nitrite-containing freshwater; nitrite levels in blood plasma of fish acclimated for 24 hours were significantly lower than in fish acclimated for 48 hours.

The levels of oxidizing agents (e.g. nitrites) and reducing agents (e.g. ascorbic acid) in the blood, and the activity of the reductase system, determine the concentration of methemoglobin in the blood. The effect of methemoglobinemia will be exacerbated when oxygen demand is high, thus it is not surprising that a wide range of toxic levels of nitrite have been reported (EIFAC 1984; MacDonald et al. 1987). One would expect nitrite toxicity and the subsequent increase in methemoglobin formation to be most apparent in those situations involving either hypoxia or high oxygen demand by the fish, e.g. at high temperatures and high levels of activity. Recovery from nitrite toxicity requires the addition of reducing agents in the blood, and/or increased red blood cell turnover.

Nitrite has also been implicated in the formation of N-nitroso compounds (Archer et al. 1971; Wolff and Wasserman 1972; Mirvish 1975; DeFlora and Arillo 1983), and nitrosamines have been shown to be carcinogenic to zebrafish (*Brachydanio rerio*), rainbow trout, and guppy (*Poecilia reticulata*) (Stanton 1965; Ashley and Halver 1968; Sato et al. 1973).

In addition to differences in species' susceptibilities to nitrite, researchers have examined the effect of water chemistry conditions on nitrite toxicity. The principal factors studied were pH, chloride, and calcium; the effects of other anions have also been addressed to a limited extent.

Effect of pH

The aqueous nitrite equilibrium is pH-dependent, as previously mentioned, and the effect of pH on nitrite toxicity was examined by Wedemeyer and Yasutake (1978) and Russo et al. (1981). As pH increases, toxicity in terms of NO_2-N decreases and toxicity in terms of HNO_2-N increases, although neither of these chemical species alone is entirely responsible for the observed toxic action. Russo et al. (1981) concluded that in the pH range 6.4- 9.0 both species contribute significantly to the total toxicity, and the toxicity of HNO_2 is higher per unit weight than that of NO_2^-.

Effect of Chloride and Other Ions

Nitrite toxicity to rainbow trout decreases with increasing chloride ion

(Cl^-) concentration, and the relationship is linear (Russo and Thurston 1977). The mitigating effect of Cl^- on nitrite toxicity has also been reported for coho salmon (Perrone and Meade 1977), steelhead trout (*Oncorhynchus mykiss*) (Wedemeyer and Yasutake 1978), and milkfish (*Chanos chanos*) (Almendras 1987). Other anions have been found to inhibit nitrite toxicity to different degrees; these include sulfate (SO_4^{2-}), phosphate (PO_4^{3-}), and nitrate (NO_3^-) (Russo et al. 1981). Crawford and Allen (1977) studied the effect of calcium (Ca_2^+) and seawater on nitrite toxicity to chinook salmon and reported that increasing the calcium concentration in both fresh water and seawater decreased the toxicity of nitrite. Other fishes, for which it has been reported that sodium or calcium chloride reduce nitrite toxicity and/or methemoglobinemia formation, include channel catfish (Tomasso et al. 1979, 1980b; Huey et al. 1980; Schwedler and Tucker 1983), labyrinth catfish (*Clarias lazera*) (Hilmy et al. 1987b), and common carp (Hasan and Macintosh 1986b).

Current evidence is that nitrite uptake is linked with chloride uptake. Gaino et al. (1984) reported that the gills of rainbow trout exposed to nitrite showed a greater number of chloride cells at all stages of development, suggesting a response by the trout to maintain physiological chloride levels in the presence of nitrite. Tomasso (1986), in noting wide differences among fish species in their susceptibility to nitrite intoxication, correlated this with measured methemoglobin and plasma nitrite levels. He concluded that differences in plasma nitrite levels in different species depend on the discriminatory ability of the fish's active transport system for chloride ions..

NITRATE

The nitrate ion (NO_3^-) is formed by complete oxidation of ammonia through the nitrification process. High concentrations of nitrate are found in eutrophic aquatic systems, and certainly can be found in aquaculture ponds. Nitrate does not form an un-ionized species in aqueous solution; that is, HNO_3 is completely dissociated to H^+ and NO_3^-. Nitrate is relatively nontoxic to fishes (Table 3), and consequently few reports on its toxicity are available. Indeed, because it is assumed that NO_3^- will not contribute additional toxicity to the system, nitrate compounds are often the chemicals of choice for assessing the toxicities of metals and other cations.

Table 3. Representative acute toxicity values for nitrate. (Values reported are ranges.)

Species	96-hour LC50 (mg/l NO3-N)	Reference
Guppy (*Poecilia reticulata*)	180-200	Rubin & Elmaraghy (1977)
Guadalupe bass (*Micropterus treculi*)	1260	Tomasso & Carmichael (1986)
Chinook salmon (*Oncorhynchus tshawytscha*)	1310	Westin (1974)
Rainbow trout (*Oncorhynchus mykiss*)	1360	Westin (1974)
Channel catfish (*Ictalurus punctatus*)	1400	Colt & Tchobanoglous (1976)
Bluegill (*Lepomis macrochirus*)	420-2000	Trama (1954)

Trama (1954) tested bluegills in solutions of different nitrate salts, and found the fish were most susceptible to potassium nitrate (KNO3); sodium and calcium nitrates ($NaNO_3$ and $Ca(NO_3)_2$), were less toxic. The fish had 100% survival when exposed for 96 hours to $NaNO_3$ solutions containing up to 1650 mg/l NO_3-N. Colt and Tchobanoglous (1976) reported the acute (96-hour) toxicity of nitrate to channel catfish at 22, 26, and 30 C to be independent of temperature. They stated that the toxicity of nitrate appears to be related to the fish being unable to maintain normal osmoregulatory ability when transferred to water with a high salt content. In a longer-term study of nitrate toxicity on channel catfish and largemouth bass (Knepp and Arkin 1973), fish were held in a culture system in which nitrate buildup gradually occurred over a 164-day period; a concentration of 90 mg/l NO_3-N was tolerated by the catfish and bass without affecting their growth and feeding activity.

Nitrates may increase the net productivity of aquatic systems. To a certain extent this may be beneficial to the aquatic community. When nitrate concentrations become excessive, however, and other essential nutrient factors are present, eutrophication and associated algal blooms can become a problem even though the nitrate concentrations per se may not be significantly toxic to fishes.

LITERATURE CITED

Alabaster, J.S., D.G. Shurben, and G. Knowles. 1979. The effect of dissolved oxygen and salinity on the toxicity of ammonia to smolts of salmon, *Salmo salar* L. Journal of Fish Biology 15:705-712.

Alabaster, J.S., D.G. Shurben, and M.J. Mallett. 1983. The acute lethal toxicity of mixtures of cyanide and ammonia to smolts of salmon, *Salmo salar* L. at low concentrations of dissolved oxygen. Journal of Fish Biology 22:215-222.

Almendras, J.M.E. 1987. Acute nitrite toxicity and methemoglobinemia in juvenile milkfish (*Chanos chanos* Forsskal). Aquaculture 61:33-40.

Anthonisen, A.C., R.C. Loehr, T.B.S. Prakasam, and E.G. Srinath. 1976. Inhibition of nitrification by ammonia and nitrous acid. Journal of the Water Pollution Control Federation 48:835-852.

Archer, M.C., S.D. Clark, J.E. Thilly, and S.R. Tannenbaum. 1971. Environmental nitroso compounds: reaction of nitrite with creatine and creatinine. Science 174:1341-1343.

Armstrong, D.A., D. Chippendale, A.W. Knight, and J.E. Colt. 1978. Interaction of ionized and un-ionized ammonia on short-term survival and growth of prawn larvae, *Macrobrachium rosenbergii*. Biological Bulletin of the Woods Hole Institute 154:15-31.

Arthur, J.W., C.W. West, K.N. Allen, and S.F. Hedtke. 1987. Seasonal toxicity of ammonia to five fish and nine invertebrate species. Bulletin of Environmental Contamination and Toxicology 38:324-331.

Ashley, L.M. and J.E. Halver. 1968. Dimethylnitrosamine-induced hepatic cell carcinoma in rainbow trout. Journal of the National Cancer Institute 41:531-552.

Bodansky, O. 1951. Methemoglobinemia and methemoglobin-producing compounds. Pharmacological Reviews 3:144-196.

Bortz, B.M. 1977. The administration of tetramethylthionine chloride as a treatment for nitrite-induced methemoglobinemia in rainbow trout (*Salmo gairdneri*). M.S. thesis, The American University, Washington, District of Columbia, USA.

Broderius, S. R., Drummond, J. Fiandt, and C. Russom. 1985. Toxicity of ammonia to early life stages of the smallmouth bass at four pH values. Environmental Toxicology and Chemistry 4:87-96.

Broderius, S.J. and L.L. Smith, Jr. 1979. Lethal and sublethal effects of binary mixtures of cyanide and hexavalent chromium, zinc, or ammonia to the fathead minnow (*Pimephales promelas*) and rainbow trout (*Salmo gairdneri*). Journal of the Fisheries Research Board of Canada 36:164-172.

Brown, D.A. and D.J. McLeay. 1975. Effect of nitrite on methemoglobin and total hemoglobin of juvenile rainbow trout. The Progressive Fish-Culturist 37:36-38.

Brown, V.M., D.H.M. Jordan, and B.A. Tiller. 1969. The acute toxicity to rainbow trout of fluctuating concentrations and mixtures of ammonia, phenol and zinc. Journal of Fish Biology 1:1-9.

Burkhalter, D.E. and C.M. Kaya. 1977. Effects of prolonged exposure to ammonia on fertilized eggs and sac fry of rainbow trout (*Salmo gairdneri*). Transactions of the American Fisheries Society 106:470-475.

Burrows, R.C. 1964. Effects of accumulated excretory products on hatchery-reared salmonids. Bureau of Sport Fisheries and Wildlife, U.S. Fish and Wildlife Service, Research Report 66, 12p.

Butler, J.N. 1964. Ionic equilibrium. Addison-Wesley Publishing Company, Inc., Reading, Massachusetts, USA.

Calamari, D., R. Marchetti, and G. Vailati. 1981. Effects of long-term exposure to ammonia on the developmental stages of rainbow trout (*Salmo gairdneri* Richardson). Rapports et Proces-Verbaux des Reunions, Conseil International pour l'Exploration de la Mer 178:81-86.

Cameron, J.N. 1971. Methemoglobin in erythrocytes of rainbow trout. Comparative Biochemistry and Physiology 40A:743-749.

Carline, R.F., A.J. Benson, and H. Rothenbacher. 1987. Long- term effects of treated domestic wastewater on brown trout. Water Research 21:1409-1415.

Chipman, W.A., Jr. 1934. The role of pH in determining the toxicity of ammonium compounds. Ph.D. thesis, University of Missouri, Columbia, Missouri, USA.

Colt, J. and G. Tchobanoglous. 1976. Evaluation of the short-term toxicity of nitrogenous compounds to channel catfish, *Ictalurus punctatus*. Aquaculture 8:209-224.

Crawford, R.E. and G.H. Allen. 1977. Seawater inhibition of nitrite toxicity to chinook salmon. Transactions of the American Fisheries Society 106:105-109.

Dabrowska, H., and T. Wlasow. 1986. Sublethal effect of ammonia on certain biochemical and haematological indicators in common carp (*Cyprinus carpio* L.). Comparative Biochemistry and Physiology 83C:179-184.

Daoust, P.-Y. and H.W. Ferguson. 1984. The pathology of chronic ammonia toxicity in rainbow trout, *Salmo gairdneri* Richardson. Journal of Fish Diseases 7:199-205.

De Flora, S. and A. Arillo. 1983. Mutagenic and DNA damaging activity in muscle of trout exposed *in vivo* to nitrite. Cancer Letters 20:147-155.

DeGraeve, G.M., R.L. Overcast, and H.L. Bergman. 1980. Toxicity of underground coal gasification condenser water and selected constituents to aquatic biota. Archives of Environmental Contamination and Toxicology 9:543-555.

Department of the Environment, U.K. 1972. Effects of pollution on fish. In: Water Pollution Research, 1971. H.M. Stationery Office, London, England. p. 37.

Downing, K.M. and J.C. Merkens. 1955. The influence of dissolved-oxygen concentration on the toxicity of un-ionized ammonia to rainbow trout (*Salmo gairdneri* Richardson). Annals of Applied Biology 43:243-246.

Emerson, K., R.C. Russo, R.E. Lund, and R.V. Thurston. 1975. Aqueous ammonia equilibrium calculations: effect of pH and temperature. Journal of the Fisheries Research Board of Canada 32:2379-2383.

Emery, R.M. and E.B. Welch. 1969. The toxicity of alkaline solutions of ammonia to juvenile bluegill sunfish (*Lepomis macrochirus* Raf.). Water Quality Branch, Division of Health and Safety, Tennessee Valley Authority, Chattanooga, Tennessee, USA.

EIFAC (European Inland Fisheries Advisory Commission). 1984. Water quality criteria for European freshwater fish. Report on nitrite and freshwater fish. EIFAC Technical Paper 46, Food and Agricultural Organization of the United Nations, Rome, Italy.

Flis, J. 1968. Anatomicohistopathological changes induced in carp (*Cyprinus carpio* L.) by ammonia water. Part II. Effects of subtoxic concentrations. Acta Hydrobiologica 10:225-238.

Gaino, E., A. Arillo, and P. Mensi. 1984. Involvement of the gill chloride cells of trout under acute nitrite intoxication. Comparative Biochemistry and Physiology 77A:611-617.

Hasan, M.R. and D.J. Macintosh. 1986a. Acute toxicity of ammonia to common carp fry. Aquaculture 54:97-107.

Hasan, M.R. and D.J. Macintosh. 1986b. Effect of chloride concentration on the acute toxicity of nitrite to common carp, *Cyprinus carpio* L., fry. Aquaculture and Fisheries Management 17:19-30.

Hazel, R.H., C.E. Burkhead, and D.G. Huggins. 1979. The development of water quality criteria for ammonia and total residual chlorine for the protection of aquatic life in two Johnson County, Kansas streams. Office of Water Research and Technology, U.S. Department of the Interior, Washington, District of Columbia, USA.

Herbert, D.W.M. 1962. The toxicity to rainbow trout of spent still liquors from the distillation of coal. Annals of Applied Biology 50:755-777.

Herbert, D.W.M. and J.M. Vandyke. 1964. The toxicity to fish of mixtures of poisons. II. Copper-ammonia and zinc-phenol mixtures. Annals of Applied Biology 53:415-421.

Hillaby, B.A. and D.J. Randall. 1979. Acute ammonia toxicity and ammonia excretion in rainbow trout (*Salmo gairdneri*). Journal of the Fisheries Research Board of Canada 36:621-629.

Hilmy, A.M., N.A. El-Domiaty, and K. Wershana. 1987a. Acute and chronic toxicity of nitrite to *Clarias lazera*. Comparative Biochemistry and Physiology 86C:247-253.

Hilmy, A.M., N.A. El-Domiaty, and K. Wershana. 1987b. Effect of sodium chloride and calcium chloride on nitrite induced methemoglobinemia in *Clarias lazera*. Water, Air, and Soil Pollution 33:57-63.

Hockenbury, M.R. and C.P.L. Grady, Jr. 1977. Inhibition of nitrification—effects of selected organic compounds. Journal of the Water Pollution Control Federation 49:768-777.

Huey, D.W. and T.L. Beitinger. 1982. Methemoglobin levels in channel catfish, *Ictalurus punctatus*, exposed to nitrite and tricaine methanesulfonate. Canadian Journal of Fisheries and Aquatic Sciences 39:643-645.

Huey, D.W., B.A. Simco, and D.W. Criswell. 1980. Nitrite-induced methemoglobin formation in channel catfish. Transactions of the American Fisheries Society 109:558-562.

Kiese, M. 1974. Methemoglobinemia: a comprehensive treatise. Chemical Rubber Company, Cleveland, Ohio, USA.

Knepp, G.L. and G.F. Arkin. 1973. Ammonia toxicity levels and nitrate tolerance of channel catfish. The Progressive Fish-Culturist 35:221-224.

Konikoff, M. 1975. Toxicity of nitrite to channel catfish. The Progressive Fish-Culturist 37:96-98.

Krous, S.R., V.S. Blazer, and T.L. Meade. 1982. Effect of acclimation time on nitrite movement across the gill epithelia of rainbow trout: the role of "chloride cells". The Progressive Fish-Culturist 44:126-130.

Lang, T., G. Peters, R. Hoffmann, and E. Meyer. 1987. Experimental investigations on the toxicity of ammonia: effects on ventilation frequency, growth, epidermal mucous cells, and gill structure of rainbow trout *Salmo gairdneri*. Diseases of Aquatic Organisms 3:159-165.

MacDonald, D.D., L.E. Fidler, and D. Valiela. 1987. Site specific water quality criteria for fish and aquatic life in the Canadian portion of the Flathead River Basin: nitrate, nitrite and ammonia. Environment Canada Conservation and Protection, Inland Waters and Lands, Pacific and Yukon Region, Vancouver, British Columbia, Canada.

Malacea, I. 1968. Untersuchungen uber die Gewohnung der Fische an hohe Konzentrationen toxischer Substanzen. (Studies on the acclimation of fish to high concentrations of toxic substances.). Archiv fur Hydrobiologie 65:74-95, (English translation).

McCormick, J.H., S.J. Broderius, and J.T. Fiandt. 1984. Toxicity of ammonia to early life stages of the green sunfish *Lepomis cyanellus*. Environmental Pollution (Series A) 36:147- 163.

McCoy, E.F. 1972. Role of bacteria in the nitrogen cycle in lakes. Water Pollution Control Research Series 16010 EHR 03/72, Office of Research and Monitoring, U.S. Environmental Protection Agency, Washington, District of Columbia, USA.

Meade, J.W. 1985. Allowable ammonia for fish culture. The Progressive Fish-Culturist 47:135-145.

Meade, J.W. and R.L. Herman. 1986. Histological changes in cultured lake trout, *Salvelinus namaycush*, subjected to cumulative loading in a water reuse system. Canadian Journal of Fisheries and Aquatic Sciences 43:228-231.

Merkens, J.C. and K.M. Downing. 1957. The effect of tension of dissolved oxygen on the toxicity of un-ionized ammonia to several species of fish. Annals of Applied Biology 45:521-527.

Ministry of Technology, U.K. 1962. Effects of pollution on fish. Toxicity of mixtures of zinc sulphate and ammonium chloride. In: Water Pollution Research, 1961. H.M. Stationery Office, London, England. pp. 90-93.

Mirvish, S.S. 1975. N-Nitroso compounds, nitrite, and nitrate: possible implications for the causation of human cancer. *In*: Proceedings of a Conference on Nitrogen as a Water Pollutant, Vol. 1 - Analysis, Sources, Public Health, August 18-20, 1975, Copenhagen. International Association on Water Pollution Research, London, England.

Mitchell, S.J. and J.J. Cech, Jr. 1983. Ammonia-caused gill damage in channel catfish (*Ictalurus punctatus*): confounding effects of residual chlorine. Canadian Journal of Fisheries and Aquatic Sciences 40:242-247.

Palachek, R.M. and J.R. Tomasso. 1984a. Toxicity of nitrite to channel catfish (*Ictalurus punctatus*), tilapia (*Tilapia aurea*), and largemouth bass (*Micropterus salmoides*): evidence for a nitrite exclusion mechanism. Canadian Journal of Fisheries and Aquatic Sciences 41:1739-1744.

Palachek, R.M. and J.R. Tomasso. 1984b. Nitrite toxicity to fathead minnows: effect of fish weight. Bulletin of Environmental Contamination and Toxicology 32:238-242.

Patrick, R., J.E. Colt, R.E. Crawford, B.A. Manny, R.C. Russo, R.V. Thurston, and G.A. Wedemeyer. 1979. Nitrates, nitrites. *In*: A Review of the EPA Red Book: Quality Criteria for Water, edited by R.V. Thurston, R.C. Russo, C.M. Fetterolf, Jr., T.A. Edsall, and Y.M. Barber, Jr. Water Quality Section, American Fisheries Society, Bethesda, Maryland, USA. pp. 158-162.

Perrone, S.J. and T.L. Meade. 1977. Protective effect of chloride on nitrite toxicity to coho salmon (*Oncorhynchus kisutch*). Journal of the Fisheries Research Board of Canada 34:486-492.

Peters, G., R. Hoffmann, and H. Klinger. 1984. Environment- induced gill disease of cultured rainbow trout (*Salmo gairdneri*). Aquaculture 38:105-126.

Randall, D.J. and P.A. Wright. 1987. Ammonia distribution and excretion in fish. Fish Physiology and Biochemistry 3:107- 120.

Redner, B.D. and R.R. Stickney. 1979. Acclimation to ammonia by *Tilapia aurea*. Transactions of the American Fisheries Society 108:383-388.

Reichenbach-Klinke, H.-H. 1967. Untersuchungen uber die Einwirkung des Ammoniakgehalts auf den Fischorganismus. (Investigations on the influence of the ammonia content on the fish organism.). Archiv fur Fischereiwissenschaft 17:122-132, (English translation).

Rice, S.D. and J.E. Bailey. 1980. Survival, size, and emergence of pink salmon, *Oncorhynchus gorbuscha*, alevins after short- and long-term exposures to ammonia. Fisheries Bulletin 78:641-648.

Robinette, H.R. 1976. Effect of selected sublethal levels of ammonia on the growth of channel catfish (*Ictalurus punctatus*). The Progressive Fish-Culturist 38:26-29.

Robinson-Wilson, E.F. and W.K. Seim. 1975. The lethal and sublethal effects of a zirconium process effluent on juvenile salmonids. Water Resources Bulletin 11:975-986.

Roseboom, D.P. and D.L. Richey. 1977. Acute toxicity of residual chlorine and ammonia to some native Illinois fishes. Report of Investigation 85. Illinois State Water Survey, Urbana, Illinois, USA.

Rubin, A.J. and G.A. Elmaraghy. 1977. Studies on the toxicity of ammonia, nitrate and their mixtures to guppy fry. Water Research 11:927-935.

Russo, R.C. 1980. Recent advances in the study of nitrite toxicity to fishes. *In*: Proceedings of the Third USA-USSR Symposium on the Effects of Pollutants upon Aquatic Ecosystems, 2-6 July 1979, Borok, Jaroslavl Oblast, USSR. Edited by W.R. Swain and V.R. Shannon. Ecological Research Series EPA-600/9-80-034, U.S. Environmental Protection Agency, Duluth, Minnesota, USA. pp. 226-240.

Russo, R.C., D.J. Randall, and R.V. Thurston. 1988. Ammonia toxicity and metabolism in fishes. In: Protection of River Basins, Lakes, and Estuaries. Edited by R.C. Ryans, American Fisheries Society, Bethesda, Maryland, USA. pp. 159- 173.

Russo, R.C., C.E. Smith, and R.V. Thurston. 1974. Acute toxicity of nitrite to rainbow trout (*Salmo gairdneri*). Journal of the Fisheries Research Board of Canada 31:1653-1655.

Russo, R.C. and R.V. Thurston. 1977. The acute toxicity of nitrite to fishes. *In*: Recent Advances in Fish Toxicology. Edited by R.A. Tubb. Ecological Research Series EPA-600/3-77-085, U.S. Environmental Protection Agency, Corvallis, Oregon, USA. pp. 118-131.

Russo, R.C., R.V. Thurston, and K. Emerson. 1981. Acute toxicity of nitrite to rainbow trout (*Salmo gairdneri*): effects of pH, nitrite species, and anion species. Canadian Journal of Fisheries and Aquatic Sciences 38:387-393.

Sato, S., T. Matsushima, N. Tanaka, T. Sugimura, and F. Takashima. 1973. Hepatic tumors in the guppy (*Lebistes reticulatus*) induced by aflatoxin B, dimethylnitrosamine, and 2-acetylaminofluorene. Journal of the National Cancer Institute 50:767-778.

Scarano, G., M.G. Saroglia, R.H. Gray, and E. Tibaldi. 1984. Hematological responses of sea bass *Dicentrarchus labrax* to sublethal nitrite exposures. Transactions of the American Fisheries Society 113:360-364.

Schulze-Wiehenbrauck, H. 1976. Effects of sublethal ammonia concentrations on metabolism in juvenile rainbow trout (*Salmo gairdneri* Richardson). Berichte der Deutschen Wissenschaftlichen Kommission fur Meeresforschung 24:234-250.

Schwedler, T.E. and C.S. Tucker. 1983. Empirical relationship between percent methemoglobin in channel catfish and dissolved nitrite and chloride in ponds. Transactions of the American Fisheries Society 112:117-119.

Sheehan, R.J. and W.M. Lewis. 1986. Influence of pH and ammonia salts on ammonia toxicity and water balance in young channel catfish. Transactions of the American Fisheries Society 115:891-899.

Shterman, L.Y. 1970. Methemoglobin in fish blood. Journal of Ichthyology 10:709-712.

Smart, G. 1976. The effect of ammonia exposure on gill structure of the rainbow trout (*Salmo gairdneri*). Journal of Fish Biology 8:471-475.

Smart, G.R. 1978. Investigations of the toxic mechanisms of ammonia to fish — gas exchange in rainbow trout (*Salmo gairdneri*) exposed to acutely lethal concentrations. Journal of Fish Biology 12:93-104.

Smith, C.E. and R.C. Russo. 1975. Nitrite-induced methemoglobinemia in rainbow trout. The Progressive Fish-Culturist 37:150-152.

Smith, C.E. and W.G. Williams. 1974. Experimental nitrite toxicity in rainbow trout and chinook salmon. Transactions of the American Fisheries Society 103:389-390.

Soderberg, R.W. 1985. Histopathology of rainbow trout, *Salmo gairdneri* Richardson, exposed to diurnally fluctuating un- ionized ammonia levels in static-water ponds. Journal of Fish Diseases 8:57-64.

Soderberg, R.W., J.B. Flynn, and H.R. Schmittou. 1983. Effects of ammonia on growth and survival of rainbow trout in intensive static-water culture. Transactions of the American Fisheries Society 112:448-451.

Sousa, R.J., T.L. Meade, and R.E. Wolke. 1974. Reduction of ammonia toxicity by salinity and pH manipulation. Proceedings of the Fifth Annual Workshop, World Mariculture Society 5:343-354.

Stanton, M.F. 1965. Diethylnitrosamine-induced hepatic degeneration and neoplasia in the aquarium fish, *Brachydanio rerio*. Journal of the National Cancer Institute 34:117-130.

Tabata, K. 1962. Suisan dobutsu ni oyobosu amonia no dokusei to pH, tansan to no kankei. (Toxicity of ammonia to aquatic animals with reference to the effect of pH and carbon dioxide.). Tokai-ku Suisan Kenkyusho Kenkyu Hokoku 34:67-74, (English translation).

Thurston, R.V., C. Chakoumakos, and R.C. Russo. 1981a. Effect of fluctuating exposures on the acute toxicity of ammonia to rainbow trout (*Salmo gairdneri*) and cutthroat trout (*S. clarki*). Water Research 15:911-917.

Thurston, R.V. and E.L. Meyn. 1984. Acute toxicity of ammonia to five fish species from the northwest United States. Technical Report No. 84-4, Fisheries Bioassay Laboratory, Montana State University, Bozeman, Montana, USA. 13p.

Thurston, R.V., G.R. Phillips, R.C. Russo, and S.M. Hinkins. 1981b. Increased toxicity of ammonia to rainbow trout (*Salmo gairdneri*) resulting from reduced concentrations of dissolved oxygen. Canadian Journal of Fisheries and Aquatic Sciences 38:983-988.

Thurston, R.V. and R.C. Russo. 1983. Acute toxicity of ammonia to rainbow trout. Transactions of the American Fisheries Society 112:696-704.

Thurston, R.V., R.C. Russo, and K. Emerson. 1979. Aqueous ammonia equilibrium - tabulation of percent un-ionized ammonia. Ecological Research Series, EPA-600/3-79-091, U.S. Environmental Protection Agency, Duluth, Minnesota, USA.

Thurston, R.V., R.C. Russo, R.J. Luedtke, C.E. Smith, E.L. Meyn, C. Chakoumakos, K.C. Wang, and C.J.D. Brown. 1984. Chronic toxicity of ammonia to rainbow trout. Transactions of the American Fisheries Society 113:56-73.

Thurston, R.V., R.C. Russo, E.L. Meyn, and R.K. Zajdel. 1986. Chronic toxicity of ammonia to fathead minnows. Transactions of the American Fisheries Society 115:196-207.

Thurston, R.V., R.C. Russo, and G.R. Phillips. 1983. Acute toxicity of ammonia to fathead minnows. Transactions of the American Fisheries Society 112:705-711.

Thurston, R.V., R.C. Russo, and C.E. Smith. 1978. Acute toxicity of ammonia and nitrite to cutthroat trout fry. Transactions of the American Fisheries Society 107:361-368.

Thurston, R.V., R.C. Russo, and G.A. Vinogradov. 1981c. Ammonia toxicity to fishes: effect of pH on the toxicity of the un-ionized ammonia species. Environmental Science and Technology 15:837-840.

Tomasso, J.R. 1986. Comparative toxicity of nitrite to freshwater fishes. Aquatic Toxicology 8:129-137.

Tomasso, J.R. and G.J. Carmichael. 1986. Acute toxicity of ammonia, nitrite, and nitrate to the Guadalupe bass, *Micropterus treculi*. Bulletin of Environmental Contamination and Toxicology 36:866-870.

Tomasso, J.R., C.A. Goudie, B.A. Simco, and K.B. Davis. 1980a. Effects of environmental pH and calcium on ammonia toxicity in channel catfish. Transactions of the American Fisheries Society 109:229-234.

Tomasso, J.R., B.A. Simco, and K.B. Davis. 1979. Chloride inhibition of nitrite-induced methemoglobinemia in channel catfish (*Ictalurus punctatus*). Journal of the Fisheries Research Board of Canada 36:1141-1144.

Tomasso, J.R., M.I. Wright, B.A. Simco, and K.B. Davis. 1980b. Inhibition of nitrite-induced toxicity in channel catfish by calcium chloride and sodium chloride. The Progressive Fish-Culturist 42:144-146.

Trama, F.B. 1954. The acute toxicity of some common salts of sodium, potassium and calcium to the common bluegill (*Lepomis macrochirus* Rafinesque). Proceedings of the Academy of Natural Sciences of Philadelphia, Pennsylvania, USA. 106:185-205.

Vamos, R. 1963. Ammonia poisoning in carp. Acta Biologica (Szeged) 9:291-297.

Wedemeyer, G.A. and W.T. Yasutake. 1978. Prevention and treatment of nitrite toxicity in juvenile steelhead trout (*Salmo gairdneri*). Journal of the Fisheries Research Board of Canada 35:822-827.

Westin, D.T. 1974. Nitrate and nitrite toxicity to salmonoid fishes. The Progressive Fish-Culturist 36:86-89.

Wolff, I.A., and A.E. Wasserman. 1972. Nitrates, nitrites, and nitrosamines. Science 177:15-19.

Wuhrmann, K. 1952. Sur quelques principes de la toxicologie du poisson. (Concerning some principles of the toxicology of fish.) Bulletin du Centre belge d'Etude et de Documentation des Eaux 15:49-60, (English translation).

Wuhrmann, K. and H. Woker. 1948. Beitrage zur Toxikologie der Fische. II. Experimentalle Untersuchungen uber die Ammoniak- und Blausaurevergiftung. (Contributions to the toxicology of fishes. II. Experimental investigations on ammonia and hydrocyanic acid poisoning). Schweizerische Zeitschrift fur Hydrologie 11:210-244, (English translation).

Wuhrmann, K. and H. Woker. 1953. Beitrage zur Toxikologie der Fische. VIII. Uber die Giftwirkungen von Ammoniak- und Zyanidlosungen mit verschiedener Sauerstoffspannung und Temperatur auf Fische. (Contributions to the toxicology of fishes. VIII. On the toxic effects of ammonia and cyanide solutions on fish at different oxygen tensions and temperatures). Schweizerische Zeitschrift fur Hydrologie 15:235-260, (English translation).

Wuhrmann, K., F. Zehender, and H. Woker. 1947. Uber die fischereibiologische Bedeutung des Ammonium- und Ammoniakgehaltes fliessender Gewasser. (Biological significance for fisheries of ammonium ion and ammonia content of flowing bodies of water). Vierteljahrsschrift der Naturforschenden Gesellschaft in Zurich 92:198-204, (English translation).

THE IMPACT OF VARIATIONS IN WATER pH ON FISH

David Randall

ABSTRACT

Variations in water pH have a marked effect on the performance of fish, and extremes outside the range pH 5 to 9 can cause death. Acid and alkaline waters reduce the swimming performance of trout, possibly due both to ammonia accumulation and an impairment of oxygen transport at low pH, and to ammonia accumulation at high pH. Acid waters inhibit sodium/hydrogen exchange at the gill surface and enhance sodium efflux, resulting in sodium loss from the fish; this response is ameliorated by elevated calcium levels in the water. Alkaline waters result in ammonia accumulation in trout and possibly other fish. This effect may be potentiated in waters of high buffering capacity.

The boundary layer of water next to the gill surface does not have the same pH as the bulk water. Carbon dioxide excretion acidifies this boundary layer, whereas ammonia excretion raises the pH. Above pH 5.4 carbon dioxide excretion has the dominant effect and the pH of the boundary layer is less than that of the bulk water. Below pH 5.0 ammonia excretion raises the pH of water next to the gills compared to the pH of the bulk water. Thus, changes in pH of water next to the gills are ameliorated in the face of changes in pH in the bulk phase. This effect decreases with increasing buffering capacity of the water.

INTRODUCTION

The accumulation of ammonia and carbon dioxide and the associated shifts in pH in water constitute a serious problem in aquaculture, especially in intensive, recirculating production systems. Acidification of water will increase the carbon dioxide/bicarbonate ratio but decrease the ammonia/ammonium ion ratio. The excretion of carbon dioxide by the fish will tend to lower, whereas the excretion of ammonia will tend to raise, water pH.

Extracellular pH in fish is on the alkaline side of neutrality and is midway between the point where pH equals pK for reactions of the two major excretory products, carbon dioxide and ammonia (Fig. 1). The pK of the NH_3/NH_4^+ reaction is about 9.5; whereas, the apparent pK of the CO_2/HCO_3^- reaction is around 6.1 (Boutilier et al. 1984). The CO_2/HCO_3^- and NH_3/NH_4^+ ratios are equal at the pH where the two lines cross in Fig. 1a, at the midpoint between the pK of the NH_3/NH_4^+ reaction and the apparent pK of the CO_2/HCO_3^- reaction. The pH at this point is similar to the pH of fish blood at the

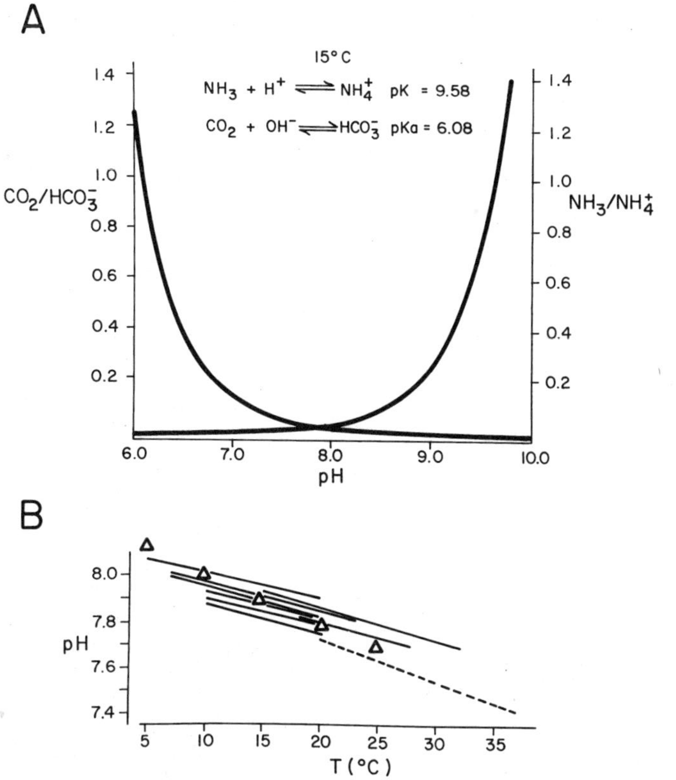

Figure 1. (A). The effect of varying pH on the CO_2/HCO_3^- and NH_3/NH_4^+ ratios in rainbow trout plasma at 15°C. (B). The pH values where the ratios are equal have been calculated at different temperatures and added to a graph of variations in plasma pH with temperature for several fish (redrawn from Randall and Wright 1989).

same temperature. The midpoint between these two pKs varies with temperature in much the same way as blood pH changes in fish (Fig. 1b).

The pH of seawater is in the same range as extracellular fluid, but freshwater systems show a much wider variation in pH, from the humic acid-laden waters of the Rio Negro in Brazil and the man-made acid waters of the Canadian shield (both around pH 4.0-5.0) to the alkaline lakes of the Kenyan Rift Valley (pH 10.0). Most living things, including fish, find extremes of pH lethal. Thus Manaus is built on the Negro rather than the Amazon because of the lack of mosquitoes at the former site. Fish have disappeared from many of the lakes of the Canadian shield due to acidification, and only *Oreochromis alcalicus grahami* can survive in the alkaline waters of lake Magadi, in the Kenyan Rift Valley (Randall et al. 1989).

There is an extensive literature on the effect of low pH environments on freshwater fish (see review by Wood and McDonald 1982). Many studies have described the physiological responses of fish to acid conditions. Among them, disturbances to acid-base state are particularly well known (Packer and Dunson 1970; Neville 1979; Packer 1979; McDonald et al. 1980). In addition, studies of the effects of acid on ionoregulation (Wood and McDonald 1982), ammonia excretion (Wright and Wood 1985; Randall and Wright 1989), carbon dioxide excretion (Lin and Randall 1989), and oxygen transport (Packer 1979) have been reported.

The resistance of various fresh-water fishes to strong alkalies (that is, sodium, potassium, and calcium hydroxides) in distilled and natural waters has been studied by many investigators (Douderoff and Katz 1958; Wright and Wood 1985; Randall and Wright 1989; Lin and Randall 1989). In general, pH levels above 9 are often lethal to freshwater fish, with only certain species surviving at pH 10. Detailed investigations of the physiological effects of alkaline water on fish, however, are more recent. Exposure of rainbow trout to alkaline conditions impaired ammonia excretion (Wright and Wood 1985), caused elevated blood ammonia levels and was lethal (Randall and Wright 1989; Lin and Randall 1989). The Tilapia, *Oreochromis alcalicus grahami* survives in alkaline lake Magadi, (pH 9-10.5) in the Kenyan Rift Valley, in part because it can detoxify ammonia by producing urea via the ornithine-urea cycle (Randall et al. 1989). *Tilapia* sp. have extremely low surface permeabilities to ions and this also may confer upon them the capacity to survive in these extreme conditions which also have high ionic content (Maloiy et al. 1978; Eddy et al. 1981). Maetz and De Renzis (1978) studied aspects of the adaptation of fish to high external alkalinity and found blood pH increased (see also Johansen et al. 1975), gill potential decreased and there was a 9-fold stimulation of Na and Cl influx in *O. mossambica*. Wright and Wood (1985), however, found that sodium influx in rainbow trout was inhibited in alkaline water following a few hours exposure.

WATER pH AND SWIMMING PERFORMANCE

Fish are often chronically exposed to, and may have to migrate through, waters with extremes of pH. Fish must swim actively in a variety of normal behaviors (feeding, avoiding predation, migration, spawning, etc.); if swimming ability is impaired in acid or alkaline water, the survival of the fish population may be in jeopardy.

Ye and Randall (1990) demonstrated that exposure of fish to either acid or alkaline waters reduced the maximum critical swimming speed of trout (Fig. 2). They reported that maximum critical speed was not appreciably

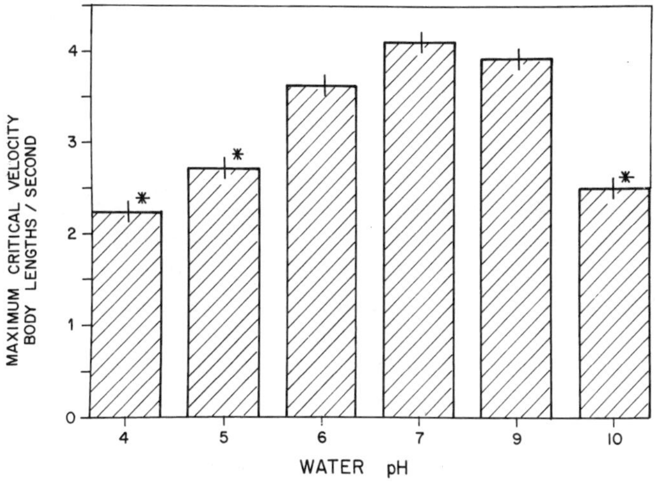

Figure 2. The effect of acid water on the maximum critical velocity in rainbow trout. * = significantly different from fish in pH 7 control water ($P < 0.05$)(from Ye and Randall 1989).

decreased in water between pH 6.0 and 9.0, but in water at pH below 5.0 and above 9.0 the maximum critical velocity decreased significantly. Swimming speed in fish exposed to waters of pH 4.0, 5.0 and 10.0 was only 54.5%, 66.5% and 61% respectively, of that recorded for fish in water of pH 7.0. Waiwood and Beamish (1978) found that critical speed was not appreciably influenced by hardness, pH, or time of exposure. The water pH range they used, however, was only from 6.0 to 8.0. Graham and Wood (1981) measured the effect of low pH on swimming in fingerling rainbow trout and found that, below pH 4.6 (soft water) or 4.4 (hard water), critical velocity declined linearly by about 4% per 0.1 pH unit. In Graham and Wood's study, the fish were exercised as soon as they were exposed to the low pH water. When Ye and Randall (1990) exercised fish as soon as they were exposed to the low pH water (pH 4.0), the reduction of maximum critical velocity was less than 20%. In 24-hour-acclimated fish, the reduction was 46%. Thus, acclimation time, as well as the pH of water, influence the maximum critical velocity of fish. The longer the fish is exposed to low pH water, the greater the stress.

Swimming in Acidic Water

There are some physiological differences between fish exercised in acid water compared to control water (Ye and Randall 1990): (1) plasma pH (pHe) and red blood cell pH (pHi) were significantly decreased in resting fish in acid water; (2) pHe was restored to resting levels after 2 hours recovery in control water, but not in acid water; (3) blood total CO_2 was significantly lowered in acid water; (4) blood lactate level increased faster in fish in acid water than in those in control water. During exercise, fish generate many acid products and, in acid water, may have problems dealing with the resulting acidotic state. In fish in neutral water, some of the H^+ ions produced during exercise are excreted across the gills, thereby restoring blood pH to resting levels 2 hours after recovery. In acid water, the large H^+ gradient between the fish body and the water probably inhibit acid excretion across the gills and impair the ability of the fish to recover from the exercise-induced acidotic state. It is possible that $Na^+/H^+(NH_4^+)$ exchange is reversed in acid water, and the increase in sodium efflux seen in acid water is due, at least in part, to a reversal of this exchange mechanism. This would result in proton uptake and ammonium ion retention by the fish under acidic conditions.

Exposure to acid conditions (pH 4.0) increased coughing and breathing frequency in rainbow trout (Ye and Randall 1989). The increase in the cough frequency in trout in pH 4.0 water may be due to the mucus accumulation on the gills. Branchial mucous secretion is a common response in fish exposed to acid water (Packer and Dunson 1972). Mucus may serve to protect the fish by limiting passive branchial ion loss and H^+ entry. On the other hand, mucous may limit oxygen uptake, which will result in an increase in breathing frequency, and a decrease of swimming ability.

It has been reported that oxygen transport is not inhibited by exercise-induced extracellular acidosis in trout (Primmett et al. 1986). Blood oxygen content is not reduced because red cell pH is maintained due to a significant increase in plasma catecholamine levels. It was proposed that adrenergic regulation of red cell pH allowed normal Hb-O_2 saturation regardless of the extent of the exercise-induced extracellular acidosis (Primmett et al. 1986).

The physiological effects in fish of swimming in acid water can be seen by comparing the effects of exposing fish to the following conditions: (1) fish chronically exposed in acid water for 24 hours under resting conditions, (2) fish forced to swim to fatigue in the respirometer in neutral water, and (3) fish forced to swim to fatigue in the respirometer in acid water. In the first condition, blood pH decreased slowly and gradually from 7.89 in control water to 7.81 after 24 hours acid exposure (Fig. 3). Although blood catecholamines increased irregularly during acid exposure, red cell pH decreased significantly from 7.46 in control water to 7.33 in acid water (Fig. 4). Since the fish were in a resting condition, the oxygen demand of the fish was low. It is assumed that the amount of oxygen available in the blood was adequate and the resting oxygen consumption was not affected (Ye et al. 1990). In the second condition, a large quantity of H^+ was produced when the fish was swum to fatigue. Blood pH decreased significantly from 7.89 in resting conditions to 7.69 at fatigue (Fig. 3), but at the same time catecholamines increased significantly (Primmett et al. 1986). As in the results of Primmett et al. (1986), red cell pH was not decreased during burst swimming (Fig. 4). Arterial blood oxygen content, therefore was maintained during the swimming and recovery periods. In the third condition, however, red cell pH was already decreased significantly in the resting fish. As a result, the blood oxygen delivery capacity was probably impaired due to Bohr and

Root effects. Red blood cell pH in fish exposed to acid water, however, was higher at fatigue and during recovery than before swimming (Fig. 4), probably as a result of the release of catecholamines into the blood.

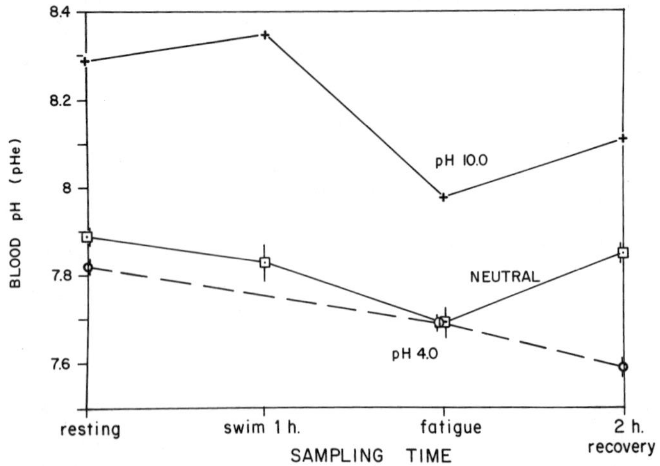

Figure 3. The effect of acid (pH 4.0) and alkaline (pH 10.0) water on blood pH (pHe) in rainbow trout during and following swimming to fatigue (from Ye and Randall 1989).

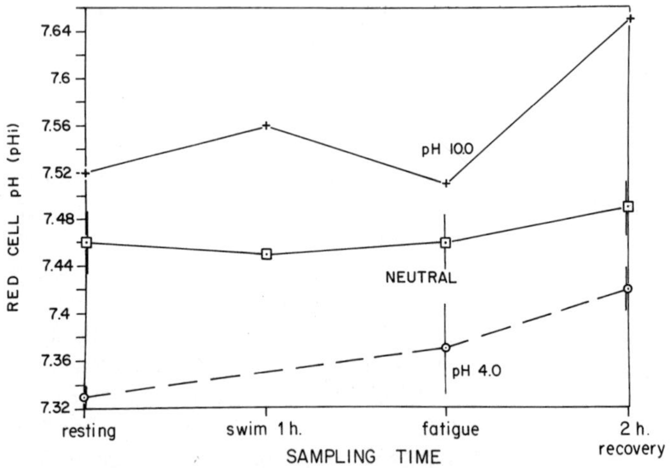

Figure 4. The effect of acid (pH 4.0) and alkaline (pH 10.0) water on red blood cell pH (pHi) in rainbow trout during and following swimming to fatigue (from Ye and Randall 1989).

Swimming in Alkaline Water

Alkaline water also impairs swimming performance, but presumably due to different causes than those in acid waters. There are some differences between fish exercised in alkaline water compared to control water (Figs. 3 and 4): (1) pHe and pHi increased in resting fish in alkaline water; (2) pHe and pHi continued to increase during routine swimming in alkaline water, but did not change in control water; (3) blood total CO_2 increased in resting fish and decreased during exercise in alkaline water, but in control water blood total CO_2 increased slightly after exercise; (4) lactate levels increased faster and higher in alkaline water than in control water. In alkaline water, the H^+ ions produced by exercise only offset the already high pHe so that the pHe decreased from 8.29 to 7.98. Clearly, the development of an acidotic state cannot account for the reduction in swimming ability under alkaline conditions, and there is no other clear reason as to why swimming is impaired. One possibility is the accumulation of ammonia in the body. Water pH has a marked effect on ammonia excretion (Wright and Wood 1985), and ammonia is retained in the body under alkaline conditions (Lin and Randall 1989).

The possible reasons of the reduction of maximum critical velocity in fish exposed to low and high water pH are manifold. A pH change in the microenvironment of tissues and cells would affect the ionization of many intermediary compounds, and therefore, affect metabolic processes. Clearly, a blood and red cell alkalosis or acidosis indicates that other tissues and cells may also be alkalotic or acidotic, which will disturb the microenvironment for many enzyme reactions in the tissues and muscles, and when fish exercise, the situation may become worse. It is easy to assume that, under such circumstances, fish will not reach the maximal critical velocity observed in normal pH water.

In conclusion, acid or alkaline conditions clearly reduce the critical swimming velocity of the fish. The farther the water pH is from neutrality and the longer the fish is exposed, the more the swimming capacity of the fish is impaired.

WATER pH AT THE GILL SURFACE

Flow of water over the secondary lamellae of the gills is unidirectional and pulsatile, but may be fairly uniform during swimming when gill water flow is dominated by ram ventilation. The surface of the gills is covered by a mucous coating. Molecular CO_2 and NH_3, along with some HCO_3^- (Perry et al. 1982) and NH_4^+ (Wright and Wood 1985), are excreted into the mucus and boundary layer of water next to the gill surface. Bicarbonate ions represent only a small portion of the total carbon dioxide excreted, probably less than 10% (see review by Randall and Daxboeck 1984), and the amount of ammonia excreted as NH_3 varies between 45 and 100% (Cameron and Heisler 1983; Wright and Wood 1985). As long as the pH is above 6.1, most of the CO_2 excreted will form HCO_3^- and acidify the water (Fig. 5). Wright et al. (1986) showed that this reaction was catalyzed by carbonic anhydrase in the mucous layer. The mucous layer contains a large number of damaged

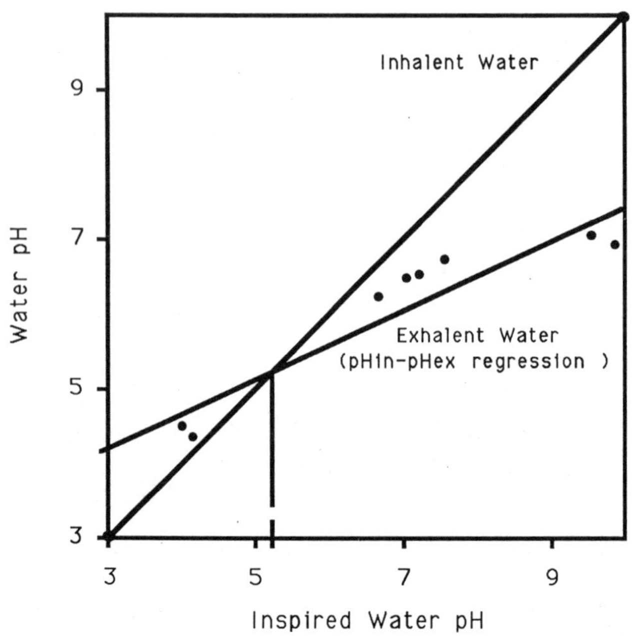

Figure 5. The relationship between inhalant and exhalent water pH in rainbow trout. The exhalent water line is the regression of the mean exhalent pH values collected over a range of inhalant pH levels. The inhalant and exhalent lines cross at pH 5.3 (from Lin and Randall 1989).

gill epithelial cells, which are known to contain high levels of carbonic anhydrase (Lacy 1983). More recently, Rahim et al. (1988) have shown that the external apical surface of the gill contains carbonic anhydrase activity. Carbon dioxide entering this boundary layer will, therefore, acidify the mucus and boundary layers (Wright et al. 1986). Ammonia excretion will have the opposite effect, but because more CO_2 than NH_3 is excreted across the gills, the overall effect will be an acidification of the boundary layer, when water pH is above 6.0.

Ammonia entering the acidified layer next to the gill surface will be converted to NH_4^+ and will diffuse out of the boundary water layer into the bulk medium. The conversion of NH_3 to NH_4^+ will reduce NH_3 levels in the water and therefore NH_3 levels in the blood (Wright et al. 1989). Starving trout have elevated blood NH_3 levels (Hillaby and Randall 1979) and this can be correlated with a decreased CO_2 excretion and, therefore, a reduced acidification of the water boundary layer.

In rainbow trout exposed to acid water (pH 3.9), carbon dioxide hydration is negligible and consumption of protons by formation of NH_4^+ results in an increase in pH of the water as it passes over the gills (Fig. 5; Lin and Randall 1989). The consumption of protons, in fact, probably exceeds NH_3 excretion, and part of the increase in pH may be due to bicarbonate dehydration as a consequence of HCO_3^-/Cl^- exchange. In alkaline water (pH 9.9), ammonium ion formation is reduced but carbon dioxide hydration is almost 100%; hence, there is a large acidification of water as it passes over the gills (Lin and Randall 1989). These processes are balanced for rainbow trout exposed to water of pH 5.3; at this pH there is no pH change as water passes over the gills (Lin and Randall 1989).

WATER pH AND CARBON DIOXIDE AND AMMONIA EXCRETION

Changes in inspired water pH have little or no effect on carbon dioxide excretion, although there are changes in water and blood partial pressures of carbon dioxide. Ammonia excretion is impaired at both high and low water pH. Ammonia excretion will be enhanced in acidic water because of ammonium ion trapping, but total ammonia excretion is depressed because acid conditions inhibit Na^+/NH_4^+ exchange (Wright and Wood 1985). Alkaline conditions impair NH_3 excretion (Lin and Randall 1989) and inhibit Na^+/NH_4^+ exchange (Wright and Wood 1985), so there is also an impairment of ammonia excretion under alkaline conditions. In fact rainbow trout die in soft alkaline water, probably because of ammonia accumulation in the body (Lin and Randall 1989). Tilapia can survive in Lake Magadi, an alkaline lake in the Kenyan Rift Valley, because they are more tolerant to ammonia than other vertebrates, and have a large capacity to detoxify ammonia by converting it to urea via the ornithine-urea cycle (Randall et al. 1989); these Lake Magadi tilapia excrete no ammonia, only urea.

Thus fish, in general, are intolerant of extremes of water pH, outside the range pH 5 to 9. There are a few species that have evolved special mechanisms to either tolerate or ameliorate the effects of these extremes in water pH, on carbon dioxide, ammonia, and ion transfer. The accumulation of molecular carbon dioxide does not appear to be a problem per se; animals can tolerate a wide range of carbon dioxide and bicarbonate levels. Animals, including fish, cannot tolerate large shifts in body pH and ammonia levels, both of which are toxic. Carbon dioxide accumulation is a problem, not so much because of any toxic effect of carbon dioxide itself, but because of its affect on pH.

LITERATURE CITED

Boutilier, R. G., T. A. Heming, and G.K. Iwama, 1984. Appendix: Physio-chemical parameters for use in fish respiratory physiology. Pages 403-430 *in* W. S. Hoar and D. J. Randall, editors. Fish Physiology 10A: Academic Press, Inc., New York.

Cameron, J. N. and N. Heisler, 1983. Studies of ammonia in the rainbow trout: physio-chemical parameters, acid-base behaviour, and respiratory clearance. Journal of Experimental Biology 105: 107-125.

Doudoroff, P. and M. Katz, 1958. Critical review of literature on the toxicity of industrial wastes and their components to fish. Sewage and Industrial Wastes 22: 1432-1458.

Eddy, F. B., O.S. Bamford, and G. M. O. Maloiy, 1981. Na^+ and Cl^- effluxes and ionic regulation in *Tilapia grahami*, a fish living in conditions of extreme alkalinity. Journal of Experimental Biology 91: 349-353.

Graham, M. S. and C. M. Wood, 1981. Toxicity of environmental acid to the rainbow trout: interactions of water hardness, acid type, and exercise. Canadian Journal of Zoology 59: 1518-1526.

Hillaby, B. A. and D. J. Randall, 1979. Acute ammonia toxicity and ammonia excretion in rainbow trout (*Salmo gairdneri*). Journal of the Fisheries Research Board of Canada 36: 621-629.

Johansen, K., G. M. O. Maloiy, and G. Lykkeboe, 1975. A fish in extreme alkalinity. Respiratory Physiology 24: 159-162.

Lacy, E. R. 1983. Histochemical and biological studies of carbonic anhydrase activity in the opercular epithelium of the euryhaline teleost, *Fundulus heteroclitus*. American Journal of Anatomy 166: 19-39.

Lin, H. and D. J. Randall, 1989. The effect of varying water pH on the acidification of the water as it passes over the gills of rainbow trout. Submitted.

Maloiy, G. M. O., G. Lykkeboe, K. Johansen, and O.S. Bamford, 1978. Osmoregulation in *Tilapia grahami*: a fish in extreme alkalinity. Pages 229-238 *in* K. Schmidt-Nielsen, L. Bolis and S. P. H. Maddrell, editors. Comparative Physiology of Water, Ions and Fluid Mechanics. Cambridge University Press, Cambridge, U.K.

McDonald, D. G., H. Hobe, and C.M. Wood, 1980. The influence of calcium on the physiological responses of the rainbow trout. *Salmo gairdneri*, to low environmental pH. Journal of Experimental Biology 88: 109-131.

Maetz, J. and G. De Renzis, 1978. Aspects of the adaptation of fish to high external alkalinity: comparison of *Tilapia grahami* and *T. mossambica*. Pages 213-228 *in* K. Schmidt-Nielsen, L. Bolis and S. H. P. Maddrell, editors. Comparative Physiology of Water, Ions, and Fluid Mechanics. Cambridge University Press, Cambridge, U.K.

Neville, C. M. 1979. Sublethal effects of environmental acidification on rainbow trout *Salmo gairdneri*. Journal of the Fisheries Research Board of Canada 36: 84-87.

Packer, R. K. 1979. Acid-base balance and gas exchange in brook trout (*Salvelinus fontinalis*) exposed to acidic environments. Journal of Experimental Biology 79: 127-134.

Packer, R. K. and W.A. Dunson, 1970. Effects of low environmental pH on blood pH and sodium balance of brook trout. Journal of Experimental Zoology 174: 65-72.

Packer, R. K. and W.A. Dunson, 1972. Anoxia and sodium loss associated with the death of brook trout at low pH. Comparative Biochemistry and Physiology 41A: 17-26.

Perry, S. F., P.S. Davie, C. Daxboeck, and D.J. Randall, 1982. A comparison of CO_2 excretion in a spontaneously ventilating blood-perfused trout preparation and saline-perfused gill preparations: contribution of branchial epithelium and red blood cell. Journal of Experimental Biology 101: 47-60.

Primmett, D. R. N., D.J. Randall, M. Mazeaud, and R.G. Boutilier, 1986. The role of catecholamines in erythrocyte pH regulation and oxygen transport in rainbow trout (*Salmo gairdneri*) during exercise. Journal of Experimental Biology 122: 139-148.

Randall, D. J. and C. Daxboeck, 1984. Oxygen and carbon dioxide transfer across fish gills. Pages 263-314 *in* W. S. Hoar and D. J. Randall editors. Fish Physiology Vol. 10A. Academic Press, New York, New York, USA.

Randall, D. J., C.M. Wood, S.F. Perry, H. Bergman, G.M.O. Maloiy,, T. Mommsen, and P.A. Wright, 1989. Urea excretion as a strategy for survival in a fish living in a very alkaline environment. Nature 337: 165-166.

Randall, D. J. and P.A. Wright, 1989. The interaction between carbon dioxide and ammonia excretion and water pH in fish. Canadian Journal of Zoology 67:2936 - 2942.

Rahim, S., J.P. Delaunoy, and P. Laurent, 1988. Identification and immunocytochemical localization of two different carbonic anhydrase isoenzymes in teleostean fish erythrocytes and gill epithelia. Histochemistry 89: 451-459.

Waiwood, K. G. and F.W.H. Beamish, 1978. Effects of copper, pH and hardness on the critical swimming performance of rainbow trout (*Salmo gairdneri* Richardson). Water Research 12: 285-287.

Wood, C. M. and D.G. McDonald, 1982. Physiological mechanisms of acid toxicity to fish. Pages 197-226 *in* R. E. Johnson, editor. Acid Rain/ Fisheries, Proceedings of an International Symposium on Acidic Precipitation and Fishery Impacts in North-Eastern North America. American Fisheries Society, Bethesda, Maryland, USA.

Wright, P. A., T.A. Heming, and D.J. Randall, 1986. Downstream pH changes in water flowing over the gills of rainbow trout. Journal of Experimental Biology 126: 499-512.

Wright, P. A., D.J. Randall, and S.F. Perry, 1989. Fish gill water boundary layer: a site of linkage between carbon dioxide and ammonia excretion. Journal of Comparative Physiology 158: 627-635.

Wright, P. A. and C.M. Wood, 1985. An analysis of branchial ammonia excretion in the freshwater rainbow trout: Effects of environmental pH change and sodium uptake blockade. Journal of Experimental Biology 114: 329-353.

Ye, X., and D.J. Randall, 1989. The effect of water pH on swimming performance in rainbow trout (*Salmo gairdneri*). Fish Physiology and Biochemistry (in press).

Ye, X., D.J. Randall, and X. He, 1990. The effect of acid water on gill potential, blood ionic and acid-base status, and resting oxygen consumption in plasma in rainbow trout (*Salmo gairdneri*). Fish Physiology and Biochemistry (in press).

EFFECTS OF SALINITY ON AQUACULTURE PRODUCTION

Robert R. Stickney

ABSTRACT

Fish vary widely in their ability to acclimate to increasing or decreasing salinity. Many of the species of aquaculture interest are stenohaline, but some, such as *Tilapia mossambica*, certain hybrid tilapias, red drum, and various salmonids, demonstrate a remarkable degree of euryhalinity. Survival can be compromised if a fish is placed in an unsuitable salinity regime, but growth, food conversion efficiency, and reproduction are also significant factors that can be influenced by salinity. For some species of aquaculture interest, responses to salinity have been well documented,but for others little or no information has been collected to date.

INTRODUCTION

Aquaculturists who practice their profession in brackish and sea water areas are well aware of the relationships between salt in the water and fish farm management. Exposure to even low salinity water may greatly increase stress in freshwater fish, while salinities below that of sea water may lead to severe stress in stenohaline marine fishes. Performance is often reduced dramatically when fish are exposed to stressful salinities, and death can result. Compounding the physiological impacts of salt concentration on aquaculture animals is the corrosive nature of salt water. Anyone who has worked around sea water is familiar with the manner in which metal objects are attacked and quickly destroyed from contact with the salts that are present.

Salinity is defined as the total weight of solid material contained in one kilogram of sea water when all the carbonate has been converted to oxide, the bromine and iodine replaced by chlorine, and all organic matter completely oxidized (Sverdrup, et al. 1942). Included in the solids present in sea water are virtually all of the elements that exist in nature. Sodium and chloride ions are dominant, but calcium, magnesium, iron, zinc, copper, and a host of other elements are present in significant concentrations. While the absolute concentrations of the various elements present in sea water vary considerably, the ratios that exist between the more abundant elements are constant. This condition, embodied in the Law of Constant Proportions, was first described by Dittmar (1884) after his collection of water samples during the famous H.M.S. Challenger expedition.

For many years, salinity was calculated using the following formula (Sverdrup, et al. 1942):

Salinity = 0.03 x 1.805 x Chlorinity.

The relationship was redefined in 1967 (Ross 1970) as:

Salinity = 1.8066 x Chlorinity.

Chlorinity is defined as the total weight of chlorine, bromine, and iodine contained in one kilogram of sea water, assuming that the bromine and iodine have been replaced by chlorine. The determination of chlorinity can be made by titration against a standard seawater sample, known as "normal water" and prepared by the Hydrographical Laboratories in Copenhagen, Denmark (Sverdrup, et al. 1942).

Precise determinations of the salinity of a water sample can be obtained through titration against "normal water" or by measurement of electrical conductivity. Other methods of measuring salinity are by determination of the density of a water sample with a hydrometer, measurement of refractive index with a refractometer, and the determination of freezing point depression. Hydrometers are made of glass and, thus, are easily broken. Hand-held refractometers, which can directly measure salinity from as little as a single drop of water and are sufficiently accurate for most aquaculture purposes, are available for a few hundred dollars.

Salinity is measured in g/kg or parts per thousand (ppt), which can be abbreviated o/oo. Full strength seawater typically measures 35 o/oo salinity, while freshwater is defined as water having a salinity of less than 0.5 o/oo.

The aquaculturist who works in saline water should maintain salinity within the optimum range for the species under culture. In water systems which receive rapid exchanges of water (net pens, cages, raceways), salinity will only change as a result of dilution from precipitation or freshwater runoff. Such changes are transient and may not impact the entire water column in the system since freshwater is less dense than saltwater and may be restricted to the surface water and only mix into the deeper waters at a slow rate. In ponds, freshwater inflow may eventually decrease salinity throughout the water column because of wind-induced mixing.

Freshwater inflow to pond aquaculture systems is generally not a significant problem to the mariculturist unless there is an inordinate amount of rainfall or runoff. More significant is the tendency of a pond to increase in salinity temporally. As salt water evaporates, the concentration of salts in the remaining liquid increases, thereby causing salinity to increase. If evaporation is replaced with sea water, there will be some reduction in salinity, but the overall salt content of the water will be higher than it was when the pond was filled because the total amount of salts will have increased (no salt is lost from the system as a result of water evaporation). Eventually, the salinity may exceed that which is desirable for the species under culture. To avoid this problem, evaporation losses should be replaced with fresh, not salt water. Thus, it is important for the mariculturist who is using static ponds to have an adequate supply of fresh or sufficiently low salinity water to replace evaporative losses.

This paper deals with the salinity tolerances and requirements of several finfish species of aquaculture interest or importance. Included is information on the salinity tolerance of some commercially important freshwater species, along with information on the optimum salinities of some marine and estuarine species. Preceding the information on salinity tolerances and requirements is a section on osmoregulation in fishes and on the very crucial parr-smolt transformation which occurs in anadromous salmonids.

OSMOREGULATION IN FISHES

The blood and tissue fluid salt concentrations in freshwater fishes are somewhat hypertonic to the external medium (10-12 o/oo salinity within the fish as compared with about 0.5 o/oo in the water). Because the gills of fish act as semipermeable membranes, there is a tendency for water in the external medium to flow into the fish and dilute the body fluids. The fish must continuously excrete water in order to maintain its internal salt concentration. The process by which this is accomplished is known as osmoregulation.

Freshwater fishes do not drink water because of the continuous intake of water primarily through the gills. The kidneys of freshwater fishes are efficient at removing water, but retaining salt. Thus, freshwater fishes produce copious amounts of dilute urine. Certain ions can be taken up by fish through the process of active transport, often through the gills. Most of the salts that are required by the animals are obtained through ion exchange systems on the gills and intestinal absorption from ingested food.

While fishes which inhabit estuaries may be able to find conditions where the external medium is isotonic with their internal salt concentrations, it is improbable that many of them would constantly remain in an isotonic medium. Most marine fishes are hypotonic to the surrounding water. Thus, the situation is the reverse of that which faces freshwater fishes. A marine fish tends to lose water from its tissues to the surrounding environment. In order to replace water, the marine fish ingests copious amounts, but the ingested water contains salts which must be excreted so tissue salt concentrations will not become excessive. The kidneys in marine fishes are adapted to excrete small amounts of urine that contain high concentrations of salt. Conte (1969) reviewed the literature and found that ingestion of water by teleosts in the marine environment ranges from 0.3 to 1.5% of body weight of the fish per hour. Some of the ingestion rates reported in that review are as follows:

American eel (*Anguilla rostrata*)	2.77	ml/hr/kg
Rainbow trout (*Salmo gairdneri*)	5.37	"
Southern flounder (*Paralichthys lethostigma*)	10.00	"
Tilapia (*Tilapia mossambica*)	234.00	"

The energy costs of maintaining proper internal salt balance in fishes can be considerable, thus growth may be reduced when nutrient energy is used for osmoregulation.

Euryhaline fishes have the ability to osmoregulate over a wide range of salinities, while stenohaline fishes are restricted to narrow ranges of salinity. Anadromous fishes, such as salmon, spawn in freshwater and later migrate to the ocean, while catadromous species, such as the American eel, spawn in the sea and migrate to freshwater. Morphological and physiological changes may occur when anadromous or catadromous fishes adapt to a new salinity regime. This concept can be extended to euryhaline species which may show seasonal patterns in such indices as glomerular filtration rate as has been shown in southern flounder (*Paralichthys lethostigma*) held for long periods at constant salinity (Hickman 1968). The changes in kidney activity seem to represent a physiological response to the normal pattern of behavior in which the fish would move into low salinity water during part of the year and migrate offshore in the winter.

Osmoregulatory control in the embryos and larvae of fishes has long been thought to be under the control of so-called chloride cells, which are rich in mitochondria. Chloride cells have been conclusively shown to transport salt in the case of the early life history of at least one important aquaculture species, *Tilapia mossambica* (Foskett and Scheffey 1982). The presence or absence and density of chloride cells is a function of age, species, and the salinity in which a fish is found. Chloride cells may first appear during early or late embryonic development and may disappear after yolk sac absorption when other regulatory tissues become developed (Alderdice 1988).

It is important to note that, with respect to egg hatching and larval survival, salinity tolerance information will be incomplete if the effect of temperature is not taken into consideration. Alderdice and Forrester (1968) demonstrated with English sole (*Parophrys vetullus*) that there is an optimum range of salinity and temperature when egg hatching and the production of normal larvae are maximized. A larger range of salinities and temperatures will result in good levels of hatching and with most of the larvae being normally developed. Outside of the latter range, various abnormalities occur and the percent of eggs hatched is limited. This concept can undoubtedly be extended to various other species.

SMOLTING IN SALMONIDS

Perhaps the most widely studied phenomenon related to acclimation from one salinity regime to another relates to smoltification or the parr-smolt transformation that occurs in anadromous salmonids. Smoltification has been the subject of a large number of scientific investigations. The topic has been reviewed by Conte (1969) and Folmar and Dickhoff (1980). Morphologically, the dark pigmented bars (parr marks) which lie perpendicular to the lateral line on the sides of the fish during the parr stage fade and the fish become silvery as they become smolts. In nature, the silvering of the fish is accompanied by a behavioral change during which the fish become increasingly tolerant of salinity and begin migrating to sea (Folmar and Dickhoff 1980).

Fish size has long been thought of as the most important variable influencing smolting. The relationship has been demonstrated in Atlantic salmon, *Salmo salar* (Huntsman and Hoar 1939; Elson 1957; Parry 1960; Johnston and Eales 1970; Hoar 1976) and is common in other species within the genus *Salmo* as well as those in the genus *Oncorhynchus* (Hoar 1976). While fish size seems to be a dominating factor, age is also important (Parry 1960) and can often be directly correlated with size. Other environmental factors that influence the onset of smoltification are light and temperature (Wagner 1974; Hoar 1976; Clarke, et al. 1978; Johnston and Saunders 1981). The relative importance of the latter seems to vary among species. In steelhead trout (*Salmo gairdneri*), for example, increasing daylength is a key factor in the regulation of smolting. Temperature seems to have little influence over a relatively broad range, and neither absolute daylength nor total time of exposure to light is important (Wagner 1974).

Physiologically, thyroid hormones have been strongly implicated in controlling the metabolic and developmental changes that occur during parr-smolt transformation (Dickhoff and Sullivan 1987). Changes in the levels of circulating thyroid hormones during smoltification have been shown by Dickhoff et al. (1978) for coho salmon (*Oncorhynchus kisutch*). Dramatic increases in plasma thyroxine (T4) occur in that species when they smolt (Folmar and Dickhoff 1981). Seven month old Atlantic salmon are more tolerant of transfer from fresh water to 32 o/oo salinity after being fed small amounts of the thyroid hormone T3 (3,5,3'-L-thyronine) according to a study by Refstie (1982).

Smolting is associated with increasing levels of Na^+-K^+ ATPase in the gills of coho and chinook salmon (*O. kisutch* and *O. tshawytscha*, respectively) and steelhead trout (Zaugg and McClain 1970, 1972; Ewing, et al. 1979; Zaugg, et al. 1985). Increases in plasma Na^+-K^+ ATPase have been demonstrated in coho salmon (Folmar and Dickhoff 1981). Gill ATPase levels will increase even in coho and chinook salmon which are held in fresh water while undergoing the parr-smolt transformation, but migrating salmon show much higher enzyme activity (Zaug, et al. 1985). In addition, there is a proliferation of chloride cells in the gills and subopercular epithelium, and involvement by various endocrine systems (Folmar and Dickhoff 1980).

Changes in hepatic mitochondrial proteins have been observed in smolting steelhead trout which may be associated with adaptation to the marine environment and to the concomitant change in nutritional status (Bradley 1987). In brown trout (*Salmo trutta*) smoltification is correlated with a high capacity for the regulation of magnesium ions (Hogstrant and Haux 1985). Smoltification stage may also be indicated by measurement of plasma sodium levels after 24 hour exposure of juvenile salmon to saltwater (Clarke and Blackburn 1977).

SALINITY TOLERANCE OF SELECTED FISHES OF AQUACULTURE IMPORTANCE

Most fishes of aquaculture importance are reared within relative narrow ranges of salinity, though several euryhaline species are of interest to fish culturists. Several important fish species and groups of species are discussed in the following subsections.

Trout

The ability of certain trout species to undergo smoltification and enter sea water has been well documented for the steelhead strain of rainbow (*Salmo gairdneri*) and searun cutthroat trout (*S. clarki*). Not all rainbow trout respond the same when introduced into seawater, so some factor other than salinity could be responsible for the differences that have been observed.

Murai and Andrews (1973) reared rainbow trout in water up to 30 o/oo salinity for several weeks after a nine day period during which the fish were adapted from freshwater. Survival and growth were considered to be excellent in that study. Another study published during the same year was conducted in Alabama, where trout were reared in cages during the winter in 20 o/oo water with good growth and survival (Tatum 1973). In a later study by Jürss et al. (1985), rainbow trout growth was inhibited in 20 o/oo water because food conversion efficiency was poor. That effect was overcome in part when the dietary protein level was increased from 36.6% to 45.2%. Tatum (1973) obtained his fish from a regional federal hatchery. The sources of fish in the other two studies were not reported. The protein level in the feed used by Murai and Andrews (1973) was not reported. Tatum (1973) fed a diet containing 40% protein.

Age is an important consideration in the transfer of trout to seawater. Underyearling brown trout and rainbow trout do not thrive in saltwater, but adapt at a later age (Houston 1961). Transfer of rainbow trout from freshwater to sea water is facilitated if the transfer is effected at the proper water temperature. In a study which compared responses of fish acclimated at 10°C and 17°C, the plasma osmolarities of fish maintained at the higher temperature and then introduced from freshwater into sea water increased much more rapidly and reached higher levels than in fish held at the colder temperature before being exposed to sea water. A higher degree of body dehydration was observed in fish maintained at 17°C (Alexis, et al. 1984). Changes in ATP-ase activity associated with the parr-smolt transformation in steelhead trout have been documented by Zaugg and Wagner (1973).

Saltwater-tolerant rainbow trout are currently being produced in marine net pens in Finland. Some of those fish, which originated from stock developed at the University of Washington by Lauren R. Donaldson in the 1940's, have recently been returned to the United States where they were introduced into net pens in Puget Sound, Washington (Jon Weiss-Hansen, personal communication). Growth and survival appear to be excellent in the net pen environment, though it should be remembered that the fish were bred for salt water tolerance and optimum performance under confinement.

Salmon

Like brown trout and rainbow trout, Atlantic salmon and coho salmon are unable to thrive in seawater as underyearlings. Chum salmon (*Oncorhynchus keta*), pink salmon (*O. gorbuscha*), and perhaps sockeye salmon (*O. nerka*) can tolerate early transfer to saltwater (Houston 1961). Chum and pink salmon can tolerate full strength sea water while in the alevin stage (Hoar 1976), and chinook salmon migrate to the sea after only a few weeks in fresh water. The alevins of chinook salmon can survive longer in seawater than those of either coho or sockeye salmon (Weisbart 1968). Tolerance by Atlantic salmon parr to sea water can be improved if the fish are fed diets containing up to 12% added salt, as compared with tolerance by fish which receive no dietary salt supplementation (Basulto 1976). Thus, there may be some ways that managers can improve osmoregulatory performance in certain salmonid species.

Overall, the ability to osmoregulate is greater in chum and pink salmon than it is in coho, chinook, or sockeye salmon. Chum and pink salmon may actually spawn near the sea, and even in the intertidal zone (Skud 1954; Clemens and Wilby 1961) and normal fertilization and development can occur at salinities as high as 18 o/oo (Rockwell 1956).

Determination of smoltification is of particular importance to the net pen salmon culturist. Incompletely smolted coho salmon may live for extended periods if introduced to salt water, but they may not grow. A saltwater challenge will not reveal whether such fish are actually smolts (Clarke and Naggahama 1977), so some other gauge of smoltification needs to be employed. As previously mentioned, various physiological parameters are indicators of parr-smolt transformation, but it may not be possible for the fish farmer to measure those parameters, and individual differences among fish may mean that a large sample size would be needed to assess the degree of smoltification reached by the majority of the population.

Size, age, and development of silvery color have often been used as the basis for transfer of salmon from freshwater to marine net pens, though size is the most commonly used criterion and is genetically linked to smoltification at least in coho salmon (Saxton et al. 1984). Coho salmon are typically reared for 18 months in freshwater hatcheries after which they are moved to marine net pens for growout. If pan-size fish are desired (200-400 g), the saltwater rearing phase may be as short as 7-12 months (Saxton et al. 1983). Mahnken (1973) recommended a minimum size of 15 g before transfer of coho salmon from freshwater to marine net pens. Commercial Atlantic salmon producers typically grade fish by size and move those that reach a particular size to net pens while retaining undersized fish in fresh water.

Presmolt coho and chinook salmon can be reared in brackish water, but growth is retarded in both species when seawater salinity is approached (Otto 1971; Kepshire and McNeil 1982). Canagaratnam (1959) found that growth of 0.4 to 0.5 g coho salmon was slower in fresh water than in water having salinity in the range of 6-12 o/oo. When the fish reached 1.0 g, best growth occurred at a salinity of 18 o/oo. While culture of coho salmon to smoltification in brackish water may be desirable from the standpoint of reducing osmoregulatory stress (Heard and Crone 1976), most net pen culturists do not have a wide range of salinities available to them and are forced to move fish from freshwater to relatively high salinity seawater without an intermediate period in brackish water.

Allee (1981) and Allee and Suzumoto (1981) reported that coho salmon which returned to an ocean ranching facility demonstrated generally poorer survival and a lower percentage of egg fertility when matured in seawater than if they were taken to a freshwater hatchery for maturation and spawning. Osmotic stress associated with maturation was hypothesized as a factor. There is also a link between the salinity at which coho salmon are matured and the thyroid as thyroxine levels are elevated in fish matured in saltwater as compared with those matured in freshwater (Sower and Schreck 1981). Wertheimer (1981a,b) reported good success with the maturation and spawning of coho salmon in saltwater net pens. In the latter instance, the net pens were located in an area that had a distinct halocline. Surface salinities in a lens of low salinity water ranged from 0-19 o/oo. The lens was from a few centimeters to about one meter in depth, below which salinities reached 24-32 o/oo. It was felt that salmon utilized the low salinity surface water lens during the maturation process. High gamete viability occurred (Wertheimer 1981b), and was shown to be inversely correlated with blood plasma osmolality in fish matured in net pens with a low salinity water lens (Wertheimer 1984).

Catfishes

In North America, channel catfish (*Ictalurus punctatus*) and blue catfish (*I. furcatus*) have been collected from salinities as high as 11.4 o/oo (Perry 1967). Those two species, along with the white catfish (*I. catus*) have been grown in brackish water ponds at salinities as high as 11 o/oo (Perry and Avault 1970). Channel catfish, for example, grow well in ponds with salinity ranging from about 4-5 o/oo (Burnside et al. 1975). The lethal concentration that will kill 50% of the fish exposed (LC_{50}) has been established for white catfish by Kendall and Schwartz (1968) at 14 o/oo with 60 hours exposure. Channel catfish eggs of three days of age or older can tolerate up to 16 o/oo salinity. This is reduced to 8 o/oo at hatching and increases slightly following yolk sac absorption. Channel catfish from six months to about one year of age will tolerate at least 12 o/oo salinity (Allen and Avault 1970). The salinity tolerance of blue catfish, like that of white catfish is 14 o/oo, slightly higher than that for channel catfish (Allen 1971). However, hybrids between channel and blue catfish can tolerate salinities of 14-15 o/oo (Stickney and Simco 1971).

Growth of channel catfish for several months in New Mexico saline groundwater has been evaluated by Turner (1988). Fingerlings were exposed to salinities up to 11 o/oo. No differences in growth, food conversion, or survival were observed in salinities below 6.5 o/oo. The ionic ratios in the New Mexico groundwater were not the same as those in modern sea water, so long-term growth of catfish in waters of different ionic composition could yield other conclusions.

Currently, there are no major efforts underway to rear ictalurid catfishes in brackish water within North America. Some advantage might be gained from exposing catfish to low salinity water. For example, the protozoan parasite *Ichthyophthirius multifiliis* can be controlled if catfish are placed in water of 2 o/oo salinity (Johnson 1976). Little information is available on the salinity tolerance of the walking catfishes. *Clarias lazera* has been found to tolerate 9.5 o/oo salinity with 94% survival (Chervinski 1984a).

Carp

The vast majority of carp production in the world is conducted in fresh water, though as is true with channel catfish, some interest has been demonstrated in rearing carp in brackishwater situations. The unacclimated tolerance of common carp (*Cyrpinus carpio*) to salinity has been placed at 15 o/oo (Payne 1983). Grass carp (*Ctenopharyngodon idella*) have a similar salinity tolerance, reportedly 14 o/oo (Maceina and Shireman 1979; Kilambi and Zdinak 1980). Growth and survival of common carp larvae is actually better in salinities up to at least 3 o/oo than it is in fresh water (Lam and Sharma 1985). Thus, brackish water culture is possible with common carp and grass carp. Various Indian carps (*Labeo rohita*, *Catla catla*, and *Cirrhinus mrigala*) can also be cultured in brackish water (Saha et al. 1984a, b, c).

Silver carp (*Hypopthalmichthys molitrix*) are somewhat less tolerant of salinity than either common carp or grass carp. Short-term tolerance of 10 o/oo has been observed (Chervinski 1977; Oertzen 1985), but high mortality will result within a few days. Maximum salinity tolerance occurs at between 18 and 22 C (Oertzen 1985). Growth is actually enhanced in low salinity water as compared with freshwater (Scildhauer 1981), with an optimum reported at 3 to 4 o/oo (Oertzen 1985).

Tilapia

Several fishes in the family Cichlidae and known by the common name tilapia are cultured in warm water around the world. Modifications in the taxonomic nomenclature proposed have led to proposed changes in the generic name of most of the species of aquaculture interest from *Tilapia* to *Sarotherodon* (Trewavas 1973) and later to *Oreochromis* (Trewavas 1982). The American Fisheries Society continues to recognize the genus *Tilapia* (Robins, et al. 1980), and that genus name is used throughout this discussion.

A review of the salinity tolerance of various tilapia species by Stickney (1986) concluded that the species most tolerant of brackish and high salinity waters are *Tilapia zilli* and *T. mossambica*. The former species holds little interest for aquaculturists, but the latter has been widely cultured throughout the tropical world. *T. mossambica* is not the most ideal species for culture, however, as it has dark external pigmentation that is objected to in some regions, reproduces as early as three months of age leading to overpopulation and stunting, and has a lower dressout percentage than some of the more desirable, but less salt tolerant species such as *T. aurea* and *T. nilotica*.

Sperm motility of *T. mossambica* is highest in distilled water and declines as the salinity of the medium is increased. Sperm from acclimated males are able to fertilize eggs in water of 35 o/oo salinity but only 0.2% of the sperm are motile (Harvey and Kelley 1984). Fertilized eggs, fry and fingerling *T. nilotica*, for example, suffer 50% mortality within 96 hours at between 15 and 20 o/oo salinity (Watanabe et al. 1985). Tempering will allow eggs and larvae of *T. nilotica to* survive somewhat higher salinities than if they are moved directly from freshwater (Al-Asgah 1984). *T. aurea* can be adapted to full strength sea water if they are properly tempered, though growth is reportedly not as great as in freshwater and disease problems have been reported (McGeachin and Wicklund 1987). Finale-Gómez and Brito-Pérez (1984) advised increasing the salinity by 15 o/oo every 24 hours. Survival of *T. mossambica* has been reported from salinities of 60, 70, and even 120 o/oo (Assem and Hanke 1979; Potts, et al. 1967; Whitfield and Blaber 1979). That species can efficiently reproduce at salinities up to 49 o/oo (Popper and Lichatowich 1975).

Red hybrid tilapia have been developed in Taiwan, the Philippines, Guam, Florida (USA), and Israel. The origins of the red hybrid tilapia have been reviewed by Galman and Avtlion (1983). Most of the red hybrids evaluated to date show a remarkable ability to tolerate sea water. This adaptability to high salinity may relate to the fact that most of the red hybrids have *T. mossambica* as one species involved in the cross. *T. hornorum* is also frequently used in crosses that produced red hybrid tilapia. That species has been found to occur naturally in brackish water (Philippart and Ruwet 1982). There is evidence that some groups of red hybrid tilapia actually grow more rapidly in brackish and full strength sea water than in freshwater (Meriwether et al. 1984; Watanabe 1985).

Red Drum and Hybrid Striped Bass

The red drum, *Sciaenops ocellatus*, has received a great deal of attention from researchers and commercial aquaculturists in the past few years. Red drum can be readily spawned in captivity through photoperiod and temperature control (Arnold et al. 1977), and the larvae can be reared in ponds to market size within a year.

Red drum juveniles are well adapted to rapid salinity changes (Wakeman and Wohlschlag 1983), though winterkill in freshwater ponds has been reported by commercial producers and may be related to osmotic stress in cold water. Embryos and larval red drum are less tolerant of fresh and low salinity water than are postlarvae and juveniles (Holt et al. 1981; Crocker et al. 1981, 1983). When the environmental temperature ranges from 18 to 30 C, larval red drum occur in nature at salinities from 16 to 34 o/oo. Juveniles, on the other hand, have been found from 0 to 34 o/oo over a temperature range of 12.5 to 32 C (Peters and McMichael 1987) and can tolerate direct transfer to fresh water from 28 o/oo salinity (Crocker et al. 1983).

Environmental calcium levels significantly affect red drum survival in both fresh and saline waters, though variations in environmental magnesium appear not to influence the performance of red drum (Wurts and Stickney 1988). In freshwater, the minimum recommended calcium concentration for red drum culture has been placed at 25 mg/l, though levels two to four times higher seem to be desirable. In 35 o/oo salinity water, calcium should be present at no less than 340 mg/l, though 400 mg/l is recommended. In water with salinities ranging from 9-35 o/oo, a calcium:sodium ratio of approximately 0.038:1 appears appropriate (Wurts and Stickney 1988).

Striped bass (*Morone saxatilis*) are anadromous, and therefore, during various stages of their lives, are fully adapted to freshwater and seawater. White bass (*M. chrysops*), on the other hand, are generally found in freshwater environments, though they have been reported in coastal streams along the Gulf of Mexico (Moyle 1976) and may, therefore, be able to tolerate brackish water. Hybrids between the two species, which have generated a good deal of interest among aquaculturists in recent years, appear to be broadly euryhaline. They are able to survive direct transfer from freshwater to 36 o/oo and back and have reportedly survived at 36 o/oo for at least a week (Wattendorf and Shafland 1982). In a recent study of hybrid striped bass performance at salinities ranging from freshwater to 35 o/oo, Smith et al. (1986) found that survival was above 90% at all measured salinities up to 28 o/oo, but that survival was only 60% at a salinity of 35 o/oo. Growth was not affected by salinity in that study.

Other Species

The white sturgeon (*Acipenser traansmontanus*) is presently being produced commercially on the west coast of the United States and has been the object of a considerable amount of research in the same region. The species spawns in fresh water but spends most of its adult life in coastal marine waters and is thus considered to be semi-anadromous (McEnroe and Cech 1987). Aquaculture is currently being practiced in fresh water, but the potential for brackish water culture would seem to exist. A related species in Asia, the sevruga (*A. stellatus*), may be more typical of freshwater sturgeons. *A. stellatus* juveniles can tolerate direct transfer to 10 o/oo salinity. Complete mortality occurs within six hours when the fish are exposed to 20 o/oo (Shikin and Lapina 1982).

Gilthead sea bream (*Sparus aurata*) are being commercially cultured in the Mediterranean region. The species can tolerate transfer from full strength sea water to salinities as low as 4 o/oo without suffering mortality. The same species can tolerate 2 o/oo if tempered to that salinity (Chervinski 1984b).

Another species of interest in the Mediterranean Sea is the sea bass, *Dicentrarchus labrax*. The eggs of that species have been shown to hatch over a relatively wide range of temperatures and salinity. At salinities from 29 to 47 o/oo, larvae hatched over a temperature range of 8 to 20 C. No hatching occurred below 4 C at any of the salinities tested (Marangos et al. 1986).

Milkfish (*Chanos chanos*) are cultured in the Far East. The species is broadly euryhaline. Larvae can be found in coastal waters with salinities ranging from 10 to 34 o/oo (Bardach et al. 1972) and will occasionally enter freshwater lakes. Net pen culture of milkfish is well established in Laguna de Bay, an extension of Manila Bay in the Philippines where very low salinities can occur (verbal reports to the author indicate that the salinity range is between freshwater and about 10 o/oo).

SUMMARY

It is clear that many freshwater fish species reared by aquaculturists can not only tolerate moderate levels of salinity, but that performance is actually improved when those fish are exposed to low salinities; levels of 2 to 4 o/oo seem to be appropriate in many cases. Most anadromous fish reared by aquaculturists are tolerant of a much broader salinity range than are the strictly freshwater species, and in some cases results obtained when rearing anadromous fishes in freshwater are similar to those obtained from sea water culture. Certain stages in the life of anadromous salmonids require tighter control on salinity, however. Prior to parr-smolt transformation and at the time of maturation, performance at high salinities is negatively impacted.

Most of the non-anadromous marine fishes under culture today are relatively euryhaline, though there are exceptions. This review concentrates on species which are of interest in the Americas, though several examples of species primarily under culture in Europe, Africa, and Asia are included.

LITERATURE CITED

Al-Asgah, N. A. 1984. The effect of salinity on eggs and larvae of *Oreochromis niloticus* (Linnaeus). Arabian Gulf Journal of Scientific Research 2: 673-681.

Alderdice, D. F. 1988. Osmotic and ionic regulation in teleost eggs and larvae. Pages 163-251 *In*: W. S. Hoar and D. J. Randall, editors. Fish Physiology, Volume XI, Part A. Academic Press, New York, New York, USA.

Alderdice, D. F. and C. R. Forrester. 1968. Some effects of salinity and temperature on early development of the English sole (*Parophrys vetulus*). Journal of Fisheries Research Board of Canada 28: 727-744.

Alexis, M. N., E. Papaparaskeva-Papoutsoglou, and S. Papoutsoglou. 1984. Influence of acclimation temperature on the osmotic regulation and survival of rainbow trout (*Salmo gairdneri*) rapidly transferred from fresh water to sea water. Aquaculture 40: 333-341.

Allee, B. J. 1981. The status of saltwater maturation of coho salmon (*Oncorhynchus kisutch*) at Oregon Aqua-Foods, Inc. Pages 1-8 *In*: T. Nosho (editor). Salmonid broodstock maturation. Washington Sea Grant, Seattle, Washington USA.

Allee, B. J. and B. K. Suzumoto. 1981. Maturation studies of the 1980 brood year coho salmon (*Oncorhynchus kisutch*) at Oregon Aqua-Foods, Inc. pp. 71-74, *In*: T. Nosho (editor) Salmonid broodstock maturation. Washington Sea Grant, Seattle, Washington USA).

Allen, K. O. 1971. Notes on the relative salinity tolerance of channel and blue catfish. The Progressive Fish-Culturist 33: 135-137.

Allen, K. O. and J. W. Avault, Jr. 1970. Effects of salinity on growth and survival of channel catfish, *Ictalurus punctatus*. Proceedings of the Southeastern Association of Game and Fish Commissioners 23: 319-331.

Arnold, C. R., T. D. Williams, F. A. Fable, J. L. Lasswell, and W. H. Bailey. 1977. Laboratory spawning and larval rearing of red drum and southern flounder. Proceedings of the Southeastern Association of Fish and Wildlife Agencies 31: 437-440.

Assem, H. and W. Hanke. 1979. Concentrations of carbohydrates during osmotic adjustment of the euryhaline teleost *Tilapia mossambica*. Comparative Biochemistry and Physiology 64A: 5-16.

Bardach, J. E., R. H. Ryther and W. O. McLarney. 1972. Aquaculture. Wiley-Interscience, New York USA. 868 pages.

Basulto, S. 1976. Induced saltwater tolerance in connection with inorganic salts in the feeding of Atlantic salmon (*Salmo salar*). Aquaculture 8: 45-55.

Bradley, T. M. 1987. Changes in hepatic mitochondrial proteins of steelhead during parr-smolt transformation. Pages 230-235 *In*: M. J. Dadswell, R. J. Klauda, C. M. Moffit, R. L. Saunders, R. A. Rulifson, and J. E. Cooper (editors). Common strategies of anadromous and catadromous fishes. American Fisheries Society, Bethesda, Maryland USA.

Burnside, M. C., J. W. Avault, Jr. and W. G. Perry. 1975. Comparison of a wild and a domestic strain of channel catfish grown in brackish water. The Progressive Fish-Culturist 37: 52-54.

Canagaratnam, P. 1959. Growth of fishes in different salinities. Journal of the Fisheries Research Board of Canada 16: 121-130.

Chervinski, J. 1977. Note on the adaptability of silver carp - *Hypopthalmichthys molitrix* (Val.) and grass carp - *Ctenopharyngodon idella* (Val.) to various salinity concentrations. Aquaculture 11: 179-182.

Chervinski, J. 1984a. Salinity tolerance of young catfish, *Clarias lazera* (Burchell). Journal of Fish Biology 25: 147-149.

Cervinski, J. 1984b. Salinity tolerance of young gilthead sea bream *Sparus aurata* L. Bamidgeh 36: 121-123.

Clarke, W. C. and J. Blackburn. 1977. A seawater challenge test to measure smolting of juvenile salmon. Fisheries and Marine Services Research Division Technical Report number 705, Department of Fisheries and Environment, Canada. 14 p.

Clarke, W. C. and Y. Nagahama. 1977. Effect of premature transfer to seawater on growth and morphology of the pituitary, thyroid, pancreas, and interrenal in juvenile coho salmon (*Onchorhynchus kisutch*). Canadian Journal of Zoology 55: 1620-1630.

Clarke, W. C., J. E. Shelbourn and J. R. Brett. 1978. Growth and adaptation to seawater in 'underyearling' sockeye (*Oncorhynchus nerka*) and coho (*O. kisutch*) salmon subjected to regimes of constant or changing temperature and daylength. Canadian Journal of Zoology 56: 2413-2421.

Clemens, W. A. and G. V. Wilby. 1961. Fishes of the Pacific coast of Canada. 2nd edition. Fisheries Research Board of Canada Bulletin 68. 443 p.

Conte, F. P. 1969. Salt secretion. Pages 241-292 *In*: W. S. Hoar and D. J. Randall (editors). Fish Physiology, Vol 1. Academic Press, New York.

Crocker, P. A., C. R. Arnold, J. A. DeBoer and J. D. Holt. 1981. Preliminary evaluation of survival and growth of juvenile red drum (*Sciaenops ocellata*) in fresh and salt water. Proceedings of the World Mariculture Society 12: 122-134.

Crocker, P. A., C. R. Arnold, J. A. DeBoer and G. J. Holt. 1983. Blood osmolality shift in juvenile red drum, *Sciaenops ocellata* L. exposed to fresh water. Journal of Fish Biology 23: 315-319.

Dickhoff, W. W. and C. V. Sullivan. 1987. Involvement of the thyroid gland in smoltification with special reference to metabolic and developmental processes. Pages 197-210 *In*: M. J. Dadswell, R. J. Klauda, C. M. Moffit, R. L. Saunders, R. A. Rulifson, and J. E. Cooper (editors). Common strategies of anadromous and catadromous fishes. American Fisheries Society, Bethesda, Maryland USA.

Dickhoff, W. W., L. C. Folmar and A. Gorbman. 1978. Changes in plasma thyroxine during smoltification of coho salmon, *Oncorhynchus kisutch*. General and Comparative Endocrinology 36: 229-232.

Dittmar, W. 1884. Report on researches into the composition of ocean water, collected by H.M.S. Challenger. Challenger Reports, Physics and Chemistry 1: 1-251.

Elson, P. F. 1957. The importance of size on the change from parr to smolt in Atlantic salmon. Canadian Fish Culturist 21: 1-6.

Ewing, R., J. Johnson, H. Pribble and J. Lichatowich. 1979. Temperature and photoperiod effects on gill (Na-K)-ATPase activity in chinook salmon (*Oncorhynchus tshawytscha*). Journal of the Fisheries Research Board of Canada 36: 1347-1353.

Finale-Gómez, E. and R. Brito-Pérez. 1984. Adaptación de la tilapia (*Oreochromis aureus*) al agua salada mediante cambios paulatino de salinidad. Revista de Investigaciones Marinas 5: 57-64.

Folmar, L. C. and W. W. Dickhoff. 1980. The parr-smolt transformation (smoltification) and seawater adaptation in salmonids. A review of selected literature. Aquaculture 21: 1-17.

Folmar, L. C. and W. W. Dickhoff. 1981. Evaluation of some physiological parameters as predictive indices of smoltification. Aquaculture 23: 309-324.

Foskett, J. K. and C. Scheffey. 1982. The chloride cell: definitive identification as the salt-secretory cell in teleosts. Science 215: 164-166.

Galman, O. R. and R. R. Avtalion. 1983. A preliminary investigation of the characteristics of red tilapias from the Philippines and Taiwan. Pages 291-301 *In*: Proceedings, International Symposium on Tilapia in Aquaculture. Tel Aviv University, Tel Aviv, Israel.

Harvey, B. and R. N. Kelley. 1984. Control of spermatozoan motility in a euryhaline teleost acclimated to different salinities. Canadian Journal of Zoology 62: 2674-2677.

Heard, W. R. and R. A. Crone. 1976. Raising coho salmon from fry to smolts in estuarine pens, and returns of adults from two smolt releases. The Progressive Fish-Culturist 38: 171-174.

Hickman, C. P. 1968. Glomerular filtration and urine flow in the euryhaline southern flounder, *Paralichthys lethostigma*. Canadian Journal of Zoology 46: 427-437.

Hoar, W. S. 1976. Smolt transformation: evolution, behavior, and physiology. Journal of the Fisheries Research Board of Canada 33: 1234-1252.

Hogstrand, C. and C. Haux. 1985. Evaluation of the sea-water challenge test on sea trout, *Salmo trutta*. Comparative Biochemistry and Physiology 82A: 261-266.

Holt, J., R. Godbout and C. R. Arnold. 1981. Effects of temperature and salinity on egg hatching and larval survival of red drum, *Sciaenops ocellata*. Fishery Bulletin 79: 569-573.

Houston, A. H. 1961. Influence of size upon the adaptation of steelhead trout (*Salmo gairdneri*) and chum salmon (*Oncorhynchus keta*) to sea water. Journal of the Fisheries Research Board of Canada 18: 401-415.

Huntsman, A. G. and W. S. Hoar. 1939. Resistance of Atlantic salmon to seawater. Journal of the Fisheries Research Board of Canada 4: 409-411.

Johnson, S. K. 1976. Laboratory evaluation of several chemicals as preventatives of ich disease. Pages 91-96 *In*: Proceedings of the 1976 Fish Farming Conference and Annual Convention of the Catfish Farmers of Texas. Texas A&M University, College Station, Texas USA.

Johnston, C. E. and J. G. Eales. 1970. Influence of body size on silvering of Atlantic salmon (*Salmo salar*) at parr-smolt transformation. Journal of the Fisheries Research Board of Canada 27: 933-987.

Johnston, C. E. and R. L. Saunders. 1981. Parr-smolt transformation of yearling Atlantic salmon (*Salmo salar*) at several rearing temperatures. Canadian Journal of Fisheries and Aquatic Sciences 38: 1189-1198.

Jürss, K., Th. Bittorf and Th. Völker. 1985. Influence of salinity and ratio of lipid to protein in diets on certain enzyme activities in rainbow trout (*Salmo gairdneri* Richardson). Comparative Biochemistry and Physiology 81B: 73-79.

Kendall, A. W., Jr. and F. L. Schwartz. 1968. Lethal temperature and salinity tolerance for white catfish, *Ictalurus catus*, from the Patuxent River, Maryland. Chesapeake Science 9: 103-108.

Kepshire, B. M., Jr. and W. J. McNeil. 1982. Growth of premigratory chinook salmon in seawater. Fishery Bulletin 70: 119-123.

Kilambi, R. V. and A. Zdinak. 1980. The effect of acclimation on the salinity tolerance of grass carp *Ctenopharyngodon idella* (Gur. and Val.). Journal of Fish Biology 16: 171-175.

Lam, T. J. and R. Sharma. 1985. Effects of salinity and thyroxine on larval survival, growth and development in the carp, *Cyprinus carpio*. Aquaculture 44: 201-212.

Maceina, M. J. and J. V. Shireman. 1979. Grass carp: effects of salinity on survival, weight loss, and muscle tissue water content. The Progressive Fish-Culturist 41: 69-73.

Mahnken, C. V. W. 1973. The size of coho salmon and time of entry into seawater: part 1. Effects on growth and condition index. Pages 30-31 *In*: Proceedings of the 24th Annual Northwest Fish Culture Conference, Wemme, Oregon.

Marangos, C., H. Yagi, and H. J. Ceccaldi. 1986. Rôle de la température et de la salinitie sur le taux de survie et la morphologenése au cours du developpement embryonnaire chez les oeuffs du loup de mer *Dicentrarchus labrax* (Linnaeus, 1758)(Pisces, Teleostei, Serranidae). Aquaculture 54: 287-300.

McEnroe, M. and J. J. Cech, Jr. 1987. Osmoregulation in white sturgeon: life history aspects. Pages 191-196 *In*: M. J. Dadswell, R. J. Klauda, C. M. Moffit, R. L. Saunders, R. A. Rulifson, and J. E. Cooper (editors). Common strategies of anadromous and catadromous fishes. American Fisheries Society, Bethesda, Maryland USA.

McGeachin, R. B. and R. I. Wicklund. 1987. Growth of *Tilapia aurea* in seawater cages. Journal of the World Aquaculture Society 18: 31-34.

Meriwether, F. H., II, E. D. Scura and W. Y. Okamura. 1984. Cage culture of red tilapia in prawn and shrimp ponds. Journal of the World Mariculture Society 15: 254-265.

Moyle, P. B. 1976. Inland fishes of California. University of California Press, Berkeley. 405 pp.

Murai, T. and J. W. Andrews. 1973. Growth and food conversion of rainbow trout reared in brackish and fresh water. Fishery Bulletin 70: 1293-1295.

Oertzen, Jörg-A. von. 1985. Resistance and capacity adaptation of juvenile silver carp, *Hypophthalmichthys molitrix* (Val.), to temperature and salinity. Aquaculture 44: 321-332.

Otto, R. G. 1971. Effects of salinity on the survival and growth of pre-smolt salmon (*Oncorhynchus kisutch*). Journal of the Fishery Research Board of Canada 28: 343-349.

Parry, G. 1960. The development of salinity tolerance in salmon, *Salmo salar* L. and some related species. Journal of Experimental Biology 37: 425-434.

Payne, A. I. 1983. Estuarine and salt tolerant tilapias. Pages 534-543 *In*: L. Fishelson and Z. Yaron (editors). International Symposium on Tilapia in Aquaculture. Tel Aviv University, Tel Aviv, Israel.

Perry, W. G. 1967. Distribution and relative abundance of blue catfish, *Ictalurus furcatus*, and channel catfish, Ictalurus punctatus, with relation to salinity. Proceedings of the Southeastern Association of Game and Fish Commissioners 21: 436-444.

Perry, W. G. and J. W. Avault, Jr. 1969. Preliminary experiment on the culture of blue, channel and white catfish in brackish water ponds. Proceedings of the Southeastern Association of Game and Fish Commissioners 22: 397-406.

Perry, W. G. and J. W. Avault, Jr. 1970. Culture of blue, channel and white catfish in brackish water ponds. Proceedings of the Southeastern Association of Game and Fish Commissioners 23: 592-605.

Peters, K. M. and R. H. McMichael, Jr. 1987. Early life history of the red drum, *Sciaenops ocellatus* (Pisces:Sciaenidae), in Tampa Bay, Florida. Estuaries 10: 92-107.

Philippart, J.-C. and J.-C. Ruwet. 1982. Ecology and distribution of tilapias. Pages 15-59 *In*: R. S. V. Pullin and R. H. Lowe-McConnell (editors). The biology and culture of tilapias. ICLARM Conference Proceedings 7, International Center for Living Aquatic Resources Management, Manila, Philippines.

Popper, D. and T. Lichatowich. 1975. Preliminary success in predator control of *Tilapia mossambica*. Aquaculture 5: 213-214.

Potts, W. I. M., M. A. Foster, P. P. Rady and G. P. Howell. 1967. Sodium and water balance in the cichlid teleost *Tilapia mossambica*. Journal of Experimental Biology 47: 461-470.

Refstie, T. 1982. The effect of feeding thyroid hormones on saltwater tolerance and growth rate of Atlantic salmon. Canadian Journal of Zoology 60: 2706-2712.

Robins, C. R., R. M. Bailey, C. E. Bond, J. R. Brooker, E. A. Lachner, R. N. Lea and W. B. Scott. 1980. A list of common and scientific names of fishes from the United States and Canada. American Fisheries Society, Bethesda, Maryland. 174 p.

Rockwell, J. J. 1956. Some effects of seawater and temperature on the embryos of the Pacific salmon, *Oncorhynchus gorbuscha* (Walbaum) and Oncorhynchus keta (Walbaum). Ph.D. dissertation, University of Washington, Seattle. 416 p.

Ross, D. A. 1970. Introduction to oceanography. Appleton-Century-Crofts, New York. 384 p.

Saha, K. C., D. N. Chakraborty, B. K. De and S. Shakraborty. 1984a. Studies on the salinity tolerance of fry of Indian major carps in captivity. Indian Journal of Fisheries 11: 247-248.

Saha, K. C., D. N. Chakraborty, B. K. Jana, B. K. De, J. N. Misra, B. K. Pal and A. K. Talapatra. 1984b. Studies on the potentiality of brackish water fish farming along Alampore coast West Bengal. Indian Journal of Fisheries 11: 256-267.

Saha, K. C., D. N. Chakraborty, A. Mahalanabish, D. K. Nag, G. C. Paul and H. B. Dey. 1984c. Studies on the potentiality of brackish water fish farming at Junput sea coast, Contai, West Bengal. Indian Journal of Fisheries 11: 249-255.

Saxton, A. M., R. N. Iwamoto and W. K. Hershberger. 1983. Smoltification in the net-pen culture of accelerated coho salmon, *Oncorhynchus kisutch* Walbaum: prediction of saltwater performance. Journal of Fish Biology 22: 363-370.

Schildhauer, B. 1981. On the salinity tolerance of silver carp (*Hypophthalmichthys molitrix* Valentin). Fisherei Forshung 19: 35-39.

Shikin, Yu. N. and N. N. Lapina. 1982. Locomotor and feeding activity of juvenile sevruga, *Acipenser stellatus* (Acipenseridae), with increasing salinity. Journal of Ichthyology 22: 138-142.

Skud, B. E. 1954. Salinity gradients in the intertidal zone of an alaskan pink salmon stream. Fishery Bulletin 22: 65-142.

Smith, T. I. J., W. E. Jenkins and R. W. Haggerty. 1986. Growth and survival of juvenile striped bass (*Morone saxatilis*) x white bass (*M. chrysops*) hybrids reared at different salinities. Proceedings of the Annual Conference of the Southeastern Association of Fish and Wildlife Agencies 40: 143-151.

Sower, S. and S. B. Schreck. 1981. Sexual maturation of coho salmon (*Oncorhynchus kisutch*) induced ovulation and serum hormone and ion levels of salmon in fresh versus salt water. Page 46 *In*: T. Nosho (ed.). Salmonid broodstock maturation. Washington Sea Grant, Seattle.

Stickney, R. R. 1986. Tilapia tolerance of saline waters: a review. The Progressive Fish-Culturist 48: 161-167.

Stickney, R. R. and B. A. Simco. 1971. Salinity tolerance of catfish hybrids. Transactions of the American Fisheries Society 100: 790-792.

Sverdrup, H. U., M. W. Johnson and R. H. Fleming. 1942. The Oceans: Their Physics, Chemistry, and General Biology. Prentice-Hall, Englewood Cliffs, New Jersey USA. 1087 p.

Tatum, W. M. 1973. Brackish water cage culture of rainbow trout (*Salmo gairdneri*) in South Alabama. Transactions of the American Fisheries Society 102: 826-828.

Trewavas, E. 1973. On the cichlid fishes of the genus *Pelmatochromis* with proposal of a new genus for *P. congicus*; on the relationship between *Pelmatochromis* and *Tilapia* and the recognition of *Sarotherodon* as a distinct genus. Bulletin of the British Museum (Natural History) Zoology 25: 1-26.

Trewavas, E. 1982. *Tilapias*: taxonomy and speciation. Pages 3-13 *In*: Pullin, R. S. V. and R. H. Lowe-McConnell (eds.). The biology and culture of tilapias. ICLARM Conference Proceedings 7, International Center for Living Aquatic Resources Management, Manila, Philippines.

Turner, P. R. 1988. Growth of channel catfish in saline groundwaters of the Pecos Valley of New Mexico. New Mexico Water Resources Research Institute Report Number 231. New Mexico State University, Las Cruces. 54 p.

Wagner, H. H. 1974. Photoperiod and temperature regulation of smolting in steelhead trout (*Salmo gairdneri*). Canadian Journal of Zoology 52: 219-234.

Wakeman, J. M. and D. E. Wohlschlag. 1983. Time course of osmotic adaptation with respect to blood serum osmolality and oxygen uptake in the euryhaline teleost, *Sciaenops ocellelatus* (red drum). Contributions in Marine Science 26: 165-177.

Watanabe, W. O. 1985. Experimental approaches to the saltwater culture of tilapias. Pages 3-5 *In*: ICLARM Newsletter, January. International Center for Living Aquatic Resources Management, Manila, Philippines.

Watanabe, W. O., C-M. Kuo and M-C. Huang. 1985. Salinity tolerance of Nile tilapia fry (*Oreochromis niloticus*), spawned and hatched at various salinities. Aquaculture 48: 159-176.

Wattendorf, R. J. and P. L. Shafland. 1982. Observations on salinity tolerances of striped bass x white bass hybrids in aquaria. Progressive Fish-Culturist 44: 148-149.

Weisbart, M. 1968. Osmotic and ionic regulation in embryos, alevins, and fry of the five species of Pacific salmon. Canadian Journal of Zoology 46: 385-397.

Wertheimer, A. 1981a. Maturation success of coho salmon and pink salmon held under different salinity regimes. Pages 75-78 *In*: T. Nosho (ed.). Salmonid broodstock maturation. Washington Sea Grant, Seattle, Washington USA.

Wertheimer, A. 1981b. Use of estuarine netpens for holding returning broodstock. Pages 9-11 *In*: T. Nosho (ed.). Salmonid broodstock maturation. Washington Sea Grant, Seattle, Washington USA.

Wertheimer, A. C. 1984. Maturation success of pink salmon (*Oncorhynchus gorbuscha*) and coho salmon (*O. kisutch*) held under three salinity regimes. Aquaculture 43: 195-212.

Whitfield, A. K. and S. J. M. Blaber. 1979. The distribution of the freshwater cichlid *Sarotherodon mossambica* in estuarine systems. Environmental Biology of Fishes 4: 77-81.

Wurts, W. A. and R. R. Stickney. 1989. Responses of red drum (*Sciaenops ocellatus*) to calcium and magnesium concentrations in fresh and salt water. Aquaculture 76:21 - 35.

Zaugg, W. S. and L. R. McClain. 1970. ATPase activity in gills of salmonids: seasonal variation and saltwater influence in coho salmon, *Oncorhynchus kisutch*. Comparative Biochemistry and Physiology 35: 587-596.

Zaugg, W. S. and L. R. McClain. 1972. Changes in gill ATPase activity associated with parr-smolt transformation in steelhead trout, coho and spring salmon. Journal of the Fisheries Research Board of Canada 29: 167-171.

Zaugg, W. S. and H. H. Wagner. 1973. Gill ATPase activity related to parr-smolt transformation in steelhead trout (*Salmo gairdneri*): influence of photoperiod and temperature. Comparative Biochemistry and Physiology 45: 955-965.

Zaugg, W. S., E. F. Prentice and F. W. Waknitz. 1985. Importance of river migration to the development of seawater tolerance in Columbia River anadromous salmonidae. Aquaculture 51: 33-47.

ENVIRONMENT-RELATED OFF-FLAVORS IN FISH

Craig S. Tucker and John F. Martin

ABSTRACT

Off-flavors in fish can be caused by feed ingredients or natural foods, post-mortem oxidative rancidity, or odorous compounds absorbed from the environment. This review examines environment-related off-flavors with particular reference to the occurrence and impact of the problem on the farm-raised channel catfish *Ictalurus punctatus* industry. Sources of environment-related off-flavors include chemical pollution and extracellular metabolites of aquatic bacteria or algae. Geosmin and 2-methylisoborneol, two "earthy-smelling" metabolites of aquatic microorganisms, are the most common causes of off-flavor in cultured fish. These two compounds are synthesized by cyanobacteria (blue-green algae) in the order Hormogonales and by various actinomycetes. Organic compounds responsible for off-flavor are rapidly absorbed by fish from water and stored in lipid-rich tissues; elimination is relatively slow. The occurrence of off-flavor in pond-cultured fish is episodic, unpredictable, and unpreventable.

Although off-flavors can be removed from fish by moving them to a clean environment, this practice is not always applicable because some off-flavor compounds are not quickly purged from fish. Earthy off-flavors can be partially masked by certain processing methods, but it is doubtful if these methods will be widely used in large-scale aquaculture industries.

INTRODUCTION

"Off-flavors" are objectionable tastes or odors in water or foods. Off-flavor in fish and other aquatic products is an important practical problem in commercial fisheries and aquaculture. Consumption of aquatic products is elective in most developed countries, and disagreeable flavors in the product, and resulting consumer dissatisfaction, may adversely affect market demand. This is particularly so for products such as channel catfish *Ictalurus punctatus*, which have not historically been a common food item for a large segment of the population. If off-flavored fish are marketed, first-time buyers might assume the flavor is inherent in the product and shun future purchases.

Off-flavor can be the result of post-mortem oxidation of fats (rancidity) in fish. These off-flavors are usually the result of prolonged or improper storage. Fish may also acquire off-flavors from certain natural foods or feed ingredients. For instance, feeds high in marine fish oil may impart an excessively "fishy" flavor to otherwise mild-flavored fish. This review addresses a third source of off-flavors — the environment.

A variety of odorous compounds can be absorbed by fish from the environment and impart undesirable flavors to the flesh. These compounds may enter the environment as pollution or they may be produced by microorganisms growing in or near the water. The most common causes of off-flavor in cultured fish are odorous metabolites of aquatic bacteria or algae. Off-flavors arising from pollution are important in many commercial fisheries, but are rare in aquaculture because fish culture facilities are usually located where they can be supplied with unpolluted water. However, some culture facilities, such as open-water net pens, are susceptible to unforeseen pollution that may cause fish to develop off-flavors.

RELATIONSHIP BETWEEN TASTES AND ODORS IN WATER AND FLAVORS IN FISH

Tastes and odors in drinking water have been intensely studied for decades. The current understanding of taste and odor problems in water and the history of the research in this area have been summarized by Mallevialle and Suffet (1987). It has long been recognized that waters with disagreeable tastes or odors may yield fish with unusual or objectionable flavors. The relationship between odors in water and flavors in fish is complicated by the differing degrees to which odorous compounds in water are bioconcentrated and metabolized by fish, as well as by the varying sensory interactions between specific odorous compounds and the natural flavors of fish. Nevertheless, Persson (1984) developed an empirical relationship that is useful for estimating the potential for a compound in water to cause flavor problems in fish. This relationship was developed from published "estimated threshold concentrations" (the lowest concentration of a compound in water that will impair the flavor of fish) and corresponding "threshold odor concentrations" (the lowest concentration of the compound that is detectable in water). Estimated threshold concentrations have been experimentally determined for only about 60 compounds, but threshold odor concentrations are known for hundreds of compounds and products (Mallevialle and Suffet 1987). The functional relationship between estimated threshold concentration (ETC) and threshold odor concentration (TOC) is the power equation:

$$ETC = 0.97\ TOC^{0.81}$$

Phenolic compounds appear to cause flavor problems in fish at lower concentrations in water than can be detected by sensory methods ($ETC < TOC$). For this class of compounds, the relationship is:

$$ETC = 0.16\ TOC^{1.14}$$

The threshold odor concentrations for a number of compounds discussed in this paper are compiled in Table 1.

OFF-FLAVORS ARISING FROM POLLUTION

Man discharges into natural waters thousands of organic compounds, many of which are highly odorous and may affect the flavor of fish. Impairment of fish flavor by man-made pollutants has been widely reported (reviewed by Shumway and Palensky 1973, Persson 1984) and it appears that the most common sources of these off-flavors are organic compounds derived from petroleum products or paper processing. For example, in a study of fish off-flavors in Finland, 41% of the flavor problems were attributed to pulp mill effluents and 24% to petroleum products (Kuusi and Suihko 1983). In waters of more industrialized countries, off-flavors related to petroleum compounds are probably most common.

Off-Flavors Related to Petroleum Compounds

Fish and other aquatic animals with an "oily", "diesel fuel", or "petroleum" flavor are commonly found in areas near oil spills or waters that

Table 1. Description of odors and threshold odor concentrations in water of some organic compounds. (from data compiled by Persson 1984, Mallevaille and Suffet 1987).

Compound	Odor	Threshold odor concentration (μg/L)
Allyl mercaptan	garlic, disagreeable	0.05
Benzyl mercaptan	sharp, unpleasant	0.2
Carbon tetrachloride	characteristic	200
Chloroform	characteristic	100
Crotyl mercaptan	skunk	0.03
Cumene	sharp	80
1,4-Dichlorobenzene	aromatic	0.3
2,3-Dichlorophenol	medicinal	30
2,4-Dichlorophenol	medicinal	40
Dimethyltrisulfide	swampy, sewage	0.01
Ethylmercaptan	rotten cabbage	0.1
Geosmin	earthy	0.02
2-Methylisoborneol	musty	0.04
α-Methylnapthalene	coal tar	12
Monochlorophenols	medicinal	10-60
Napthalene	coal tar	5
Phenol	phenolic	5,000
Pyridine	irritating	4
Thiophenol	putrid, nauseating	0.06
Toluene	sour, burnt	170
2,4,6-Trichlorophenol	phenolic	30
Xylene	sweet, solvent	20

receive wastes from petroleum refineries or petrochemical manufacturing plants. Petroleum products may also enter waters in runoff containing motor oils and fuels, or from wastes and spills of boat fuels.

Crude oils and refined petroleum products such as diesel fuel, kerosene, and gasoline are complex mixtures of hydrocarbons and other organic compounds. Crude oils are usually at least 75% hydrocarbons with the remainder being various sulfur-, nitrogen-, and oxygen-containing organic compounds. Alkanes and alkylcyclohexanes constitute a highly variable proportion (50 to 98 %) of crude oil. Aromatic hydrocarbons (benzene, toluene, xylene, and polycyclic aromatic compounds such as napthalene) comprise from less than 1 to over 15% of crude oil. Crude oil is then distilled and processed into a variety of products such as diesel fuels, kerosenes, and gasolines.

Threshold odor concentrations for crude oils in water vary from 0.5 to over 350 µg/L (Mallevialle and Suffet 1987). Refined petroleum products have somewhat higher threshold odor concentrations because highly odorous sulfur-containing compounds are removed during refining and processing; minimum concentrations in water that impair the flavor of fish range from less than 0.1 mg/L for crude oil to over 5 mg/L for kerosene (Persson 1984).

Most components of crude or refined petroleum are highly bioconcentrated from the water into fish. The variety of compounds isolated from fish is characteristic of the product to which the fish is exposed (Shipton et al. 1970; Mackie et al. 1972). However, not all the compounds found in petroleum-tainted flesh contribute to off-flavor. Unsaturated alkanes, aromatic hydrocarbons, and sulfur-containing organic contaminants are primarily responsible for off-flavors in fish exposed to petroleum products (Motohiro 1983). Saturated aliphatic hydrocarbons are not as odorous and apparently do not contribute greatly to flavor problems in fish.

Off-Flavor Caused by Pulp and Paper Mill Wastes

Untreated effluents from pulp mills contain a variety of polymeric lignins, phenolic compounds, mercaptans, terpenes, and other residues derived from chemically digested wood chips. If the pulp is bleached, the untreated effluent will contain chlorinated derivatives of these substances as well as degradation products of chlorolignins such as chloroform, carbon tetrachloride, and chloroethylenes (Paasivirta et al. 1983). The concentrations and relative amounts of these substances released into the environment depends upon the extent to which the effluents are treated.

Fish exposed to untreated pulp-mill effluents develop flavors described as "sewage" (Shumway and Chadwick 1971), "phenolic", "sulfide-like", and "pulp-mill" (Kuusi and Suihko 1983). The flavor is considered quite distinctive. Exposure of fish to untreated effluent at concentrations of 1-2% (v/v) induced off-flavors in coho salmon *Oncorhynchus kisutch* (Shumway and Chadwick 1971) and yellow perch *Perca flavescens* (Cook et al. 1973). However, some components of pulp-mill effluent, particularly those from chlorobleaching processes, are highly odorous and may impair the flavor of fish at much lower concentrations. Monochlorophenols and dichlorophenols have threshold odor concentrations of 2 to 60 μg/L (Table 2); fish exposed to 2,4-dichlorophenol at concentrations as low as 0.4 μg/L developed off-flavors (Shumway and Palensky 1973). Some mercaptans present in pulp-mill waste have threshold odor concentrations of less than 0.1 μg/L.

Biological treatment of pulp-mill waste before discharge greatly reduces the potential for flavor-impairment of fish exposed to the effluent. Coho salmon exposed to 1% untreated pulp-mill effluent developed a strong off-flavor, but fish exposed to 3% biologically treated effluent (the highest level tested) did not develop an off-flavor after 96 hours of exposure (Shumway and Chadwick 1971) .

Other Anthropogenic Off-Flavors

Mallevialle and Suffet (1987) discuss possible sources of odors and tastes in drinking water and list specific compounds and threshold odor concentrations associated with these sources. Four general sources are identified: dispersed urban pollution (such as solvents, gasoline, and motor oils); municipal wastewater effluents (which may contain odorous chlorinated and sulfur-containing organic compounds); industrial wastewater effluents; and organic pollutants from agricultural activities (such as odorous pesticides). At times, fish living in waters polluted by these wastes may accumulate sufficient quantities of these compounds to cause off-flavors, but documentation is meager.

Reviews of off-flavors in fish caused by water pollutants are provided by Shumway and Palensky (1973), Smith (1974), and Persson (1984).

ENVIRONMENTAL OFF-FLAVORS OF MICROBIAL ORIGIN

A causal relationship between certain odors in surface waters and aquatic microorganisms was suspected before the turn of the century. This relationship was suggested by the similarity of odors in water and odors produced in cultures of certain bacteria. The most common odor produced by these cultures has been variously described as "musty", "muddy", or "earthy". Similar off-flavors have been noticed in fish for centuries (Lovell et al. 1978).

The first complete description of the origin and nature of earthy off-flavors in fish was provided by Thayson (1936) and Thayson and Pentelow (1936). They associated the cause of off-flavor in Atlantic salmon *Salmo salar* from certain rivers in Scotland with actinomycetes present in decaying beds of submerged reeds along the river bank. Thayson (1936) speculated that the "earthy" compound produced by these actinomycetes is absorbed across the gills and transported to various tissues via the bloodstream. The earthy flavor was also experimentally produced in fish exposed to concentrated distillates from cultures of actinomycetes (Thayson and Pentelow 1936). The off-flavor was rapidly removed by transferring the fish to clean water.

Aschner et al. (1969) provided the first complete account of earthy off-flavors in fish produced by cyanobacteria (blue-green algae). They reported that *Oscillatoria tenuis* growing in fish culture ponds in Israel was responsible for earthy off-flavors in carp *Cyprinus carpio* and suggested that the fish acquired the off-flavor either by absorbing the compound from the water or by ingesting masses of cyanobacteria. They also reported that the problem could be alleviated either by killing the offending organism with the algicide copper sulfate, or by moving the off-flavored fish to clean water.

Earthy off-flavors in pond-raised channel catfish were initially described by Lovell and Sackey (1973) and Maligalig et al. (1973). The incidence of off-flavor in channel catfish has increased dramatically as culture practices have changed in efforts to increase fish yields (Brown and Boyd 1982). Off-flavors in catfish are reviewed in detail later.

Considerable research has been conducted on earthy flavors in rainbow trout *Oncorhynchus mykiss* cultured in prairie pot-hole lakes in central Canada. Two highly odorous compounds, geosmin and 2-methylisoborneol, were confirmed as causes of the flavors, and cyanobacteria capable of producing these compounds were isolated and identified (Yurkowski and Tabachek 1974; Tabachek and Yurkowski 1976). These two compounds also caused off-flavor in walleye *Stizostedium vitreum*, cisco *Coregonus artedii*, lake whitefish *Coregonus clupeaformis*, and northern pike *Esox lucius* from Cedar Lake, an important commercial fishery in Manitoba, Canada (Yurkowski and Tabachek 1980).

Kuusi and Suihko (1983) surveyed off-flavors in fish from Finland from 1969 to 1981. Most of the off-flavors were related to man-made pollution, but earthy off-flavors were common, particularly in carp. Earthy off-flavors in bream *Abramis brama* from a low salinity brackish-water bay in the Gulf of Finland have been intensively studied by Persson (1978, 1979b, 1981). The flavor is caused by absorption of geosmin apparently produced by the cyanobacterium *Oscillatoria agardhii*.

Reports of earthy off-flavors are limited to aquatic animals from freshwater or brackish waters. Apparently, the microorganisms that produce earthy-smelling metabolites do not thrive in saline waters. For example, marine shrimp cultured in Ecuador develop geosmin-related off-flavors only after periods of excessively heavy rainfall which causes the salinity of the culture ponds to drop from the usual 10 to 35 ppt to less than 10 ppt (Lovell and Broce 1985).

Compounds Causing Off-Flavors

Dozens of volatile, odorous compounds have been isolated from cultures of actinomycetes (Gerber 1979), cyanobacteria (Slater and Blok 1983), or algae (Juttner 1983). Several of these compounds have odors similar to those described in off-flavored fish (Table 2), but only two, geosmin and 2-methylisoborneol, have been consistently reported as causes of off-flavor in fish.

Table 2. Examples of compounds isolated from cultures of aquatic microorganisms.

Compound	Odor	Organism	Reference
geosmin	earthy	*Streptomyces*	Gerber 1968
2-methylisoborneol	musty	*Streptomyces*	Gerber 1969
β–cyclocitral	tobacco	*Microcystis*	Juttner 1984
2-isopropyl-3-methoxypyrazine	musty	*Streptomyces*	Gerber 1979
furfural	putrid	*Streptomyces*	Gerber 1979
cadin-4-ene-1-ol	woody	*Streptomyces*	Gerber 1971
dimethylsulfide	fishy	*Asterionella*	Juttner and Muller 1979
β–ionone	floral	*Synura*	Juttner 1981
isopropylmercaptan	onion	*Microcystis*	Jenkins et al. 1967
1-octen-3-one	mushroom	*Anabaena*	Mohren and Juttner 1983
hexanal	lettuce	*Stephanodiscus*	Juttner 1981
heptanal	rancid	*Cryptomonas*	Juttner 1983

Geosmin

Production of earthy odors by cultures of actinomycetes has been recognized since before 1900. Early attempts to isolate the chemical compounds responsible for the odor were limited by the lack of powerful analytical separation methods. Gerber and Lechevalier (1965) analyzed an odor concentrate from an actinomycete culture by gas chromatography and found a single sharp peak corresponding to the earthy odor detected at the exit port. They named the compound geosmin, from the Greek "ge", for earth, and "osme", for odor. Subsequent work elucidated the chemical structure of geosmin (Fig. 1) and its dehydration product, argosmin (Gerber 1968; Rosen et al. 1968). Argosmin can be produced in the laboratory by reacting geosmin with acid. Argosmin is considered to be odorless, although there is some evidence that certain argosmin isomers may also cause earthy flavors in fish (Martin, unpublished).

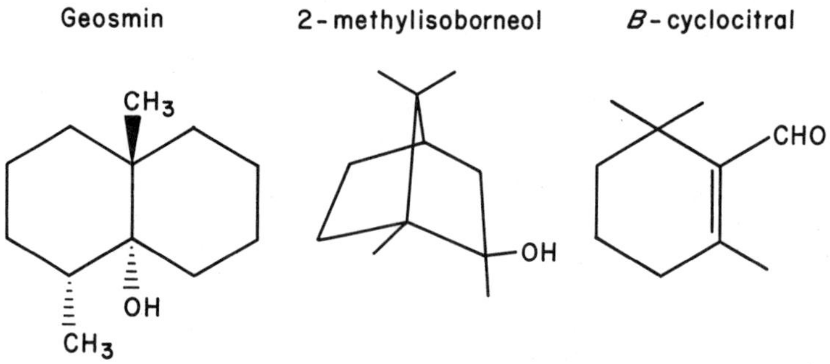

Figure 1. Structures of geosmin, 2-methylisoborneol, and beta-cyclocitral.

Once the identity of geosmin was known and authentic geosmin standards became available, its presence in soils and a variety of aquatic environments was quickly established. Geosmin has been identified as a flavor component of some vegetables and other food products and has been implicated as a cause of taste and odor problems in water supplies and freshwater fish worldwide.

The threshold odor concentration of geosmin in water is about 0.02 µg/L (Buttery et al. 1976; Persson 1980). The sensory threshold concentration of geosmin in fish ranges from 0.6 to 6.5 µg/kg (Yurkowski and Tabachek 1974; Persson 1980). Sensory threshold flavor concentrations vary with the species tested; geosmin is more easily detected in mild-flavored fish than in fish that have a fairly strong natural flavor, such as rainbow trout.

2-Methylisoborneol

Soon after the identity of geosmin was known, another earthy-smelling metabolite of actinomycetes was described (Gerber 1969; Medsker et al. 1969). This compound, 2-methylisoborneol (Fig. 1), was previously known as a synthetic product prepared by methylation of camphor. The odor of 2-methylisoborneol is musty or earthy in dilute solutions, but camphorous in concentrated solutions. This compound has been isolated from soils and freshwaters worldwide.

The threshold odor concentration of 2-methylisoborneol is about 0.04

μg/L in water (Persson 1979a). This is slightly higher than the threshold odor concentration for geosmin in water. However, the flavor imparted to fish by 2-methylisoborneol appears to be detected by taste panels at concentrations about an order of magnitude lower than geosmin. The sensory threshold concentration in pike is about 0.09 μg/kg, and in rainbow trout, about 0.6 μg/kg (Persson 1980).

2-Methylenebornane and 2-methyl-2-bornene

These two musty-smelling compounds are dehydration products of 2-methylisoborneol (Fig. 2). They were known only from laboratory synthesis until Martin et al. (1988b) isolated both from chronically off-flavored channel catfish. The compounds were found in the flesh at concentrations up to 4 mg/kg and the concentration of 2-methylenebornane was always higher than the concentration of 2-methyl-2-bornene. The two dehydration products have been isolated from fish with and without concurrent detectable levels of 2-methylisoborneol.

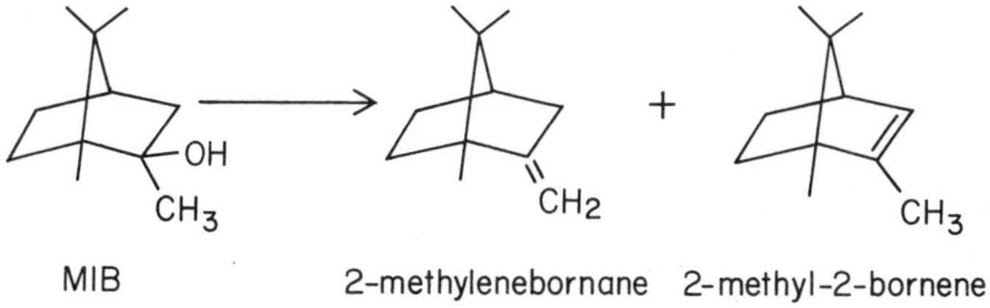

Figure 2. Dehydration products of 2-methylisoborneol (MIB).

The source of the dehydration products is unknown. Martin et al. (1988b) speculated that when fish are exposed to 2-methylisoborneol for long periods, the compound is dehydrated in the flesh and stored as the highly lipophilic, hydrocarbon products. However, recent investigations indicate the presence of 2-methylenebornane and 2-methyl-2-bornene in the water of some catfish culture ponds at low (about 10 ng/L) concentrations. It is possible, then, that the two compounds are produced in the environment from 2-methylisoborneol and then absorbed by the fish. The threshold odor concentrations in water and the sensory threshold concentrations in fish for the two dehydration products of 2-methylisoborneol are not known.

β-cyclocitral

This compound (Fig. 1) is a nor-carotenoid with an odor described as tobacco-like and has been isolated from cultures of *Microcystis aeruginosa* and from lake water during a bloom of *M. wesenbergii* (Juttner 1984). β–cyclocitral has also been isolated from channel catfish with an off-flavor described as "celery-like" or "spicy" (Martin, unpublished). The threshold odor concentrations of β–cyclocitral in water and fish are not known.

Products of decomposition

Large amounts of decaying organic matter can cause taste and odor problems in water by providing substrate for the growth of actinomycetes which then produce metabolites such as geosmin or 2-methylisoborneol. Objectionable tastes or odors can also result directly from the odorous products released during the bacterial decomposition of proteinaceous organic matter. These products include noxious and highly odorous sulfur compounds such as methylmercaptan, isobutylmercaptan, dimethylsulfide, and trimethylsulfide (Slater and Blok 1983). Channel catfish collected from culture ponds immediately following massive die-offs of phytoplankton blooms often have off-flavors described as "sewage", "rotten", or "fishy". These flavors are somewhat characteristic of the sulfur compounds mentioned above, but their presence in fish has not been confirmed. Some of these compounds are also produced during microbial spoilage of fish and crustaceans (Whitfield and Freeman 1983) which complicates the interpretation of their etiology.

Microbiology of Geosmin and 2-Methylisoborneol

Geosmin and 2-methylisoborneol are synthesized via the isoprenoid biosynthetic pathway (Bentley and Meganathan 1981). Geosmin is probably derived from a C-15 sesquiterpenoid by loss of an isopropyl side group. 2-Methylisoborneol appears to be derived from methyl addition to a monoterpenoid (Fig. 3). Intermediate products and mechanisms controlling biosynthesis are unknown. The biosynthetic pathways for geosmin and 2-methylisoborneol are presumably common to all organisms found to produce these compounds.

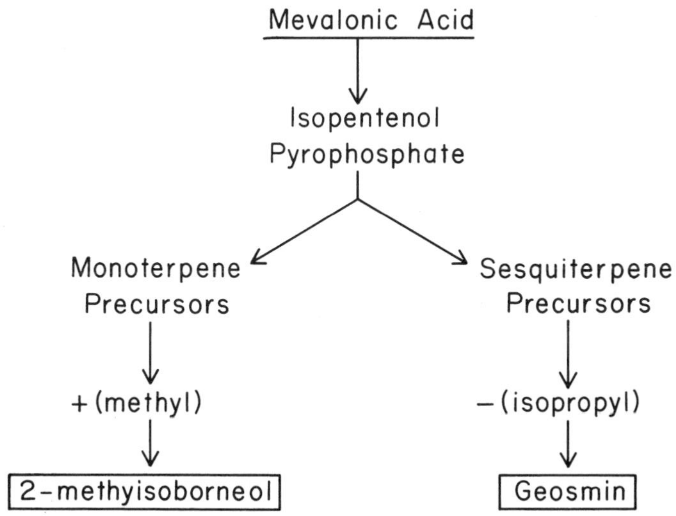

Figure 3. Biosynthesis of 2-methylisoborneol and geosmin.

Since the isolation of geosmin from cultures of *Streptomyces griseus* (Gerber and Lechevalier 1965), production of geosmin has been confirmed in numerous actinomycetes, most notably species of *Streptomyces* and *Nocardia*. 2-Methylisoborneol was initially isolated from cultures of *S. antibioticus*, *S. praecox*, and *S. griseus* (Medsker et al. 1969), and has also been found in cultures of several other actinomycetes. The actinomycetes that produce these compounds are aerobic heterotrophs, and are common inhabitants of soils and surface waters.

Actinomycetes are more abundant in eutrophic lakes or ponds than in nutrient-poor waters. Abundance of actinomycetes increases dramatically following development of phytoplankton blooms. This suggests that actinomycetes are important in the decomposition of organic matter produced by phytoplankton (Silvey and Wyatt 1970). Actinomycetes also flourish in littoral areas where decomposing plants serve as a nutrient source. Actinomycetes are often more abundant in rivers than in lakes or ponds, particularly after periods of high runoff (Cross 1981). This is thought to be the result of large numbers of propagules washed into the river from soils and the influx of organic matter and nutrients that serve as substrate for bacterial growth.

Production of geosmin, and also presumably 2-methylisoborneol, by actinomycetes in culture varies with carbon source, temperature, and pH. Geosmin production by a strain of *Streptomyces* was greatest at 25 C and a culture pH of 10 (Weete et al. 1977). Geosmin was not produced at temperatures below 15 C and geosmin production was markedly decreased at pH-values less than 7. Growth of *Streptomyces* and subsequent production of geosmin is greatest in environments with high dissolved oxygen concentrations (Blevins 1980; Wood et al. 1983).

Production of geosmin by cyanobacteria was confirmed shortly after the initial identification of the compound in actinomycete cultures (Safferman et al. 1967; Medsker et al. 1968). Production of 2-methylisoborneol by cyanobacteria was first reported by Tabachek and Yurkowski (1976) who recovered the compound from cultures of *Lyngbya cryptovaginata* isolated from a lake in Manitoba, Canada, used to culture rainbow trout. Subsequent work has indicated geosmin production by at least six genera (in two families of the Hormogonales: Oscillitoriaceae and Nostocaceae) and 2-methylisoborneol production by at least three genera of Oscillatoriaceae (Table 3). As with the actinomycetes, the production of odorous metabolites by cyanobacteria appears to be dependent on environmental conditions, but little information is available describing these interactions.

Cyanobacteria occur in a wide variety of environments (Carr and Whitten 1982; Reynolds 1984), but the abundance of cyanobacteria, and the dominance of the microbial community by this group, is markedly increased in nutrient-rich environments, particularly those environments with high ratios of phosphorus to nitrogen. Many of the known geosmin-producing cyanobacteria are notorious for their occurrence in aquaculture ponds and other eutrophic bodies of water. Species such as *Anabaena circinalis* and *Oscillatoria agardhii* are particularly noxious due to the formation of gas vesicles within the cells of the trichome, a bouyancy-regulating mechanism which often results in massive surface blooms. The occurrence of these blooms is often seasonal, with maximum standing crops attained in late summer and early autumn. Despite the general selection for cyanobacteria in eutrophic bodies of water, certain species, particularly benthic or epiphytic forms, also are commonly found in oligotrophic waters.

The relative importance of actinomycetes and cyanobacteria in the

Table 3. Cyanobacteria reported to be producers of geosmin and 2-methylisoborneol.

Compound	Genus	Species	Reference
Geosmin	*Anabaena*	*circinalis*	Henley 1970
		scheremetievi	Izaguirre et al. 1982
		macrospora	Yagi et al. 1983
	Aphanizomenon	*gracile*	Juttner et al. 1986
	Lyngbya	*aestuarii*	Tabachek and Yurkowski 1976
		species	Tabachek and Yurkowski 1976
	Oscillatoria	*agardhii*	Persson 1979b
		brevis	Berglind et al. 1983
		bornetii	Berglind et al. 1983
		cortiana	Tabachek and Yurkowski 1976
		prolifica	Tabachek and Yurkowski 1976
		simplicissima	Izaguirre et al. 1982
		splendida	Tabachek and Yurkowski 1976
		tenuis	Tabachek and Yurkowski 1976
		variabilis	Tabachek and Yurkowski 1976
	Schizothrix	*mulleri*	Kikuchi et al. 1973
	Symploca	*muscorum*	Medsker et al. 1968
2-methyl-isoborneol	*Lyngbya*	*cryptovaginata*	Tabachek and Yurkowski 1976
	Oscillatoria	*curviceps*	Izaguirre et al 1982
	Oscillatoria	*tenuis* var. *levis*	Izaguirre et al. 1982
	Phormidium	*tenue*	Yagi et al. 1983

production of odorous metabolites in natural waters is in some dispute. It is quite clear that representatives of either group are capable of producing these compounds and, in certain specific instances, the causative organisms are known. For instance, Silvey and Roach (1975) present good evidence for actinomycete involvement in taste and odor problems in reservoirs in the southwestern United States. However, some research has suggested that actinomycetes are relatively inactive in many aquatic environments (Johnston and Cross 1976; Cross 1981) and that the contribution of actinomycetes to odor problems is primarily the result of drainage of the bacteria or metabolites from shoreline areas (Persson 1982; Wood et al. 1983).

The relative role of actinomycetes and cyanobacteria in the production of odorous metabolites is more than an academic question. Possible approaches to off-flavor abatement in cultured fish will depend upon identification of the causative organism. Obviously, treatment of a fish culture pond with an algicide to kill cyanobacteria may be ineffective if actinomycetes are responsible for the off-flavor episode. In fact, such a treatment may aggravate the problem by supplying actinomycetes with nutrients (dead cyanobacteria and algae) for further growth.

Uncertainty about which organisms cause off-flavor results primarily from difficulties encountered in conducting research in the aquatic environment. This has led to many vague and misleading statements in the literature. A causal relationship is difficult and time-consuming to prove and most statements regarding this relationship are based on presumptive evidence, such as correlations between incidence of an odor and the most abundant organism. Mallevialle and Suffet (1987) discuss problems associated with analyzing cause-and-effect relationships in taste and odor problems in water and offer a minimal set of "rules of evidence" that should be met to establish causal relationships. Loosely interpreted these rules are:

1. Identify and quantify the organisms present during the taste and odor problem, and identify and quantify the compounds responsible for the problem;
2. Isolate and culture the organisms and identify the metabolites produced in culture;
3. Correlate the compounds and organisms found in the environment with compounds identified in culture.

Research based on these criteria is rare (but see Tabachek and Yurkowski 1976; Izaguirre et al. 1982), and the organisms responsible for most episodes of odors and tastes in water and off-flavor in fish are not known with certainty. It appears that the majority of odor and taste episodes in eutrophic lakes and ponds (such as aquaculture ponds) are caused by proliferation of cyanobacteria rather than actinomycetes (Persson 1981, 1982, 1983). However, much more research is required for validation.

Environmental Fate of Geosmin and 2-Methylisoborneol

The environmental fates of geosmin and 2-methylisoborneol are largely unknown and much contradictory information exists in the literature. Volatilization and biodegradation appear to be the major route of loss of geosmin and 2-methylisoborneol from natural waters, but the relative importance of these two mechanisms is not known.

At 25 C, 50 to 100% of the geosmin in sterile solutions was lost within 24 h, and the rate of loss increased as the surface area to volume ratio of the solution increased (Lelana 1987). Aeration also increased rates of geosmin loss. On the other hand, data presented by Lalezary et al. (1984) indicate that geosmin and 2-methylisoborneol are not readily volatilized from aqueous solutions. Although the importance of volatilization of odorous compounds to the atmosphere from natural bodies of water is unknown, it is generally accepted that aeration or air-stripping is not a practical method of removal of geosmin or 2-methylisoborneol from drinking water supplies (Mallevialle and Suffet 1987).

Lelana (1987) also found rapid rates of biodegradation of geosmin. Within 72 h, 98% of the geosmin in a non-sterile solution of pond water was apparently metabolized by the naturally occurring microorganisms. Rapid rates of geosmin biodegradation have been attributed to the activity of certain *Bacillus* species (Narayan and Nunez 1974). However, Danglot et al. (1983) could not demonstrate rapid geosmin degradation and were unable to document the presence of a postulated exoenzyme "geosminase" from cultures of *Bacillus subtilis*. Geosmin degradation did occur in the presence of lysates from cultures of *Streptomyces* indicating that such an enzyme may, however, exist. Their work also demonstrated that geosmin is rapidly adsorbed onto glass vessels, indicating a possible source of serious experimental errors in work of this sort.

2-Methylisoborneol appears to be much more resistant to biodegradation than geosmin. Izaguirre et al. (1988) isolated predominantly *Pseudomonas* species from enrichment cultures that degraded 2-methylisoborneol. Degradation of 2-methylisoborneol occurred only in cultures consisting of three or more different species of bacteria. Complete degradation required 5 days to more than two weeks, with a long lag time before significant 2-methylisoborneol degradation was detected. Isoborneol, on the other hand, was degraded rapidly by mixed cultures of bacteria, indicating that the addition of the methyl group in 2-methylisoborneol accounts for the recalcitrance of the compound, probably by steric hindrance of the reactive hydroxyl group.

UPTAKE AND ELIMINATION OF ODOROUS COMPOUNDS

Sites of Uptake

Fish accumulate chemicals from water or food. The absorption of organic compounds from water can occur at the gills, through the skin, or across the intestinal epithelium from water swallowed while drinking or incidentally while feeding. From and Horlyck (1984) placed the geosmin-producing cyanobacterium *Symploca muscorum* at isolated locations on live rainbow trout and recorded the time required for development of earthy off-flavor in fillets obtained after various exposure periods. The most rapid absorption occurred through the gills, with only 6 min of exposure required to obtain an earthy flavor. Absorption was slower through the skin (1.5 h), small intestine (4 h), and stomach (7 h). Lovell and Sackey (1973) and Maligalig et al. (1975) also suggest that gills are the major route of uptake of odorous compounds from water. However, Persson and York (1978) noted that total uptake of 2-methylisoborneol from water was higher in fish that were fed compared to those not fed. They speculated that the alimentary tract may be a significant route of uptake of odorous compounds and that these compounds are absorbed from water incidentally swallowed during feeding. From and Horlyck (1984) suggested that stomach acids may partially react with geosmin to yield odorless argosmin isomers. This reaction may diminish the importance of the alimentary tract as a route of uptake for this particular compound.

Fish can also develop off-flavor by ingesting the organisms that synthesize the odorous compounds. Lovell and Sackey (1973) placed channel catfish in aquaria with masses of geosmin-producing strains of *Oscillatoria tenuis* or *Symploca muscorum*, and in aquaria without the cyanobacteria, but supplied with glass wool-filtered water from cultures of the organisms. Fish in both sets of aquaria developed earthy off-flavors that were qualitatively similar, but fish in aquaria with the cyanobacterial masses developed a more intense flavor, and developed the flavor more rapidly than fish exposed to cell-free filtrates of cyanobacterial cultures. However, the concentration of geosmin in the filtered water may have been lower than in the unfiltered water because geosmin is adsorbed by glass wool filters (From and Horlyck 1984).

Ingestion of organisms is probably of minor importance in the development of off-flavor in intensively cultured fish because virtually all nutriment is derived from feeds. The major route of uptake is absorption of the compounds from the water across the gills. Wild fish and fish in extensive fish culture systems may, however, absorb significant quantities of these compounds by ingesting the organisms responsible for their synthesis. The cells are lysed in the gastrointestinal tract and release intracellular stores of the odorous compounds which are then absorbed.

Rates of Uptake

As a rule, uptake of odorous compounds by fish is rapid and elimination is relatively slow. Persson (1984) summarized the times required to produce and eliminate off-flavors in fish by several organic compounds that may be present in polluted waters. Generally, only minutes to hours are required for accumulation of quantities sufficient to impair flavor.

The uptake of 2-methylisoborneol at two concentrations in water by small (5-10 g) channel catfish is depicted in Figure 4 (Martin et al. 1988a). At steady-state, the concentration of 2-methylisoborneol in fish muscle was about 10 times greater than in water. Assuming a sensory threshold concentration of about 0.1 µg/kg, these fish would have developed a detectable off-flavor in less than 2 h of exposure to either concentration of 2-methylisoborneol. Highest concentrations of 2-methylisoborneol were found in lipid-rich tissues such as skin and visceral fat; concentrations of 2-methylisoborneol in visceral fat were almost 100-fold greater than in water.

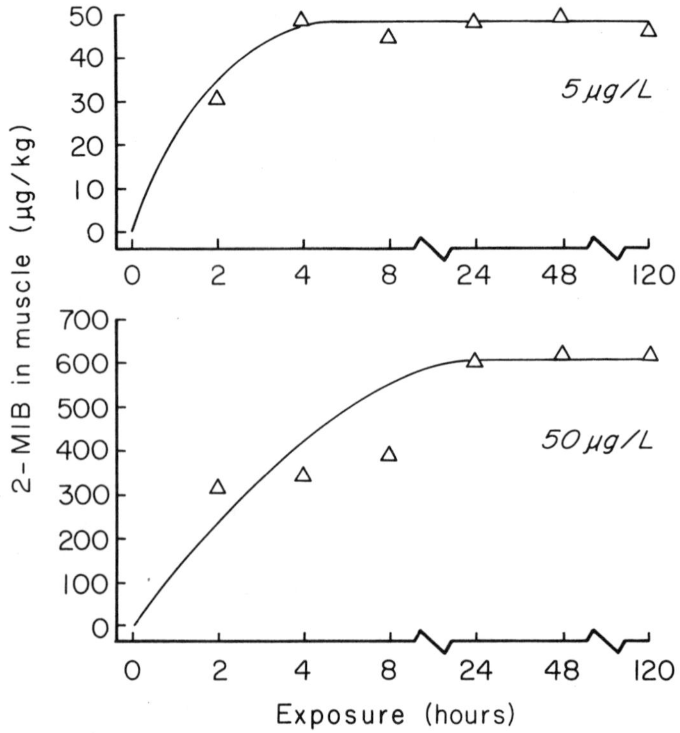

Figure 4. Accumulation of 2-methylisoborneol in channel catfish muscle (Martin et al. 1988a). Fish were exposed to either 5 or 50 µg/L 2-methylisoborneol in water.

Lelana (1987) developed an empirical equation describing the rate of geosmin uptake by channel catfish. At 25C, the geosmin concentration in the muscle of catfish is described by the equation:

$$G_1 = 0.55 + 2.38\ G_2 + 0.23\ T + 2.68\ G_2T$$

where: G_1 = geosmin in the tissue (µg/kg)
G_2 = geosmin in the water (µg/L)
T = exposure time (h)

This equation apparently describes an initial uptake rate because steady-state conditions are never reached. As an artifact of the regression used to derive the equation, a background geosmin concentration in fish of 0.55 µg/kg is predicted and should be disregarded. Despite these shortcomings, the equation is useful for describing short-term relationships. If the concentration of geosmin in water were 1 µg/L, only about 2 h would be required for the fish to absorb enough geosmin to reach the sensory threshold concentration (estimated to be 8 µg/kg). If the concentration of geosmin in the water were 3 µg/L, the time to reach the sensory threshold is reduced to only about 2 min.

Uptake of hydrocarbons from petroleum products is rapid because these compounds are highly lipophilic. Concentrations of toluene, xylene, and polycyclic aromatic hydrocarbons from crude oil reach steady-state conditions in fish tissues after less than 24 h of exposure (Anderson et al. 1974). Because these compounds have high bioconcentration factors and uptake rates, fish exposed to petroleum products probably develop detectable off-flavors very rapidly.

Shumway and Chadwick (1971) found significant impairment of the flavor of coho salmon after 72 h of exposure to untreated pulp-mill effluent (1% by volume), but the minimum time to develop detectable off-flavor was not determined. Rainbow trout (100-250 g) developed detectable off-flavors after only 15 min exposure to *o*-cresol (900 µg/L), butanethiol (100 µg/L), 2,4-dichlorophenol (100 µg/L), or pyridine (100 µg/L); maximum flavor impairment occurred after about 1.5 h exposure to each of the four compounds at the concentrations listed (Shumway and Palensky 1973).

Elimination of Odorous Compounds

Nonpolar, lipophilic compounds are eliminated from fish by passive diffusion across the gills or skin, or by metabolism to more polar compounds which are excreted from the kidney or secreted in gallbladder bile. The relative importance of these processes for most odorous compounds is not known. Elimination through the gills is probably the major route, but this may not be true for strongly lipophilic substances that are not readily partitioned from lipid stores into the blood.

Most information currently available on rates of elimination of odorous compounds from fish is based on sensory analysis. The flavor of rainbow trout exposed to 2,4-dichlorophenol improved "substantially" 6 h after transferal to clean water and disappeared completely after 33 h (Shumway and Palensky 1973). The time required to purge fish of earthy flavors varied from 6 to 18 days for channel catfish (Lovell and Sackey 1973; Maligalig et al. 1973) and 5 to 16 days for rainbow trout (Iredale and York 1976). The chemical causes of the earthy off-flavors were not identified in these studies. Lovell (1983b) also used sensory analysis to evaluate the effect of temperature on rate of loss of an "earthy-musty" off-flavor from channel catfish (Table 4). Disappearance of the off-flavor was most rapid at the highest water temperature.

Table 4. Average flavor scores for "earthy-musty" off-flavor channel catfish held in clean, flowing water at three temperatures. A score of 10 = no off-flavor, 8 = slight, 6 = distinct, 4 = intense, 2 = extreme (Lovell 1983b).

Water Temperature (°C)	Flavor Score				
	0 day	3 days	6 days	10 days	15 days
16	4.7	6.1	7.1	7.4	9.1
22	4.9	7.0	8.7	9.3	9.6
26	5.0	7.7	8.7	9.6	9.6

Elimination of geosmin from large (about 400 g) channel catfish held in clean, flowing water is illustrated in Figure 5 (Lelana 1987). About 6 days was required for geosmin concentrations to decline below the estimated sensory threshold concentration (about 8 µg/kg). This was substantiated by sensory analysis which also indicated that after 6 days, flavor scores were in the range considered "acceptable". Geosmin appears to be eliminated somewhat more slowly from rainbow trout than from channel catfish, probably because of the effect of cooler water temperatures on the rate of elimination. Yurkowski and Tabachek (1974) found that 5 to 7 days was required to remove a muddy flavor caused by geosmin in rainbow trout. Although this is similar to the time found by Lelana (1987) to purge channel catfish of geosmin-related flavor, the initial geosmin concentration in rainbow trout (11 µg/kg) was much lower than in channel catfish (90 µg/kg).

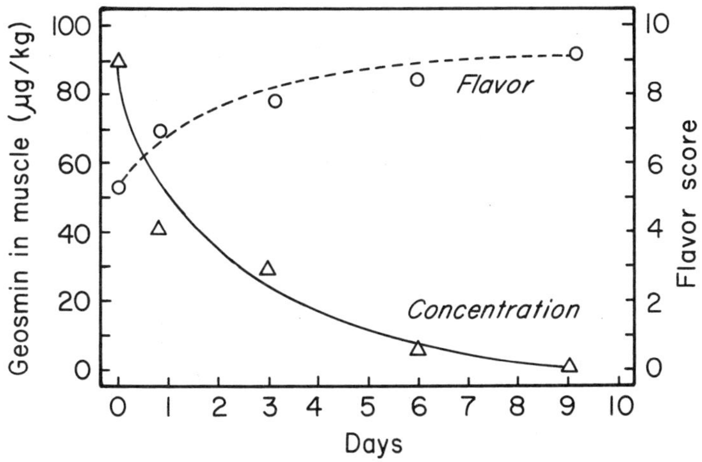

Figure 5. Elimination of geosmin and changes in flavor scores for channel catfish held in clean, flowing water (Lelana 1987). A score of 10 = no off-flavor, 0 = intense off-flavor.

Elimination of 2-methylisoborneol from fish appears to be more rapid than for geosmin. Martin et al. (1988a) exposed small (5-10 g) channel catfish to 2-methylisoborneol at concentrations of 5 and 50 μg/L. Fish were exposed for 7 days and steady-state conditions were apparently achieved. Concentrations of 2-methylisoborneol in both groups of fish declined to less than 2 μg/kg over a 48 h period after the fish were moved to water free of the compound.

OFF-FLAVORS IN POND-RAISED CHANNEL CATFISH: A CASE STUDY

Pond culture of channel catfish is the largest aquaculture industry in the United States. In 1988, over 150 million kg of fish were produced. About 80% of the total was produced in a limited geographical area of west-central Mississippi. Producers of channel catfish consistently identify environment-related off-flavors as their major production problem. The following account summarizes the importance of the problem to the industry and the current understanding of its causes and solutions.

Fish Production Practices

Most channel catfish are reared in ponds contained within earthen levees and filled with ground water. The ponds average between 1 to 2 m in depth and are 5 to 15 ha in size. Several different production schemes are used by producers (Busch 1985). In the most common production scheme, fingerlings weighing 20 to 50 g each are stocked at densities ranging from 10,000 to 20,000 fish/ha. Fish are fed a high quality diet once or twice daily at rates that often exceed 75 kg/ha per day. Market-sized fish (about 0.5 to 0.7 kg) are selectively harvested from ponds with a large mesh seine. Small fish escape through the mesh and remain in the pond for further growth. Fingerlings are then added to replace the fish removed by seining and another cycle begins. In this production system, ponds are operated for years without draining and there is little or no water exchange during the growing season; ground water is only added periodically to replace evaporative losses. Fish production varies greatly depending upon management expertise, but averages about 3500 kg/ha per year.

Prior to harvest, a sample of one or two fish is submitted to quality control personnel at the processing plant for evaluation of flavor acceptability. If the flavor is deemed acceptable, and the processing schedule for the plant permits, fish are tentatively approved for purchase. One or two additional samples for flavor evaluation may be required, usually on the day before or day of harvest. A final sample for flavor-testing is usually taken from the transport truck immediately before the live fish are unloaded at the processing plant. If the flavor of the fish that are sampled at any time is unacceptable, the fish are rejected for processing; fish are either not harvested or returned to the pond from the transport truck.

Phytoplankton Communities

The continuous addition of large amounts of nutrients in the form of fish metabolic wastes makes channel catfish ponds ideal habitats for aquatic plants. Although filamentous algae and vascular plants at times become established and grow luxuriantly, the most common plant form is phytoplankton.

The concentration of chlorophyll *a* (an indicator of phytoplankton standing crop) in the water of channel catfish ponds is generally greater than 100 µg/L throughout the year. During summer months, chlorophyll *a* concentrations increase in response to increased solar radiation and the higher amounts of fish feed (and hence fish metabolic waste) added to the pond. Chlorophyll *a* concentrations average 200-400 µg/L at this time (Fig. 6). Assuming an average depth of 1 m for these ponds, this translates into a chlorophyll *a* content per unit surface area of 200-400 µg/m^2. Similar values (180 to 300 µg/m^2) have been estimated to be the maximum that occur in nature (Wetzel 1975).

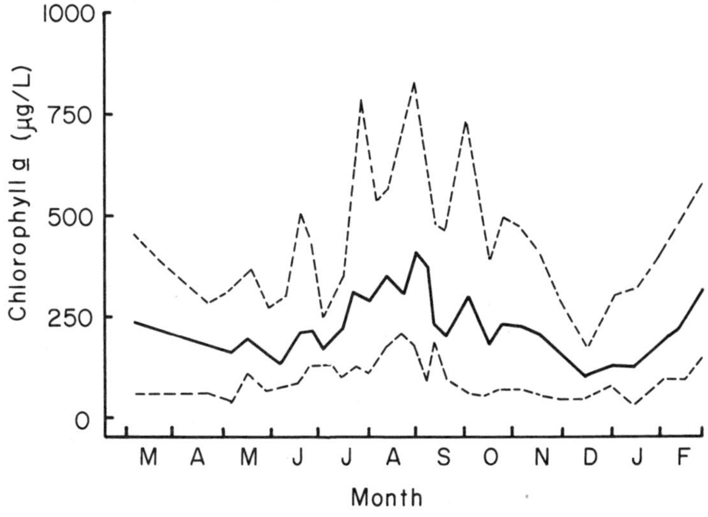

Figure 6. Chlorophyll *a* concentrations in six experimental channel catfish culture ponds in Mississippi (Tucker and Lloyd 1984). Lines represent minimum, average, and maximum values.

The composition and seasonal occurrence of phytoplankton taxa found in channel catfish ponds are typical of temperate, eutrophic bodies of water (Tucker and Lloyd 1984; Tucker 1985). Phytoplankton density and taxonomic composition differs among ponds at any given time, but, in general, phytoplankton associations are dominated by cyanobacteria during the summer months and diatoms (Bacillariophyceae) during the winter months (Fig. 7). Communities are markedly more diverse in the spring and fall. The

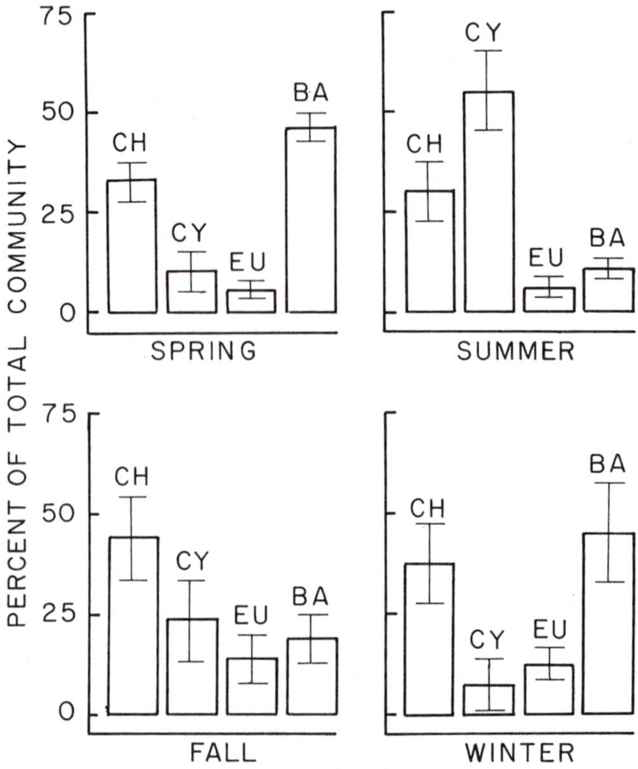

Figure 7. Composition of phytoplankton communities in experimental channel catfish ponds in Mississippi (Tucker and Lloyd 1984). CH = Chlorophyta (green algae); CY = Cyanobacteria (blue-green algae); EU = Euglenophyta (euglenoid algae); BA = Bacillariophyceae (diatoms).

most common cyanobacteria found during the warmer months are species of *Microcystis*, *Oscillatoria*, *Rhaphidiopsis*, and *Anabaena*. Benthic algal communities and planktonic and benthic bacterial communities have not been studied in channel catfish ponds.

Types of Off-Flavor

Dozens of terms have been used by fish producers and quality-control personnel to describe the types of off-flavors encountered in cultured channel catfish. Some attempt to compile and standardize these terms has been made by Lovell (1983a) and Johnsen et al (1987). Lovell (1983a) collected channel catfish from culture ponds in Alabama and Mississippi over a 60-day period in spring and summer. Fish from 24 ponds were judged to be off-flavor and unmarketable. Fish from these ponds were subjectively evaluated for flavor description and intensity by a trained sensory panel. Most fish had more than one off-flavor; however, eight dominant flavors were described (Table 5). The chemical causes of these off-flavors were not identified.

Table 5. Off-flavor descriptors for pond-cultured channel catfish (Lovell 1983a).

Descriptor	Comments
Sewage	This flavor was variously described as "fecal", "lagoon", or "sewage", and was often found in combination with other flavors.
Stale	This flavor is best characterized as severely lacking in freshness, and was also commonly found in combination with the other flavors.
Earthy-musty	This flavor was generally the most intense of the flavors described.
Rancid	This flavor is similar to that of fish which have undergone oxidation after prolonged frozen storage.
Metallic	This flavor was also described by the terms "paint", "chemical", or "astringent", and was often found in combination with other flavors.
Moldy	This flavor is somewhat similar to the earthy-musty and stale flavors, and was also described as "cobweb", "moldy-hay", and "wet-paper".
Petroleum	This flavor was quite objectionable. The flavor was also described by panel members as "diesel", "gasoline", "kerosene", and "motor oil".
Weedy	This flavor was also described as "grassy", "plant-like", and "algae-like", and was sometimes found in combination with the metallic flavor.

A lexicon (Table 6) of flavor descriptors for pond-raised channel catfish was compiled by Johnsen et al. (1987) with the explicit aim to standardize terminology. This lexicon included positive flavor notes as well as off-flavors and is also notable for separating the earthy-musty flavors into

Table 6. Flavor descriptors for channel catfish (Johnsen et al. 1987).

Descriptor	Definition
AROMATICS	
Nutty/Pecan	The flavor associated with fresh pecans and other hard-shell nuts.
Boiled Chicken/Buttery	The flavor associated with sweet boiled chicken meat; the combination of sweet protein and some oil/fat character.
Grainy/Corn/Green Vegetable	The flavor associated with grains or corn husks. This flavor is a favorable characteristic at low levels, but may be objectionable at higher levels or when the character is more vegetative and green than just grain-like.
Lagoon/Bluegreen	2-Methylisoborneol is the reference.
Musty/Woody	Geosmin is the reference.
Putrid	The flavor associated with decaying protein.
Rotten Plants	The flavor associated with slightly oxidized fats and oils and reminiscent of wet cardboard.
Painty	The flavor associated with linseed oil and oil based paints.
TASTES	
Sweet	The taste associated with sugars.
Salty	The taste associated with sodium ions.
FEELING FACTORS	
Astringent	The feeling on the tongue described as puckering/dry and associated with tannin and alum.
Metallic	The feeling on the tongue described as flat; associated with iron and copper.

distinct flavor notes with known chemical reference standards, geosmin and 2-methylisoborneol. Chemical etiologies of the other off-flavors are not known. The panel that developed this lexicon consisted of eighteen representatives of catfish producers, processors, and researchers from five states in the southeastern United States. It was believed that all common off-flavors, with the exception of "petroleum", were represented.

Incidence and Cause of Off-Flavors

Sensory scores for channel catfish sampled from four experimental ponds at Stoneville, Mississippi (Tucker and Martin, unpublished) illustrate the dynamic nature of the incidence and severity of off-flavor (Fig. 8). Note

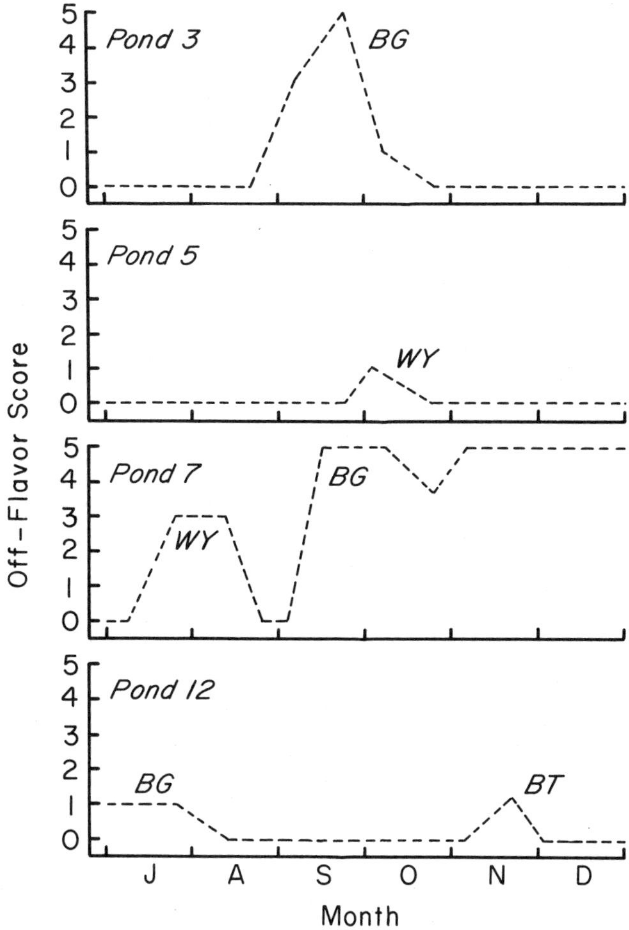

Figure 8. Flavor scores of channel catfish sampled from four experimental culture ponds. A score of 0 = no off-flavor, 5 = intense off-flavor. BG = "blue-green" flavor, WY = "woody" flavor, BT = "bitter" taste.

that the flavor-intensity scoring system used in this figure differs from that used in Table 4 and Figure 5. On the initial sampling date in July, all four ponds were judged to contain acceptably flavored fish; in early October, three of the four samples were judged off-flavor to varying degrees. Off-flavor episodes varied from two weeks to over three months. These ponds were part of a larger study involving 14 ponds that were sampled from June through February. The highest incidence of off-flavor occurred in September and October with samples from 10 of the 14 ponds judged off-flavor. The most intense off-flavors also occurred at this time; the flavor described as "blue-green" (confirmed as being caused by 2-methylisoborneol) was most common and the most intense. The lowest incidence of off-flavor occurred in December, with samples from five ponds judged off-flavor.

Studies of catfish culture ponds in Alabama (Brown and Boyd 1982; Armstrong et al. 1986; Lovell et al. 1986) showed that fish from 50 to 75% of the ponds sampled in late summer or early fall were off-flavor and unacceptable for harvesting. The predominant off-flavor was "earthy-musty". Geosmin was confirmed in 80% of the fish with earthy-musty off-flavors; 2-methylisoborneol was not detected in any fish sampled (Lovell et al. 1986). The total incidence of ponds with off-flavor fish, and the incidence of earthy-musty flavors, increased during the summer growing season as phytoplankton biomass increased in response to higher water temperatures and amounts of feed added to ponds.

The incidence of various types of off-flavor in channel catfish raised commercially in Mississippi has not been studied. Anecdotal reports indicate that the overall incidence of off-flavor is similar to that found in Alabama, and that earthy-musty off-flavors are also the most common. However, 2-methylisoborneol and its dehydration products, rather than geosmin, are usually the cause of these off-flavors in Mississippi (Martin et al. 1988b,c). Off-flavor related to 2-methylisoborneol dehydration products may be particularly prevalent during the winter months. These highly lipophilic compounds are slowly eliminated from fish, particularly when metabolism is slowed at colder water temperatures (Martin et al. 1988b). The reasons for the different chemical etiologies of earthy-musty off-flavors in Alabama and Mississippi are not known, although differences in the density and composition of channel catfish pond phytoplankton communities in the two states have been reported (Tucker and Boyd 1985).

Lovell (1983b) reported no consistent relationship between earthy-musty off-flavors in fish from Alabama and the presence of odor-producing actinomycetes; however, distinct earthy-musty off-flavors did seem to be associated with the presence of *Anabaena circinalis*, a known producer of geosmin. Brown and Boyd (1982) found that ponds in Alabama with low fish densities and low maximum daily feed allotments generally yielded the best tasting fish. No strong relationship was found between either phytoplankton density or type of phytoplankton and the degree of off-flavor, but in several ponds with intense off-flavor, the cyanobacterium *Lyngbya* sp. was abundant. Similar results were obtained in another study conducted in Alabama by Armstrong et al. (1986). Regression analyses between the incidence of off-flavor and several environmental variables, including abundance of cyanobacteria and actinomycetes, failed to yield consistent or strong correlations. Cyanobacteria, especially species of *Oscillatoria* and *Microcystis* were abundant in many ponds with off-flavor fish, but were also present in ponds containing fish without off-flavors. Species of green algae (Chlorophyta) were dominant in some ponds with intensely off-flavored fish. Species of *Oscillatoria* are present in many Mississippi ponds containing fish with off-flavors caused by 2-methylisoborneol. Martin et al. (1988c) noticed that this cyanobacterium is also present in ponds without off-flavored fish. The onset of off-flavor appeared to be related to the senescence, and eventual death, of the populations of *Oscillatoria* sp.

Cost of Off-Flavor to the Producer

The cost effects of off-flavor are difficult to evaluate because the incidence is episodic and dynamic. Keenum and Waldrop (1988) attempted to estimate the effects of off-flavor on production costs of Mississippi pond-raised channel catfish. They used three simple scenarios with the only variable being the time of year when off-flavor occurred. In each scenario it was assumed that 25% of the production would be held in inventory for 3 months due to off-flavor.

When off-flavor occurred during the winter (assumed to be the non-growing season), the only additional cost charged was an opportunity cost for delayed income. For this scenario, an additional cost of 3.85 cents/kg of catfish harvested was calculated. When off-flavor occurred during the summer growing season, interest was charged on the quantity of fish that could not be sold. Additionally, net income from foregone sales must be forfeited by the producer because restocking and growth of the restocked fingerlings cannot be accomplished. Because off-flavor fish held during the growing season continue to grow, additional revenue offsets part of the interest and foregone net income. For this scenario, an additional cost of 4.10 cents/kg of catfish harvested was added to production costs. The cost of off-flavor occurring during part of both the non-growing and growing season was estimated as the sum of the costs for the first two scenarios, or 7.95 cents/kg of catfish harvested. Using these three estimates and an annual production of 150 million kg of channel catfish in 1988, the added production costs of off-flavor for that year was between 5.8 and 12 million dollars.

The above estimates are likely conservative, but other potential costs accruing from off-flavor are difficult to estimate. Fish may be lost to infectious diseases or poor water quality while the fish are in inventory waiting for off-flavor to abate. This represents loss of investment in fish that should have been marketed. As fish grow, they convert feed less efficiently and compete with smaller fish for feed allotments. This lowers the overall feed conversion rate and growth rate of the pond population. Additional labor may also be required for sampling and management programs. Market constraints also impact the effect of off-flavor on actual production costs. Market constraints (not being able to sell fish when they reach market size, also called quotas) increase production costs in a manner similar to off-flavor. This cost may be added to or subtracted from the cost of off-flavor because fish may not have been marketable regardless of flavor.

MANAGEMENT AND TREATMENT OF OFF-FLAVORS

Three approaches have been used to deal with off-flavor in pond-raised fish: manage around off-flavor episodes, prevent off-flavor episodes, or remove the flavor from fish once they have developed it. Removing the off-flavor from fish can be accomplished by treating the pond with a biocide to kill the organism producing the odorous substance or by moving the fish to a "clean" environment. Some work has also been conducted on methods to process off-flavored fish in a manner that will mask the flavor. Little systematic research has been conducted on management of off-flavors in cultured fish. Much of the following discussion is anecdotal.

Managing Around Off-Flavor Episodes

Off-flavor episodes in ponds are dynamic and when several ponds are considered, not all are likely to contain off-flavor fish at the same time. Ponds that contain a sufficient quantity of market-sized fish should be checked frequently for off-flavor and sold promptly if the fish are suitable. Market constraints may impact the practicability of this procedure, but every effort should be made to harvest and market fish free of off-flavor as soon as possible.

Preventing Off-Flavor Episodes

A significant correlation exists between amounts of feed added to ponds and flavor scores of channel catfish (Brown and Boyd 1982). This implies that the incidence and severity of off-flavor can be decreased by lowering fish stocking rates and adding less feed to ponds. However, for pond-raised channel catfish, it appears that stocking and feeding rates would have to be lower than those currently considered profitable for this procedure to have a significant effect. Costs of channel catfish production are very sensitive to harvested yield (Keenum and Waldrop 1988), and although off-flavor decreases long-term yield, savings in reduction of the incidence of off-flavor probably would not offset decreased revenue from lower yields. Nevertheless, this approach may be useful in other fish culture situations.

Earthy-musty flavors related to geosmin or 2-methylisoborneol are the most common off-flavors in pond-raised fish. Because cyanobacteria are frequently implicated as the major source of these compounds in eutrophic waters, most research in preventing off-flavor has concerned the use of algicides to reduce cyanobacterial abundance. Only one algicide, Solricin 135™, is marketed in the United States with specific claims that its use will prevent or eliminate off-flavor problems associated with blooms of cyanobacteria. The active ingredient in Solricin 135™ is potassium ricinoleate, derived from saponfication of castor oil. Potassium ricinoleate is similar in structure to naturally occurring allelopathic compounds isolated from certain aquatic angiosperms. In laboratory tests, potassium ricinoleate inhibits the growth of cyanobacteria to a greater extent than that of algae (Van Aller et al. 1985). However, when tested in ponds, Solricin 135™ did not reduce the abundance or percentage of cyanobacteria in phytoplankton communities and did not prevent severe episodes of off-flavors in channel catfish from treated ponds (Tucker and Lloyd 1987). The failure of Solricin 135™ to inhibit cyanobacteria in ponds was attributed to either rapid metabolism of the compound by bacteria or precipitation as a sparingly soluble calcium salt.

Copper sulfate, a relatively non-selective algicide, has been used in attempts to prevent episodes of off-flavor by applying the chemical frequently (weekly or so) at low concentrations (less than 0.2 mg/L as copper). This practice appears to be most common among channel catfish producers in Alabama. Although producers report some success using this technique, there is no sound evidence that this procedure is successful in reducing the incidence or severity of off-flavor. Frequent treatment of aquaculture ponds with non-selective algicides also cannot be recommended because chronic deterioration in water quality causes decreased fish yields (Tucker and Boyd 1978, 1979; Tucker et al. 1983; Smith and Piedrahita 1988).

Torrens and Lowell (1987) reported a decrease in the incidence of off-flavor in channel catfish polycultured with the planktonivorous fish, blue tilapia *Oreochromis aurea*. Apparently, the feeding habits of tilapia altered the environment in a manner that did not favor the growth of odor-producing microorganisms. However, Tucker and Martin (unpublished) failed to demonstrate a similar benefit of tilapia polyculture on off-flavor in channel catfish. Furthermore, with the onset of cold water temperatures, the tilapia died and catfish developed a severe "rotten fish" or "putrid" flavor from feeding on the decaying tilapia. Nevertheless, the use of other planktonivorous fish, such as silver carp *Hypothalmichthys molotrix*, may hold some promise as a biological control agent and should be further investigated.

Eliminating Off-Flavors from Fish

Some compounds responsible for off-flavor, such as geosmin and 2-methylisoborneol, are eliminated fairly rapidly from fish once uptake of the compound from water ceases. Other compounds are more slowly eliminated, and several weeks may be required for off-flavors to dissipate. Examples of these compounds include petroleum substances and, probably, the microbial products β–cyclocitral and the hydrocarbon dehydration products of 2-methylisoborneol. Obviously, the nature of the compound responsible for off-flavor will influence the practicability of procedures undertaken to dissipate the flavor.

Algicides are sometimes applied to ponds with off-flavor fish in an attempt to kill the organism responsibly for producing the odorous compound. Success has been equivocal, probably because no effort is usually made to identify the chemical and biological cause of the problem. Limited field observations suggest that treatment of ponds with a quick-acting algicide such as copper sulfate has some merit when off-flavors are due to geosmin or 2-methylisoborneol and there is good reason to suspect a specific cyanobacterium as the causative organism. This procedure often fails when compounds other than geosmin and 2-methylisoborneol are involved, possibly because other compounds are either eliminated from fish too slowly or they originate from sources other than cyanobacteria or algae. This treatment procedure should be contrasted to the procedure where frequent, low doses of an algicide are used to "thin" phytoplankton communities and prevent off-flavor. Frequent treatments affect water quality for extended periods and reduce fish growth; a single treatment immediately before harvest has little effect on fish growth. However, treatment of fish culture ponds with algicides may cause oxygen depletion.

Removal of fish from the "off-flavor pond" to a clean environment is a proven procedure for dissipating certain off-flavors. Pond-cultured fish are sometimes simply moved to an adjacent pond, but there is a risk that the new pond will also contain or develop a community of odor-producing microorganisms. To obviate this problem, a special facility, such as tanks or small ponds, may be used to hold fish while the off-flavor dissipates. Water exchange or algicides are used to prevent phytoplankton blooms from developing during the short period the fish are held. These treatments add to the cost of production because of the labor involved in harvesting and moving the fish.

Fish moved to tanks or ponds and held at high densities often are not offered feed and may not accept feed even if offered. Weight loss by fish in these facilities is proportional to the length of time fish are held and water temperature (Table 7). Because fish may lose considerable weight, this procedure will be useful only for those off-flavors that are quickly dissipated.

Table 7. Percent weight loss of channel catfish held, without feeding, in clean, flowing water at three temperatures (Lovell 1983b).

Water Temperature (°C)	Percent Weight Loss				
	0 day	3 days	6 days	10 days	15 days
16	0	5	7	8	9
22	0	10	10	13	15
26	0	13	15	17	18

Processing Off-Flavored Fish

Limited research has been conducted to develop methods of processing fish to mask or eliminate earthy off-flavors. Smoke-processing earthy-flavored catfish improved the flavor, but the earthy flavor was still noted by trained judges. Similar results were obtained by Iredale and Shaykewich (1973) for "muddy-flavored" rainbow trout. Smoke-processed trout were rated as acceptable by a consumer panel, but the muddy flavor was still evident to a trained panel. Iredale and Shaykewich (1973) also found that certain canning methods could significantly reduce the intensity of muddy flavors in trout. Steam precooking fillet strips with subsequent addition of either vegetable oil or smoke-flavored oil resulted in a highly acceptable canned product.

Canned and smoked products constitute a negligible proportion of processed channel catfish, and it is unlikely that these methods will ever be used to any extent to process off-flavored catfish. However, canning or smoking may be useful for fish where these products are traditionally accepted or in situations where the volume of off-flavored fish to be processed is small.

LITERATURE CITED

Anderson, J. W., J. M. Neff, B. A. Cox, H. E. Tatus, and G. M. Hightower. 1974. Effects of oil on estuarine animals: uptake and depuration, respiration. Pages 285-310 in F. J. Vernberg and W. B. Vernberg, editors. Pollution and Physiology of Marine Organisms. Academic Press, New York, New York, USA.

Armstrong, M. S., C. E. Boyd, and R. T. Lovell. 1986. Environmental factors affecting flavor of channel catfish from production ponds. Progressive Fish-Culturist 48:113-119.

Aschner, M., C. Leventer, and I. Chorin-Kirsch. 1969. Off-flavor in carp from fish ponds in the coastal plain and the Galil. Bamidgeh 19:23-25.

Bentley, R. and R. Meganathan. 1981. Geosmin and 2-methylisoborneol biosynthesis in streptomyces: evidence for an isoprenoid pathway and its absence in non-differentiating isolates. Federation of European Biochemical Societies 125:220-222.

Berglind, L., H. Holtan, and O. M. Skulberg. 1983. Case studies on off-flavors in some Norwegian lakes. Water Science and Technology 15:199-210.

Blevins, W. T. 1980. Geosmin and other odorous metabolites of microbial origin. Pages 350-357 in F. E. Guthrie and J. J. Perry, editors. Introduction to Environmental Toxicology. Elsevier, New York, New York, USA.

Brown, S. W. and C. E. Boyd. 1982. Off-flavor in channel catfish from commercial ponds. Transactions of the American Fisheries Society 111:379-383.

Busch, R. L. 1985. Channel catfish culture in ponds. Pages 13-84 in C. S. Tucker, editor. Channel Catfish Culture. Elsevier, Amsterdam, The Netherlands.

Buttery, R. G., D. G. Guadagni, and L. C. Ling. 1976. Geosmin, a musty off-flavor of dry beans. Journal of Agricultural and Food Chemistry 24:419-420.

Carr, N. G. and B. A. Whitten, editors. 1982. The Biology of Cyanobacteria. University of California Press, Berkeley, California, USA.

Cook, W. H., F. A. Farmer, O. E. Kristense, K. Reid, J. Reid, and R. Rowbottom. 1973. The effect of pulp and paper mill effluents on the taste and odor of receiving water and the fish therein. Pulp and Paper Magazine Canada 74:97-106.

Cross, T. 1981. Aquatic actinomycetes: a critical survey of the occurrence, growth and role of actinomycetes in aquatic habitats. Journal of Applied Bacteriology 50:397-423.

Danglot, C., G. Amar, and R. Vilagines. 1983. Ability of *Bacillus* to degrade geosmin. Water Science and Technology 15:291-300.

From, J. and V. Horlyck. 1984. Sites of uptake of geosmin, a cause of earthy flavor, in rainbow trout (*Salmo gairdneri*). Canadian Journal of Fisheries and Aquatic Sciences 41:1224-1226.

Gerber, N. N. 1968. Geosmin from microorganisms is trans -1,10-dimethyl-trans-9-decalol. Tetrahedron Letters 25:2971-2974.

Gerber, N. N. 1969. A volatile metabolite of actinomycetes, 2-methylisoborneol. Journal of Antibiotics 22:508-509.

Gerber, N. N. 1971. Sesquiterpenoids from actinomycetes: candin-4-ene-l-ol. Phytochemistry 12:185-189.

Gerber, N. N. 1979. Volatile substances from actinomycetes: their role in the odor pollution of water. Critical Reviews in Microbiology 7:191-194.

Gerber, N. N. and H. A. Lechevalier. 1965. Geosmin, an earthy-smelling substance isolated from actinomycetes. Applied Microbiology 13:935-938.

Henley, D. E. 1970. Odorous metabolites and other selected studies of Cyanophyta. Ph.D. Thesis. North Texas State University, Denton, Texas, USA.

Iredale, D. G. and K. J. Shaykewich. 1973. Masking or neutralizing muddy flavor in rainbow trout (*Salmo gairdneri*) by smoking and canning processes. Journal of the Fisheries Research Board of Canada 30:1235-1239.

Iredale, G. D. and R. K. York. 1976. Purging a muddy-earthy flavor taint from rainbow trout (*Salmo gairdneri*) by transferring to artificial and natural holding environments. Journal of the Fisheries Research Board of Canada 33:160-166.

Izaguirre, G., C. J. Hwang, S. W. Krasner, and M. J. McGuire. 1982. Geosmin and 2-methylisoborneol from cyanobacteria in three water supply systems. Applied and Environmental Microbiology 43:708-714.

Izaguirre, G., R. L. Wolfe, and E. G. Means. 1988. Degradation of 2-methylisoborneol by aquatic bacteria. Applied and Environmental Microbiology 54:2424-2431.

Jenkins, D., L. L. Medsker, and J. F. Thomas. 1967. Odorous compounds in natural water: some sulfur compounds associated with blue-green algae. Environmental Science and Technology 1:731-735.

Johnsen, P. B., G. V. Civille, and J. R. Vercellotti. 1987. A lexicon of pond-raised catfish flavor descripters. Journal of Sensory Studies 2:85-91.

Johnston, D. W. and T. Cross. 1976. Actinomycetes in lake muds: dormant spores or metabolically active mycelium? Freshwater Biology 6:464-469.

Juttner, F. 1981. Detection of lipid degradation products in the water of a reservoir during a bloom of *Synura uvella*. Applied and Environmental Microbiology 41:1001-06.

Juttner, F. 1983. Volatile odorous excretion products of algae and their occurrence in the natural aquatic environment. Water Science and Technology 15:247-257.

Juttner, F. 1984. Dynamics of the volatile organic substances associated with cyanobacteria and algae in a eutrophic shallow lake. Applied and Environmental Microbiology 47:814-820.

Juttner, F. and H. Muller. 1979. Excretion of octadienes and octatrienes by a freshwater diatom. Naturwissenschafter 66:363-364.

Juttner, F., B. Haflacher, and K. Wurster. 1986. Seasonal analysis of volatile organic biogenic substances (VOBS) in freshwater phytoplankton populations dominated by *Dinobryon*, *Microcystis*, and *Aphanizomenon*. Journal of Phycology 22:169-172.

Keenum, M. E. and J. E. Waldrop. 1988. Economic analysis of farm-raised catfish production in Mississippi. Mississippi Agricultural and Forestry Experiment Station Technical Bulletin 155. Mississippi State, Mississippi, USA.

Kikuchi, T., T. Mimura, K. Harimaya, H. Yano, T. Arimoto, Y. Masada, and T. Inove. 1973. Odorous metabolites of blue-green alga: *Schizothrix muelleri* Naegali, collected in the southern basin of Lake Biwa. Chemical and Pharmaceutical Bulletin 21:2342-2343.

Kuusi, T. and M. Suihko. 1983. Occurrence of various off-flavors in fish in Finland from 1969 to 1981. Water Science and Technology 15:47-58.

Lalezary, S., M. Pirbazari, M. J. McGuire, and S. W. Krasner. 1984. Air stripping of taste and odor compounds from water. Journal of the American Water Works Association 55:169-176.

Lelana, I. Y. 1987. Geosmin and off-flavor in channel catfish. Ph.D. Thesis, Auburn University, Auburn, Alabama, USA.

Lovell, R. T. 1983a. New off-flavors in pond-cultured channel catfish. Aquaculture 30:329-334.

Lovell, R. T. 1983b. Off-flavors in pond-cultured channel catfish. Water Science and Technology 15:67-73.

Lovell, R. T. and D. Broce. 1985. Cause of musty flavor in pond-cultured penaeid shrimp. Aquaculture 50:169-174.

Lovell, R. T. and L. A. Sackey. 1973. Absorption by channel catfish of earthy-musty flavor compounds synthesized by cultures of blue-green algae. Transactions of the American Fisheries Society 102:774-777.

Lovell, R. T., R. O. Smitherman, and E. W. Shell. 1978. Progress and prospects in fish farming. Pages 261-292 in A. M. Altshul and H. L. Wilke, editors. New Protein Feeds, Vol. 3. Academic Press, New York, New York, USA.

Lovell, R. T., I. Y. Lelana, C. E. Boyd, and M. S. Armstrong. 1986. Geosmin and musty-muddy flavors in pond-raised channel catfish. Transactions of the American Fisheries Society 115:485-489.

Mackie, P. R., A. S. McGill, and R. Hardy. 1972. Diesel oil contamination of brown trout (*Salmo trutta* L.). Environmental Pollution 3:9-16.

Maligalig, L. L., J. F. Caul, and O. W. Tiemeier. 1973. Aroma and flavor of farm-raised channel catfish: effects of pond condition, storage, and diet. Food Products Development 7:86-92.

Maligalig, L. L., J. F. Caul, R. Basette, and O. W. Tiemeier. 1975. Flavoring channel catfish experimentally: effects of concentration and exposure time. Journal of Food Science 40:1242-1245.

Mallevialle, J. and I. H. Suffet, editors. 1987. Identification and Treatment of Tastes and Odors in Drinking Water. American Water Works Association, Denver, Colorado, USA.

Martin, J. F., L. W. Bennet, and W. H. Graham. 1988a. Off-flavor in channel catfish (*Ictalurus punctatus*) due to 2-methylisoborneol and its dehydration products. Water Science and Technology 20:99-105.

Martin, J. F., T. H. Fisher, and L. W. Bennett. 1988b. Musty odor in chronically off-flavored channel catfish: isolation of 2-methylene-bornane and 2-methyl-2-bornene. Journal of Agricultural and Food Chemistry 36:1257-1260.

Martin, J. F., C. P. McCoy, C. S. Tucker, and L. W. Bennett. 1988c. 2-methylisoborneol implicated as a cause of off-flavor in channel catfish, *Ictalurus punctatus* (Rafinesque), from commercial culture ponds in Mississippi. Aquaculture and Fisheries Management 19:151-157.

Medsker, L. L., D. Jenkins, and J. F. Thomas. 1968. Odorous compounds in natural waters. An earthy-smelling compound associated with blue-green algae and actinomycetes. Environmental Science and Technology 2:461-466.

Medsker, L. L., D. Jenkins, J. F. Thomas, and C. Kock. 1969. Odorous compounds in natural waters. 2-Exohydroxy-2-methylbornane, the major odorous compound produced by several actinomycetes. Environmental Science and Technology 3:476-477.

Mohren, S. and F. Juttner. 1983. Odorous compounds of different strains of *Anabaena* and *Nostoc* (Cyanobacteria). Water Science and Technology 15:221-228.

Motohiro, T. 1983. Tainted fish caused by petroleum compounds—a review. Water Science and Technology 15:75-84.

Narayan, L. V. and W. J. Nunez. 1974. Biological control: isolation and bacterial oxidation of the taste and odor compound geosmin. Journal of the American Water Works Association 66:532-536.

Paasivirta, J., J. Knuutinen, J. Tarhanen, T. Kuokkonen, K. Surma-aho, R. Paukku, H. Kaariainen, M. Lahtipera, and A. Veiganen. 1983. Potential off-flavor compounds from chlorobleaching of pulp and chlorodisinfection of water. Water Science and Technology 15:97-104.

Persson, P.-E. 1978. Muddy off-flavor in bream (*Abramis brama*) from the Porvoo area, Gulf of Finland. Internationale Vereinigung for Theoretische und Angewandte Limnologie Verhandlungen 20:2098-2102.

Persson, P.-E. 1979a. Notes on muddy flavor. IV. Variability of sensory response to 2-methylisoborneol. Aqua Fennica 9:48-52.

Persson, P.-E. 1979b. The source of muddy odor in bream (*Abramis brama*) from Porvoo sea area (Gulf of Finland). Journal of the Fisheries Research Board of Canada 36:883-890.

Persson, P.-E. 1980. Sensory properties and analysis of two muddy odour compounds, geosmin and 2-methylisoborneol, in water and fish. Water Research 14:1113-1118.

Persson, P.-E. 1981. The etiology of muddy odour in water and fish. Finnish Fisheries Research 4:1-13.

Persson, P.-E. 1982. Muddy odour: a problem associated with extreme eutrophication. Hydrobiologia 86:161-164.

Persson, P.-E. 1983. Off-flavors in aquatic ecosystems. Water Science and Technology 15:1-12.

Persson, P.-E. 1984. Uptake and release of environmentally occurring odorous compounds by fish. Water Research 18:1263-1271.

Persson, P.-E and R. K. York. 1978. Notes on muddy odor: I. Sensory properties and analysis of 2-methylisoborneol in water and fish. Aqua Fennica 8:83-88.

Reynolds, C. S. 1984. The Ecology of Freshwater Phytoplankton. Cambridge University Press, Cambridge, England.

Rosen, A. A., R. S. Safferman, C. I. Mashni, and A. H. Romano. 1968. Identity of odorous substance produced by *Streptomyces griseoluteus*. Applied Microbiology 16:178-179.

Safferman, R. S., A. A. Rosen, C. I. Mashni, and M. E. Morris. 1967. Earthy-smelling substance from a blue-green alga. Environmental Science and Technology 1:429-430.

Shipton, J., J. H. Last, K. E. Murray and G. L. Vale. 1970. Studies on a kerosene-like taint in mullet (*Mugil cephalus*). II. Chemical nature of the volatile constituents. Journal of the Science of Food and Agriculture 21:433-436.

Shumway, D. L. and G. G. Chadwick. 1971. Influence of Kraft mill effluent on the flavor of salmon flesh. Water Research 5:997-1003.

Shumway, D. L. and J. R. Palensky. 1973. Impairment of the flavor of fish by water pollutants. United Sates Environmental Protection Agency, Ecological Research Series EPA-R3-73-010.

Silvey, J. K. G. and A. W. Roach. 1975. The taste and odor producing aquatic actinomycetes. CRC Critical Reviews in Environmental Control 5:233-273.

Silvey, J. K. G. and J. T. Wyatt. 1970. The interrelationship between freshwater bacteria, algae and actinomycetes in southwestern reservoirs. Pages 249-275 in J. Cairns, editor. The Structure and Function of Freshwater Microbial Communities. Research Division Monograph 3, Virginia Polytechnic Institute, Blacksburg, Virginia, USA.

Slater, G. P. and V. C. Blok. 1983. Volatile compounds of the cyanophyceae — a review. Water Science and Technology 15:181-190.

Smith, A. L. 1974. The effects of effluents from the Canadian petrochemical industry on aquatic organisms. Environment Canada, Fisheries and Marine Service, Technical Report No. 472.

Smith, D. W. and R. H. Piedrahita. 1988. The relationship between phytoplankton and dissolved oxygen in fish ponds. Aquaculture 68:249-265.

Tabachek, J. L. and M. Yurkowski. 1976. Isolation and identification of blue-green algae producing muddy odor metabolites, geosmin and 2-methylisoborneol, in saline lakes in Manitoba. Journal of the Fisheries Research Board of Canada 33:25-35.

Thayson, A. C. 1936. The origin of an earthy or muddy taint in fish. I. The nature and isolation of the taint. Annals of Applied Biology 23:99-104.

Thayson, A. C. and F. T. K. Pentelow. 1936. The origin of an earthy or muddy taint in fish II. The effect on fish of the taint produced by an odoriferous species of Actinomyces. Annals of Applied Biology 23:105-109.

Torrens, L. and F. Lowell. 1987. Effects of blue tilapia/channel catfish polyculture on production, food conversion, water quality, and channel catfish off-flavor. Proceedings of the Arkansas Academy of Science 41:82-86.

Tucker, C. S. 1985. A checklist of phytoplankton (exclusive of diatoms) from Mississippi channel catfish ponds. Mississippi Agricultural and Forestry Experiment Station Technical Bulletin 126. Mississippi State, Mississippi, USA.

Tucker, C. S. and C. E. Boyd. 1978. Consequences of periodic applications of copper sulfate and simazine for phytoplankton control in catfish ponds. Transactions of the American Fisheries Society 107:316-320.

Tucker, C. S. and C. E. Boyd. 1979. Effects of simazine treatment on channel catfish and blue-gill production in ponds. Aquaculture 15:345-352.

Tucker, C. S. and C. E. Boyd. 1985. Water quality. Pages 135-227 in C. S. Tucker, Editor. Channel Catfish Culture. Elsevier, Amsterdam, The Netherlands.

Tucker, C. S. and S. W. Lloyd. 1984. Phytoplankton communities in channel catfish ponds. Hydrobiologia 112:137-141.

Tucker, C. S. and S. W. Lloyd. 1987. Evaluation of potassium ricinoleate as a selective blue-green algicide in channel catfish ponds. Aquaculture 65:141-148.

Tucker, C. S., R. L. Busch, and S. W. Lloyd. 1983. Effects of simazine treatment on channel catfish production and water quality in ponds. Journal of Aquatic Plant Management 21:7-11.

Van Aller, R. T., G. F. Pessoney, V. A. Rogers, E. J. Watkins, and H. G. Leggett. 1985. Oxygenated fatty acids: a class of allelochemicals from aquatic plants. Pages 387-400 in A. C. Thompson, editor. The Chemistry of Allelopathy. American Chemical Society, Washington, D.C., USA.

Weete, J. D., W. T. Blevins, G. R. Wilt, and D. Durham. 1977. Chemical, biological, and environmental factors responsible for the earthy odor in the Auburn city water supply. Alabama Agricultural Experiment Station Bulletin 490. Auburn, Alabama, USA.

Wetzel, R. G. 1975. Limnology. W. B. Saunders Co. Philadelphia, Pennsylvania, USA.

Whitfield, F. B. and D. J. Freeman. 1983. Off-flavors in crustaceans caught in Australian coastal waters. Water Science and Technology 15:85-96.

Wood, S., S. T. Williams, W. R. White, and F. Jones. 1983. Factors influencing geosmin production by a streptomycete and their relevance to the occurrence of earthy taints in reservoirs. Water Science and Technology 15:191-198.

Yagi, M., M. Kajino, U. Matsua, K. Ashitani, T. Kita, and T. Nakamura. 1983. Odor problems in Lake Biwa. Water Science and Technology 15:311-322.

Yurkowski, M. and J. L. Tabachek. 1974. Identification, analysis, and removal of geosmin from muddy-flavored trout. Journal of the Fisheries Research Board of Canada 31:1851-1858.

Yurkowski, M. and J. L. Tabachek. 1980. Geosmin and 2-methylisoborneol implicated as a cause of muddy odor and flavor in commercial fish from Cedar Lake, Manitoba. Canadian Journal of Fisheries and Aquatic Sciences 37:1449-1450.

REVIEW OF WATER REUSE SYSTEMS - WATER REUSE IN HATCHERIES

Ronald D. Mayo

ABSTRACT

Aquaculture water reuse encompasses a variety of system designs ranging from relatively simple aeration systems to 100% water containment for complete environmental control. This spectrum of systems design is presented and examined. The economic factors governing the selection of system components include the relative capital and operating costs of these components. The history and current status of water reuse are reviewed as are the major limitations to, and prospects for, future development of recirculating aquaculture systems.

WATER REUSE IN HATCHERIES - A PERSONAL AND HISTORICAL PERSPECTIVE

This is not the first time the author has been asked to review water reuse in a formal presentation. In 1972 he discussed various aspects as they relate to salmonid hatcheries (Liao and Mayo 1972) and, in 1974, related water reuse to pollution control (Liao and Mayo 1974). The 1976 review presented in Kyoto, Japan was broadened to include elements of the economic justification and, better yet, fairly detailed technical description of 17 systems then in service in major North American facilities (Mayo 1976). In 1980, in Stavanger, Norway, the review presented was focused on North America and made an attempt to regularize the terminology (Mayo 1981).

Insofar as possible, each of these documents followed appropriate editorial conventions for professional papers and took a wide technical view. Thus it would not be unreasonable that they serve as a model and foundation for this presentation. In fact, such a use would have made these efforts less demanding, an idea that is quite compelling.

However, instead of following these earlier models, we are going to step back into the first person and attempt this review from our personal perspective in order that those of you who have just become involved in aquaculture or those of you who are working in a fairly narrow technical field can understand better water reuse in the context of aquacultural development. We will use specific projects to illustrate its evolution in aquaculture in terms of its justification and intent rather than simply its existence. We will focus on projects that we have been involved in as we know the basis on which the planning decisions were made.

TERMINOLOGY

The Stavanger review and other papers presented at the same meeting helped focus on the issue of terminology and contributed to the decision to prepare on EIFAC/FAO technical paper "Flow-through and recirculation systems, Report of the working group on terminology, format and units of measurement" by a task force convened by Harald Rosenthal (EIFAC Technical Paper 49, 1986). This paper should be required reading for all involved in reuse systems. It contains some very fundamental and useful concepts that are of great use in clear communication in this field. It also points out the difficulty of achieving consistency in the terminology used in this field and suggests that each author attempt a "clear description of the actual mode of operation" being referred to. As I describe various projects, I will attempt to do this, but with only faint hope of even partial success. For example, terms like "simple reuse" and "complex reuse" may be found in discussions of various systems. These are misleading terms because a simple process can require a complex facility and vice versa. (For example, a simple process such as adding oxygen to water can be accomplished by anything from water dropping into a pool to a complex set of pumps and generators, and either can be justified economically under certain conditions.) These terms are used in the interest of brevity and not clarity, and for this we apologize to our friend, Harald Rosenthal.

In this paper I will talk about two processes, oxygen injection and the use of ozone disinfection, which have major implications for water reuse though the examples provided do not involve reuse in the sense that the water is reused several times. However, I will provide a definition of the term "reuse system" that will explain why all of the processes described here are appropriate for inclusion.

A reuse system is one that subjects the water to one or more processes that improve its quality in a way that will allow greater fish production than would otherwise be possible.

Many of you will recognize this as a very wide definition for a reuse system and in fact would include many systems not thought of as "reuse" systems by their operators. Thus you have discovered the first law of water reuse: More people have reuse systems than know they do.

OXYGEN ADDITION

Parkview Hatchery - New Mexico

This project is nearing completion at a site near the town of Chama River near the Colorado border. It is the reconstruction of a facility originally built in 1935 at the same site. It is intended that this project produce 150,000 pounds of rainbow trout each year for planting in lakes and streams for harvest by sportsmen (Mayo and Associates 1986). It will also maintain the trout broodstock for this and other hatcheries. The water supply is a series of springs producing approximately 2,200 gpm of water at 48 °F, year around. The water is of exception quality, but the elevation of 7,300 feet above sea level significantly reduces the oxygen carrying capacity of the water. (At 48 °F, saturation at 7,300 feet is 8.9 mg/l, at sea level it is 11.6 mg/l. Thus, if we assume a minimum oxygen level criteria of 5 mg/l, the available oxygen at this site is only 59% of that of a sea level site.)

The owner, the State of New Mexico, wanted to maximize use of the spring resource for production as similar quality water supplies are limited. At the same time, the owner expressed a strong bias against major pumping or mechanical systems.

As planners, we went quickly to the examples provided by the original designers who constructed the first facility at this site in the 1930's. Though the site is flat, some fall is available so they linked their ponds in series and achieved some reaeration by a series of drops from raceway to raceway. In our design for reconstruction we followed a similar pattern but with considerable elaboration.

The rearing raceways are built in three parallel groups. At the head of each group are the fry raceways. They receive "first water" only from the head box/aeration structure. Water leaving the fry raceways, plus additional water from the head box entered the upper pair of fingerling raceways. That water then dropped through four more sets of raceways and finally into the broodstock system.

Of primary concern was the reaeration of water as it passed from raceway to raceway. This is accomplished by passing water through an aluminum screen as it leaves the raceway and again as it enters the next raceway. These screens were designed to make the greatest possible use of the available head to accomplish the aeration function. While each "fall" of the water is fairly short, the repetition is expected to accomplish a 1.5 mg/l increase between each pair of raceways. The basis for this estimate is contained in a brief paper by Soderburg, (1982).

With the low oxygen carrying capacity problem resolved by a number of separate aeration steps the water quality issue of concern became ammonia. In this facility, we used un-ionized ammonia as the controlling criteria as follows: For fry and small fingerlings the criteria for maximum un-ionized ammonia was 0.006 mg/l; for fingerlings it was 0.010 mg/l, and for brood stock it was 0.015 mg/l. The use of these ammonia criteria, which may have been first proposed by Burrows in the 1960's, is not satisfactory to all with an interest in this subject. Other, more complex ways to determine the carrying capacity of water where oxygen is not a factor are being proposed. However, as these criteria reflect feeding levels, pH's and water temperatures, and as they have some merit by long use, they were used in this instance.

Thus as indicated above, at Parkview, the criterion which defines the projected production levels is ammonia. The possibility of some form of ammonia removal, through biological filtration, ion exchangers, or ozone treatment was given consideration and some design decisions reflected the possibility of some such future attempts being made. However, in this facility, reuse is restricted to levels that could be accomplished by reaeration. The impact of reaeration is significant. In general terms, the use of reaeration increased the capacity of this site by a factor of 3, and, because of the particular site conditions, was accomplished at a fairly low level of mechanical complexity.

Parkview is an example of atmospheric reaeration to accomplish reuse. Though the process is fairly simple and the cost low, the planning and execution was refined as compared to the original 1935 project. However, that our best efforts did not produce radical changes in concept would suggest that all that is known about trout hatcheries is not new.

Other Examples of the Use of Atmospheric Aeration

Other examples of planned aeration for reuse include the Rathbun Hatchery in Iowa and the Alsea Hatchery in Oregon. At Rathbun, each concrete rearing pond was equipped with an airlift system that removed water from the center of the pond where a central drain was located and, after reaeration, returned the water to the pond inlet. This system had several features: The recycled water was taken from near the surface so as to avoid the recirculation of solids. The water was recirculated only to the pond from which it was removed as a disease control measure.

At Alsea a similar system was planned with a slightly different approach. Here raceways were "paired" in the summer when water supplies where low and less than half of the raceways required. The water being recirculated was captured at the lower end of the raceway (at the bottom of the water column) and "air lifted" into the adjacent, empty raceway. Water was "air lifted" again from the other end of the empty raceway and placed in the original raceway carrying the fish load. This tended to allow solids to settle more completely in the unused raceway, thus providing a water quality benefit.

The Use of Pure Oxygen

Atmospheric reaeration is not the only way to put oxygen back into the water in a hatchery environment. Over the last 15 years, the use of injected pure oxygen has become an important tool in extending the usefulness of a given water supply for rearing fish (Colt 1987).

Typically such systems will consist of a source of oxygen, a device for dissolving oxygen into the water, most often either under high pressure or with counterflow reactors, and the plumbing necessary to inject the oxygen rich water into the fish rearing systems. The oxygen source can either be mechanical systems which make the oxygen at the site or liquid oxygen purchased from commercial sources and stored on site.

The justification for using oxygen is typically an economic one. We have this example from a planning study for a large private salmon rearing farm north of Santa Cruz, California. Unlimited water was available for the price of the intake system and the pumping energy. The water requirement, based on oxygen criteria, was 90,000 gpm. Based on ammonia criteria it was 42,000 gpm. The choices were to pump 90,000 gpm at an energy cost of U$0.80/kg of oxygen or to pump 42,000 gpm and provide the rest of the oxygen from stored liquid oxygen costing U$0.22/kg (Mayo Associates 1986).

Examples of systems in place in the US include the Oregon Aquafoods facility and the Anadromous facility, both in Oregon. These are very large salmonid production facilities and for both injected oxygen is vital. These facilities were developed 8 to 12 years ago (Mayo Associates 1988). Other examples include the British Oxygen trout facilities in Scotland, a marine research facility on Guam, and various commercial facilities in Germany.

It is interesting however, that examples of the use of pure oxygen in large production facilities in the US is limited almost entirely to commercial facilities though serious consideration of this technology is beginning in state and federal facilities. For example, construction of a prototype project is being funded by the Bonneville Power Administration in Central Oregon.

Dworshak National Fish Hatchery

Typically, when water reuse in hatcheries is discussed, the example of the Dworshak Hatchery in Idaho comes to mind. When constructed, it was the largest and most complex salmonid facility in the US and possibly the world. It may still be.

The engineering planning for the construction of the Dworshak Steelhead Hatchery started in 1966. This hatchery, like many built in the last 25 years, was constructed in an attempt to mitigate the damage done to the natural runs of anadromous salmonids, in this case the anadromous rainbow trout, the steehead. It was intended to raise for release over 8 million fish per year weighing 50 grams each. The major rearing units are 140 concrete tanks each with a volume of 3,200 cubic feet. The water supply pump station has a capacity of 75,000 gpm. In the beginning, 25 of the rearing tanks were connected to a reuse water treatment system that provided for the treatment of water as it left these tanks and its partial reuse in these same tanks. In terminology then current, it was a 10-pass complex reuse system.

Reuse was considered in this application because during the first 5 years of hatchery operation the water supply, the north fork of the Clearwater River, was excessively warm in the summer and cold in the winter. After completion of the dam, the problem of warm water would be eliminated.

The heart of the reuse system was a series of 8 filters, 75 feet long, 22 feet wide and 7 feet deep. The water was distributed on top of the filter bed and removed from the bottom. The filter consisted of 12 inches of crushed oyster shell and 3 feet of crushed rock. After leaving the filter, the water was aerated by being pumped into a headtank through a series of aspirating nozzles. Water used in this system was mechanically cooled in the summer and warmed in the winter. All water entering the system was filtered in pressure sand filters and subjected to ultraviolet sterilization.

The design of these filters was based on prototype units constructed by Burrows and Combs (1968) at a federal hatchery in Washington State. Their developmental work, intuitive science at its best, was limited by the schedule imposed by the construction of the dam that was the basis for construction of the hatchery.

As engineers for the initial planning, we were frustrated by the lack of a basic understanding of the process elements. We were given a description of an entire system, including the rearing raceways and told that we had to build a system "just like that one, except bigger". The application rates throughout the system were given in exact terms but without knowing if changes in loading would result in changes in results. The effects of different temperatures were not considered nor were the differing patterns of metabolic waste production quantified. It was kind of like being asked to design an airplane without knowledge of aeronautics. It can be done but you shouldn't expect too much.

The primary justification for providing for reuse was to reduce the energy and capital costs associated with cooling, heating and disinfecting the water supply. Though the initial installation was for only part of the hatchery, it was conceived as a large scale prototype for things to come.

Since initial construction took place, the Dworshak Hatchery has been the center of reuse research and development for the federal government. The original filter system has long since been replaced by other systems and they in turn have been replaced (May 1981). Filters of plastic and sand have been tried with some success though no one involved in the work would suggest that a final solution has been achieved.

OZONE AS A DISINFECTION AND TREATMENT PROCESS

The Evolution

For the last several years, the staff at Dworshak has been involved in prototype testing of ozone disinfection systems at Dworshak for application in the Clearwater Hatchery soon to be constructed across the river (Pratschner et al. 1985). In this work, the effectiveness of ozone in the control of Infectious Hematopoietic Necrosis (IHN) for steelhead has been investigated. Though this work related to a "flow-through" system, it has implications for reuse systems in two ways.

1. In order that reuse systems operate effectively, especially the "complex systems" involving numerous reuses of the water, it is generally considered necessary that the water supply be relatively disease free. Ozone can accomplish this.

2. There are those who believe that ozone has special promise as a reuse treatment process for the removal of ammonia, nitrate, nitrite and biochemical oxidation demand (Colberg 1978).

Work at Dworshak has attempted to look into the effectiveness of ozone for reuse treatment but as the available supplies of ozone (after other demands) were small, the results were inconclusive.

As we will describe below, it would appear that ozone will be an important tool for the disinfection of water to hatcheries and perhaps as a reuse treatment process. However, we should not overlook other disinfection systems.

Chlorination followed by dechlorination has been an important part of the operation of Oregon Aquafood's facility in Oregon. They have demonstrated, on a large scale, that toxic effects on the fish can be avoided by conservative design and careful operation.

Ultraviolet sterilization (UV) has been used in hatcheries for some time with good effect. While there have been system failures, the failure has often been in the excessive expectations rather than the UV technology.

As examples of ozone applications in the near future, I would like to describe two projects I am planning in the State of Washington.

Merwin Trout Hatchery, Washington State

At the other end of the spectrum from the initial Dworshak design, in terms of time and approach, is the Merwin Trout Hatchery. This facility, now in design for Pacific Power & Light Co., will rear 67,000 pounds of trout for release each year. It will have 16 concrete raceways with volumes of 2,250 cubic feet each and will have a maximum water supply of 4,500 gpm.

Disinfection of the supply is considered necessary, as testing has indicated IHN virus is present and experience in nearby facilities has determined this will impact production. A secondary reason for disinfection is to prevent this hatchery from becoming a source of infection for existing salmon downstream on the Lewis River.

While the design is still evolving, we anticipate that this facility will be equipped with an ozone disinfection system that is fed from stored liquid oxygen and will treat up to 3,600 gpm for a significant part of the year. The water will pass through contractor vessels, injected with ozone, and then pass through stripping towers to reduce the ozone to acceptable levels. The ozone that is removed will be catalyzed to oxygen gas and may be used to provide oxygen enrichment for the rearing water in the fingerling raceways. The presence of oxygen storage facilities will make it desirable to use oxygen enrichment all year around, which could reduce water requirements and the related use of ozone.

At first glance, it would appear disinfection of a water supply alone would not fit our earlier definition of reuse systems. However, without the disinfection of the water supply from Lake Merwin to kill the IHN virus, it is likely that the quantity of fish released would be reduced and the quality would be such that their return rate would be diminished. Thus, this is a system that subjects the water to a process that improves the quality in a way that will allow greater fish production than would otherwise be possible.

The basis for the Merwin Hatchery design, in addition to a considerable background in ozone disinfection for domestic and industrial water, is research carried out over the last several years at Dworshak (see above) and the Cowlitz Trout Hatchery as discussed below. At these facilities, pilot and prototype studies were conducted that determined that ozone can both disinfect water that is infected with *Ceratomyxa shasta* and IHN and it can be treated in such a way as to effectively remove the ozone prior to fish contact. While time has not allowed a determination of return rates from fish planted for ocean migrations, it is clear that their health and size are better than fish in control groups and their survival is much better in the hatchery environment.

There are those who will argue that fish grown in a "sterile" environment may somehow be less likely to perform well after release. This may have merit. However, we have yet to see a manager of a hatchery with perfect spring water infect the water with a view towards growing stronger fish. Nor have we seen any adults return from the ocean that had died as a fingerling in the hatchery.

Cowlitz State Trout Hatchery, Washington

Planning for the Cowlitz Trout Hatchery started at the same time as planning for the Dworshak Hatchery, in 1966. But in many ways these two facilities differed. At the Cowlitz hatchery heating and cooling weren't required because the river temperatures were satisfactory (with the use of wells) for fry rearing and for some tempering. The rearing of fingerlings in raceways made use of very large quantities of river water that appeared to be adequately disease free.

However, as years passed the changing river environment and the presence of a very large salmon hatchery upstream saw increasing problems with Infectious Hematopoietic Necrosis (IHN) and *Ceratomyxosis shasta*. It reached the point that the State Department of Wildlife directed the agency responsible for this facility, Tacoma City Light, to take steps to reduce the disease problem. As this project was built under agreement to mitigate fish losses from a major hydroelectric project, the State's direction has considerable force.

Tacoma City Light undertook several years of research (Tipping 1987; Vu 1987) and in 1988 signed an agreement to construct a 9,000 gpm ozone disinfection system at Cowlitz. Design work for that project has begun. It will be similar in nature to the Merwin facility, except that it will be constructed as a modification to an existing facility and will be somewhat larger.

EXAMPLE OF THE ECONOMIC IMPLICATIONS OF OZONE SYSTEMS

Reuse processes can be very simple and the only cost for their installation is a little common sense. An example might be the use of available fall to increase dissolved oxygen. Certainly some of our examples indicate that reuse is not always costly. On the other hand, a major ozone system can be costly. However, if fish reared in an ozone system return at a significantly higher level the added cost may well be offset. And if this added ozone cost makes a release program possible as opposed to high hatchery mortalities and minimal returns, then ozone may not be costly.

There are also situations where a costly process may be justified to take advantage of the "economies of scale". To illustrate this, we are going to set up an example based on a purely hypothetical situation and some fairly real costs.

a. A state has two sites they can use to construct hatcheries on. The water supplies are similar, each exposed to some diseases, each with some other quality problems.

b. Each site can support 200,000 pounds of trout production using standards applicable to these rather marginal water supplies.

c. Either site could support the space for 400,000 pounds of production if water quality were better.

d. Research has suggested that if the existing water supplies are treated with ozone, the water quality will be improved in terms of chemical quality, oxygen levels and pathogens. It is determined that these improvements would make it possible to rear 400,000 pounds of improved quality fish on a single site.

However, the boss says "too expensive." Is he right? Let's look at some prior work on the cost of trout hatcheries:

a. In a study for the State of California (TMA 1988), a formula, based on earlier studies for the State of Washington, was put forth as follows:

Capital Cost of a hatchery in Millions ($1988) = 0.0037 x Rearing Volume in Cubic Feet to the power of 0.67

b. Data from the same study, after some manipulation, produced these relationships relative to the State's eight larger trout hatcheries (production range/year 300,000 to 700,000 pounds):

Rearing Volume in Cubic Feet = Annual Production in Pounds x 0.133 + 58,600

Annual Operating Costs in \$ = Annual Production in Pounds x \$0.25 + \$500,000

Peak Required Water Supply in gpm = Annual Production in Pounds divided by 42

Average Required Water Supply in gpm = Annual Production in Pounds divided by 70

c. Data from the preliminary work on the Merwin Hatchery, suggests these relationships:

Dosage required to accomplish disease control, etc. = 1.5 mg/l

Energy to convert Oxygen to Ozone = 10 KWHr/Pound

Cost of Energy = \$0.05/KWHr

Cost of Oxygen for feed to Ozone System = \$0.072/pound

Capital Cost for a 300 pound per day ozone system = \$1,500,000, the cost of other systems assumes a scaling exponent of 0.67.

Ozone produced per Pound of Oxygen = 4% of Oxygen

Efficiency of Ozone Use = 85%

O&M Cost (Excluding Oxygen and Power) for ozone system = 2% of capital costs

Given all of the above, we have these relationships:

The production cost for two - 200,000 pound/year hatcheries without an ozone system is \$4.24/pound. For a single 400,000 pound facility, the cost is \$2.96/pound.

At another scale, the production cost for two - 500,000 pound/year hatcheries without an ozone system is $2.02/pound. For a single 1,000,000 pound facility, the cost is $1.71/pound.

These values are illustrated in Figure 1.

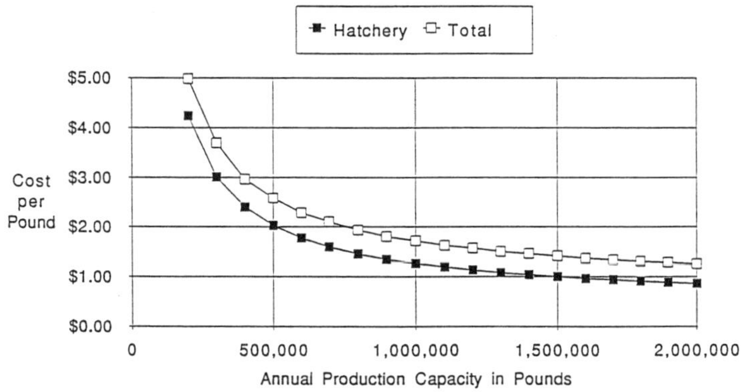

FIGURE 1. Production Cost for Example Trout Hatcheries of Various Sizes. Hatchery = Standard Facility without Ozone Treatment System. Total = Hatchery with Ozone Treatment of Supply @ 1.5 mg/l

Thus, in response to the question of cost, we would say that in this example, the production costs of a single large facility with an ozone system are less than those at two smaller facilities without expensive ozone systems.

However, just because we have set up this example doesn't make ozone a good tool for every case, or fewer, bigger hatcheries a good answer in every case, or data from California or Washington representative of what will happen everywhere. However, it does suggest that complexity does not necessarily cost more and that, at a minimum, each case should be considered on its own merit.

To summarize the economic implications of ozone technology, "It can make you money or it can cost you money, but it can't be ignored."

SUMMARY

I have described a range of projects that involve water reuse and why they were built in the first place. As I warned you at the beginning, this is primarily in the context of my own experience. Not because there are no other people working in this field, but because I think it is appropriate that you understand what shapes a planner's thinking in selecting water reuse processes. As I try to summarize our thoughts about reuse, it becomes clear that water reuse as a separate technical concept from other aspects of hatchery planning does not exist. I will try to explain what I mean.

Reuse systems, even those that appear to be physically very simple, are combinations of very complex biological and chemical processes. But these individual processes are within our ability to understand, and as has been demonstrated at this meeting, are better understood every day.

In this we have come a long way from the hurried, underfunded and largely intuitive research that was the basis for the initial Dworshak design. We have certainly gone beyond the need to depend on people selling "black box" solutions. We are moving from an art to an understanding of predictable processes that can be combined with confidence into working systems.

However, as we look at these individual processes that let us build better reuse systems, we should see that not one of them is unique to water reuse. Each concept and relationship has significance to any fish culture project and perhaps it is time, after 25 years to stop reviewing water reuse as if it is a separate thing.

I will finish this summary with a last review of the basic laws of water reuse:

The first law is: More people have reuse systems than know they do.

The second law is: You might make a reuse system work without understanding it. To keep it working is harder.

The third law is: People who announce they have perfect reuse systems in a little black box may not be totally sincere.

And finally: As quality water supplies diminish, and they will, all fish growing systems will be reuse systems and the term will disappear from use and no more reuse review papers will be written. This is probably a good thing.

LITERATURE CITED

Burrows, R. and R. Combs. 1968. Controlled Environments for Salmon Propagation. Progressive Fish-Culturist, Volume 30 (3).

Burrows, R. 1964. Effects of Accumulated Excretory Products on Hatchery Reared Salmonids. U.S. Sports Fisheries and Wildlife Research Report 66.

*Carey and Kramer. 1966. Fish Hatchery Design Memorandum No. 14.1, Dworshak Dam and Reservoir. Prepared for the U.S. Army Engineering District, Walla Walla, Washington.

Colberg, P. 1978. Effect of Ozone on Microbial Fish Pathogens, Ammonia, Nitrate, Nitrite, and BOD in Simulated Reuse Hatchery Water. Idaho Agriculture Experiment Station Research Paper No. 7759.

*Colt, J. and Watten. 1987. Pure Oxygen in Fish Culture. U.S. Trout Farmers Association Annual Convention, Feb. 24-28.

EIFAC Technical Paper No. 49. 1986. Flow-through and Recirculation Systems. Report of the working group on terminology, format and units of measurement, by a task force convened by Harald Rosenthal.

Liao, Paul B. and Ronald D. Mayo. 1972. Salmonid hatchery water reuse systems. Aquaculture 1: 317-335.

Liao, Paul B. and Ronald D. Mayo. 1974. Intensified fish culture combining water reconditioning with pollution abatement. Aquaculture 3: 61-85.

*Mayo Associates. 1986. Planning Report Amended for the Parkview Hatchery Reconstruction Project. Prepared for the State of New Mexico Department of Game and Fish (with Leedshill-Herkenhoff, Inc.).

*Mayo Associates. 1987. A facility development program for SilverKing Oceanic Farms of Santa Cruz, California.

*Mayo Associates. 1988. An assessment of private salmon ranching in Oregon. Prepared for the Oregon Coastal Zone Commission.

*Mayo Associates. 1989. The California hatchery evaluation study, options, public and private, for meeting fish management needs in the year 2000. Presented at the World Aquaculture Society's annual meeting, Los Angeles, California, February 12-16.

Mayo, Ronald D. 1976. A technical and economic review of the use of reconditioned water in aquaculture. Presented at FAO Conference on Aquaculture, Kyoto, Japan, May 26-June 2.

Mayo, Ronald D. 1981. Recirculation systems in northern America. Prepared for the World Symposium on Aquaculture in Heated Effluents and Recirculation Systems. Stavanger 38-30, May 1980, Vol. II, Berlin 1981.

Pratschner, G. Owsley, D., et al. 1985. Final Report, A test of ozone for disinfection of a hatchery water supply to prevent waterborn transmission of infectious hematopoietic necrosis (IHN) virus. U.S. Fish and Wildlife Service Report.

Soderberg, R. 1982. Aeration of water using gravity aerators. Progressive Fish-Culturist.

The Mayo Associates (TMA). 1988. California Hatchery Evaluation Study. Prepared for State of California Fish and Game Commission.

Tipping, Jack. 1987. Use of ozone to control ceratomyxosis in steelhead trout (*Salmo gairdneri*). Ozone Science and Engineering, Volume 9.

*Vu., Chaun. 1987. Engineering aspects of the ozone pilot system at Cowlitz Trout Hatchery. Tacoma City Light Report.

*Certain of the above citations are not available through the normal literature sources. This is typical of an applied, though fairly narrow, science like hatchery bio-engineering. The author will attempt to make copies or partial copies available on request.

AERATION IN INTENSIVE CULTURE

John Colt and Christina Orwicz

ABSTRACT

Gas transfer may be required in intensive culture to add or remove specific gases, such as oxygen, nitrogen, or carbon dioxide. The addition of oxygen (aeration) is the most common gas transfer process used in aquaculture. Aerators are rated under standard conditions and their actual operating performance in an aquatic system depends strongly on the desired gas concentrations and the ambient temperature and pressure.

Aeration can increase the carrying capacity of an aquatic system when dissolved oxygen is the most limiting factor. The actual increase in carrying capacity will depend primarily on pH and temperature. Pure oxygen aeration systems are widely used in intensive culture due to their ability to economically saturate or supersaturate water with dissolved oxygen and reduce nitrogen gas. Gas transfer systems for removal of carbon dioxide or gas supersaturation are designed quite differently from aeration systems.

The "best" aeration system for a given application depends on site conditions, production schedules, layout of the rearing units, and operational procedures. General design procedures for aeration systems are presented.

NOMENCLATURE

a = Area of interfacial contact between gas and liquid (m^2)

A = Constant for each gas and equal to 760/1000K. Equal to 0.5318 for oxygen ((mm Hg•l)/(mg•atm))

AF = Aeration effectiveness; used to rate gravity aerators and equal to the amount of oxygen added divided by the total deficit; expressed at a base temperature of 20 C (%)

BP	=	Barometric pressure (mm Hg)
C	=	Dissolved gas concentration (mg/l)
COC	=	Cumulative oxygen consumption in a rearing unit or series of rearing units (mg/l)
C_{in}	=	Influent dissolved oxygen concentration to an aerator (mg/l)
C_{out}	=	Effluent dissolved oxygen concentration from an aerator (mg/l)
C_o	=	Dissolved gas concentration at time zero in a unsteady-state test (mg/l)
C_T	=	Total analytical concentration of an acid-base system (mol/l)
C^*	=	Saturation dissolved gas concentration at a given temperature, pressure, and mole fraction (mg/l)
C^*_{760}	=	Saturation dissolved gas concentration at a given temperature, 1 atm total pressure, and for air (mg/l)
C^*_z	=	Saturation dissolved gas concentration corresponding to a given water depth z (mg/l)
$\overline{C}^*_\infty$	=	Saturation dissolved gas concentration attained at an infinite time in a unsteady-state test (mg/l)
$\overline{C}^*_{\infty f}$	=	$\overline{C}^*_\infty$ evaluated under field conditions at local barometric pressure and temperature (mg/l)
$\overline{C}^*_{\infty s}$	=	$\overline{C}^*_\infty$ evaluated under standard conditions (clean water, temperature = 20 C, and barometric pressure = 760 mm Hg) (mg/l)

$\overline{C}^*$	=	Effective average saturation dissolved gas concentration; depends on type of aerator, aeration basin depth, and depletion of oxygen in the gas phase (mg/l)
DC	=	Dissolved carbon dioxide (mg/l)
D_e	=	Equivalent aeration depth expressed as a percent of basin depth (%)
DN	=	Dissolved nitrogen + argon (mg/l)
DO	=	Dissolved oxygen (mg/l)
DO_{in}	=	Influent dissolved oxygen to a rearing unit (mg/l)
DO_{out}	=	Effluent dissolved oxygen from a rearing unit (mg/l)
DO_{min}	=	Minimum dissolved oxygen criteria (mg/l)
DO_{max}	=	Maximum dissolved oxygen criteria (mg/l)
FAE	=	Field aeration efficiency; oxygen transfer rate under field conditions divided by the power input (kg oxygen/kW•hr)
G/L	=	Ratio of gas flow at standard conditions to water flow (dimensionless)
K	=	Ratio of molecular weight to molecular volume and equal to 1.42903 for oxygen (mg/ml)
K_L	=	Overall liquid phase mass-transfer coefficient (m/hr)
K_La	=	Volumetric mass transfer coefficient (1/hr)

M	=	Mass of fish in an individual rearing unit (kg)
M_e	=	Metabolizable energy; the energy available from food less exogenous excretion (fecal + nonfecal) (kcal/g feed)
$\dot{m}$	=	Mass flow rate of oxygen (kg/hr)
N	=	Symbol used for Field Aeration Efficiency (FAE) in older literature
N_o	=	Symbol used for Standardized Aeration Efficiency (SAE) in older literature
OFR	=	Oxygen-Feed Ratio; the amount of oxygen required per mass of feed ingested by a culture animal (kg oxygen/kg of dry feed)
OFR_d	=	Oxygen-Feed Ratio used for design purposes; may include the effects of other organisms in the system
OTE_o	=	Standardized oxygen transfer efficiency; equal to SOTR/$\dot{m}$ (decimal fraction)
OTE_f	=	Field oxygen transfer efficiency; equal to $OTR_f/\dot{m}$ (decimal fraction)
OTR_f	=	Field oxygen transfer rate (kg/hr)
P	=	Pressure (mm Hg)
P_w	=	Vapor pressure of water (mm Hg)
P_{bot}	=	Pressure at the bottom of an aeration basin (mm Hg)
PP	=	Partial pressure of a gas (mm Hg)

Q_w = Water flow (m^3/s)

Q_{ox} = Oxycalorific equivalent; the number of calories released by a substrate per mass of oxygen consumed (kcal/g of oxygen)

R = Total ration (kg/d)

SAE = Standardized aeration efficiency; oxygen transfer rate in clean water at standard conditions divided by power input (kg oxygen/kW•hr)

SOTR = Standardized oxygen transfer rate in clean water at standard conditions (kg/hr)

t = Time (hr)

T = Temperature (C)

TGP = Total gas pressure (mm Hg)

V = Volume of an individual rearing unit (m^3)

Z = Depth (m)

Z_{bot} = Depth of an aeration basin (m)

Z_e = Equivalent aeration depth (m)

Greek Symbols

α' = Ratio of K_La in a dirty water application to K_La in clean water at equivalent conditions of temperature, geometry, mixing, etc. (dimensionless)

α_g = Mole fraction of "free" gas such as carbon dioxide, hydrogen sulfide, or ammonia (dimensionless)

β'	=	Ratio of C^* in a dirty water application to C^* in clean water at equivalent conditions of temperature and partial pressure (dimensionless)
β	=	Bunsen coefficient for a gas (liter/(liter•atm))
ΔP	=	Difference between total gas pressure and the local barometric pressure measured by the membrane-diffusion method (mm Hg)
ΔP_{uncomp}	=	Uncompensated ΔP is the difference between total gas pressure and the local pressure including hydrostatic pressure (mm Hg)
Θ	=	Temperature adjustment factor (dimensionless)
ρg	=	Specific weight of water (73.42 mm Hg/m at 20 C)
ΔDO	=	Change in dissolved oxygen (mg/l)
ΔDN	=	Change in dissolved nitrogen + argon (mg/l)
X	=	Mole fraction of a gas (%/100) (dimensionless)

Superscripts and Subscripts

g	=	Gas phase
l	=	Liquid phase
~	=	A time and position dependent variable

INTRODUCTION

In intensive culture, many of the important water quality parameters are dissolved gases such as oxygen, carbon dioxide, hydrogen sulfide, ammonia, and nitrogen. Other dissolved gases such as ozone, chlorine, or sulfur dioxide are commonly used in disinfection processes. The maintenance of environmental quality requires the control of dissolved gas levels. In many cases, an aeration system will be designed to add oxygen and strip out other gases (Watten et al., 1989). While addition of oxygen (aeration) is the most common gas transfer process (Sowerbutts and Forster, 1981), the design of a gas transfer system must consider the potential impact of all dissolved gases (Colt and Tchobanoglous, 1981; Speece, 1981).

For some types of aeration equipment, standardized testing and rating procedures are available. The actual performance under aquaculture conditions may be quite different than under standard conditions. Some of the standardized procedures may not be useful for typical aquatic systems due to differences in system configuration or operation mode.

The design and operation of aeration equipment depends strongly on the water quality criteria for the various dissolved gases, hydraulics of rearing units, and mode of operation. For some gases such as oxygen, both maximum and minimum water quality criteria are needed.

This paper suggests water quality criteria for major dissolved gases, reviews aeration equipment testing procedures, describes the performance and rating of aeration equipment, and discusses the design and operational characteristics of aeration systems as a function of rearing unit type, configuration, and mode of operation. The scope of this article is limited to the transfer of oxygen, except when other gases have a significant impact on normal design and operation. It is hoped that this information will result in the design and operation of more efficient systems.

TYPES AND CONFIGURATION OF AERATORS

Aeration devices can be classified as surface, subsurface, or gravity. If the source of oxygen is enriched or "pure" oxygen gas rather than air, the units are called pure oxygen aerators.

Types of Aerators

Surface aerators spray or splash water into the air and transfer oxygen from the air into the water. The major types of surface aerators are shown in Figure 1.

a)

b)

c)

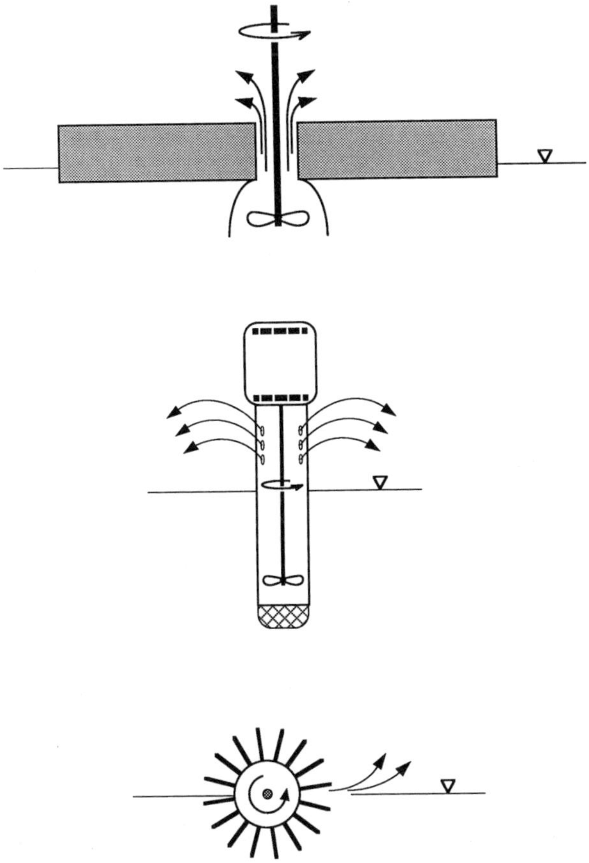

Figure 1. Typical surface aerators (a) floating aerator, (b) surface aerator with draft tube, (c) brush, rotor, or paddlewheel aerator

Subsurface aerators mix water and air together in an aeration basin and transfer oxygen from the air bubbles into the water. The major types of submerged aerators are shown in Figure 2.

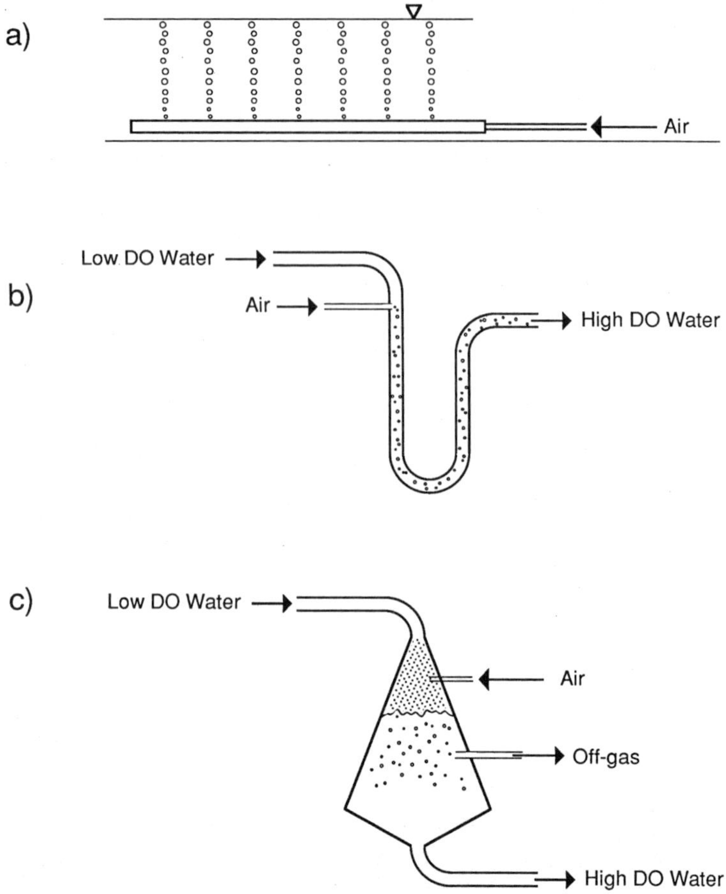

Figure 2. Typical submerged aerators (a) diffused (b) u-tube, (c) aerator cone, (continued next page) (d) static tube, (e) air-lift, (f) venturi, (g) nozzle

d)

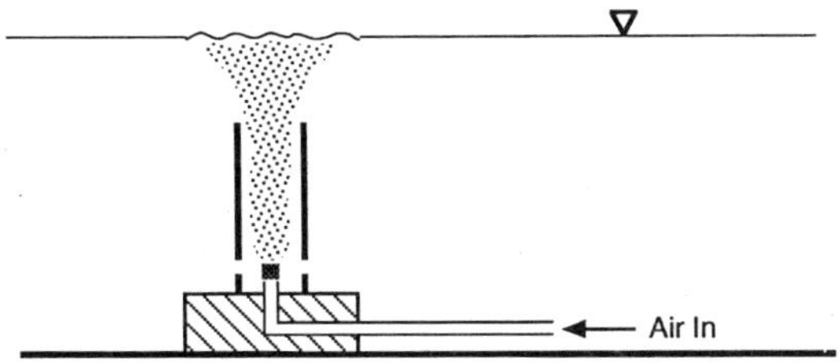

e)

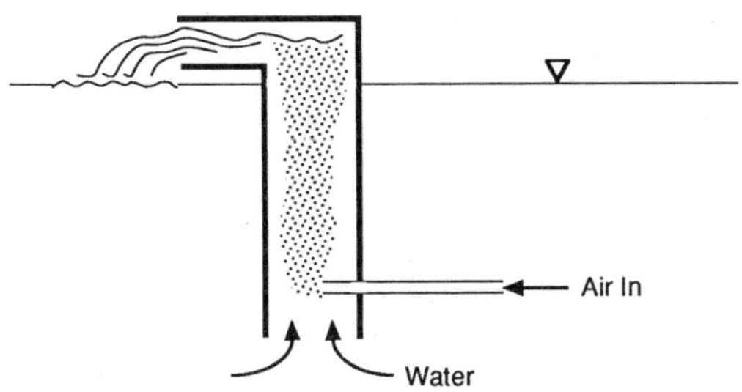

f)

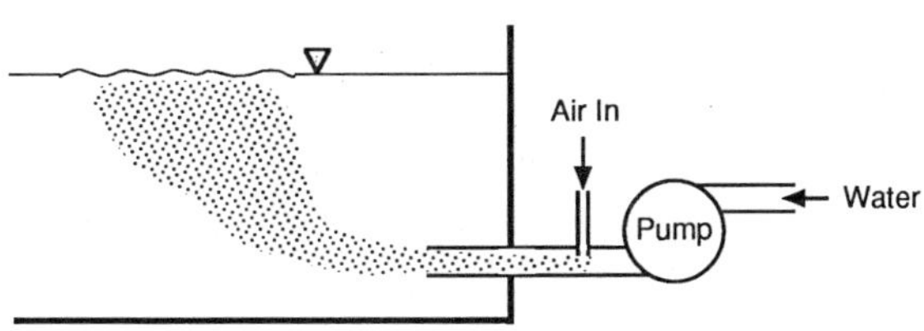

g)

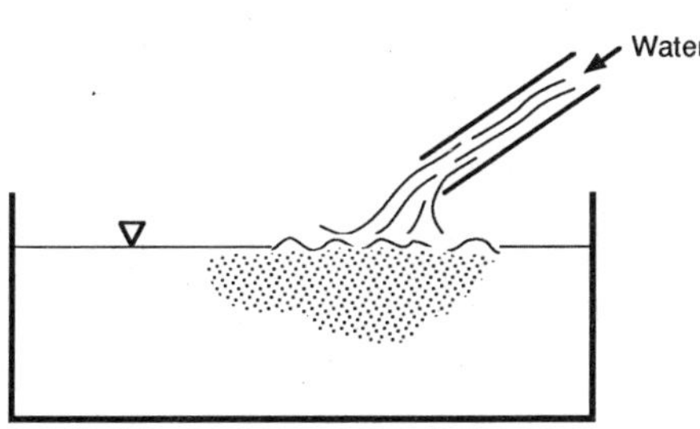

Gravity aerators are a special type of surface aerator that uses gravity rather than mechanical power to transfer oxygen. This type of aerator is commonly used in flow-through systems where adequate head is available. The major types of gravity aerators are shown in Figure 3.

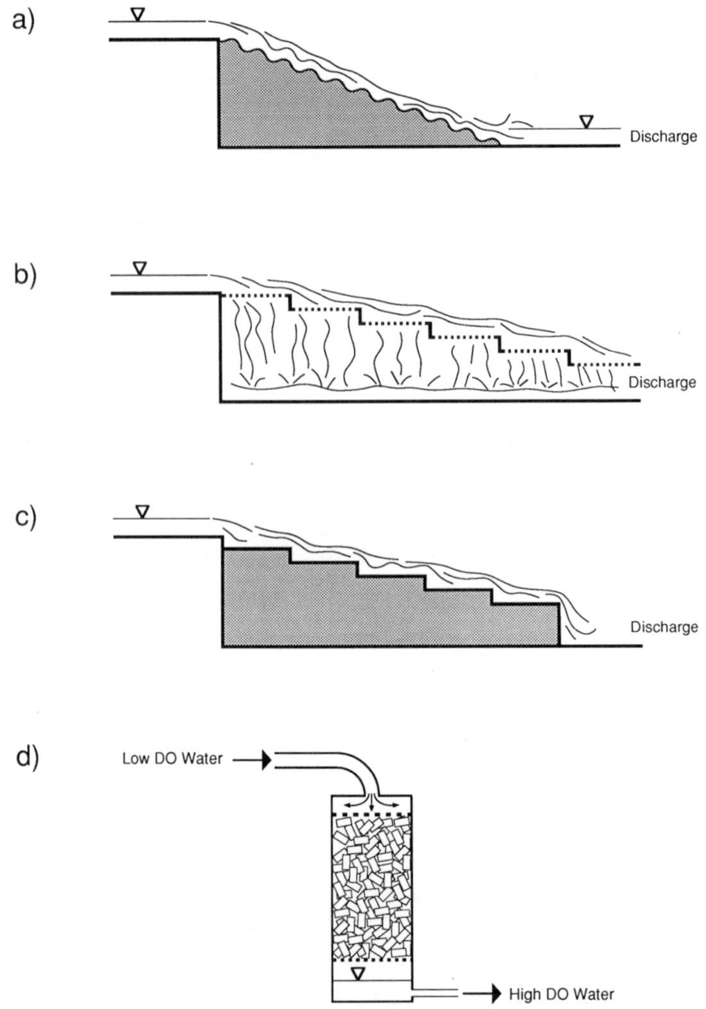

Figure 3. Typical gravity aerators (a) corrugated inclined plane (b) lattice, (c) cascade, (d) packed column, (continued next page) (e) spray column, (f) tray or screen

e)

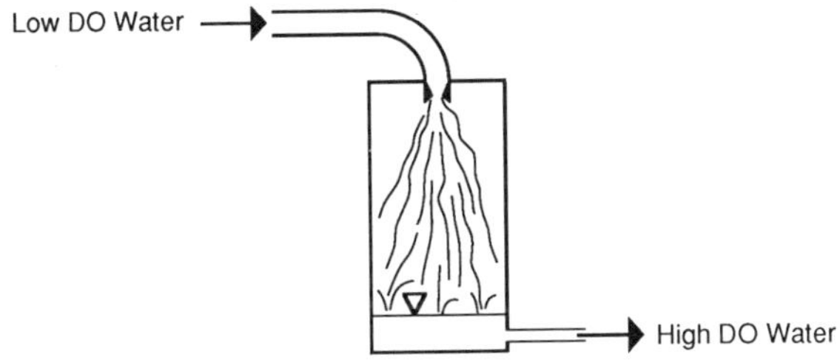

f)

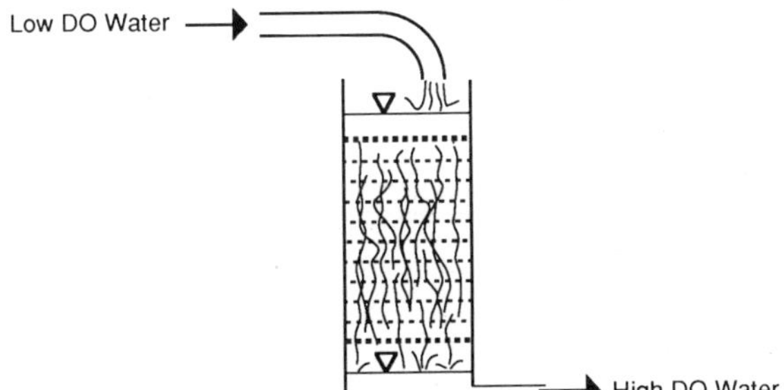

Pure oxygen aerators use enriched oxygen gas (85-98%) rather than air. With some modifications, almost all types of aerators can be used. The most commonly used types of aerators are the u-tube, packed column, spray column, aeration cone, and high pressure oxygen injection (Colt and Watten, 1988). Pure oxygen aeration using diffused aerators is commonly employed in the transport of aquatic animals. The u-tube aerator may be modified to recycle off-gas and increase the transfer efficiency (Fig. 4a). The high pressure oxygen injection is especially useful when there is adequate head between the water source and hatchery (Fig. 4b).

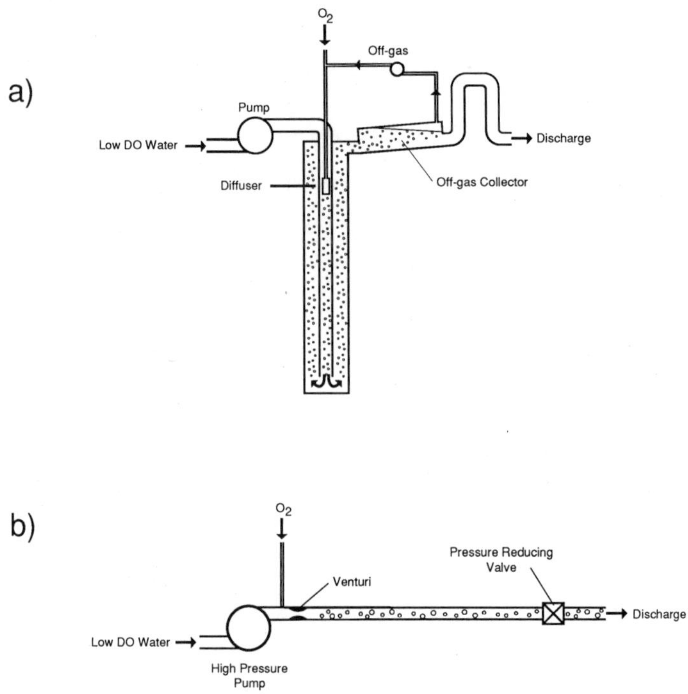

Figure 4. Pure oxygen systems (a) u-tube with recycle, (b) high pressure oxygen injection

System Configuration

Some of the common system configurations are presented in Figure 5. The single-pass (Fig. 5a) and series-reuse (Fig. 5c) configurations are the most common. The series-parallel configuration (Fig. 5d) can allow installation of aeration in the common connecting lines.

Aerator Configuration

Depending on the system configuration or operational limitations, aerators may be placed at different locations (Fig. 6). The aerators may be a)

b)

c)

d)

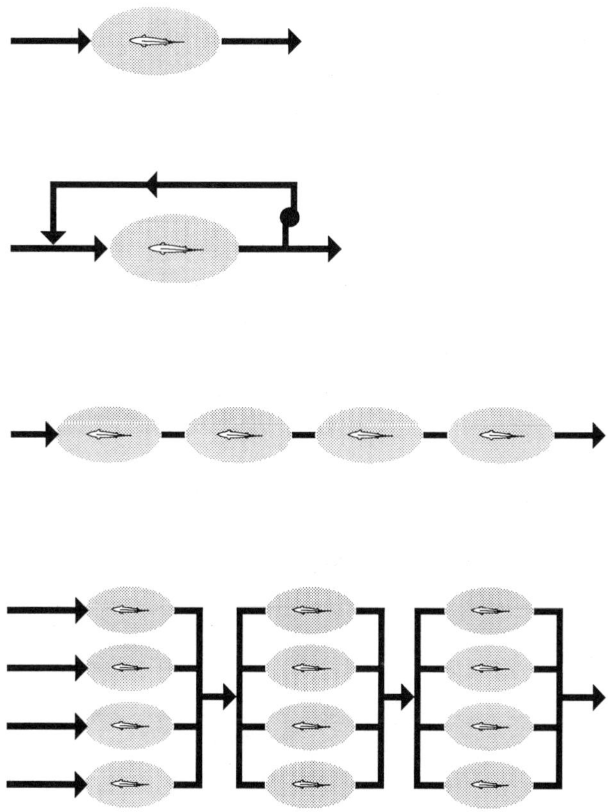

Figure 5. Configuration of rearing units (a) single-pass, (b) recycle, (c) series–reuse, and (d) series-parallel

located in the influent stream (Figs. 6a, 6b, and 6c), recycle stream (Figs. 6e and 6f), or rearing unit (Fig. 6d). In the recycle configuration (Figs. 6e and 6f), the recycle flow may be several times larger than the influent flow. In a side-stream system (Figs. 6b and 6c), only a part of the water flow passes through the aeration system.

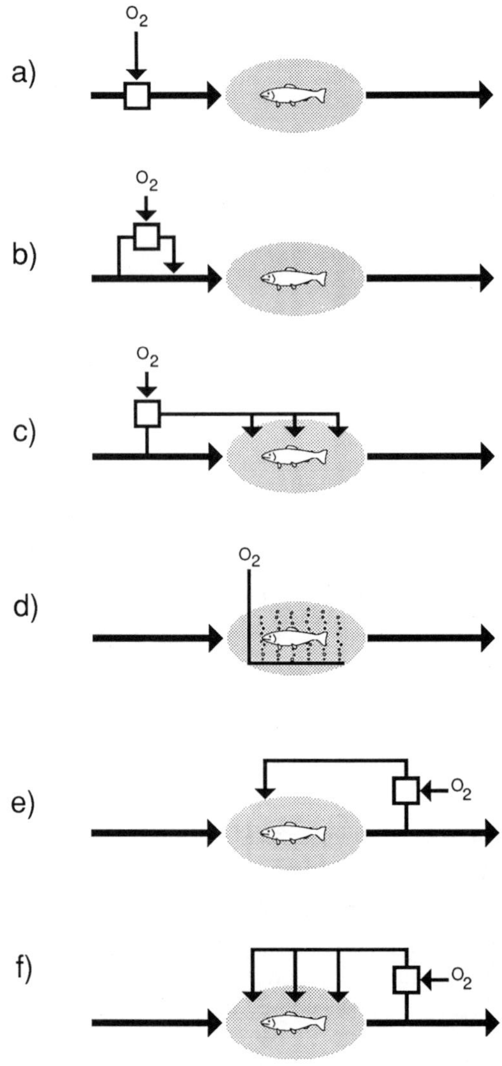

Figure 6. Location of aerators (a) influent, (b) influent, side-stream mode with single point return, (c) influent, side-stream mode with multi-point return, (d) in-unit, (e) recycle, single point return, and (f) recycle, multi-point return

Rearing Unit Hydraulics

The hydraulics of the individual rearing unit may be classified as a plug-flow reactor (PFR), continuous-flow stirred-tank reactor (CFSTR), or some combination (Colt and Tchobanoglous 1981). The standard rectangular raceway can be modeled as a PFR, with water moving through the system with little longitudinal mixing. Therefore, the DO has a significant longitudinal gradient. A circular tank can be modeled as a CFSTR. The dissolved gas concentrations of an ideal circular tank are well-mixed and equal to the concentration in the effluent. The hydraulics of the rearing units have a highly significant impact on the design and operation of aeration systems.

SOLUBILITY OF GASES IN WATER

The solubility of a gas depends on its temperature, salinity, gas composition, and total pressure as related by Henry's law (Colt, 1984):

$$C^{*} = 1000K\beta X(P - P_{w})/760 \quad (1)$$

K is a constant for each gas and P_w varies slightly with temperature and salinity, so gas solubility depends primarily on the Bunsen coefficient (β), mole fraction (X), and total pressure (P). Non-ideal gas corrections to Henry's law (Benson and Krause 1984; Weiss 1974) can be ignored for typical computations. The atmospheric solubility of a gas depends on both the solubility of the pure gas and the mole fraction of the gas. The solubility of the major atmospheric gases at 15 C in freshwater are presented in Table 1 as a function of pressure and gas composition.

The air solubility of gases at 1 atm (C^{*}_{760}) as a function of temperature is listed in standard references and may be adjusted to other barometric pressures by the following equation:

$$C^{*} = C^{*}_{760} \frac{(BP - P_{w})}{(760 - P_{w})} \quad (2)$$

Equation 2 is restricted to the solubility of air while Equation 1 can be used to compute the solubility of a mixture of arbitrary composition.

Table 1. Solubility of major atmospheric gases (mg/l) as a function of pressure and gas composition at 15 C (Benson and Krause 1984; Weiss 1970; 1974).

Pressure (atm)	Gas Composition	Gas Solubility (mg/l) Nitrogen	Oxygen	Argon	Carbon Dioxide
1	air	16.36	10.08	0.62	0.69[a]
1	pure gas	20.95	48.14	65.94	1992.53
2	air	33.00	20.32	1.24	1.38[a]
2	pure gas	42.26	97.02	133.01	3979.36

[a] Based on a moles fraction of 0.000350 for air.

Reporting of Dissolved Gas Levels

Dissolved gases are typically measured in mg/l, but may be reported in a variety of other units, including percent of saturation:

$$\text{Percent of saturation} = \left[\frac{C}{C^*}\right]100 \quad (3)$$

For gas supersaturation or oxygen toxicity considerations, it is necessary to express the gas concentration in terms of partial pressures, which is the pressure of an individual gas if it occupied the entire gas volume by itself. The partial pressure in the gas phase, based on Dalton's Law, is equal to (Colt, 1984):

$$PP^g = X(BP - P_w) \quad (4)$$

The partial pressure in the liquid phase is commonly called gas tension:

$$PP^l = \left[\frac{C}{\beta}\right]A \quad (5)$$

Gas supersaturation is described by total (dissolved) gas pressure (TGP):

$$\text{Total Gas Pressure} = PP^l_{N2} + PP^l_{O2} + PP^l_{CO2} + P_W \quad (6)$$

or

$$\Delta P = TGP - BP \qquad (7)$$

The ΔP can be measured directly with several types of membrane-diffusion instruments (Bouck 1982; D'Aoust and Clark 1980). The actual ΔP that an animal experiences is equal to the difference between the total gas pressure and the local pressure (barometric + hydrostatic pressures). This is the uncompensated ΔP and is equal to:

$$\Delta P_{uncomp} = \Delta P - \rho g Z \qquad (8)$$

Effects of Environmental Variables

Temperature, salinity, pressure, and mole fraction all have a significant effect on the solubility of gases. The computation of C^* from Equations 1 or 2 may not be straight forward due to temporal and spatial changes of these parameters within the aerator unit.

Temperature and Salinity

The solubility of a gas expressed in concentration units (mg/l) decreases with increasing temperature and salinity (Colt, 1984). For example, the atmospheric solubility of oxygen in freshwater decreases from 11.29 mg/l at 10 C to 7.56 mg/l at 30 C. At 35 ‰ salinity, oxygen solubility decreases from 9.02 mg/l at 10 C to 6.24 mg/l at 30 C.

Pressure

For surface aerators, gas transfer takes place at atmospheric pressure, and the local barometric pressure can be used in Equations 1 or 2. Increasing the total pressure increases the solubility of the all gases proportionally (Equation 1). The total pressure can be increased by pressurization of the aeration system, and for aerators with uniform and constant pressures (for example, pressurized packed columns), the resulting absolute pressure can be used directly in Equation 1.

In many aeration systems, hydrostatic head is used to increase the pressure at which gas transfer occurs. At 20 C, hydrostatic head is equal to 73.42 mm Hg/m submergence. The C^* corresponding to a depth Z is equal to:

$$C_z^* = \frac{C^*(BP + \rho g Z - P_w)}{(760 - P_w)} \tag{9}$$

A depth of approximately 10 m will double the solubility of a gas.

As a bubble rises to the surface, the ambient pressure experienced by the bubble decreases from P_{bot} at the bottom of the basin to the barometric pressure at the surface (Figure 7). Therefore, the C^* varies from C_z^* (Equation 9, $Z=Z_{bot}$) at the bottom of the basin to the normal atmospheric value C^* (Equation 9, Z=0) at the surface.

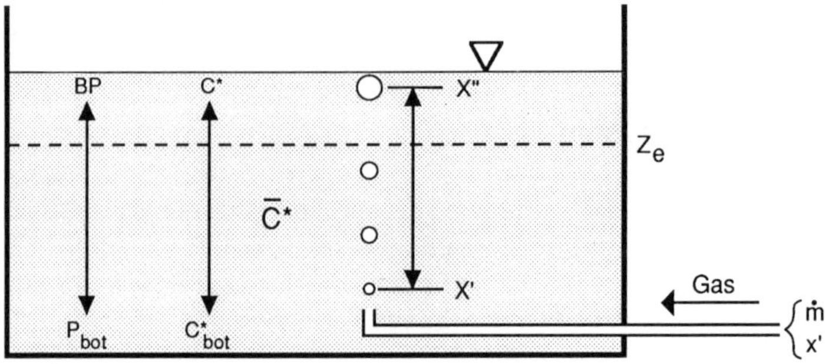

Figure 7. Definition sketch for $\overline{C}^*$

For submerged aerators, the effective average saturation concentration (Baillod, 1979) can be defined:

$$\overline{C}^* = \frac{\int_0^{Z_{bot}} C^* \, dZ}{Z_{bot}} \tag{10}$$

It is important to note that C^* in Equation 10 may be difficult to compute from either Equations 1 or 2 because the mole fraction of the rising bubble is changing due to the transfer of gases both into and out of the bubbles. The value of C^* after a long period of aeration ($\overline{C}^*$) can be measured directly.

Since dC/dt = 0, the value of $\overline{C}^*_\infty$ is reproducible for a given type of submerged aerator.

The equivalent aeration depth Z_e is found by equating the measured $\overline{C}^*_\infty$ to the right side of Equation 9 and solving for Z_e:

$$\overline{C}^*_\infty = \frac{C^*(BP + \rho g Z - P_w)}{(760 - P_w)} \qquad (11)$$

The equivalent aeration depth may be expressed as a percent of the total basin depth:

$$D_e = \frac{Z_e}{Z_{bot}}(100) \qquad (12)$$

Typically, D_e will range from 24 to 61 % of the total basin depth for different types of subsurface aerators (Colt and Westers 1982; Ewing et al. 1979). The value of D_e usually remains relatively constant over a wide range of depths for a given type of aerator.

Mole Fraction

In dry air, the mole fraction of nitrogen, oxygen, argon, and carbon dioxide is 0.78084, 0.20946, 0.00934 and 0.000350, respectively (Gleuckauf 1951; Machta 1983). The mole fraction of carbon dioxide in the atmosphere is increasing at an average rate of 1-1.5 ppm by volume/year (Machta 1983). If the mole fraction of oxygen is increased from 0.20946 (air) to 1.00 (pure oxygen) at a total pressure of 1 atmosphere, the solubility of oxygen is increased from 10.08 mg/l to 48.14 mg/l (Table 1).

Chemical Reaction of Dissolved Gases

Several important gases have significant chemical reactions in the liquid phase:

Carbon dioxide:

$$CO_2 + H_2O = H_2CO_3 = H^+ + HCO_3^- = 2H^+ + CO_3^{-2}$$

Hydrogen sulfide:

$$H_2S = H^+ + HS^- = 2H^+ + S^{-2}$$

Ammonia:

$$NH_3 + H^+ = NH_4^+$$

These reactions do not have any effect on the equilibrium concentration of gases in solution, as this depends only on Equation 1. However, these reactions have a critical impact on the rate of gas transfer and toxicity considerations.

The concentration of the "free" carbon dioxide gas (Stumm and Morgan 1981), for example, is equal to:

$$\text{"Free" } CO_2 = \alpha_g C_T \tag{13}$$

The value of α_g depends on pH and temperature. Therefore, the amount of gas that must be added or removed to result in a given concentration of "free" gas depends on pH and temperature.

DISSOLVED GAS CRITERIA IN INTENSIVE CULTURE

Dissolved gas criteria are needed in production conditions for dissolved oxygen, carbon dioxide, nitrogen + argon, and gas supersaturation. Most of the published criteria (USEPA 1976) are based on laboratory experiments and are formulated to protect a wide range of species and life-stages. Even the hatchery water quality criteria (for example, SECL 1983) are biased toward governmental facilities that typically have abundant water. Many public facilities and most private sector facilities routinely exceed published water quality criteria. Additional research is needed to define dissolved gas criteria for production conditions.

Dissolved Oxygen (low)

The critical DO concentrations for the incubation and hatching of salmonid eggs are highest just before hatching (Rombough 1986) and decrease after hatching. At high temperatures and elevations, the critical DO for eggs may be above ambient air saturation concentrations.

The growth of fish is not affected until the dissolved oxygen concentration drops below a critical concentration. This critical concentration is influenced by temperature and feeding level and ranges from 5 to 6 mg/l for salmon and trout and from 3 to 4 mg/l for warmwater fish such as channel catfish under laboratory conditions (Andrews et al. 1973; Brett 1979).

Under intensive culture conditions, DOs $\geq$ 6.5 - 7.0 mg/l may improve growth (Christensen 1989; Poston and Williams 1988). The use of oxygen supplementation can also increase survival and improve fish health and quality (Severson et al. 1987; Westers et al. 1987). Some of the beneficial effects of oxygen supplementation may be due to the stripping of chronic levels of gas supersaturation.

Dissolved Oxygen (High)

The maximum allowable DO level depends on several factors, including oxygen toxicity, physiological dysfunctions, and developmental problems. While oxygen is required for the survival of aerobic organisms such as fish, some of the by-products of oxygen metabolism are highly toxic and can overwhelm biochemical defense mechanisms. Colt et al. (1989) has suggested a preliminary criterion of 300 mm Hg. This corresponds to a DO_{max} = 21 mg/l at 12 C and 16 mg/l at 25 C. While aquatic animals appear to be able to tolerate oxygen pressures up to approximately 400 mm Hg, the effects of changes in oxygen pressure and the resulting changes in blood pH and ion transfer are poorly documented but may be an important consideration. High dissolved oxygen concentrations can have significant developmental effects on eggs and larvae (Colt et al. 1989). In general, oxygen levels up to approximately 500 mm Hg had little effect on hatching and larval development, but there was increased mortality of some species between 500-700 mm Hg. At DOs above approximately 550-700 mm Hg, the number of erythrocytes was reduced, and above 725 mm Hg, no erythrocytes were reported. Typically, the maximum dissolved oxygen concentration used in production systems is in the range of 15 - 20 mg/l (200 - 400 mm Hg).

Carbon Dioxide

High concentrations of carbon dioxide (hypercapnia) may reduce the capacity of the blood to transport oxygen and cause formation of calcareous deposits in the kidneys. Relatively high levels of carbon dioxide (20-40 mg/l) are required to significantly reduce oxygen transport (Basu 1959). The prevalence and severity of nephrocalcinosis depends primarily on the carbon dioxide concentration and to a lesser degree on the mineral composition of the diet (Smart et al. 1979). The growth and feed conversion ratios of rainbow trout reared at 12 and 24 mg/l of carbon dioxide were similar over a 275 day exposure. Fish reared at 55 mg/l of carbon dioxide showed reduced growth and higher feed conversion ratios, but it was only after about 330 days at this exposure that growth was seriously impaired (Smart et al. 1979). The accumulation of carbon dioxide can be a serious problem in the transport of fish (Haskell and Davies 1958).

Carbon dioxide concentrations should be maintained below 20 mg/l in culture systems for salmon and trout (SECL 1983). Specific criteria for warmwater species (especially under pond conditions) are lacking.

Nitrogen + Argon

Biologically, nitrogen and argon gases are inert and do not chemically react. Criteria for these two gases will be presented in the section on gas supersaturation.

Gas Supersaturation

The biological response to gas supersaturation at a given ΔP depends on species, life stage, water quality, and the animal's depth in the water column (pressure). Little comparative information is available for chronic exposure to gas supersaturation. Based on the limited available data and personal observations of several researchers, the following fish may be more sensitive to chronic gas supersaturation: Lake trout (*Salvelinus namaycush*), Atlantic salmon (*Salmo salar*), striped bass (*Morone saxatilis*), and mullet (*Mugil cephalus*). In general, eggs and sac fry are less sensitive to gas supersaturation than feeding fry (Colt et al. 1986).

At a given ΔP, the biological effects of gas supersaturation are reduced as the nitrogen/oxygen ratio is decreased (Nebeker et al. 1976). Tolerance to supersaturation is not altered by the addition of low levels of carbon dioxide

(Nebeker et al. 1976). The animal's position in the water column has a significant impact on the tolerance to gas supersaturation by reducing the ΔP_{uncomp} the animal experiences by 73.42 mm Hg/m at 20 C (Colt 1984). The estimation of potential protection by hydrostatic pressure depends on species behavior and available water depth.

If specific species/life stage information is not available, a hatchery criteria of 10 mm Hg is suggested (Colt et al. 1989). This criteria will protect the most sensitive life stages of the less tolerant species, but may be overly conservative for some species, older fish, and deep production systems. A higher criteria may be possible for pure oxygen systems with nitrogen/oxygen ratios less than the atmospheric value of 3.77 (based on partial pressure), but specific recommendations are not available at this time.

Interaction With Other Water Quality Parameters

The intensity of a rearing system has typically been described in terms of volumetric density (mass/volume) or loading (mass/flow). In high intensity flow-through systems, the total cumulative oxygen consumption (mg/l) may be a more important parameter as it is independent of both temperature and fish size (Meade 1988).

The utilization of oxygen produces both carbon dioxide and ammonia. The depletion of oxygen may not always be the most limiting parameter and when ammonia or carbon dioxide is more limiting, aeration will have have little effort on carrying capacity. Maximum cumulative oxygen consumption (COC) based on both carbon dioxide and ammonia limitations are presented in Figure 8 for water quality criteria typical of salmon and trout culture.

Assuming that the respiratory quotient of fish is equal to 1.0 (Kutty 1968), alkalinity = 1 meq/l, and no transfer of carbon dioxide occurs across the water surface, the cumulative oxygen consumption for carbon dioxide (COC_{CO2}) is limited to approximately 14-15 mg/l (Fig. 8) over pHs of 6.0-8.5, but increases at high pHs. Assuming a peaking factor of 1.44 for oxygen consumption and carbon dioxide production, COC_{CO2}s greater than approximately 9-10 mg/l will require carbon dioxide removal to maintain the carbon dioxide criteria during the period of maximum metabolic activity (Colt et al. 1989).

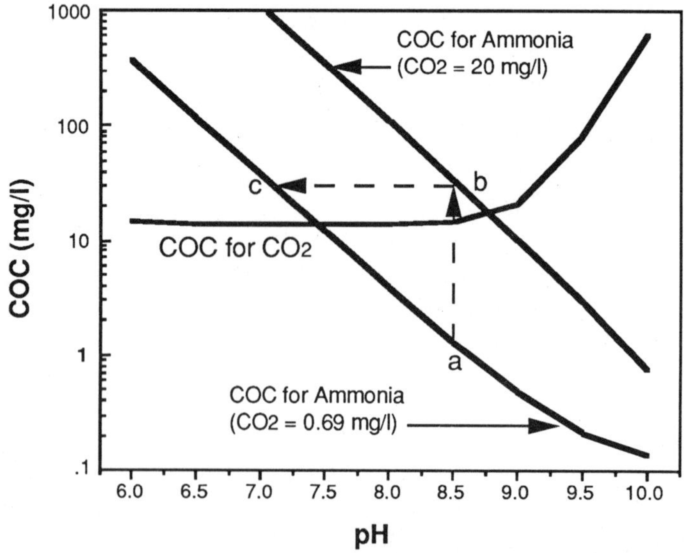

Figure 8. Maximum cumulative oxygen consumption (mg/l) as a function of pH. Separate curves are presented for ammonia, and carbon dioxide limitations (temperature = 15C, ammonia criteria = 16 µg/l NH_3-N, oxygen criteria = 6.0 mg/l, free carbon dioxide criteria = 20 mg/l, influent DO = 10.072 mg/l, influent un-ionized ammonia concentration = 0, respiratory quotient = 1.0, ammonia production = 30 g/kg feed, and oxygen consumption = 250 g/kg feed)

The maximum $COC_{ammonia}$ is presented in Figure 8 for waters in equilibrium with the atmosphere (350 parts per million carbon dioxide by volume) and waters that have accumulated 20 mg/l "free" carbon dioxide due to metabolic activities. The reduction in pH caused by the buildup of metabolic carbon dioxide significantly increases the allowable $COC_{ammonia}$. At pH 8.5, the maximum $COC_{ammonia}$ increases from 1.3 mg/l at equilibrium pH to 40 mg/l at the carbon dioxide criteria (point a to b in Fig. 8). The pH is also reduced from 8.50 to 7.04.(point b to c in Figure 8) Design of aquaculture systems for waters with pHs greater than 8.5 will require special care. Many groundwaters and surface water have a pH > 8.5, especially geothermal waters .

Many water quality criteria are strongly interrelated (Colt et al. 1989). This is especially critical in systems using pure oxygen due to the ability to maintain supersaturated dissolved oxygen levels and the potential effects of high dissolved oxygen levels on water quality criteria for both ammonia and carbon dioxide.

GAS TRANSFER

Slightly Soluble Gases

The rate at which a slightly soluble gas such as oxygen is transferred into water is proportional to the area of the gas-liquid interface and the gradient between the saturation and existing concentration of the gas in the water (Lewis and Whitman, 1924):

$$\frac{d\tilde{C}}{dt} = \tilde{K}_L \cdot \tilde{a}(\tilde{C}^* - \tilde{C}) \qquad (14)$$

A positive gradient ($\tilde{C}^* - \tilde{C}$) transfers gas into the liquid phase, and conversely, a negative gradient transfers gas into the gas phase. The transfer rate can be increased by increasing the value of $\tilde{K}_L$, $\tilde{a}$, or $\tilde{C}^*$. Pressurization of the entire system or use of hydrostatic pressure increases $\tilde{C}^*$. In a pure oxygen system, $\tilde{C}^*$ is additionally increased due to the higher mole fraction of oxygen in the gas phase.

The overall mass transfer rate (Baillod 1979) can be evaluated by integrating the local transfer rate over the volume V:

$$\int_V \frac{d\tilde{C}}{dt} dV = \int_V \tilde{K}_L \cdot \tilde{a}(\tilde{C}^* - \tilde{C}) dV \qquad (15)$$

For most aerators, it is impossible to determine $\tilde{K}_L$ and $\tilde{a}$ independently and the two parameters are lumped into an overall $K_L\,\tilde{a}$ parameter.

In general, it is assumed that $\tilde{K}_L \tilde{a}$ is constant over the whole volume of the aeration basin. For surface aerators, the $\tilde{C}^*$ equals the atmospheric equilibrium concentration. In submerged aeration systems, $\tilde{C}^*_\infty$ must be used. For other systems, both $\tilde{C}^*$ and $\tilde{C}$ may vary both temporally and spatially within the aeration device and depend on operating conditions.

Even though the value $\widetilde{C}^*$ may be constant in many systems, its computation may not be simple because of changes in pressure or mole fraction. Computation of C^* is difficult in pure oxygen systems where gas costs require high absorption efficiencies and the gas phase composition is much different than the influent gas composition. Computation is further complicated by simultaneous transfer of nitrogen, argon, and carbon dioxide gas into or out of the liquid phase (Colt and Westers 1982; Watten and Beck 1985; Watten et al. 1989).

Highly Soluble Gases

Highly soluble gases such as hydrogen sulfide, chlorine, or ammonia (carbon dioxide is intermediate in solubility) rarely present absorption problems, but may be very difficult to strip out of solution. For those gases with acid-base reactions, the value $\widetilde{C}$ of in Equations 14 and 15 must be the concentration of actual or "free" gas in solution (Howe and Lawler 1989).

Packed columns operated in the counter-current mode have been commonly used for gas stripping. The equilibrium between the gas-phase and the influent liquid limits the maximum gas-phase concentration. G/L ratios in the range of 100-500 may be needed for stripping highly soluble gas (Watten et al. 1989). These G/L ratios are much higher than those used for aeration (Hackney and Colt 1982) or degassing of nitrogen (Colt and Bouck 1984). pH adjustment may be necessary to increase liquid phase concentrations for ammonia or hydrogen sulfide stripping (Howe and Lawler 1989; Metcalf and Eddy 1979).

STANDARDIZED AERATOR TESTING UNDER CLEAN WATER CONDITIONS

Two types of testing procedures can be used to rate aerators: unsteady-state and steady-state procedures. Standardized testing and rating procedures for in-basin aerators have been developed for wastewater applications (Boyle 1979; Brown and Baillod 1982). While these standards are helpful, their use in aquaculture systems may not always be either convenient or valid.

Unsteady-State Testing

The unsteady-state testing procedures are conducted under standard conditions using an experimental test basin. Some test basins are as large as 3000 to 6000 m^3 and can be used to test aerators as large as 50 to 100 kW. Typically, a solution of sodium sulfite and cobalt chloride is used to deoxygenate the water by chemical oxidation. The aerator is then started and the dissolved oxygen is measured periodically until the saturation concentration is approached. This testing procedure is termed the unsteady-state test as the amount of oxygen transferred and the dissolved oxygen concentration are changing during the test.

Volumetric Mass Transfer Coefficient (K_La)

Fundamental to the rating of in-basin aerators is the computation of K_La (Equation 14). The K_La value is estimated from the change of dissolved oxygen with time. In exponential form, the integrated form of Equation 15 can be written as:

$$C = \overline{C}^*_\infty + (\overline{C}^*_\infty - C_0)\exp(-K_La t) \tag{16}$$

The recommended method of data analysis (Brown and Baillod 1982) requires a computer or advanced scientific calculator and uses a non-linear least squares procedure to estimate $\overline{C}^*_\infty$, C_0 , and K_La directly. The value of $\overline{C}^*_\infty$ is equal to the effective average saturation concentration when dC/dt = 0. The effective average saturation concentration is reduced at the start of the aeration test due to the gas phase depletion. As a result, the computed K_La value is an apparent value rather than the true value. For a discussion of differences between the two parameters, Brown and Baillod (1982) should be consulted. For surface aerators, the $\overline{C}^*_\infty$ value estimated from Equation 16 is very close to the atmospheric saturation value (C^*) after correction for temperature and barometric pressure. In the United States, all K_La values are reported at 20 C, 760 mm Hg, and tap water.

Standardized Oxygen Transfer Rate (SOTR)

The maximum rate of oxygen transfer occurs when the dissolved oxygen in the basin is equal to zero (Equation 14). This is called the standardized oxygen transfer rate (SOTR) and is equal to:

$$SOTR = (K_La)\,(\overline{C}^*_{\infty s})\,(V \times 10^{-3}) \qquad (17)$$

In older literature, this parameter is referred to as N_o. The value of $\overline{C}^*_{\infty s}$ can be computed from Equation 11 by setting $C^* = C^*_s$ (9.092 mg/l).

Standardized Aeration Efficiency (SAE or N_o)

The standardized aeration efficiency is equal to

$$SAE = \frac{SOTR}{\text{Power input in kW}} \qquad (18)$$

and is expressed in units of kg O_2/kW•hr. The power input should be the measured wire power, that is, actual power used by the (a) motor, drive, and blower or (b) motor, coupling, and gearbox (Yunt 1979). For aquaculture applications, the majority of units are supplied with an attached motor. The reporting of SAE based on only the power delivered by the blower or gearbox as is done by some manufacturers has no justification other than to increase SAE for marketing purposes.

Standardized Oxygen Transfer Efficiency (OTE_o)

The standardized oxygen transfer efficiency is equal to the oxygen transferred into the water divided by the mass flowrate of oxygen supplied to the aerator:

$$OTE_o = \frac{SOTR}{\dot{m}} \qquad (19)$$

SOTR, OTE_o, and SAE are reported for aerators and can be used to compare different types or brands of aerators. The standardized procedure for the determination of K_La reduces the uncertainty in aerator rating and allows more meaningful comparisons between units. Unfortunately, the standards for the rating of aerators changes from country to country. In Europe, the standard temperature used in the computation of SOTR and SAE is 10 C rather than the 20 C used in the United States. Additionally, in the United Kingdom, a commercial detergent is added to change the properties of tap water to more closely approach the properties of wastewater (Stenstrom and Gilbert 1981). The desirability of this modification for aquatic culture applications is unknown. The standard state for air flow and pure oxygen measurements in the United States is 20 C, 1 atm, and 36% relative humidity (1.205 kg/m^3) and 20 C, 1 atm, and 0 relative humidity (1.331 kg/m^3), respectively (Colt and Watten 1988; Yunt 1979). The standard state for air flow measurements may also be different in other parts of the world (Fisette 1979) and may introduce a potential error as large as 18% when comparing aeration units.

Steady-state Testing

For a number of gravity aerators such as packed columns, both the input and effluent oxygen concentrations can be directly measured. For these types of aerators, the field oxygen transfer rate, field aeration efficiency, and field oxygen transfer efficiency can be directly computed.

Field Oxygen Transfer Rate (OTR_f)

$$OTR_f = 3.6 Q_w (C_{out} - C_{in}) \tag{20}$$

Field Aeration Efficiency (FAE or N)

$$FAE = \frac{OTR_f}{\text{Power input in kW}} \tag{21}$$

Field Oxygen Transfer Efficiency (OTE_f)

$$OTE_f = \frac{OTR_f}{\dot{m}} \tag{22}$$

The values of the field parameters under standard conditions can be computed using the equations presented in the "Performance and Rating of Aeration Systems Under Field Conditions" section.

Aerator Effectiveness (AF)

For some types of atmospheric gravity aerators, $\dot{m}$ can not be measured and the aeration effectiveness (Chesness and Stephens, 1971) has been used as a rating parameter:

$$AF = \left[\frac{C_{out} - C_{in}}{C^{*} - C_{in}}\right] \qquad (23)$$

Testing of Pure Oxygen Units

Presently, there are no standards for the reporting of transfer information for pure oxygen systems. Generally, only FAE and OTR_f are reported for pure oxygen systems under a specific set of conditions. While the SOTR can be computed from Equation 17, the value of C^* used in this equation can vary by a factor of 2-4 by changing the system pressure and oxygen mole fraction. Therefore, comparison of SOTR between different pure oxygen units is meaningless except for marketing purposes.

The FAE values for pure oxygen systems should include only the wire power for the aerator and exclude the power for oxygen generators (if used). Inclusion of the power for the oxygen generators will bias the comparison between systems using liquid oxygen and on-site oxygen generation. An additional FAE value can be presented for the total system power input (pumping + off-gas recycle pumping + oxygen generator compressors), if desired. For some types of high pressure bubble generators, significant oxygen transfer may occur in the rearing units because of the very fine bubbles produced. Therefore, it is important to report both the performance of the contact system and the overall performance of the contact + rearing system. Other testing techniques such as the tracer or off gas methods (Mueller and Boyd 1988) may be more appropriate for pure oxygen systems.

PERFORMANCE AND RATING OF AERATION SYSTEMS UNDER FIELD CONDITIONS

Values of SOTR, SAE, and OTE_o, are computed for 20 C, 760 mm Hg, tap water, and zero dissolved oxygen and therefore can not be used directly for design of aquatic culture systems. The actual performance under field conditions depends primarily on the required dissolved oxygen concentration (C) and to a lesser extent on the temperature, pressure, and water characteristics. The rating of aerators under field conditions requires the computation of the field oxygen transfer rate (OTR_f), field aeration efficiency (FAE), and field oxygen transfer efficiency (OTE_f).

The computation of these parameters assumes that the field installation is identical to the unit tested under standard conditions. This includes not only the size, type, and physical characteristics of the aerators, but the number and orientation of the units and depth of the system. In general, changing one or more of these parameters will greatly decrease the accuracy of the estimated field parameters. Many submerged aerators are tested in 4-5 m water depths typical of aeration basins used in wastewater applications. Therefore, many published SOTR, SAE, and OTE_0 will significantly over-estimate field values for 1-2 m water depths common in aquaculture.

Characteristics of Culture Water

The values of α', β', and Θ have a significant effect on the oxygen transfer rate under field conditions. Drastic over or under-design can result if inaccurate values are used (Stenstrom and Gilbert 1981).

Alpha (α')

The effects of water characteristics on oxygen transfer are corrected for by the alpha factor:

$$\alpha' = \frac{K_La \text{ in field conditions}}{K_La \text{ in clean water conditions}} \tag{24}$$

The value of α' depends primarily on the concentration of surfactants in the water. Surfactants reduce the surface tension of the water and decrease bubble size in diffused aeration. Reduced gas transfer occurs because of reduced surface renewal around the slowly rising bubble and the formation of adsorbed surfactant film on the bubble (Stenstrom and Gilbert 1981). Laboratory-scale systems are typically used to determine α'. Because α' depends on both the level of turbulence and type of aerator used, the test system should resemble the full-scale system as much as possible.

In production catfish ponds, alpha ranged from 0.66 to 1.07 and averaged 0.94 (Shelton and Boyd 1983). In recycle systems, alpha values as low as 0.36 have been measured following feeding (Weaver 1981). The depression of alpha appears to depend on the leaching of soluble compounds from feed or from compounds produced by algae. For a given water, a lower α' may be produced in packed columns as compared to either diffused or mechanical surface aerators (Weaver 1981).

Salinity has a major effect on alpha that can not be explained by conventional aeration theory. The transfer of both oxygen and air increases in seawater (Bouck and King 1983; Kils 1976/1977). This appears to be due to a decrease in the bubble size. The magnitude of this effect depends on the particular type of aerator used. At 40 ‰ salinity, the SAE of a rotating disc turbine was 10.5 times larger than in freshwater. For a venturi and diffused aerator, the seawater values were increased 2.9 and 1.8 times, respectively (Kils 1976/1977). Due to the large and variable effect of salinity, it may be necessary to determine the SOTR or SAE at the salinity of interest.

Beta (β')

The effects of water characteristics on oxygen solubility are corrected for by the beta factor:

$$\beta' = \frac{C_s^* \text{ in field condition}}{C_s^* \text{ in clean water conditions}} \qquad (25)$$

The beta factor is influenced primarily by dissolved solids and to a lesser extent by dissolved organics and suspended solids. In wastewater, beta values typically range from 0.95 to 1.00 (Metcalf and Eddy, Inc., 1979; Stenstrom and Gilbert, 1981). Beta values for aquaculture conditions are unavailable. It should be noted that beta can not be measured with a dissolved oxygen probe since this instrument measures the activity of oxygen rather than the concentration (Stenstrom and Gilbert, 1981). At a given temperature and pressure, the output of a dissolved oxygen probe will read a constant value in saturated water regardless of the salinity or concentration of dissolved organics and suspended solids. Until further information is available, a value of 0.95 can be used for ponds and recycle systems and 1.00 for flow-through systems.

Theta (Θ)

The theta factor is used to correct K_La for changes in viscosity, surface tension, and diffusion constants as a function of temperature (Stenstrom and Gilbert, 1981). The temperature variation of K_La is:

$$K_La(T) = K_La(20\ C)\Theta^{(T - 20\ C)} \quad (26)$$

where 20 C is the standard temperature and T is the field temperature. A value of 1.024 is recommended in freshwater, although values ranging from 1.008 to 1.047 can be found in the literature. Theta values in seawater appear to be larger than 1.024, especially for temperatures greater than 20 C (Kils 1976/1977).

Field Oxygen Transfer Rate (OTR_f)

The field oxygen transfer rate (OTR_f) is equal to:

$$OTR_f = \frac{\alpha'(SOTR)\,\Theta^{(T-20)}(\beta'\overline{C}^*_{\infty f} - C)}{\overline{C}^*_{\infty s}} \quad (27)$$

The computation of OTR_f requires (1) SOTR, (2) α' and β' for the particular water condition, and (3) local temperature. For surface aerators, it is generally assumed that $\overline{C}^*_{\infty f}$ is equal to the atmospheric saturation concentration and $\overline{C}^*_{\infty s}$ = 9.092 mg/l. For submerged aerators, the values of $\overline{C}^*_{\infty s}$ and $\overline{C}^*_{\infty f}$ can be computed from Equation 11 (for $\overline{C}^*_{\infty s}$, C^* equals $\overline{C}^*_s$ (9.092 mg/l) and for $\overline{C}^*_{\infty f}$, C^* is evaluated at the actual field temperature and pressure). The computation of $\overline{C}^*_{\infty s}$ and $\overline{C}^*_{\infty f}$ requires either Z_e (Equation 11) or D_e (Equation 12) and Z_{bot}. When $\alpha' = \beta' = 1$, temperature = 20 C, $\overline{C}^*_{\infty f}$ = 0.0, and $\overline{C}^*_{\infty s}$ = , Equation 27 reduces to SOTR .

Since gravity aerators and other in-line aerators are tested using steady-state methods, OTR_f is measured directly and Equation 27 can be used to compute SOTR. For atmospheric gravity aerators at one atmosphere pressure (760 mm Hg), C_{out} is limited to 9.092 mg/l (C^*_s) and the maximum SOTR can not exceed:

$$SOTR \leq 3.6(9.092)Q_w \tag{28}$$

Equation 28 assumes that the G/L ratio is high enough to prevent gas phase oxygen depletion. But if gas phase oxygen depletion occurs, the performance of the aerators will be reduced.

Field Aeration Efficiency (FAE)

The field aeration efficiency (FAE) is equal to:

$$FAE = \frac{\alpha'(SOTR)\,\Theta^{(T-20)}(\beta'\overline{C}^*_{\infty f} - C)}{\overline{C}^*_{\infty 6}} \tag{29}$$

In older literature, this parameter is referred to as N. Procedures for computing $\overline{C}^*_{\infty}$ and $\overline{C}^*_{\infty 6}$ were discussed in the last section (Field Oxygen Transfer Rate). The impact of a change in a given parameter on the field aeration efficiency is summarized in Table 2. The overall resulting change in FAE due to simultaneous changes in a number of parameters will have to be computed directly.

SAE for an atmospheric pressure gravity aerator can not exceed:

$$SAE \leq \frac{3.6(9.092)Q_w}{\text{Power input in kW}} \tag{30}$$

Field Oxygen Transfer Efficiency (OTE_f)

Once the OTR_f is computed from Equation 27, the OTE_f can be computed from Equation 22.

Table 2. Effects of various parameters on field aeration efficiency (FAE)

Parameter	Comment
An Increase in these Parameters Decreases FAE	
Temperature	If the desired dissolved oxygen concentration is specified in an absolute manner (mg/l)
C	This parameter has the greatest impact on FAE
An increase in these Parameters Increases FAE	
Temperature	If the desired dissolved oxygen concentration is specified as a percent of saturation
α'	Effects of a given level of surfactant may vary between different types of aerators
β'	May not be significant parameter in aquaculture systems
$\overline{C}^{*}_{\infty f}$	May be increased by selection of an aerator with high $\overline{C}^{*}_{\infty s}$ or increasing the depth of the aeration basin
BP	Increases as elevation increases
SAE	Specified by manufacturer, but methods of computation may vary
P	Increasing pressure increases $\overline{C}^{*}_{\infty f}$ in air systems. Pressure is limited by gas supersaturation problems
X	Increasing mole fraction of pure oxygen in pure oxygen system increases $\overline{C}^{*}_{\infty f}$

Field Aerator Effectiveness

If an aerator's performance is described in terms of aerator effectiveness (Equation 23), the effluent DO as a function of influent DO can be computed from:

$$C_{out} = C_{in} + \frac{(C^{*} - C_{in})(AF)}{100}\,\Theta^{(T-20)} \qquad (31)$$

Equation 31 assumes that the reported AF value was corrected to a base temperature of 20 C.

Pure Oxygen Aerators

As discussed in the previous section, there are no testing standards for pure oxygen systems and therefore it is difficult to specify either SOTR or SAE. In addition, the value C_s^* of is not a constant and depends on C. The functional relationships between the C and other performance parameters are shown in Figure 9 for a pure oxygen packed column; the relationships may be different for other pure oxygen systems. It is important to note that increasing the required effluent DO (C) requires an increase in the C^* concentration (Fig. 9a) and actually increases the OTE_f (Fig. 9b). This is at the expense of increasing the G/L ratio (Fig. 9c) and variable costs (Fig. 9d). The required G/L depends on both the influent and effluent gas concentrations and the physical characteristics of the system; therefore, the design will be very site-specific.

Rating of pure oxygen systems will require detailed system performance data such as those shown in Figure 9. Due to the large number of different conditions that may be experienced, it may be more convenient to work directly with a multi-component mass transfer model of the specific aeration process (Colt and Watten 1988; Nirmalakhandan et al. 1988; Watten and Beck 1986).

PROCESS SELECTION AND DESIGN

A wide range of aeration devices are available for aquaculture. The actual type selected will depend on a variety of factors related to the characteristics of the aerator, the culture system, site conditions, and system operations. It should be noted that aerator selection may require serious trade-offs between some of these parameters. Many of these tradeoffs may be difficult to quantify, especially if the long-term objectives of the culture system are not well-defined or the oxygen demand of the system changes significantly over the production cycle.

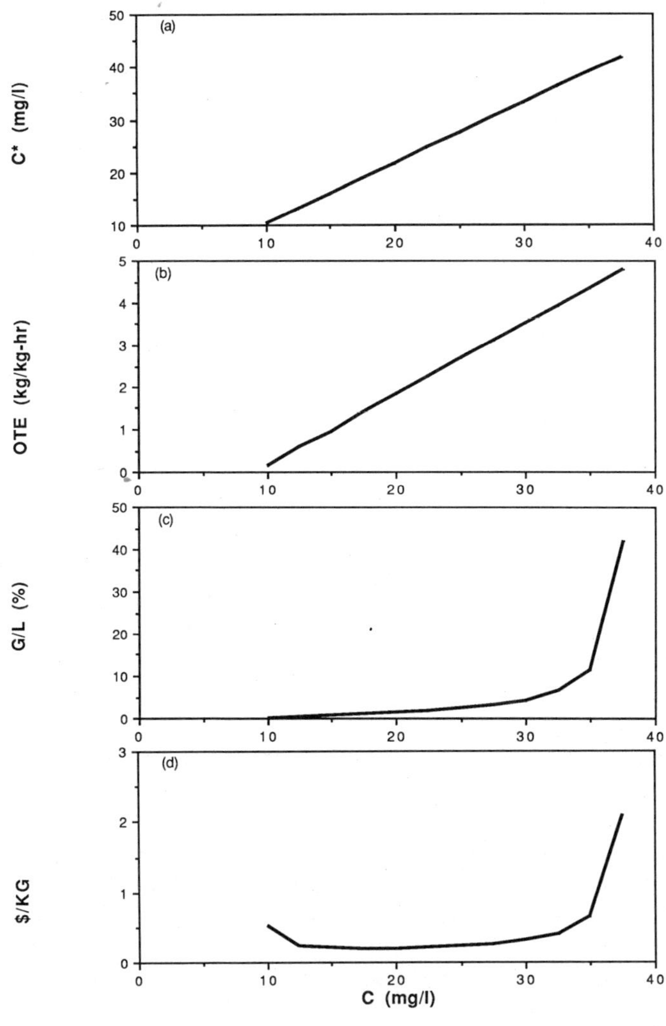

Figure 9. Impact of C in a pure oxygen packed column on (a) C^*, (b) oxygen transfer efficiency, (c) gas-to-liquid ratio, and (d) variable costs. Based on the model developed by Watten et al., 1989. (temperature = 20 C, barometric pressure = 760 mm Hg, influent DO = DN = DC = 100%, column height = 1.0 m, 1 1/2 inch pall rings, oxygen purity = 99%, $\alpha'=\beta'=1$, oxygen cost = $0.40 per 100 standard cubic feet, power costs = $0.06 per kW•hr, additional head losses in system = 0.30 m , and pump efficiency = 60%)

Efficiency

The selection of an aerator will depend strongly on the efficiency of the unit, especially for aerators designed to operate over a significant part of the production cycle. The efficiency can be measured by the field aeration efficiency (FAE) or the aeration effectiveness (AF). The second parameter is more important in systems using gravity aerators that do not require pumping.

Field Aeration Efficiency

Standardized aeration efficiency (SAE) for aerators commonly used in aquatic systems are listed in Table 3. The SAE will typically range from 1.0 to 2.6 kg O_2/kW•hr. The SAE of some types of subsurface aerators may range as high as 3.2 to 3.5 kg O_2/kW•hr. If 0.5 to 1.0 m of head is available, the SAE of the U-tube and packed column aerators can range as high as 40 - 80 kg O_2/kW•hr. The only power required is for injection of air into the u-tube or low pressure fans for the packed columns. The SAE values in Table 3 are for standard conditions and the field aeration efficiency (FAE) must be computed from Equation 29. The FAE values for aquaculture systems will be significantly less than the listed SAE values due primarily to the necessity of maintaining a dissolved oxygen concentration of 5 to 7 mg/l. For a given location and water supply, only SAE, C, and $\overline{C}^*_{\infty f}$ can be specified. The selection of an aerator with a high SAE will result in a high FAE value. For typical aeration systems, increasing the C value significantly decreases the FAE value. The maintenance of very high C values may be more economic with pure oxygen systems. High values of $\overline{C}^*_{\infty f}$ can be achieved by selecting an aerator with a high $\overline{C}^*_{\infty s}$ value or by the use of deep aeration basins to increase Z_e. In submerged aeration systems using air, aerators with high SAE and $\overline{C}^*_{\infty f}$ may result in lethal dissolved gas levels and gas bubble disease (Colt 1986) and therefore limit the use of many highly efficient subsurface aerators in aquatic systems (Colt and Westers 1982).

Field Aeration Effectiveness

Most gravity aerators and some submerged aerators can be designed to operate with no power input if 1.0 m or more head is available. Typical values

Table 3. Typical standardized aerator efficiency (SAE) and equivalent aeration depths (D_e) for aerators used in aquaculture.

Type	SAE ($kg\ O_2/kW\cdot hr$)	De (%)	Reference
	SURFACE AERATORS		
low-speed surface	1.2 - 2.4	0	Metcalf and Eddy, Inc. 1979
Low-speed surface with draft tube	1.2 - 2.4	0	
High-speed surface	1.2 - 2.4	0	
Paddlewheel			
Triangular blades	2.7 - 2.9	0	Ahmad and Boyd 1988
PVC pipe blades	1.2 - 1.9	0	Boyd et al. 1988
Tractor powered	1.3 - 2.0	0	Busch et al. 1984
	GRAVITY AERATORS		
Cascade weir (45°)	1.5 - 1.8	0	Tebbutt 1972
Corrugated inclined plane (20°)	1.0 - 1.9	0	Chesness et al. 1971;1972
Horizontal screens	1.2 - 2.6	0	Hartman 1983
Lattice aerator	1.8 - 2.6	0	Chesness et al. 1971;1972
Packed column			
Zero head	1.2 - 2.4	0	Hackney and Colt 1982
0.5 - 1.0 m head	10 - 80[a]	0	
	SUBSURFACE AERATORS		
Aeration cone	2.5	—	Speece et al 1971
Air-lift pump	2.0 - 2.1	30[b]	Nagy 1979;Cornacchia and Colt 1984; Reinemann and Timmons 1989
Diffused-air			
fine-bubble	1.2 - 2.0	36[b]	Metcalf and Eddy Inc. 1979; Colt and Westers 1982
medium-bubble	1.0 - 1.6	32[b]	Metcalf and Eddy 1979
coarse-bubble	0.6 - 1.2	28[b]	
Nozzle aerator	1.3 - 2.6	—	Chesness et al. 1973; Dijkstra et al. 1979
Propeller-aspirator-pump	1.7 - 1.9	—	Boyd and Martinson 1984
Static tube	1.8 - 2.4	29[b]	Ball and Campbell 1974; Colt and Wester 1982
U-tube			
zero head	0.72 - 2.3	—	Speece 1970; Speece et al. 1969
0.5 - 1.0 m head	10 - 40[a]	—	
Venturi aerator	2.0 - 3.3	51[b]	Toerber and Mandt 1979; Colt and Westers 1982

[a] Does not include pumping power

[b] Only an estimate

of the aeration effectiveness (AF) are presented in Table 4. Many gravity aerators have very low SAE and OTE_o values but may be useful in some applications due to their simplicity of construction and operation.

Table 4. Aeration effectiveness (AF) of typical gravity aerators.

Type Height (cm)	AF(%)	Reference
	Low-head types	
Cascade (45°)		
25	22 - 26	Tebbutt 1972
50	36 - 38	
Corrugated inclined plane (20°)		
30	18 - 29	Chesness and Stephens 1971
60	30 - 50	
Horizontal perforated trays		
110	95 - 100	Strasburg 1964
Lattice aerator		
30	29 - 37	Cheness and Stephens 1971
60	48 - 61	
Simple weir		
30	7 - 10	Haskell et al 1960
Splash board		
30	23 - 25	Chesness and Stephens 1971
60	36 - 41	
Packed Column		
30	94 - 96	Hackney and Colt 1982
60	96 - 98	
	High-Head Types[a]	
Alfalfa gate	61	Moore and Boyd 1984
Ell aspirator	83	
Gate vavle (half-open)	76	
Screen	59	
Screen cover with rocks	63	
Screen cover	52	
Screen extension	51	
Slotted cap	65	
Splashboard	51	
Splashboard with holes	53	
Straight pipe	25	
Tee aspirator	72	

[a] Pressure head ranged from 9.9 to 17 cm.

Field Oxygen Transfer Rate

In a number of systems, oxygen transfer rate is more important than efficiency. Tractor-powered paddlewheel aerators have been widely used in catfish ponds for emergency aeration (Boyd and Tucker 1979) and can easily be moved from pond to pond when needed. Diffused aeration with pure oxygen is widely used in transport systems (Carmichael and Tomasso 1988) and emergency systems for high intensity systems due to its ability to transfer large amounts of oxygen without any power input.

Utilization Efficiency

The utilization efficiency is the percent of the total oxygen passing through the system that is used by the aquatic animal. For a single pass system without in–basin aeration, the utilization efficiency is equal to:

$$\text{Utilization efficiency (\%)} = \left[\frac{DO_{in} - DO_{out}}{DO_{in}}\right]100 \qquad (32)$$

The utilization efficiency in a non-aerated system will often range from 20 - 40%, depending primarily on temperature and elevation.

In flow-through systems using aeration, the utilization efficiency is equal to:

$$\text{Utilization efficiency(\%)} = \left[\frac{(Q_w)\,DO_{in} + OTR_f - Q_w\,DO_{out}}{(Q_w)\,DO_{in} + OTR_f}\right]100 \qquad (33)$$

The utilization efficiency in aerated systems ranges from 50 -70 % and depends primarily on the layout of the rearing units. In high intensity systems using pure oxygen, the utilization efficiency may approach 95 %.

Dissolved Gas Concentrations and Pressures

The dissolved gas concentrations (or pressures) in the effluent from the aeration system are critical parameters in the design. These parameters include oxygen, carbon dioxide, and gas supersaturation.

Oxygen

The dissolved oxygen available to the animals is equal to $DO_{in} - DO_{min}$; therefore, increasing DO_{in} increases the carrying capacity of the rearing unit. High DO_{in} also reduces the size of the absorber unit, pumps, and piping.

If a high DO_{in} is needed or the site water temperature and elevation are high, the resulting dissolved oxygen gradient ($C^* - C$) and SAE are low. Under these conditions, a pure oxygen system may be needed.

Carbon Dioxide

Because carbon dioxide is highly soluble in water, the carbon dioxide removal characteristics of an aerator is a strong function of the gas flow or G/L ratio. Surface or gravity aerators have the best performance characteristics for carbon dioxide removal. The carbon dioxide stripping effectiveness for gravity and end-of-pipe aerators is in the range of only 2-13% (Moore and Boyd 1984). Pure oxygen systems typically remove a minimal amount of carbon dioxide (Watten et al. 1989). The majority of fish transport systems use both pure oxygen diffused aerators and surface aerators for gas transfer and mixing (Carmichael and Tomasso 1988).

Gas Supersaturation

A variety of physical and biological processes produce gas supersaturation (Colt 1986). High levels of gas supersaturation may be present in surface and ground waters or produced within the hatchery due to poor design of water treatment processes or operational problems. Submerged aeration systems using air or pure oxygen aeration systems operated at high absorption efficiencies can also produce high levels of gas supersaturation. The characteristics of an aeration system to remove gas supersaturation may be described in terms of the change in ΔP or the oxygen:nitrogen stripping ratio:

$$DO/DN = \Delta DO/\Delta DN \qquad (34)$$

Note that the addition of oxygen and stripping of nitrogen produces a negative DO/DN value.

The most common type of degassing or stripping system for influent water is the packed column (Colt and Bouck 1984; McLean and Boreham 1980; Owsley 1981) and can reliably reduce ΔPs to the range of 20-30 mm Hg. To reduce the ΔP below this level either vacuum stripping (Fuss 1983) or pure oxygen systems are needed (Severson et al. 1987; Westers et al. 1987). Due to problems with oxygen loss in vacuum stripping columns, waters low in dissolved oxygen should be first pre-aerated in a packed column to increase the DO (Marking et al. 1983).

Gas supersaturation will be more of a problem in static systems or serial reuse systems where the concentration of nitrogen and argon can approach $\overline{C}^*_{\infty f}$ (Colt and Westers 1982). When the saturation concentrations of all the gases are equal, the ΔP is equal to

$$\Delta P = BP\left[\frac{\overline{C}^*_{\infty f}}{C^*} - 1\right] \quad (35)$$

For example, a venturi aerator at 4 m (BP = 760 mm Hg, temperature = 20 C) will have a $\overline{C}^*_{\infty f}$ equaling 10.91 mg/l or 120 % of the surface concentration. The ΔP computed from Equation 35 will be equal to ΔP = 152 mm Hg, an acutely lethal value. In larval culture systems, a diffused aerator or air-lift pump with a diffuser submerged greater than 1 to 2 m can be lethal to sensitive life stages (Cornacchia and Colt 1984).

In pure oxygen systems, the allowable effluent DO (or ΔDO) determines the required DO/DN ratio (Colt and Watten 1988; Watten et al. 1989). The DO/DN ratio is a function of the physical design, oxygen mole fraction (or G/L ratio), and system pressure. Control of effluent ΔP may limit the absorption efficiency and increase the G/L ratio, especially for raceways.

Computation of Oxygen Demand and Supplemental Requirements

Sizing of aeration systems requires an estimate of the total oxygen demand by the aquatic animals and other organisms, the available oxygen supplied by water flow, and the supplemental oxygen requirement. Since the total oxygen demand of the fish depends on the number, size, and temperature, it is necessary to estimate these parameters over the whole production cycle on a weekly or monthly basis. Temperatures should be based both on average, extreme maximum, and extreme minimum values.

Average Daily Oxygen Demand

The average daily oxygen demand (Haskell 1955; Willoughby 1968) is proportional to the total ration:

$$\text{Average Daily Oxygen Demand (kg/d)} = (\text{OFR})R \qquad (36)$$

The oxygen requirement to process a given mass of feed depends on animal size, feeding rate, composition of the ration, digestibility of the feed components, and moisture content and can be described by the oxygen:feed ratio (OFR).

The oxygen/feed ratio at maintenance (no growth) is equal to:

$$\text{OFR} = \frac{M_e}{Q_{ox}} \qquad (37)$$

Assuming a value of 3.24 kcal/g oxygen for the Q_{ox} (Brett and Groves 1979), a feed containing 3.67 kcal/g feed of metabolizable energy would require 1.13 kg oxygen/kg dry feed or 0.93 kg oxygen/kg wet feed at 17.75% moisture for a maintenance feeding level. At a given fish size, the OFR decreases as the ration size increases (Fig. 10). For a single growing animal, the OFR appears to be relatively constant for a given feeding regime, even though the feeding level as a percent of body weight decreases (Hogendoorn 1983). As animal weight increases, the OFR vs feeding level curve (Fig. 10) will be shifted to the left.

It is important to differentiate between a laboratory OFR for a species and the design value (OFR_d) for an overall system. The actual design OFR_d will depend strongly on the type of system, other sources of oxygen, operational efficiency, and other organisms present in the system. Because of the critical impact of this parameter on system design, better documentation on the variation of OFR as a function of species, feed type, system configuration, and temperature is needed.

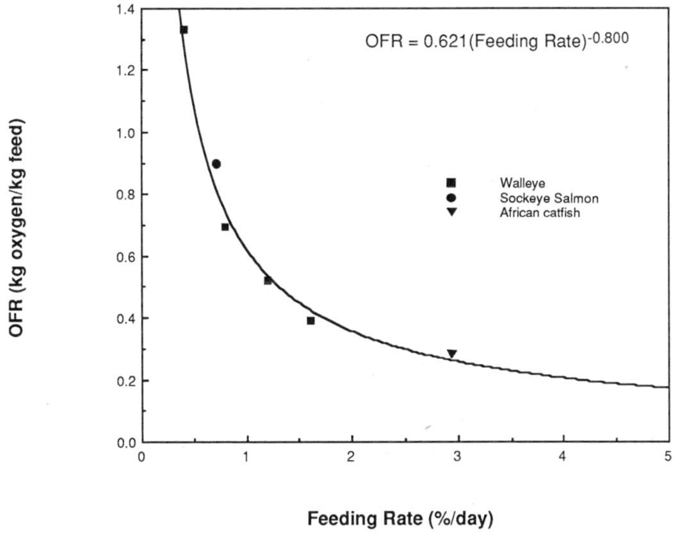

Figure 10. Variation of oxygen feed ratio (OFR) with ration size for fish in the 20-40 g size (based on Beamish and MacMahon 1988; Brett and Zala,1975; Hogendoorn, 1983)

Flow-through systems. This type of system is characterized by high flowrate and minimal algal and bacterial activity within the rearing units. In production salmon and trout systems, OFR_ds ranging from 0.20 to 0.22 kg oxygen/kg wet feed have been reported (Westers 1981; Willoughby 1968).

High intensity pure oxygen systems. This type of system is characterized by a low flowrate and a high density of animals. In commercial high density warmwater fish culture, a value of OFR_d = 1.00 kg oxygen/kg wet feed is commonly used (Anthonie Schuur, personal communication). The higher value of OFR_d for warmwater fish may be due to higher levels of M_e in the feed, lower moisture levels in the feed, lower reaeration across the water surface, higher bacteria oxygen demand from oxidation of organics and ammonia, and differences in activity and feeding behavior.

Recycle systems. Limited data is available for OFR_d in recycle systems. The oxygen demand from bacterial oxidation of organic compounds, ammonia, and solids strongly depends on the unit processes and their operation. The upper bound for OFR_d equals the ultimate Biochemical Oxygen Demand of the feed, which for channel catfish feed is equal to 1.1 kg O_2/kg dry feed (Harris 1971). Careful feeding and rapid removal of solids from the system can significantly reduce the OFR_d.

Ponds. Due to the minor impact of culture animals on the whole pond respiration, computation of OFR_d under pond conditions may not be particularly important. The computation of the variation of DO and aeration demand can be computed by the techniques presented by Boyd et al. (1978), Madenjian et al. (1987), or Meyer and Brune (1982).

Maximum Daily Oxygen Demand

On a daily basis, the maximum oxygen consumption occurs 4 to 6 hours following feeding in a flow-through system. Westers (1981) suggested a peaking factor of 1.44 to account for the maximum daily oxygen consumption rate:

$$\text{Maximum Daily Oxygen Demand (kg/d)} = (1.44)(\text{OFR})R \quad (38)$$

The peaking factor can be reduced by increasing the number of feedings per day. In recycle systems, the impact of feed type and feeding on α' must be carefully evaluated (Weaver 1981).

Oxygen Supplied by Flow

The amount of available oxygen supplied by the flow is equal to:

$$\text{Oxygen Supplied by Flow (kg/d)} = 86.4(Q_w)(DO_{out} - DO_{min}) \quad (39)$$

In systems where this source of oxygen is important, the design should be based on at least one year of DO and flow data collected at weekly intervals.

Supplemental Oxygen Requirement

The amount of supplemental oxygen is equal to the difference between Equations 38 and 39. For design purposes, the supplemental oxygen requirement should be based on a weekly or monthly biomass and feeding level. Depending on the harvest schedule and temperature, the maximum supplemental oxygen requirement may occur prior to the end of the production cycle. If there is a large variation in biomass between the various rearing units, it may be necessary to compute the supplemental oxygen requirement for each rearing unit.

Mixing Characteristics of Aerators

The mixing characteristics of aerators are an important design consideration. In ponds, it may be desirable to prevent thermal and chemical stratification during the summer months. Mixing during the day may evenly distribute the oxygen produced by algae and decrease the loss of oxygen across the air-water interface (Busch and Goodman 1983; Rogers and Fast 1988). The use of static mixers or surface aerators with draft tubes or other systems that pull water from the bottom may result in better mixing.

For emergency aeration of ponds, aerators that produce a zone of high dissolved oxygen can be used to produce a refuge zone for the animals and prevent mortality. Due to extremely high DO consumption rates that occur in ponds during emergency conditions, uniformly distributing the oxygen throughout the ponds would have little impact.

In cage systems, low-speed propeller units have been used (Aase et al. 1985). These units can increase the DO inside the cages by increasing water flow through the cages. Increased flow through the cage is only useful if the outside water has a high DO. When the surface water is warm and low in DO, a "flow developer" can be mounted with a vertical draft tube to circulate the colder (higher DO) bottom water through the cages.

The mounting of surface or submerged aerators within the cage has also been tried. When the surface waters are low in DO, aerators that produce limited water movement are needed (Kils 1979) to allow the added oxygen to remain within the cage.

Strong aeration and the resulting turbulence reduces larval mortality of fish (Barahona-Fernandes 1978) and increases the growth rate of larval

Table 5. Common site considerations.

Item	Considerations
Head	If enough head is available, operation of a gravity aerator may be possible without the use of external power.
Power	If electrical power is unavailable or unreliable, it may be necessary to use a motor-generator system or engine-powered aerators.
Location	In remote locations, the lack of spare parts and trained personnel may favor simple systems.
Subsurface	In areas with rocky or unstable soils, excavation may be expensive.
Layout	Retro-fitting existing hatcheries may involve careful consideration of problems associated with installing additional electrical lines, piping, and pumps between or around existing structures and utilities. A side-stream pure oxygen system may be easier to retro-fit at some sites.

oysters (Helm and Spencer 1972). Aeration may aid in the normal inflation of the swim bladder of fish (Doroshev and Cornacchia 1979).

System Characteristics

The selection of aerators will also be based on the physical characteristics of the (a) site, (b) number, size, and configuration of the rearing units, (c) hydraulics of the rearing units, and (d) mode of operation. In many cases, the system may be completed before the need for an aeration system is realized. Therefore, it is commonly necessary to design or retrofit an aeration system around a given system, rather than design a complete culture system from scratch.

Site

Table 5 presents common site considerations. The use of gravity aerators requires careful design of the rearing systems with respect to grade. Pumped systems may have increased costs, but may be more flexible for siting, design, and operation. In areas lacking electrical power, gasoline powered paddlewheel aerators may be more economic if the cost of bringing electrical power to site is considered.

Number, Size, and Configuration of Rearing Units

The number, size, and configuration of rearing units (Fig. 5) is important in the selection process. In large tanks or ponds, individual mechanical or floating aerators may be used. In the aquarium trade where a large number of small tanks must be aerated, diffused aeration is commonly used. Although the amount of air available is relatively fixed, the air can be inexpensively distributed to a large number of individual units and the flow to an individual unit may be varied easily. In raceway systems, a single centralized aerator may be used to aerate the influent water. A centralized system may be less expensive to build and operate but is much less flexible if the basic operation is changed.

The full-flow configuration (Fig. 6a) has been used when existing hatcheries are retro-fitted with packed column (Westers et al. 1987) or jet aerators (Colt and Westers 1982). This configuration is ideal for sites with enough head to operate the aeration system without an external power input. Available dissolved oxygen (DO_{in} - DO_{min}) limits the maximum carrying capacity of a full-flow configuration.

The side-stream configuration (Figs. 6b, c, e, f) is much more flexible than the full-flow configuration as the oxygen rich process water can be mixed in at several points (Figs. 6c and f) to reduce problems with gas supersaturation. It is also possible to increase the oxygen capacity of an individual rearing unit by construction of additional oxygen absorption units, something that can not be done with the full-flow configuration. In production systems, the cost of distribution piping for a side stream system may be significant.

Hydraulics of Rearing Units

The design of aeration systems also depends strongly on the hydraulics of the rearing units. For atmospheric aerators, the FAE value for a circular rearing unit may be significantly larger than for an equivalent raceway due to a lower C value (Larmoyeux et al. 1973).

For a given water flow, a circular rearing unit has a greater carrying capacity than a raceway. This is due to (1) the absence of dissolved gas gradients which limit the maximum DO concentration in the effluent from the absorber, and (2) rapid mixing and dilution that reduces bubble formation. The maximum DO and ΔP in a raceway must be based on the influent concentrations to the rearing unit rather than the effluent concentrations as in the case with a circular tank. Since a higher DO concentration can be obtained, the size of the absorption unit and piping in a circular rearing unit can be reduced. Less nitrogen stripping will be required since both the influent DN and DO are rapidly diluted to values equal to the effluent values from the rearing unit. Reduction of nitrogen stripping requirements allows higher oxygen absorption efficiencies and reduces the loss of oxygen in the off-gas (Colt and Watten 1988).

Purpose or Mode of Operation

Depending on the total oxygen demand and the oxygen supplied by flow, an aeration system can be designed to operate continuously over the production cycle, operate only over the peak oxygen demand period, or operate only as an emergency system. Due to the relative importance of capital and operating costs for these three modes of operation, different types of aeration systems may be needed for different operating modes.

Interactions With Culture Animals And Operational Staff

Interactions of the aerators with the culture animals and operational staff are important design considerations that are often not adequately considered.

Physical Damage

Some submerged and surface aerators tend to "eat" fish and require some type of screening to prevent entry of fish into the rotating components. Clogging with vegetation or other material can require a significant amount of time for cleaning and may result in damage to the shaft or motor if the debris is allowed to accumulate.

Sound Production

Some types of surface or submerged aerators may produce high levels of sound. Under laboratory conditions, noise can have a significant effect on the hatching and growth rate of fish and crustaceans (Banner and Hyatt 1973; Lagardère, 1982). The effects of this noise under production conditions are not well-defined, but may be significant in high density systems where the animals are closely held.

Personnel Safety

Electrical motors, engines, rotating shafts, fuel supplies, and pure oxygen supplies all represent some degree of risk to personnel. None of these items should present a significant risk if common sense and normal safety procedures are followed. Personnel must be properly trained to operate and maintain the equipment, have the required tools available, and actually follow the correct procedures.

Under freezing conditions, the build-up of ice due to spray can present a serious hazard to those walking in the area or if personnel must physically remove the ice. Prevention of icing problems may favor certain types of aerators or require installation of the aerators in a heated building.

Control

The oxygen demand of a production system has a significant diel and seasonal variation (Fig. 11). In addition, the oxygen demand from a single raceway or raceway series can change when transferring or harvesting of fish. The amount of oxygen transfer can be adjusted by turning on another pump or blower or changing the pure oxygen flow rate (G/L ratio). The degree of control depends on the total number of aeration units, the aerators operational characteristics, and the layout of the rearing units.

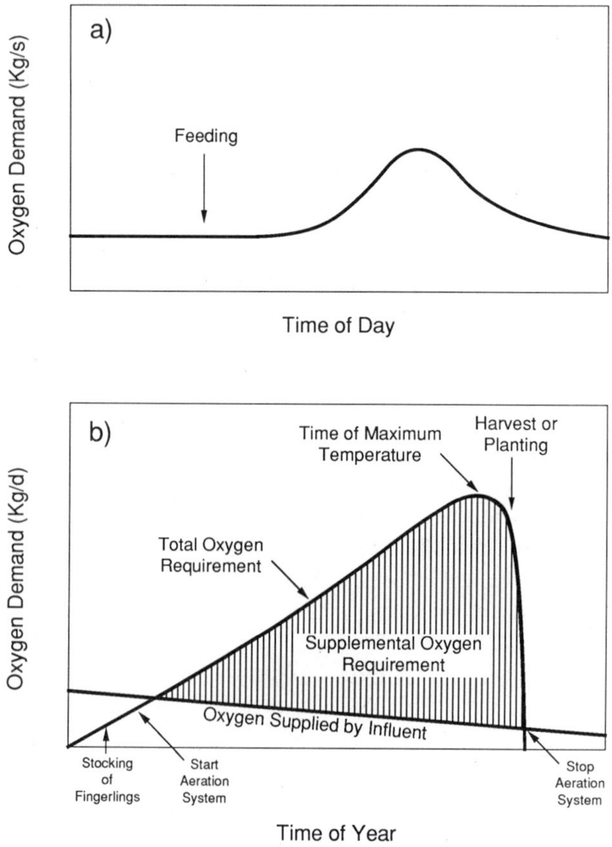

Figure 11. Variation of oxygen demand in a flow-through system with (a) time of day (b) production cycle

Increasing the degree of control over the aeration process increases the capital costs for process control equipment and maintenance and may reduce the over-all reliability of the whole system. A simple control system with a lower FAE and OTE_f may be more desirable than a complex system with better operational characteristics.

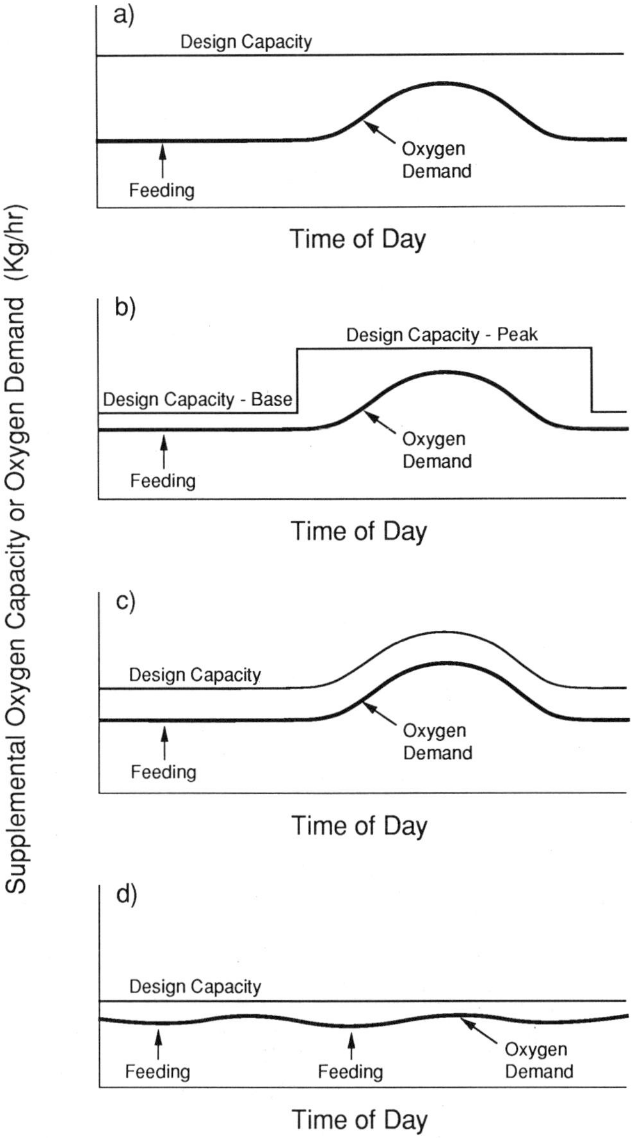

Figure 12. Control strategies (a) peak demand, (b) step control, (c) proportional control, and (d) reduced peak demand

Maximum Output Control

The simplest control strategy for diel changes in oxygen demand is to design for the maximum oxygen demand (Fig. 12a). This may result in low FAE values over much of the day.

Step Control

Step control (Fig. 12b) uses one system to provide base capacity and a second system to provide peak capacity. In raceways, surface aerators are commonly used to provide additional aeration when the biomass is high and the output of the gravity aerators is insufficient. Surface aerators can also be used to increase the oxygen transfer rate following feeding when the oxygen consumption of the culture animals increase. In subsurface aerators, the oxygen transfer can be changed by changing the air flow to the unit. A number of smaller units that can be turned on when needed may be more efficient than running a single large blower continuously.

Proportional Control

With a pure oxygen system, a control system could increase the ΔDO to track the change in oxygen demand (Fig. 12c). Pure oxygen systems have a much wider dynamic range, due primarily to the ability to adjust effluent DO with both the G/L ratio or system pressure. In a centralized side-stream system, it may be possible to adjust the available dissolved oxygen by adjusting the individual flows to the rearing units. In production systems, the biomass within individual rearing units may vary by a factor of 5 or more. It will be necessary to design the oxygen absorber unit to supply the maximum value to each rearing unit. It also may be necessary to operate all the rearing units at the maximum rate if the operational staff is unable to adjust G/L ratios (or other control parameters) on a frequent basis or ensure even distribution of animals within rearing units. This may increase the total oxygen supply capacity by 2-3 times over the computed supplemental oxygen requirement for the whole system but greatly reduce operational problems. The series/parallel configuration (Fig. 5d) allows mixing of dissolved oxygen from the individual rearing units in the common effluent manifold and may reduce control problems. This configuration is undesirable from a fish health standpoint due to its poor isolation characteristics. The use of proportional controls generally requires on-line monitoring of dissolved oxygen.

Reduced Oxygen Demand Control

Total required oxygen capacity may be minimized by staggering the feeding times within each raceway series to reduce the peak oxygen demand following feeding (Fig. 12d). This may also have the additional benefit of eliminating the need for continuous DO monitoring and on-line control.

Continuous Monitoring And Control

Optimization of aerator operation requires accurate and continuous information on dissolved gas concentrations in the rearing units. Continuous gas monitoring information can be used to adjust the feeding schedule or aerator operation. Computer controlled monitoring systems are highly reliable; but, all DO probes require maintenance on a regular basis.

Ideally, a separate oxygen monitoring probe should be installed in each rearing unit. An alternative is to use a single DO probe at a central site and switch between the different process waters (Kinghorn 1982). This last option may have reduced capital costs, but requires extensive piping and regular line cleaning to prevent significant oxygen consumption in the piping by bacteria. While continuous monitoring and computer control of DO concentrations may allow optimal control of aeration, it also increases the risk if the system fails.

DOs in ponds are highly inhomogeneous in both the horizontal and vertical directions. Until less expensive and more reliable DO probes are available, visual observation of animals or boom-mounted DO probes attached to a vehicle may be more effective and reliable (Anonymous 1983).

Operational Considerations

Aerators that interfere with the normal operations of a culture system or require extensive maintenance will probably not be used too long. Operational personnel may lack the time, knowledge, or tools to operate and repair some types of aerators. Aerators such as gravity aerators or paddlewheels that can be constructed and repaired on-site may be better choices for some operations even if their overall efficiency is lower than other types of aerators. The operational characteristics of different types of aerator systems are presented in Table 6.

Table 6. Operational considerations.

Parameter	Considerations
Fouling of Diffusers	Diffusers (airstones) may foul due to growth of algae or bacteria. Fine bubble diffusers may require special air filters and non-metal air lines to prevent clogging due to rust and scale. Diffusers may foul rapidly if not operated continuously.
Icing	Surface aerators and some types of gravity aerators produce enough spray to develop ice on walkways and roads. This may present a safety hazard to personnel.
Safety	Electrical lines, fuels tanks, or rotating shafts may present a safety hazard to personnel. Diffused aeration systems may be safer than other systems because the electrical lines are only required for the central blower unit. Electrical safety is a major concern in marine systems due to the high conductivity of seawater.
Harvesting/feeding	In ponds, static tube or surface aerators may need to be removed prior to harvesting. Aerators should not interfere with the daily operations of the facility.
Repair	The ease of repair may be an important consideration in remote locations. This includes both the skills and tools required and local availability of spare parts.
Reliability	A simple and highly reliable aerator is desirable. When adequate head is available, gravity aerators will operate during power failures.

Reliability

Although aeration systems can significantly increase production, significant to total mortality may occur if they fail. Some of the most common failures are presented in Table 7. While good design practices and operating

Table 7. Common types of aerator failures.

Component/Impact
Diffusers
Clogging of diffuser from scale or biological growth
Air, Water, and Electrical Lines
Gunshot damage
Vehicle damage
Backhoe damage
Pumps and motors
Gunshot damage
Loss of electrical power or loss of one leg in 3 phase power
Poor regulation of frequency or voltage
Power transients
Poor alignment of motor, shafts, and pumps
Debris lodged in the impeller or against the shaft
Splashing water leaks resulting in shorting of controls or motors
Inadequate lubrication or maintenance
Blowers
Flooding with water when shut down
Excessive pressure resulting in surging and damage
Electrical supply
Tree or animal damage to power lines
Flooding of conduit and shorting of lines
Wind or storm damage to power lines
Failure of the standby electrical power system after failure of the primary power source
Conrols
Failure of controller or solenoid valve in oxygen generator resulting in a significant reduction in oxygen purity
Failure of a regulator due to clogging with scale
Poor adjustment of operating water level, oxygen flow rate, or pressure may result in decreased performance or system failure

procedures can reduce the incidence of failure, a well-designed alarm system is useful and commonly employed. At a minimum this would include the following alarms:

Loss of electrical power to site
Loss of water pressure or flow
Differential pressure in oxygen feeder lines (for pure oxygen systems)
Dissolved oxygen meter and gasometer in rearing units

These alarms should be connected to a central control panel with audible and visual signals or automatic phone dialers. In many production facilities using pure oxygen, on-site personnel will be required 24 hours a day.

DESIGN PROCEDURE

Due to differences in site conditions, operational modes, and specific aerator units, it is impossible to suggest a single rigid design procedure. Some general guidelines for design are presented below (Colt and Watten 1988). It may be necessary to iterate several times through steps 3 to 8 to design an acceptable system.

1. Define the Site Conditions
   Elevation or barometric pressure
   Water temperature (average monthly and extreme monthly)
   Water flow (average monthly and extreme monthly)
   Available head or grade across the site

2. Set the Water Quality Criteria
   DO criteria (both maximum and minimum, mg/l)
   DO criteria (maximum partial pressure, mm Hg)
   DC criteria (maximum, mg/l)
   ΔP or TGP criteria (mm Hg or %)
   Un-ionized ammonia criteria (µg/l)

3. Estimate Oxygen Demand For Production Objectives
   Production objectives
   Production schedule (monthly)

Feeding levels and total feed consumption (monthly)
Total oxygen demand (monthly)
Available DO provided by flow (monthly)
Supplemental hourly oxygen requirement (maximum and minimum)
Supplemental monthly oxygen requirement (maximum and minimum)

4. Select Flow and DO Concentration

5. Select Aerator Type(s)

6. Select Operating Point
   Field oxygen transfer rate (OTR_f)
   Field aeration efficiency (FAE)
   Field oxygen transfer efficiency (OTE_f)
   Stripping ratio ($\Delta DO/\Delta DN$)
   Utilization efficiency
   Water flow, head, and power requirement

7. Compute Operating Point Under Daily and Monthly Variations

8. Compute Over-all Economics
   Capital costs ($)
   Operating costs ($/month)
   Total costs ($/kg of oxygen, $/kg of fish)
   Facility profit or loss

OPERATION OF AERATION SYSTEMS

While aerators are only a small part of any hatchery or production facility, they are critical to its operation. When an aeration system fails, significant to total mortality can result within minutes to hours. Operating policies and procedures to increase the reliability of aerators are discussed in the following sections.

Assignment of Responsibility

Depending on the size of the facility, the operation may be divided into different functional parts, each under the direction of an individual who reports to the hatchery manager. Commonly, the person responsible for servicing, maintenance, and repair activities may work for several functions or departments but should be accountable only to the hatchery manager. In this way, the correct priorities can be assigned and long-term but critical aspects such as preventive maintenance, scheduled overhauls, and an adequate spare parts inventory can be maintained. With all the problems that routinely occur in hatchery operations, the service functions are often ignored until they become critical, usually by unfortunate but not unexpected equipment failure.

Redundancy and Spares

Commonly, redundant equipment is designed into a rearing facility. Areas for such redundancy often include spare aerators, blowers, oxygen sources, and backup power sources. Redundant equipment gives time for repair without loss of service. Unfortunately, convincing management to purchase expensive equipment when there is no immediate problem with the current equipment is very difficult. What often happens in practice is that the backup equipment is brought on-line to increase the service; it then is no longer redundant but primary. Equipment failure under these conditions can be especially disastrous.

The availability of spare parts is critical to efficient aerator operation. If spare parts are readily and quickly available then a minimum of on-hand spare parts is needed. If spare parts are difficult and time consuming to acquire, a large inventory is required. Usually some spare parts are ordered with the basic equipment and when used should be immediately replaced as these are the ones most likely to be needed again. This is often put off in the presence of more immediate problems. Management must be convinced of the importance of adequate on-hand spare parts.

Preventative Maintenance

Much of the philosophy relating to spares and redundancy also applies to preventative maintenance. Equipment that is not properly maintained will fail. Even if properly maintained, it, or at least some parts of it, may wear out with extended service. Routine maintenance, equipment overhaul, and replacement should occur while the equipment is operating normally without immediate problems. All these activities require personnel time, resources, and sometimes appreciable money.

Monitoring and Control

Monitoring and control are critical hatchery activities. Monitoring and control does not necessarily imply monitoring/alarm systems and automated equipment, although these can be helpful. The most valuable inputs of the operational status of a hatchery are often sounds or rather changes in sounds. An experienced person will react to even minor sound changes and promptly investigate. Having all personnel check visually and listen during their normal daily activities and report anything unusual may be more effective than expensive monitoring equipment.

It is important not only to know something is wrong, but that information gets to someone who can act quickly. In an emergency, actions may require sacrifice of some activities to save others. While the normal chain of command for such decision making is usually clear, problems can develop in the absence of the hatchery manager or at odd hours (like 3 am on a Sunday morning). There must be an identifiable individual and a backup available at all times who are responsible for reacting to emergencies. These people must be provided with decision making guidelines by the hatchery manager and access to technical information, communications, and keys that may be required. Problems must be anticipated before they occur.

COSTS

The total costs associated with the operation of an aeration system comprise four components: (1) the cost of bringing power to the individual aerator, (2) construction and capital costs, (3) operating costs, and (4) maintenance costs. The costs of bringing power to the site and individual aerators may be significant for ponds and undeveloped sites. In some locations, this factor may require on-site generators or the use of gasoline or diesel-powered aerators. Construction costs are difficult to estimate as many systems can be constructed and installed by on-site personnel during the off-season period. The capital costs of the more commonly used 1–2 kW aerators range from $600 to $1300/kW. Operating costs (primarily energy) can be estimated from the maximum oxygen demand of the system and appropriate FAE value. The yearly maintenance costs may range from 2 to 10 % of the capital costs.

Continuous aeration with electrical paddlewheel aerators can produce greater income with less risk than use of emergency aeration (Engle and Hatch 1988). When electricity is not available for a pond, a reduction in stocking rate can more effectively reduce risk than use of emergency aeration.

Cost information for pure oxygen systems is very site specific, depending strongly on both the scale of operation and mode of operation. Many commercial operators are not willing to release operational information or cost data. For existing facilities, the use of pure oxygen can increase the production by 1.7 to 3.0 times (Boersen and Chesney 1987; Collins et al. 1984; Gowan 1987; Severson et al. 1987). Compared to increasing production by pumping more water, the use of pure oxygen is 2.5 to 5.5 times less expensive (Severson et al. 1987).

Capital costs associated with the use of pure oxygen are in the range of $0.44 to $2.20/kg ($0.20 to $1.00/lb) of annual production. As a comparison, construction costs of Federal hatcheries in the Pacific Northwest of the United States are in the range of $110 to $150/kg of annual production (Gerald Bouck, personal communication). Depending on the mode of operation and other factors, the total operating costs for pure oxygen are in the range of $0.11 to $0.44/kg of fish produced ($0.05 to $0.20/lb). The total capital costs of a pure oxygen production system may range from $3,000 to greater than $200,000, depending on site and operating conditions.

LIERATURE CITED

Aase, H., T. H. Riple, E. Boge, A. M. Aarseth, and P. Jensen. 1985. Effect of the use of flow developers in fish-rearing cages for salmon. Fiskeridirektoratets Havforskningsinstitutt, Bergen, Norway (in English).

Ahmad, T. and C. E. Boyd. 1988. Design and performance of paddle wheel aerators. Aquacult. Eng., 7:39-62.

Anonymous. 1983. CFA Innovation award winner invents "oxygen buggy", Aquacult. Magazine, 9:33.

Andrews, J. W., T. Murai, and G. Gibbons, 1973. The influence of dissolved oxygen on the growth of channel catfish. Trans. Am. Fish. Soc., 102:835-838.

Baillod, R. C. 1979. Review of oxygen transfer model refinements and data interpretation. In Proceedings: Workshop Towards an Oxygen Transfer Standard, W. A. Boyle (ed), EPA-600/9-78-021, Cincinnati, Ohio, pp.17-26.

Ball, R. O. and H. J. Campbell, Jr. 1974. Static aeration systems problems and performance. Proceedings of the 29th Purdue Industrial Waste Conference, Purdue University, pp.328-337.

Banner, A. and M. Hyatt. 1973. Effects of noise on eggs and larvae of two estuarine fishes. Trans. Am. Fish. Soc., 102:134-136.

Barahona-Fernandes. M. H. 1978. Effects of aeration on the survival and growth of sea bass (*Dicentrarchus labrax* L.) larvae: A preliminary study. Aquaculture, 14:67-74.

Basu, S. P. 1959. Active respiration of fish in relation to ambient concentrations of oxygen and carbon dioxide. J. Fish. Res. Bd. Can., 16:175-212.

Beamish, F. W. H. and P. D. MacMahon. 1988. Apparent heat increment and feeding strategy in walleye, *Stizostedion vitreum vitreum*. Aquaculture, 68:73-82.

Benson, B. B. and D. Krause. 1984. The concentration and isotopic fraction of oxygen dissolved in freshwater and seawater in equilibrium with the atmosphere. Limn. Oceanogr., 29:620-632.

Boersen, G. and J. Chesney. 1987. Engineering considerations in supplemental oxygen. In Papers on the Use of Supplemental Oxygen to Increase Hatchery Rearing Capacity in the Pacific Northwest, Bonneville Power Administration, Portland, Oregon, pp 17-24.

Bouck, G. R. 1982. Gasometer: an inexpensive device for continuous monitoring of dissolved gases and supersaturation. Trans. Am. Fish. Soc., 111:505-516.

Bouck, G. R. and R. E. King. 1983. Tolerance to gas supersaturation in fresh water and sea water by steelhead trout, *Salmo gairdneri* Richardson. J. Fish Biol., 23:293-300.

Boyd, C. E. and D. J. Martinson. 1984. Evaluation of propeller-aspirator-pump aerators. Aquaculture, 36:283-292.

Boyd, C. E., T. Ahmad, and Z. La-fa. 1988. Evaluation of plastic pipe, paddle wheel aerators. Aquacult. Eng., 7:63-72.

Boyd, C. E. and C. S. Tucker. 1979. Emergency aeration of fish ponds. Trans. Am. Fish. Soc., 108:299-306.

Boyd, C. E., R. P. Romaire, and E. Johnston. 1978. Predicting early morning dissolved oxygen concentrations in channel catfish ponds. Trans. Am. Fish. Soc., 107:484-492.

Boyle, W. A. (ed). 1979. Proceedings: Workshop Towards an Oxygen Transfer Standard. EPA-600/9-78-021, U. S. Environmental Protection Agency, Cincinnati, Ohio.

Busch, C. D. and R. K. Goodman. 1983. Fishpond management without emergency aeration. Trans. Amer. Soc. Agricult. Eng., 26:290-292,296.

Busch, R. L., C. S. Tucker, J. A. Steeby, and J. E. Reames. 1984. An evaluation of three paddlewheel aerators used for emergency aeration of channel catfish ponds. Aquacult. Eng., 3:59-69.

Brett, J. R. 1979. Environmental factors and growth. In Fish Physiology. Vol. 8, W. S. Hoar, D. J. Randall, and J. R. Brett (eds.), Academic Press, New York, pp.599-675.

Brett, J. R. and T. D. D. Groves. 1979. Physiological energetics. In Fish Physiology. Vol. 8, W. S. Hoar, D. J. Randall, and J. R. Brett (eds.), Academic Press, New York, pp.279-352.

Brett, J. R. and C. A. Zala. 1975. Daily pattern of nitrogen excretion and oxygen consumption of sockeye salmon (*Oncorhynchus nerka*) under controlled conditions. J. Fish. Res. Bd. Can., 32:2479-2486.

Brown, L. C. and C. R. Baillod. 1982. Modeling and interpretation of oxygen transfer data. J. Env. Eng. Div., ASCE, 108:607-628.

Carmichael, G. J. and J. R. Tomasso. 1988. Survey of fish transportation equipment and techniques. Prog. Fish-Cult., 50:155-159.

Chesness, J. L., L. J. Fussell, and T. K. Hill. 1973. Mechanical efficiency of a nozzle aerator. Trans. Am. Soc. Agricult. Eng., 16:67-68,71.

Chesness, J. L. and J. L. Stephens. 1971. A model study of gravity flow aerators for catfish raceway systems. Trans. Am. Soc. Agricult. Eng., 14:1167-1169,1174.

Chesness, J. L., J. L. Stephens, and T. K. Hill. 1972. Gravity flow aerators for raceway culture systems. Research Report 137, Agricultural Experiment Station, University of Georgia.

Collins, C. M., G. L. Burton, and R. L. Schweinforth. 1984. Evaluation of liquid oxygen to increase channel catfish production in heated water raceways. Tennessee Valley Authority, TVA/ONRED/AWR-84/8, Gallatin, Tennessee.

Colt, J. 1984. Computation of Dissolved Gas Concentrations in Water as Functions of Temperature, Salinity, and Pressure. Special Publication No. 14, American Fisheries Society, Bethesda, Maryland.

Colt, J. 1986. Gas supersaturation - impact on the design and operation of aquatic systems. Aquacult. Eng., 5:49-85.

Colt, J and Bouck, G. 1984. Design of packed columns for degassing. Aquacult. Eng., 3:251-273.

Colt, J., K. Orwicz, and G. Bouck. 1989. Water quality criteria for high density fish culture using supplemental oxygen. Fisheries Bioengineering Symposium, American Fisheries Society, Bethesda, Maryland, (in press).

Colt, J. and G. Tchobanoglous. 1981. Design of aeration systems for aquaculture. In Proceeding of the Bio-Engineering Symposium for Fish Culture, L. J. Allen and E. C. Kinney (eds), American Fisheries Society, Bethesda, Maryland, pp. 138-148.

Colt, J. and B. Watten. 1988. Application of pure oxygen in fish culture. Aquacult. Eng., 7:397-441.

Colt, J. and Westers, H. 1982. Production of gas by aeration. Trans. Am. Fish. Soc., 111:342-360.

Christensen, P. M. 1989. Pure oxygen for increased production. Fisheries Bioengineering Symposium, American Fisheries Society, Bethesda, Maryland, (in press).

Cornacchia, J. W. and J. E. Colt. 1984. The effects of dissolved gas supersaturation on larval striped bass *Morone saxatilis* (Walbaum). J. Fish Dis., 7:15-27.

D'Aoust, B. G. and M. J. R. Clark. 1980. Analysis of supersaturated air in natural waters and reservoirs. Trans. Amer. Fish. Soc., 109:708-724.

Dijkstra, F. H. Jennekens, and P. Nooren. 1979. The development and application of water-jet aeration for waste water treatment. Prog. Water Tech., 11:181-191.

Doroshev, S. I. and J. W. Cornacchia. 1979. Initial swim bladder inflation in the larvae of *Tilapia mossambica* (Peters) and *Morone saxatilis* (Walbaum). Aquaculture, 16:57-66.

Engle, C. R. and U. Hatch. 1988. Economic assessment of alternative aquaculture aeration strategies. J. World Aquacult. Soc.,19:85-96.

Ewing, L., D. T Redmon, and J. D. Wren. 1979. Testing and data analysis of diffused aeration equipment. J. Wat. Poll. Contr. Fed., 51:2384-2400.

Fisette, G. R. 1979. An aeration equipment manufacturer's view of an oxygen transfer testing standard. In Proceedings: Workshop Towards an Oxygen Transfer Standard, W. A. Boyle (ed). EPA-600/9-78-021, Cincinnati, Ohio, pp.3–6.

Fuss, J. T. 1983. Effective flow-through vacuum degasser for fish hatcheries. Aquacult. Eng., 2:301-307.

Gowan, R. 1987. Use of supplemental oxygen to rear chinook in seawater. In Papers on the Use of Supplemental Oxygen to Increase Hatchery Rearing Capacity in the Pacific Northwest, Bonneville Power Administration, Portland, Oregon, pp 35-39.

Gleuckauf, E. 1951. The composition of atmospheric air. In: Compendium of Meteorology, American Meteorology Society, Boston, pp.3-11.

Hackney, G. E. and J. E. Colt. 1982. The performance and design of packed column aeration systems for aquaculture. Aquacult. Eng., 1:275-295.

Harris, J. C. 1971. Pollutional characteristics of channel catfish culture. M.S. Thesis, Environmental Health Engineering, University of Texas, Austin, Texas.

Hartman, J. 1983. Performance and operation of Alaska Department of Fish and Game screen decks. Gas supersaturation in hatcheries, causes, effects, and solutions. Bio-engineering Section, American Fisheries Society, Bethesda, Maryland, unpublished.

Haskell, D. C. 1955. Weight of fish per cubic foot of water in hatchery troughs and ponds. Prog. Fish-Cult., 17:117-118.

Haskell, D. C. and R. O. Davies. 1958. Carbon dioxide as a limiting factor in trout transport. New York Fish and Game J., 5:175-183.

Haskell, D. C., R. O. Davies, and J. Reckahn. 1960. Factors in hatchery pond design. New York Fish and Game J., 7:112-129.

Helm, M. M. and B. E. Spencer. 1972. The importance of the rate of aeration in hatchery cultures of the larvae of *Ostrea edulis* L., J. Cons. Int. Explor. Mer, 34:244-255.

Hogendoorn, H. 1983. Growth and production of the African catfish, *Clarias lazera* (C. & V.). III. Bioenergetic relations of body weight and feeding level. Aquaculture, 35:1-17.

Howe, K. J., D. F. Lawler. 1989. Acid-base reactions in gas transfer: a mathematical approach. J. Am. Wat. Works, 81:61-66.

Kils, U. 1976/1977. The salinity effect on aeration in mariculture. Meeresforsch., 25:210-216.

Kils, U. 1979. Oxygen-regime and artificial aeration of net-cages in mariculture. Meeresforsch., 27:236-243.

Kinghorn, B. P. 1982. Central system for monitoring the quality of water from many sources. Prog. Fish-Cult., 44:30-32.

Kutty, M. N. 1968. Respiratory quotients in goldfish and rainbow trout. J. Fish. Res. Bd. Can., 25:1689-1728.

Lagardère, J. P. 1982. Effects of noise on growth and reproduction of *Crangon crangon* in rearing tanks. Mar. Biol., 71:177-185.

Larmoyeux, J. D., R. G. Piper, and H. H. Chenoweth. 1973. Evaluation of circular tanks for salmonid production. Prog. Fish-Cult., 35:122-131.

Lewis, W. K. and W. C. Whitman. 1924. Principles of gas adsorption. J. Ind. Eng., 16:1215-1220.

Machta, L. 1983. The atmosphere. In: Changing Climate — Report of the Carbon Dioxide Assessment Committee, National Academy Press, Washington, DC., pp 242-51.

Madenjian, C. P., G. L. Rogers, and A. Fast. 1987. Predicting night time dissolved oxygen loss in prawn ponds of Hawaii: Part II. A new method. Aquacult. Eng., 6:209-225.

Marking, L. L., Dawson, V. K. and Crowther, J. R. 1983. Comparison of column aerators and a vacuum degasser for treating supersaturated culture water. Prog. Fish-Cult., 45:81-83.

McLean, W. E. and Boreham, A. L. 1980. The design and assessment of aeration towers. Fisheries and Oceans, Vancouver, British Columbia, Canada (unpublished).

Metcalf and Eddy, Inc. 1979. Wastewater Engineering: Treatment, Disposal, Reuse. McGraw-Hill, New York.

Meyer, D. I. and D. E. Brune. 1982. Computer modelling of the diurnal oxygen levels in a stillwater aquaculture pond. Aquacult. Eng., 1:245-261.

Moore, J. M. and C. E. Boyd. 1984. Comparison of devices for aerating inflow of pipes. Aquaculture, 38:89-96.

Mueller, J. A. and W. C. Boyle. 1988. Oxygen transfer under process conditions. J. Wat. Poll. Contr. Fed., 60:322-341.

Nagy, Z. 1979. The air-lift aerator and its application in sewage treatment. Prog. Water Tech., 11:101-109.

Nebeker, A. V., G. R. Bouck, and D. G. Stevens. 1976. Oxygen and carbon dioxide and oxygen-nitrogen ratios as factors affecting fish in air-supersaturated water. Trans. Am. Fish. Soc., 105:425-429.

Nirmalakhandan, N., Y. H. Lee, and R. E. Speece. 1988. Optimizing oxygen absorption and nitrogen desorption packed towers. Aquacult. Eng., 7:221-234.

Owsley, D. E. 1981. Nitrogen gas removal using packed columns. In Proceeding of the Bio-Engineering Symposium for Fish Culture, L. J. Allen and E. C. Kinney (eds), American Fisheries Society, Bethesda, Maryland, pp. 71-82.

Poston, H. A. and R. C. Williams. 1988. Interrelations of oxygen concentration, fish density, and performance of Atlantic salmon in an ozonated water reuse system. Prog. Fish-Cult., 50:69-76.

Reinemann, D. J. and M. B. Timmons. 1989. Prediction of oxygen transfer and total dissolved gas pressure in airlift pumping. Aquacult. Eng., 8:29-46.

Rogers, G. L. and A. W. Fast. 1988. Potential benefits of low energy water circulation in Hawaiian prawn ponds. Aquacult. Eng., 7:155-165.

Rombough, P. J. 1986. Mathematical models for predicting the dissolved oxygen requirements of steelhead (*Salmo gairdneri*) embryos and alevins in hatchery incubators. Aquaculture, 59:119-137.

Severson, R. F., J. L. Stark, and L. M. Poole. 1987. Use of oxygen to commercially rear coho salmon, In Papers on the Use of Supplemental Oxygen to Increase Hatchery Rearing Capacity in the Pacific Northwest, Bonneville Power Administration, Portland, Oregon, pp 25-34.

Shelton, Jr., J. L. and C. E. Boyd. 1983. Correction factors for calculating oxygen-transfer rates of pond aerators. Trans. Am. Fish. Soc., 112:120-122.

SECL [Sigma Environmental Consultants.Limited] 1983. Summary of Water Quality for Salmonid Hatcheries, SECL 8067. Prepared for Department of Fisheries and Oceans, Vancouver, British Columbia, Canada.

Smart, G. R., D. Knox, J. G. Harrison, J. A. Ralph, R. H. Richards, and C. B. Cowey. 1979. Nephrocalcinosis in rainbow trout *Salmo gairdneri* Richardson; the effect of exposure to elevated CO_2 concentrations. J. Fish Dis., 2:279-289.

Sowerbutts, B. J. and J. R. M. Forster. 1981. Gas exchange and reoxygenation. In Aquaculture in Heated Effluents and Recirculation Systems Vol.1,. K. Tiews (ed.), Heenemann Verlagsgesellschaft, Berlin, pp. 199-217.

Speece, R. E. 1981. Management of dissolved oxygen and nitrogen in fish hatchery waters. In: Bio-engineering Symposium for Fish Culture. eds. L. J. Allen and E. C. Kinney, American Fisheries Society, Bethesda, Maryland, pp. 53-62.

Speece, R. E. 1970. Design of u-tube aeration systems. J. Sanit. Eng. Div., ASCE, 96:715-725.

Speece, R. E., J. L. Adams, and C. B. Wooldridge. 1969. U-tube aeration operating characteristics. J. Sanit. Eng. Div., ASCE, 95:563-574.

Speece, R. E., M. Madrid, and K. Needham. 1971. Downflow bubble contact aeration. J. Sanit. Eng. Div., ASCE, 97:433-441.

Stenstrom, M. K. and R. G. Gilbert. 1981. The effects of alpha, beta, and theta factors upon the design, specification and operation of aeration systems. Wat. Res., 15:643-654.

Strasburg, D. W. 1964. An aerating device for salt well water. In Sea-water Systems for Experimental Aquariums, J. R. Clark and R. L. Clark (eds), T.F.H. Publications, Jersey City, New Jersey, pp. 161-167.

Stumm, W. and J. J. Morgan. 1981. Aquatic Chemistry. 2nd Ed., John Wiley & Sons, New York.

Tebbutt, T. H. Y. 1972. Some studies on reaeration in cascades. Wat. Res., 6:297-304.

Toerber, E. D. and M. G. Mandt. 1979. Greater oxygen transfer with jet aeration system. Water and Sewage Works, 126:71-75.

USEPA [United States Environmental Protection Agency]. 1976. Quality Criteria for Water, Washington, D.C.

Watten, B. and L. T. Beck. 1985. Modeling gas transfer in a u-tube oxygen adsorption system: effects of off-gas recycling. Aquacult. Eng., 4:271-297.

Watten, B., J. Colt, and C. Boyd. 1989. Impact of dissolved nitrogen and carbon dioxide on the operation of pure oxygen systems. Fisheries Bioengineering Symposium, American Fisheries Society, Bethesda, Maryland, (in press).

Weaver, D. E. 1981. The impact of fish feed on oxygen transport. Presented at 12th Annual Meeting of the World Mariculture Society, March 8-10, 1981, Seattle, Washington (unpublished).

Weiss, R. F. 1970. The solubility of nitrogen, oxygen and argon in water and seawater. Deep-sea Res., 17:721-735.

Weiss, R. F. 1974. Carbon dioxide in water and seawater: The solubility of a non-ideal gas. Mar. Chem., 2:203-215.

Westers, H. 1981. Fish culture manual for the state of Michigan. Michigan Department of Natural Resources, Lansing, Michigan (unpublished).

Westers, H., V. Bennett, and J. Copeland. 1987. Michigan's experience with supplemental oxygen in salmonid rearing. In Papers on the Use of Supplemental Oxygen to Increase Hatchery Rearing Capacity in the Pacific Northwest, Bonneville Power Administration, Portland, Oregon, pp 12-16.

Willoughby, H. 1968. A method for calculating carrying capacities of hatchery troughs and ponds. Prog. Fish-Cult., 30:173-174.

Yunt, F. W. 1979. Gas flow and power measurements. In Proceedings: Workshop Towards an Oxygen Transfer Standard, W. A. Boyle (ed). EPA-600/9-78-021, Cincinnati, Ohio, pp.105-127.

FIXED FILM NITRIFICATION FILTERS FOR AQUACULTURE

Fred Wheaton, John Hochheimer, and G. Ellen Kaiser

ABSTRACT

Despite many years of research, design of biofilters remains as much art as science. Progress toward developing biofilter design methodologies is discussed by reviewing the biofilter kinetics, chemistry and physics. The variables that affect filter performance are enumerated and the current state of knowledge regarding their influence on filter performance is reviewed. Finally, the type of data needed to develop a rational biofilter design process is discussed and the role of mathematical models in the design process is considered.

INTRODUCTION

One major limiting factor influencing the operation of efficient, reliable and profitable recirculating aquaculture systems is the filtration system. Biological filters (biofilters), a major component in aquaculture filtration systems, use microorganisms to process waste water effluent from the aquatic culture systems. Biofilters rely upon numerous microorganisms that metabolize organic waste products from the animals, feed and other organic debris, and the inorganic waste, ammonia. When the biofilter is operating efficiently and managed properly, the effluent has extremely low concentrations of ammonia and nitrite, two chemicals which are very toxic to aquatic organisms. After processing, the "clean" water can then be recycled to the animals being cultured.

The task of designing more low-cost, efficient biofilters is not a trivial problem. There are a multitude of factors such as pH, temperature, water flow rate, alkalinity, ammonium, dissolved oxygen (DO) and other mineral concentrations, etc. which influence the dynamic microbial population, hence the filter performance.

This paper will focus on the problem of removing ammonia from the culture water using fixed film biofilters. The purpose of the paper is to present up to date information on factors that influence biofilter performance and to contribute to development of a much needed mathematical model. Currently, designers must depend on available models and design methods which are based on very limited information, hence are not very reliable under many different conditions, or they must rely on trial and error. The result of both approaches tends to be over designed, inefficient and thus more expensive biofilters. Ultimately, a mathematical model which describes the ammonia removal rate as a function of the various environmental parameters influencing the biofilter should be developed to enable biofilter designers and operators to produce and manage more efficient, economical filtration systems for any aquaculture application.

MATHEMATICAL MODELING OF BIOFILTERS

A mathematical model is an equation or set of equations that describe a process or phenomenon. The first step in developing a mathematical model for biofilters is to determine what process to define. For example, the process may be the ammonia removal rate or efficiency of a particular biological filter type. The next step is to determine all the inputs which effect the process. For this example, the inputs would include water pH; temperature; flow rate; ammonium, DO and other chemical concentrations; alkalinity; salinity; light; and the biological slime layer thickness in the fixed film biofilter.

The last and most difficult step of the modeling procedure is to determine how all of these inputs simultaneously effect the ammonia removal rate or efficiency. There are a few different approaches to this difficult step. One approach is to use accepted chemistry, physics and biological theory to develop equations that describe the process. Therefore, to rigorously describe the system using theoretical methods, the following equations would be used as well as other theoretical considerations to develop a set of non-linear simultaneous equations and unknowns to be solved: chemical equilibria relations for each chemical species in the water, an electroneutrality equation, chemical mass balances, equations describing the rate of diffusion of each chemical and combination of chemicals through the bacterial slime layer, relationships describing the bacterial metabolism of ammonium as a function of every other factor and combination of factors affecting metabolism, etc. The reason these equations must be solved simultaneously is that all of these processes are occurring concurrently. An accurate model developed through theoretical analysis would be extremely valuable because it means that the process is well enough understood and that the effect of input changes can be predicted with confidence. However, for a process such as this one which is so complex and not well understood, there is little chance that an accurate mathematical model can be obtained from the theoretical approach.

Biofilter models can trace their roots to waste water treatment. There have been many models developed to mathematically describe the workings of biofilters, but the majority have concentrated on the removal of BOD from water. Even when models were used to describe nitrification, the initial and final concentrations of ammonia were considerably higher than those acceptable for aquaculture. Most biofilter models describe kinetics of nutrient removal, mass transfer characteristics of a particular filter configuration, or both.

Kinetic models use the relationship first described by Monod (1942) (see Equation 5). This model can be used to fairly accurately describe steady state biofilter kinetics, but falls short under unsteady conditions such as during start-up or when a filter is shock loaded. Monod kinetics have been used to evaluate biofilter performance under various operating conditions including: temperature, pH, oxygen, and the effects of many toxins (Painter 1970; Haug and McCarty 1971).

Another approach to biofilter modeling involves mass transport theory. Weatherly (1982) used dynamic response analysis to model ammonia variations in aquaculture tanks under steady state and unsteady state conditions. Weatherly concluded that this approach was valid for aquaculture if further study was done in the area of biochemical kinetics and liquid mixing. There are other examples of fluid transport models for sewage treatment found in the literature, including those by Benefield and Molz (1985) and Grady and Lim (1980).

Combining biochemical kinetic models with fluid transport models is a relatively new technique that has been made possible by the computational capabilities of modern computers. Combined models are complex and require numerical solutions that are now possible with the use of a programmable computer. Combined process models for waste water treatment have been proposed by Grady and Lim (1980) and Roberts (1985).

Another possible approach is to develop models using empirical or experimental methods that treat the biofilter as a "black box". To do this, a known change of one variable is applied to the filter while holding each of the other variables at a constant level. Then, the transient (when the filter is readjusting) and steady state (when the filter is stable) response (ammonia removal rate) is recorded. This same procedure could be repeated for each significant variable on each filter type to obtain a set of simultaneous equations and unknowns that would describe the process for that filter type. The advantage of this method is that many different investigators could combine their experimental results together to give one model that could be updated and improved with new discoveries. For instance, one might develop the ammonia removal rate as a function of pH, another the ammonia removal rate as a function of temperature. If both kept all other significant variables constant during their trials, the results could be combined to estimate the ammonia removal rate as a function of pH and temperature. A disadvantage of this method is that interactions between two or more parameters would not be included in the model. These interactions are probably very important under certain conditions.

Another experimental approach is to conduct a multi-dimension factorial experiment where many levels of each variable are tested in a way that allows the interactions between variables to be included in the model. This would be a very difficult and time consuming (several years) experiment to conduct due to the many factors involved.

Most models used to describe biofilters in aquaculture have been developed as empirical models rather than based on bacterial kinetics or fluid mass transport theory. Haug and McCarty (1971) and Liao, et al. (1972) used empirical data to develop models of nitrification. These models have been of limited usefulness in developing design equations for biofilters as the data is extremely limited in the range of applicability.

The use of models to describe biofilter performance is a key step in understanding biofilter behavior under various conditions. Ideally, models could be used to evaluate the consequences of differing operating procedures on biofilter performance. Models, when they become available, will also be important design tools for the aquaculture system designer. Unfortunately, the biofilter models that most accurately describe filter performance were developed for waste water treatment biofilters that operate under much different conditions than filters for aquaculture. Ammonia and BOD levels of input streams are orders of magnitude higher than those in aquaculture systems. Models need to be evaluated at operating parameter values outside of the range of existing models before they can be used reliably for aquaculture.

In the sections that follow, how various factors affect the biofilter operation will be presented. Some of the relationships are derived theoretically, some experimentally, and some are a combination of both. Where appropriate, it will be pointed out how the information might be used in a mathematical model describing ammonia removal rate or ammonia removal efficiency as a function of the various inputs.

AQUACULTURE WASTE WATER

There are four primary sources of nitrogenous wastes in the aquatic environment: ammonia, urea, uric acid and amino acids excreted by living organisms, organic debris from dead and dying organisms, uneaten food and feces, and nitrogen gas from the air. Ammonia and feces are the principal excretory products from aquatic animals. Generally, if the ammonia concentration is within acceptable levels, the biological oxygen demand (BOD) concentration, a measure of the organic matter concentration, is usually also within acceptable levels (Klemetson and Cohen 1984).

BIOFILTER ECOLOGY

Aquaculture filters possess many species and strains of microorganisms, each feeding on different components in the aquaculture waste water. Historically, most attention has been focused on the role of heterotrophic and autotrophic bacteria in the aquatic waste treatment process. Heterotrophic bacteria metabolize the organic matter in the waste water, reducing them to inorganic nitrogen compounds of which ammonia predominates. Ammonia is oxidized to nitrate in a two step process by autotrophic bacteria. Although there are many species of nitrifying bacteria, Cutter and Crump (1983) found that ammonia conversion rates for most of the bacteria studied were very low when compared to *Nitrosomonas*. For this paper, *Nitrosomonas* sp. will be considered able to convert ammonia to nitrite and *Nitrobacter* sp. to convert nitrite to nitrate.

It is accepted that nitrifiers exhibit relatively slow growth compared to most heterotrophic bacteria (Wheaton 1977). Thus biofilters that mature in water containing high concentrations of organics tend to have high populations of heterotrophic bacteria. Although Lomax (1976) was not able to statistically prove that solids loading of biofilters reduced nitrification rates in the filter, trends in his data, and other data in the literature (U.S. EPA 1975), indicate heterotrophs and autotrophs compete for space in a biofilter.

Chemical Theory

The chemical equations which partially describe the oxidation of ammonia are (EPA, 1975):

$$NH_4^+ + 1.5O_2 + 2HCO_3^- \xrightarrow{Nitrosomonas} NO_2^- + H_2CO_3 + H_2O \quad (1)$$

$$NO_2^- + 0.5O_2 \xrightarrow{Nitrobacter} NO_3^- \quad (2)$$

It is clear from equations 1 and 2 that ammonia, nitrite, carbonate or oxygen may become rate limiting substances.

Cell growth in nitrifiers has been described by several authors using various chemical equations (WPCF 1983; Meade 1974; Anon 1971). Equation 3 gives one equation used to describe cell growth of nitrifiers (WPCF 1983):

$$NH_4^+ + 1.83O_2 + 1.98HCO_3^- \longrightarrow 0.021C_5H_7O_2N + 0.98NO_3 + 1.041H_2O + 1.88H_2CO_3 \quad (3)$$

The stoichiometric requirements for oxygen and alkalinity and the biomass production can be calculated using Equation 3. These values are listed in Table 1.

Table 1. Stoichiometric requirements for the oxidation of 1 mg of NH_4^+-N based on Equation 3 (WPCF 1983).

Compound	Amount Used (mg)	Amount Produced (mg)
O_2	4.2	
H^+		1.98
Alkalinity as $CaCO_3$	7.14	
$C_5H_7O_2N$ (cells)		0.17
NO_3^-		4.34
N_2O		1.34
H_2CO_3		8.59

Obviously, these equations represent only a limited amount of the overall process and so only provide a small portion of information. For example, reaction rates as a function of various parameters and combination of parameters cannot be obtained by examining these equations.

Another important chemical relationship is the ammonia (NH_3) / ammonium (NH_4^+) equilibrium. The percent of the total ammonia (NH_3 plus NH_4^+) as NH_3 is a function of the pH and temperature (Wheaton 1977). An increase in pH and/or temperature increases the proportion of total ammonia in the ammonia (NH_3) form. Because the NH_3 form is the most toxic to aquatic organisms and NH_4^+ is the form that the microorganisms use for food, culture water is usually maintained near neutral or slightly alkaline.

Nitrite in aqueous solution is in equilibrium with nitrous acid. Nitrous acid is more toxic to aquatic organisms than nitrite. Hence a shift to lower pH values, which increases the nitrous acid concentration relative to nitrite, increases the risk of toxicity due to nitrous acid.

Physical Theory

Figure 1 is a diagram of a single nitrifying bacteria emersed in water containing ammonia. Typically the cell is surrounded by a cell membrane. Because nitrifiers produce slime, they often are also surrounded by a thin, but definable, slime layer. If the water is moving relative to the bacteria, there is also a layer of stagnant water often referred to as the boundary layer in fluid dynamics, surrounding the cell. In aquaculture systems, the ammonia is present in the surrounding bulk aqueous solution in relatively low concentrations. Ammonia conversion to nitrite or nitrite conversion to nitrate occurs within the cell. Thus, an ammonia molecule must pass through the boundary layer, slime layer, and cell membrane. Transfer through the boundary layer is primarily by diffusion, which is a function of the ammonia concentration gradient across the boundary layer. The low ammonia concentration in the culture water leads to low diffusion rates. Diffusion is also probably the primary transfer method for moving ammonia across the slime layer. Transfer of ammonia across the cell membrane is probably at least partially active. *Nitrosomonas* expels nitrite which must traverse this path in reverse. *Nitrobacter* requires nitrite to move into the cell and nitrate to exit the cell through similar paths.

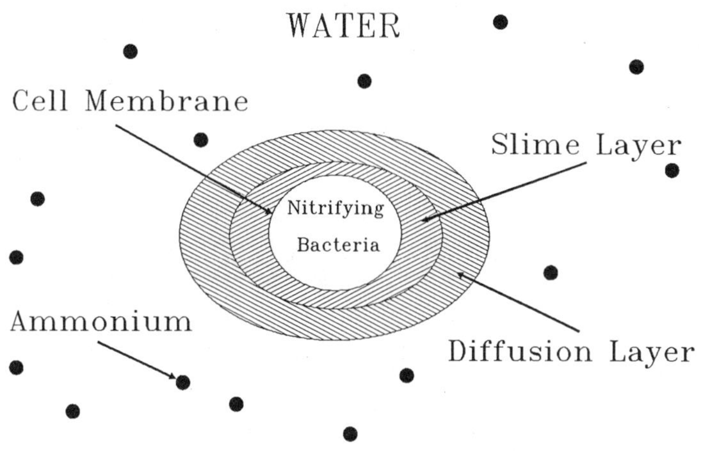

Figure 1. Diagram of a single nitrifying bacteria in water containing ammonium.

The boundary layer, and probably the slime layer to a more limited degree, can be reduced in thickness by increasing the relative velocity between the nitrifier cell and the water. Reduced thickness of these layers decreases resistance to diffusion and increases transfer rates as predicted by Frick's Law:

$$J_1 = \frac{D_1(C_1 - C)}{L_1} \qquad (4)$$

where

- J_1 = the flux across the diffusion layer
- D_1 = the diffusion coefficient
- C_1 = the concentration in the water
- L_1 = the thickness of the diffusion layer
- C = the concentration on the membrane side of the boundary layer

If the diffusion rate is limiting transfer, then increasing the particle-water relative velocity will increase nitrification rates. Work by Malone (1987) suggests diffusion may be a limiting factor. Unfortunately, relatively little research has been done to define transfer into and out of the nitrifier cells.

To understand the ammonia removal process more completely, this same analysis should simultaneously be applied to every other chemical nutrient of the bacteria.

The above discussion has focused on the limited chemical and physical theory that has been applied to biofilters. In the remainder of this paper, empirical results of studies on the effects of various physical and chemical parameters on biofilter performance will be discussed.

NITRIFICATION REACTION KINETICS

The rate of ammonia oxidation, or the kinetics of nitrification, can be described by the empirically derived Monod equation (Monod 1942):

$$u_n = \frac{u_n' N}{N + K_n} \qquad (5)$$

where

u_n = specific growth of nitrifers, day $^{-1}$

u_n' = maximum specific growth rate of microorganisms, day $^{-1}$

N = NH_4^+-N or NO_2^-–N concentration, mg/L

K_n = half saturation constant, which is equal to the growth limiting substrate concentration at half the maximum growth rate, mg/L of NH_4^+–N or NO_2–N.

Lawrence and McCarty (1969) modified the Monod expression to:

$$\frac{ds}{dt} = K\frac{SX}{K_s + S} \qquad (6)$$

where,

$\frac{ds}{dt}$ = rate of substrate utilization,

X = bacterial biomass concentration in a system in which

S = limiting substrate concentration

K = maximum observed reaction velocity when the substrate (S) is in excess, and,

K_s = half rate constant.

This relationship clearly shows that the higher the bacterial biomass in a system, the higher the ammonia removal rate. Thus, in order to promote the maximum nitrification rate, it is important to build up and maintain as large a population of nitrifiers as possible. Practically, however, there probably is a limit above which increased biomass will not increase nitrification. Increased biomass means that multiple bacterial layers must build up over the surface once a single layer has formed. Diffusion of ammonia and nitrite into the cells in the lower layers must follow a longer pathway than for surface layers. With each additional layer of bacteria, the growth rate of the lower layers will decrease due to slower diffusion of nitrogen and oxygen to these cells. The existence, or the quantitative size of this limit, have not been defined either theoretically or experimentally.

These equations also predict that nitrification will proceed at a rate determined by the most limiting substrate. This is also known as Leibig's Law of the Minimum. The biofilter should be designed such that ammonia is the limiting nutrient. However, there are several other factors that can limit growth including oxygen concentrations,temperature, pH, micronutrient concentrations, bicarbonate concentrations and others. Several of the more important variables in this group are discussed below.

Oxygen

The nitrification requires about 4.57 kg of oxygen for every kg of ammonia converted to nitrate. Much of the literature indicates that biofilters can operate efficiently down to about 2 ppm of oxygen. However, several investigators have found higher oxygen levels are needed to achieve maximum nitrification rates. Oxygen levels of 4 to 8 mg/L were found to increase nitrification rates by 10 percent over 2 mg/L (Anonymous 1965) in activated sludge plants. Schwer (1971), while working with sewage, found that 2 mg/L of dissolved oxygen (DO) gave only 40 percent of maximum nitrification rate while 4 mg/L gave 80 percent of the maximum rate. Kumar (1984) and Mathe et al. (1984) found oxygen was often a limiting factor in submerged biofilters in crab shedding systems.

Nagel and Haworth (1969) empirically derived a Monod expression relating nitrification rate to DO concentration:

$$r_n T_n = \frac{0.19(DO)}{2.0 + DO} \quad (7)$$

where

r_n = ammonia –N oxidation rate, Kg ammonium –N/Kg MLSS –day
DO = dissolved oxygen concentration, mg/l

pH

Culture system pH can strongly influence biofilter performance. It is well known that in culture systems the CO_2 produced by the cultured animals, the hydrogen ion produced by the nitrifiers, and other acids produced by the organisms in the system will lower system pH if no compensatory action is applied. For example, Collins, et al. (1975a), found a pH decline from 7.26 to 5.88 in 33 days in an unconditioned, unbuffered, culture system. A review of many studies on the effect of pH on nitrification is summarized by Kaiser and Wheaton (1983).

Table 2 gives the pH range that several authors found gave the best nitrification rates. It is obvious that the optimal pH range varies significantly but generally tends to be slightly on the basic side. The ability of bacteria to adapt within a reasonable pH range explains much of the variability. Thus, bacteria adapted to a specific environment will usually perform best under those conditions.

Table 2. The pH ranges giving the best nitrification rates for *Nitrosomonas* (NS) and *Nitrobacter* (NB).

Species	Range Tested	pH Optima	Source
NS		8.0 - 9.0	Hofman & Lees 1952
NS	7.0 - 9.0	8.0	Engle & Alexander 1958
NS		6.0 - 9.0	Winogradsky & Winogradsky 1933
NS	5.0 - 10.0	9.0	Kawai, et al. 1965
NS		7.2 - 7.8	Loveless & Painter 1969
NB		8.3 - 9.3	Mererhof 1917
NB	7.0 - 8.6	7.8	Boon & Laudelout 1962
NB		6.3 - 9.4	Winogradsky & Winogradsky 1933
NB	5.0 - 10.0	9.0	Kawa et al. 1965
BOTH	7.0 - 9.5	8.5	Kholdenbarin & Oertli1977

Temperature

Temperature affects bacterial growth rates probably more than any other variable. The temperature at which a biofilter operates is usually determined by the needs of the fish or other cultured crop, not by the nitrifying bacteria's needs. Although crop needs can vary anywhere from near 0 C to 38 C, it is more common for temperature needs to range from about 7 to 35 C. Generally, any single crop has an even narrower temperature range.

When oxygen and micro-nutrients are in sufficient quantities, the growth limiting substance for *Nitrosomonas* is ammonium and for *Nitrobacter* is nitrite (Equation 6). Table 3 gives the half saturation constants (Michaelis) for *Nitrosomonas* and *Nitrobacter*. Since nitrification proceeds from ammonia oxidation to nitrite oxidation, and *Nitrobacter* grows faster than *Nitrosomonas*, the overall kinetics of nitrification is usually controlled by ammonia oxidation (WPCF 1983).

Table 3. Half saturation constants (Kn) for *Nitrosomonas* (NS) and *Nitrobacter* (NB) at various temperatures.

Species	Temperature (C)	Kn (mg N/L)	Source
NS	30	10.0	Hofman & Lees, 1953
NS	25	3.5	Ulken, 1963
NS	20	1.0	Loveless & Painter, 1969
NB	30	6.0	Lees & Simpson, 1957
NB	28	5.0	Gould & Lees, 1960
NB	32	8.4	Laudelout & Van Tichelen, 1960
NB	25	5.0	Ulken, 1963

Maximum growth rates for Nitrosomonas have been found experimentally and are summarized in Table 4. The growth rates are related to temperature and can be linearly interpolated within the range of 10 to 30 C (WPCF 1983).

Table 4. Estimates for maximum growth rates (u_n') for *Nitrosomonas* (WPCF, 1983).

Temperature (C)	u_n' (days^{-1})
10	0.30
20	0.65
30	1.20

From Tables 3 and 4 and Equation 6, a relationship describing ammonia removal rate as a function of temperature could be derived. Another equation that could be used to predict ammonia oxidation rate as a function of temperature is provided by Knowles et. al (Kaiser and Wheaton, 1983).

Jones and Morita (1985) isolated an ammonia oxidizing bacteria capable of nitrification and growth at temperatures of -5 C. Optimal growth occurred at 22 C for cells grown at 5 C, and lethal temperatures were about 29 C. Cells grown at 25 C had optimal growth temperatures of 30 C, and lethal temperatures of 38 C (Jones & Morita, 1985). Jones & Morita (1985) concluded that temperature acclimation is possible outside of the normally accepted temperature ranges and that environmental temperatures should be used for studying nitrifiers. Temperature optima and inhibitory temperatures were found by many authors and are shown in Table 5.

Table 5. Temperature optima for *Nitrosomonas* (NS) and *Nitrobacter* (NB).

Temperature (C)				
Species	Range	Optimum	Inhibitory	Source
NS	30-36	30-36		Buswell et al. 1953
NS	10-40	30-35		Kawai et al. 1965
NS			<5	Buswell et al. 1953
NB	8-24	28		Nelson 1931
NB	4-45	34-35		Deppe & Engle 1960
NB		42		Laudelout et al. 1960
NB	10-40	30-35		Kawai et al. 1965
NB			<4	Deppe & Engle 1960
NB			>45	Deppe & Engle 1960

CHEMICAL/NUTRIENT ENVIRONMENT

Nitrifiers require more than ammonia or nitrite and carbon dioxide to survive. Obviously, there is need for micro-nutrients and other materials. There also are many compounds that suppress growth or kill nitrifiers. Usually fish culture water contains the necessary nutrient compounds for nitrifiers and few, if any, of the toxic compounds, except ammonia and nitrous acid. However, this is not always true.

It is well known that pH of culture water must be controlled and one common method of doing this is to use a carbonate based rock as a filter media. As noted earlier, it is also known that alkalinity is destroyed by nitrification. Malone and Burden (1987) have concluded that bicarbonate is also a necessary nutrient for biofilters. It supplies an additional carbon source and it buffers the water.

Anthonisen et al. (1976) concluded that free ammonia (NH_3) and free nitrous acid inhibit nitrification. Free ammonia (FA) inhibits *Nitrosomonas* at concentrations higher than those that inhibit *Nitrobacter* (10 - 150 mg/L FA compared to 0.1 to 1.0 mg/L FA). Nitrous acid (FNA) was inhibitory at concentrations of 0.22 to 218 mg/L. Acclimation to FA and FNA, temperature, and the number of active nitrifiers may also influence the inhibitory effects of FA and FNA (Anthonisen, et al. 1976).

Clark & Schmidt (1967a, b) evaluated growth response and uptake of various amino acids and vitamins by nitrifiers. L- glutamic acid, L-aspartic acid, L-serine and L-glutamine were found to be growth stimulating; whereas L-Lysine, L-histidine, L- threonine, L-valine, L-methionine, and L-arginine were inhibitory (Clark & Schmidt 1967a, b). The effects of various mineral ions, and various organic and inorganic compounds were reviewed by Kaiser and Wheaton (1983). Table 6 gives additional data on toxicity of various compounds to *Nitrosomonas* and *Nitrobacter*.

Light

Olson (1981) examined the inhibitory effects of light on both ammonia and nitrite oxidizing bacteria, and found light intensities less than 1% of sunlight intensities, 200 - 300 Lux, were inhibitory (full sunlight is about 100,000 Lux). Light is believed to oxidize cytochrome C in both species of bacteria. *Nitrobacter* is more sensitive to light because it contains less cytochrome C than *Nitrosomonas* (Olson 1981). Horrigan et al. (1981) found similar results for light inhibitors and concluded complete darkness was superior to diurnal cycling of light regimes.

Presence of light in a biofilter also encourages growth of photosynthetic algae which may clog certain filter types and may compete for media space with nitrifying bacteria. However, many species of algae assimilate ammonia and/or nitrite directly, thereby helping to reduce the toxic load. Use of algae versus nitrifying bacteria to control ammonia and nitrite in a culture system appears to be an unresolved question. However, the vast majority of systems in use today depend on nitrifying bacteria for ammonia and nitrite removal.

Table 6. Growth requirements and toxicity of various compounds to *Nitrobacter* (NB) and *Nitrosomonas* (NS).

Species	Compounds	Concentrations	Inhibition	Sources
NS	Ni	0.25 mg/l	mod	1
NS	Cr	0.25 mg/l	mod	1
NS	Cu	0.1 mg/l	slight	1
NS	Cu	0.5 mg/l	complete	1
NS	Stainless EN58B	presence	none	1
NS	Fe	<0.5 mg/l	limits growth	1
NS	L-Arginine	$5x10^{-3}M$	slight	2
NS	L-Histidine	$5x10^{-3}M$	complete	2
NS	L-Phenylalanine	$5x10^{-3}M$	slight	2
NS	L-Tyrosine	$5x10^{-3}M$	slight	2
NS	DL-Methinoninesulphoxide	$5x10^{-3}M$	slight	2
NS	Creatinine	$5x10^{-3}M$	slight	2
NS	Glycocyamine	$5x10^{-3}M$	slight	2
NS	2,4 Dini tophenol	$5x10^{-3}M$	none	2
NS	Ethylurethane	$5x10^{-3}M$	moderate	2
NS	Calcium	50 mg/l	stimulated	1
NB	Iron	7 mg/l	stimulated	3
NB & NS	Phosphate	310 mg/l-P	stimulated	4
NB	Magnesium	5 mg/l	stimulated	3
NB	Phosphorus	5 mg/l	stimulated	3
NB	Molybdenum	1 mg/l	stimulated	3
NB	Zinc	1 mg/l	stimulated	3
NB	Molybdenum	0.1 mg/l- 1mg/l	stimulated	5
NS	Magnesium	5.0 mg/l	stimulated	1
NS	Phosphate	0.28 mg/l	stimulated	1
NS	Iron	2 mg/l	stimulated	1
NS	Ni	0.1 mg/l	stimulated	1
NS	Cr	0.1 mg/l	stimulated	1
NS	Co	1.0 mg/l	stimulated	1
NS	Mn	1.0 mg/l	stimulated	1
NS	Zn	1.0 mg/l	stimulated	1

1 Skinner & Walker 1961
2 Lees 1952
3Aleem 1959
4 Laudelout, et al. 1967
5 Finstein 1965

Ammonia Loading

Biofilters are usually designed based on influent ammonia concentration or a given removal rate at the design temperature. However, Brune and Gunther (1981) demonstrated that for a fixed inlet ammonia concentration between 0.02 and 0.05 mg NH_3-N/L, the ammonia removal rate is higher at short filter detention times, which is equivalent, in this case, to a higher flow rate through the filter (Figure 2). The high ammonia loading rates may have been a reason for the high removal rates. If this is the case, ammonia loading rate may be more important than filter influent ammonia concentration.

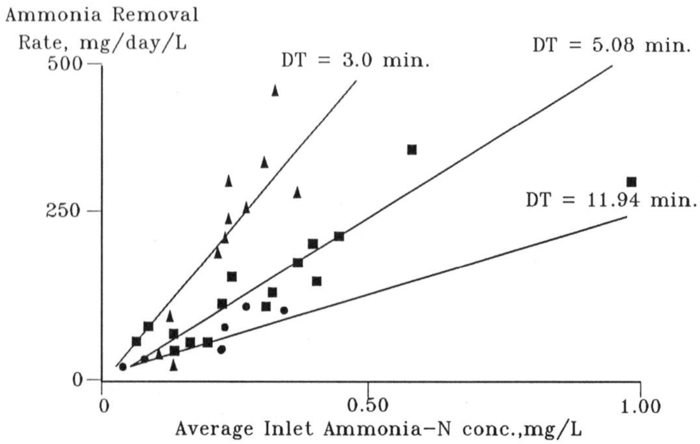

Figure 2. Rate of ammonia oxidation or removal at varying ammonia concentrations and detention times (DT) (see Brune and Gunther 1981 for original plot).

Studies at Oklahoma State University seemed to confirm this principle. The BOD removal in a biological tower was found to be a function of the total biofilter BOD loading rather than the inflow rate or concentration. With the same total organics applied, BOD removal efficiency was the same regardless of whether loading was accomplished by a high flow rate at low waste concentration, or a low flow rate at a high waste concentration (Kincannon and Gaudy 1978). Kaiser and Wheaton (1983) present a more detailed discussion on this topic.

Salinity

Biofilters can operate in a wide range of salinities ranging from fresh water to above full strength sea water. However, biofilters acclimated to one salinity require time to reacclimate if the salinity is changed over about 5 ppt. Fresh water nitrifiers are completely inhibited in saline water. Kawai et al. (1965) found that nitrification in saline filters was maximum when they were operated at constant salinities.

BIOFILTER CONFIGURATIONS

There is an almost endless number of fixed film biofilter configurations that can be used. Fortunately, most configurations can be grouped into one of six general types: submerged, trickling, reciprocating, rotating biological contactors, rotating drums, and fluidized beds.

Submerged Filters

The distinguishing feature of submerged filters is that all the filter media is maintained below the water level. The result is that all oxygen required by the filter must be supplied by the water. This leads to design conflicts that can often only be resolved by compromises in the design.

Water flow may be from top to bottom of these filters (down- flow) or from the bottom up (upflow). Upflow filters tend to have less clogging problems than downflow filters, but clogging is a constraint in either configuration. Clogging results from solids loaded on the filter and dead biomass getting caught between the solid media and plugging the passage-ways between them.

Submerged filters are simple to construct and are thus widely used. Oxygen supply, pressure drop across the filter, clogging, and backwashing requirements are their major limitations.

Trickling Filters

Trickling filters consist of solid media on which the nitrifying bacteria grow and over which the culture water is trickled. The trickling water keeps the filter bacteria wet so they can survive, but does not fill the spaces between the solid media. Thus, air passes through the filter supplying oxygen directly to the bacteria. Because oxygen does not come only from the water, design constraints on flow rate are eased.

Trickling filters are also widely used in sewage treatment as both nitrification and BOD (Biochemical Oxygen Demand) removal systems. BOD removal also occurs in aquaculture biofilters, but BOD removal is usually not as critical as nitrification in aquaculture systems.

Reciprocating Filters

Reciprocating filters are a special case of either submerged or trickling filter. In reciprocating filters, the filter is alternately flooded and drained. Depending on cycle time, oxygen may be available from the air drawn into the filter during draining or from the water. Paller and Lewis (1982) compared reciprocating and submerged filters operating on small- scale fish holding units. They found the reciprocating filter supported 29 % more fish in numbers, 33 % more fish in terms of weight, and a 29 % greater feeding rate. Superior performance of the reciprocating filter appeared to result from resistance to clogging and improved aeration in the filter.

Rotating Biological Contactor

The rotating biological contactor (RBC) consists of a series of disks mounted on a horizontal shaft that are approximately one-half submerged in a tank of culture water. Water flows in a direction parallel to the disk axis. Bacteria grow on the disk surfaces and, as the disk are rotated, the bacteria are alternately dipped into the water and then exposed to air. As long as air exposure is sufficient, the bacteria will metabolize the ammonia and nitrite while drawing oxygen from the air. The plentiful oxygen supply in air eliminates oxygen limitations on growth rate. Clogging is not a problem in these filters.

Rotating Drums

The rotating drum (RD) operates very similarly to the RBC's. The primary difference being that the RD's consist of a mesh covered cylinder that is filled with some type of media on which the nitrifiers can grow. The media can be almost anything that has a large surface area and is not excessively heavy, such as an open cell sponge, ping-pong balls, plastic rings, etc. The RD is half submerged in water and is slowly rotated about its axis, usually by an electric motor. As the RD rotates, water and air (depending on whether that section of the drum is above or below the water level) passes into and out of the drum alternately supplying nutrients and oxygen to the bacteria.

Rotation of the RD causes considerable stirring in the tank in which it operates. If the RD is located in the culture tank, the stirring action may, in some cases, provide all the circulation needed in the culture tank.

Fluidized Bed Biofilters

Fluidized bed filters consist of a vertical tube partially filled with media, typically sand. Water flows upward through the sand at a sufficient velocity to suspend (fluidize) the sand in the water column. Nitrifying bacteria grow on the sand particles. Because the water velocity is high, and the sand is fluidized, the boundary layers are minimal in thickness and any dead bacteria are scraped off the sand particles. Thus the nitrification rate in fluidized beds is usually quite high. Clogging is not a problem with these filters. They are very compact in size, and thus are useful in areas having space limitations.

Comparison of Biofilter Types

Kaiser and Wheaton (1983) compared 9 aquaculture recycle systems which included submerged upflow, RBC, trickling filter (TF), and biodrum filter types. They concluded that the biodrum and RBC type filters performed best. Miller and Libey (1985) tested an RBC, TF and fluidized bed reactor (FBR) simultaneously. The RBC provided the best nitrification efficiencies (74 to 82%), the TF was next (23 to 51%) and the FBR ranked last (8 to 32%). Rogers and Klemetson (1985) tested an RBC, biodrum, TF, and submerged filter simultaneously. The RBC provided the best ammonia removal (over 90%) for hydraulic loading rates up to 0.06 $m^3m^{-2}day^{-1}$ and influent ammonia concentrations up to 1.3 mg/L. The biodrum, which contained over 2 times greater surface area than did the RBC, removed over 80% for hydraulic loading rates up to 0.05 $m^3m^{-2}day^{-1}$ and influent ammonia concentrations up to 0.03 mg/L. The TF, with over 4.5 times the surface area of the RBC removed 50% for hydraulic loading rates of 0.012 $m^3m^{-2}day^{-1}$ and ammonia concentrations up to 2.36 mg/L. Klemetson and Cohen (1984) determined the required biological filter areas versus hydraulic loading rates for RBC, biodrum and TF type filters. Again, RBC's were the most efficient with 70.5 to 99.5% efficiency corresponding to hydraulic loading rates of 0.102 to 0.004 $m^3m^{-2}day^{-1}$. Biodrums ranked second with 85.2 to 98.6% efficiency corresponding to hydraulic loading rates of 0.035 to 0.004 $m^3m^{-2}day^{-1}$. Trickling filters ranked last with efficiencies between 34.5 to 58% for hydraulic loading rates between 0.035 and 0.008 $m^3m^{-2}day^{-1}$. Also, the total costs as a function of unit costs were calculated and the RBC's were least expensive followed by the biodrum, then the trickling filter.

MEDIA TYPE

Nearly all biological filters utilize a fixed or solid media to provide surface area on which the bacteria can attach and grow. A wide variety of materials have been tried including sand, rocks, glass balls, plastic rings and balls, oyster and clam shells, dolomite, etc. A review of the influence of some media types on filter performance is presented by Kaiser and Wheaton (1983).

Media selection for column filters has two conflicting design requirements: 1) provide as much surface area as possible per unit volume to provide space for nitrifier growth, and 2) provide large open space to prevent clogging. The smaller the particle size, the greater surface area per unit volume, but the greater is the clogging problem. Experience indicated particle size of 1.5 to 2 cm or larger in nominal diameter with few, if any fines, works well in most recirculating culture systems.

Media type also influences the void ratio - the volume of air spaces in the dry filter divided by the total media volume in the filter. Large void volumes tend to reduce clogging and increase water residence time (and hence contact time between the water and nitrifiers) in the filter. Increased residence times tend to favor increased removal rates (Laio et al. 1972).

Although solid media are nearly always used as an attachment area for nitrifiers, solid media are not necessarily a requirement for the bacteria. In fact, evidence exists to indicate nitrifiers are more efficient in suspension than when attached to solid media. Audic et al. (1984) measured the oxygen consumption of *Nitrobacter* in freely suspended bacteria as compared to bacteria attached to a surface. The specific respiratory activity, oxygen consumption for 1010 bacteria, was over two times greater; 760 uL O2/1010 bacteria for the attached bacteria and 1700 uL 02/1010 bacteria for the freely suspended bacteria (Audic et al. (1984)). More research needs to be conducted in this area.

FILTER START-UP

Biofilters have to be "broken-in" when first started or when significant load changes occur. This process of establishing the microbial community within the filter results in a typical ammonia, nitrite, and nitrate curve (Figure 3). The ammonia concentration increases due to production by the cultured

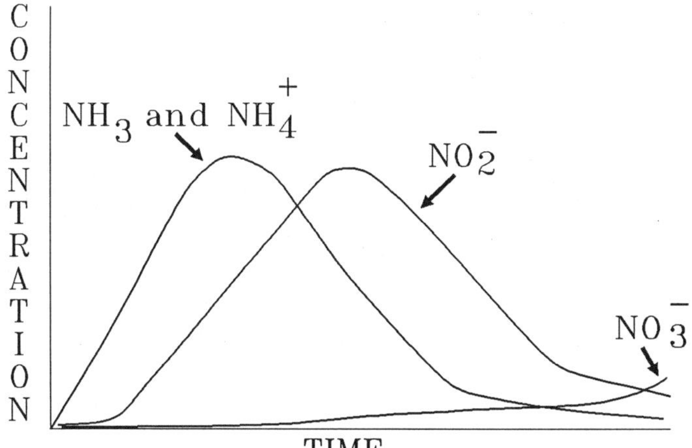

Figure 3. Typical break-in curves for a biofilter.

organisms. This stimulates growth of *Nitrosomonas* which converts the ammonia to nitrite. When the *Nitrosomonas* population is sufficiently high the ammonia concentration falls. At the same time the nitrite concentration increases, stimulating growth of *Nitrobacter*. When the *Nitrobacter* population is sufficiently high it removes nitrite faster than it is produced and the nitrite concentration falls. In a stable, "broken-in," filter both ammonia and nitrite concentrations are low and nitrate concentration slowly increases unless there is a method of removing nitrates from the culture system (e.g., denitrification or plant uptake).

Stabilizing a new biofilter typically takes 20 to 35 days. Stabilizing time for an operating filter after a significant load change will depend on the magnitude of the load change, but often takes 10 days or longer. The slow response to load changes and the long start-up time are major disadvantages of biofilters. Developing methods that reduce filter response time to either start-up or load changes would relieve biofilter managers of a major problem.

Malone and Manthe (1985) investigated biofilter performance at start-up and compared media size, commercial bacterial additives, and chemical additives. Media studies revealed a start- up time of 37 days for three different sizes of media (2 to 5 mm limestone, 5 to 10 mm limestone, and 13 to 20 mm limestone). The authors concluded media size had no effect on acclimation time. Filters were started, run for twelve days, and a commercial bacterial additive was added followed by crabs to load the system. The addition of bacteria did not reduce the start-up time (36 days) significantly. Malone and Manthe (1985) predicted that seeding biofilters with media from established biofilters exposed to similar salinities would probably be more successful. The addition of ammonium chloride alone did not reduce the acclimation time, but did eliminate the need for animals to start-up the system. Finally, Malone and Manthe (1985) concluded that ammonium chloride and sodium nitrite additions to biofilters in addition to commercial bacterial mixes significantly reduced filter acclimation time, by about 10 days. Malone and Manthe (1985) indicated that further research was needed in the area of artificial start-up of biofilters, and that reliable and reduced acclimation times of crab shedding biofilters holds great promise.

DESIGN OF BIOFILTERS

Biofilter design in its simplest terms reduces to balancing the biofilter removal rate with the ammonia production rate of the culture system. This balance varies with many factors including temperature, oxygen availability, fish size and species, total weight of fish or shellfish in the system, amount and type of feed fed, frequency of feeding, water circulation rate, percent of water recycled, types and efficiencies of other filters in the system (e.g. settling basins, foam fractionator, etc.) and other variables. The design must consider all these factors and provide a design that will keep the ammonia and nitrite concentrations at all times below lethal levels. An occasional short term rise in ammonia or nitrite concentrations into the stress level can be tolerated, but is quite undesirable.

The first information needed for design is the waste production expected from the cultured organisms over the major stages of their life cycle, at least that portion of the life cycle the organisms spend in the culture system. From this data the designer must determine the greatest load the filter will experience. The heaviest load is usually when the cultured organisms are at the largest size they ever achieve in the culture system. However, depending on the system management used, the heaviest ammonia load could occur at a different time. For example, if the fish are raised in high density and are periodically thinned to a lower density as they grow, the highest load may occur just before thinning of small fish because small fish generate more ammonia per unit weight than do larger fish.

Early in the design process a decision needs to be made as to whether all or only part of the culture system flow will pass through the filter. The filter type to be used (i.e., submerged, trickling, etc.) also needs to be decided early in the design. For any specific design the filter type must be selected early. However, the filter type may be finalized late in the design stage after possible designs for several filter types are developed. These designs can then be compared and the best selected. The filter type must be selected because, for example, the flow rates of submerged filters are often determined by filter oxygen needs whereas oxygen needs of a trickling filter or a rotating drum are independent of water flow rate.

The ammonia, or occasionally nitrite, removal rates are needed next. This value is a function of temperature, nutrients available and other factors. This data is often not very precise because there is limited data available and exactly what the real value will be in the final filter depends on many variables. Thus, designer judgement plays a significant role in selection of this parameter.

Once the production and removal rates are known, the design must balance the two by selecting the necessary media surface area. The surface area is, of course, a function of the media size, type, void volume and other factors. Thus, a filter media needs to be selected.

Detention time in the filter is somewhat controversial. Figure 2 shows that detention times as low as 3.0 min give good removal rates. In fact, low detention times may give better removal rates than longer ones, Figure 2. However, other investigators suggest removal efficiencies are directly dependent on detention time in the filter (Liao et al. 1972).

Because biofilter kinetics and the filter response to changes in a number of variables are not experimentally defined, filter design requires considerable judgement on the part of the designer. Considerable data has been presented in this paper and in many of the references that appear to be directly applicable to design of biofilters. Although some of the data is usable for biofilter design, all of it put together is not sufficient to design a filter. The data available are taken from individual studies, many of which were not set-up to develop biofilter design data. Data taken from studies on biofilters often suffers from limitations that make the data unusable for developing a general biofilter design process model. For example, a study may give ammonia removal rates for a biofilter but omit the values of pH, temperature, etc. that were used in the study. Such data may be valid for the biofilter used when operating under the study conditions, but if the conditions are not all known, the results cannot be used to make any general statement about filter performance. In fact, the data cannot even be used as one point on a n-dimensional (where n = the number of variables influencing the filter performance) response surface that, if known, could form a reliable design methodology. Thus, to develop a good design methodology, a multivariate response surface having n dimensions is needed. These relationships can also be considered as a biofilter model which expresses ammonia removal, the dependent variable, as a function of a group of independent variables, the variables affecting ammonia removal in the filter. The reason much of the experimental data reported in the literature is unusable for design is that only a few of the variables affecting filter operation are even considered in most experiments. Those variables not considered usually are not controlled or in many cases not measured or reported. Such data cannot be placed on a multivariate response surface. Thus, from the standpoint of developing a design methodology for biofilters, such data is of little use.

In conclusion, if a design methodology is to be developed for biofilter design, experiments must be designed and executed such that each experiment can be plotted on an n-dimensional surface (i.e. every investigator monitors and controls each significant variable affecting filter performance) or, a long range research program must be organized and funded to obtain the needed data. Because the first alternative is probably not viable, there is a great need for long term research.

LITERATURE CITED

Aleem, M. I. H. 1959. The physiology and chemo-autotrophic metabolism of *Nitrobacter agilis*. Thesis. Cornell University, New York.

Anonymous. 1971. Notes on water pollution, No. 52. Water Pollution Research Laboratory, Stevenage, England.

Anonymous. 1965. Water Pollution Research, London:HMSO. Original not available for examination; cited in U.S. EPA (1975).

Anthonisen, A. C., R.C. Loehr, T.B.S. Prakasam, and E. G. Srinath. 1976. Inhibition of nitrification of ammonia and nitrous acid., Journal of the Water Pollution Control Federation 48(5):835-852.

Audic, J. M., G.M. Faup, and J. M. Navario. 1984. Specific activity of nitrobacter through attachment on granular media. Water Research 18(6):745-750.

Benefield, L. and F. Molz. 1985. Mathematical simulation of a biofilm process. Biotechnology and Bioengineering 27:921-931.

Boon, B. and H. Laudelout. 1962. Kinetics of nitrite oxidation by *Nitrobacter winogradskyi*. Biochem. J. 85:440-447.

Brune, D. E. and D. C. Gunther. 1981. The design of a new high rate nitrification filter for aquaculture water reuse. Journal of World Mariculture Society 12(1):20-31.

Buswell, A.M., T.Shiota, N. Lawrence, and I. Van Meter. 1953. Laboratory Studies on the kinetics of the growth of *Nitrosomonas* with the relation to the nitrification phase of the BOD test. Applied Microbiology 2:21-25.

Clark, C. and E. L. Schmidt. 1967a. Growth response of *Nitrosomonas europea* to amino acids. J. of Bacteriology 93(4)1307-1308.

Clark, C. and E.L. Schmidt. 1967b. Uptake and utilization of amino acids by resting cells of *nitrosomonas europea*. Journal of Bacteriology 93(4):1309-1315.

Collins, M. T., J. B. Gratzek, D. L. Dawe, L. M. Campbell, D. R. Senn. 1975a. Nitrification in an aquatic recirculating system. Journal of the Fisheries Research Board of Canada. 32(11).

Collins, M. T., J. B. Gratzek, D. O. Dawe, and T. G. Nemetz. 1975b. Effects of parasiticides on nitrification. Journal of the Fisheries Research Board of Canada. 32(11).

Collins, M. T., J. B. Gratzek, D. L. Dawe, and T. G. Nemetz. 1976. Effects of antibacterial agents on nitrification in aquatic recirculating systems. Journal of the Fisheries Research Board of Canada. 33:215-218.

Cutter, D. W. and L. M. Crump. 1983. Some aspects of the physiology of certain nitrite forming bacteria. Annals of Applied Biology 20:291-296.

Deppe. K.J. and M.S. Engel. 1960. Untersuchunger iiberidie temperatuvabhang der nitratbildung durch Nitrobacter wimogradski buch. Bei undehem und gehemmtem washtum. Zentlol. Bakt. Parasitkde II: 113:561-568.

Engel, M. S. and M. Alexander. 1958. Growth and autotrophic metabolism of *Nitrosomonas europea*. J. Bact. 76:217-222.

Fenchel, T. 1979. Bacteria and Mineral Cycling, Academic Press, Inc. London.

Finstein, M. S. and C.C. Delwiche. 1965. Molybdenum as a micro nutrient for *Nitrobacter*. J. Bact. 89:123-128.

Gould, G. W. and H. Lees. 1960. The isolation and culture of the nitrifying organisms. Part I. *Nitrobacter*. Can. J. Microbiol. 6:299-307.

Grady, C.P.L. and H.C. Lim. 1980. Biological wastewater treatment. Marcel Dekker Inc. New York.

Haug, R. and P. McCarty. 1971. Nitrification with the submerged filter. Technical Report No. 149, Department of Civil Engineering, Stanford Univ., Palo Alto, California.

Hofman, T. and H. Lees. 1952. The biochemistry of the nitrifying organisms. Biochem. 52:140-142.

Horrigan, S. G., A.F. Carlucci, and P. M. Williams. 1981. Light inhibitions of nitrification in sea-surface films. J. Marine Research 39(3):557-565.

Jones, R. D. and R.Y. Morita. 1985. Low temperature growth and whole cell kinetics of a marine ammonium oxidizer. Marine Ecology Progress Series 21:239-243.

Kaiser, G. E. and F. W. Wheaton. 1983. Nitrification filters for aquatic culture systems: state of the art. Journal of the World Mariculture Society 14:302-324.

Kawai, A., Y. Yoshida and M. Kimata. 1965. Biochemical studies on the bacteria in the aquarium with circulating system. II. Nitrifying activity of the filter sand. Bull. Japanese Society of Scientific Fisheries 31:65.

Kawai, A., Yoshida, Y., and M. Kimata. 1965. Biochemical studies on the bacteria in aquarium with circulating system - I. Changes of the qualities of the breeding water and bacterial population of the aquarium during fish cultivation. Bull. Jap. Soc. Sci. Fish. 31(1)65-71.

Kholdebarin, B. and J.J. Oertli. 1977. Effect of pH and ammonia on the rate of nitrification of surface water. Journal of the Water Pollution Control Federation 49:1688-1692.

Kincannon, D. F. and A. F. Gaudy, Jr. 1978. Functional design of aerobic biological waste water treatment processes. Bioenvironmental Engineering School of Civil Engineering, Oklahoma State University, Stillwater, Oklahoma.

Klemetson, S. L. and D. Cohen. 1984. System analysis and design of biofilters for the three components of *Macrobrachium* production. Final Report, Project Number US-60-80 of BARD (U.S.-Israel Binational Agricultural Research and Development Fund). Washington, D.C.

Kumar, S. 1984. Closed recirculation nitrification systems for soft shell crabs. Unpublished Master's Thesis. Louisiana State University, Baton Rouge, Louisiana.

Laudelout, H. and L. Van Tichelen. 1960. Kinetics of the nitrite oxidation by *Nitrobacter winogradsky*. J. Bacteriology 79:39- 42.

Lawrence, A. and P. McCarty. 1969. A kinetic approach to biological wastewater treatment design and operation. Technical Report No. 23, Cornell University, Water Resources and Marine Center, Ithaca, New York. Original not available for examination; cited in Srna and Baggaley (1975).

Lees, H. 1952. The biochemistry of the nitrifying organism. Biochem. J. 52:134-139.

Lees, H. and J. R. Simpson. 1957. The biochemistry of the nitrifying organisms. Biochem. J. 65:297-305.

Liao, P. B., R. D. Mayo, and W. Williams. 1972. A study for development of fish hatchery water treatment systems. Contract No. DACW68-71-0059, U.S. Department ofthe Army, Walla Walla District, Walla Walla, Washington.

Lomax, K. M. 1976. Nitrification with waste treatment on a closed cycle catfish culture system. Unpublished Ph.D. Dissertation, University of Maryland, College Park, Maryland.

Loveless J. E., and H. A. Painter. 1969. The influence of metal ion concentration and pH value on the growth of a nitrosomonas strain isolated from activated sludge. J. Gen Microbiol. 52:1-14.

Malone, R. F. 1987. Professor, Civil Engineering Department, Louisiana State University, Baton Rouge, Personal Communication.

Malone, R.F. and D.G. Burden. 1987. High rate nitrification in recirculating blue crab shedding systems. Paper presented at 2nd National Symposium on Soft-shell Blue Crab Fisheries. Gloucester Point, Va.

Malone, R. F. and D. P. Manthe. 1985. Chemical addition for accelerated nitrification of biological filters in closed blue crab shedding systems. National Symposium on the Soft-Shelled Blue Crab Fishers, Ocean Springs, Mississippi.

Manthe, D. P., R.F. Malone and S. Kuman. 1984. Limiting factors associated with nitrification in closed blue crab shedding systems. Aquacultural Engineering 3:119-140.

Meade, T. L. 1974. The technology of closed system culture of salmonids. Marine Technological Report 30. Animal Science/NOAA Sea Grant, Narragansett Bay, University of Rhode Island, Narragansett, Rhode Island.

Meyerhof, O. 1917. Untersuchunger uber den atmungsuourgong nitrifizierenden bakterien. Pflugers Archges Physiol. 166:240-280.

Miller, G. E. and G. S. Libey. 1985. Evaluation of three biological filters suitable for aqaculture applications. J. World Mariculture Society 16:158-168.

Monod, J. 1942. Reserches sur la croissance des cultures. Hermann and Cic, Paris. Original not available for examination; cited in U.S. EPA (1975).

Nagel, C. A. and J. G. Haworth. 1969. Operational factors affecting nitrification in the activated sludge process. Presented at the 42nd Annual Conference of the Water Pollution Control Federation, Dallas, Texas. Original not available for examination; cited in U.S. EPA(1975).

Nelson, D. H. 1931. Isolation and characterization of *Nitrosomonas* and *Nitrobacter*. Zenthol Bakt Parasit. Abt II 83:280-311.

Olson, R. J. 1981. Differential photoinhibition of marine nitrifying bacteria, J. Marine Research 39(2):227-238.

Painter, H. A. 1970. A review of literature on inorganic nitrogen metabolism in microorganisms. Water Research 4:393-450.

Paller, M. H. and W. M. Lewis. 1982. Reciprocating biofilter for water reuse in aquaculture. Aquacultural Engineering 1:139- 151.

Roberts, J. 1985. Mathematical Models for the trickling filter process. Pages 243-324. *In*: S.E. Jorgensen and M.J. Gromiec, editors. Mathematical models in biological wastewater treatment. Elsevier, Amsterdam.

Rogers, G. L. and S. L. Klemetson. 1985. Ammonia removal in selected aquaculture water reuse biofilters. Aquacultural Engineering 4:135-154

Schwer, A. D. 1971. Letter communication to D. S. Parker - Metropolitan Sewer District of Greater Cincinnati. Original not available for examination; cited in U.S. EPA (1975).

Skinner, F. A. and N. Walker. 1961. Growth of *Nitrosomonas europea* in batch and continuous culture. Archiv fur Mikrobiologie 38:339-349.

Ulken, A. 1963. Die Herkunft des nitrotsin der elbe. Arch Hydrobiol. 59:486-501. Original not available, cited in: U.S. Environmental Protection Agency. 1975. Process design manual for nitrogen control. A design manual prepared for the Office of Technology Transfer of the U.S. EPA

Water Pollution Control Federation. 1983. Nutrient Control, Manual of Practice. No. FD-7, Washington, DC.

Weatherly, L.R. 1982. Application of simple dynamic response analysis to a recirculating aquaculture system; a review. Aquaculture Engineering 1: 93-103.

Wheaton, F. W. 1977. Aquacultural Engineering, John Wiley and Sons. New York.

Winogradsky, S. and H. Winogradsky. 1933. Etudes sur la microbiologie clu sol. Annls Inst. Pasteur, Paris 50:350-432.

ADVANCES IN ON-LINE WATER QUALITY MONITORING

Charles R. Gibbs

ABSTRACT

On-line water quality monitoring is a field undergoing rapid change. Process control has been recognized as a key means of productivity improvement but is dependent upon on-line analysis. Two main classes of continuous water analyzers are covered: those using sensors but no reagents and wet-chemistry analyzers. Current instrument technology and developments in a number of areas of research are reviewed. Trends and future requirements for practical analyzers and broader application of on-line water quality monitoring are discussed. As understanding of the importance of water quality to aquaculture increases, the demand for better on-line monitoring tools will increase. This paper discusses key concepts of on-line water quality monitoring, the main areas of research activity, and significant developments in the field.

BACKGROUND AND THEORY

Aquaculture can be seen as a process with inputs, outputs and internal cycles as shown in Figure 1. The quality of the medium in which the process occurs—water—affects the process and must be measured throughout the cycle. Water testing is currently performed manually on a grab sample basis in the vast majority of cases but interest in on-line monitoring is growing.

The purpose of on-line measurements is not just in data collection but in optimizing the process though feedback and feedforward mechanisms to adjust and control treatment as conditions change (Fig. 2). On-line analyzers are a key part of the trend toward automated control systems (Sherman 1988; Yeh 1986). These improvements are required to meet growing needs in the areas of efficiency, product quality, environmental protection and the reuse and recycling of water (Borman 1987).

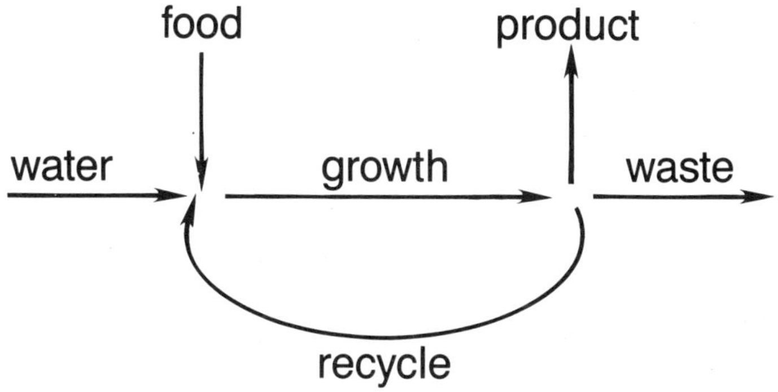

Figure 1: Aquaculture can be viewed as a process with inputs, outputs and loops, like other production processes.

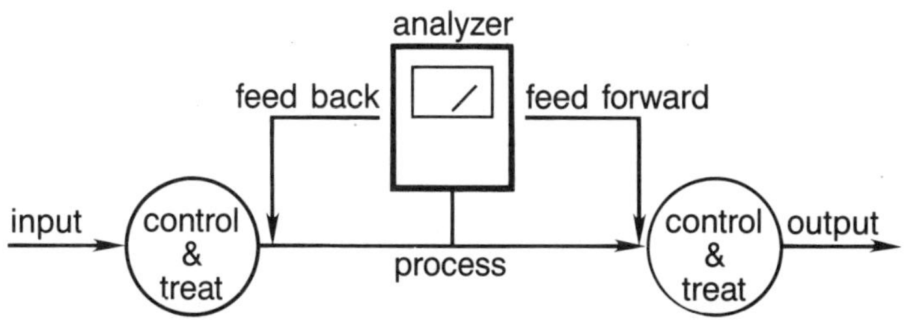

Figure 2: The purpose of on-line analysis is to provide the data needed for improved control and treatment efficiencies through feed back and feed forward loops.

In addition, recognition of the need for on-line testing is growing as a result of demands for: (1)real time control of processes, (2)reduction of variation for improvements in productivity and quality, (3)quicker response to accommodate the faster changes and worse consequences as aquaculture systems are pushed closer to the limits of instability by more intensive production, and (4)automation to reduce labor yet yield more analytical results at a lower cost. In certain industry segments, such as power generation, these needs have resulted in widespread use of automated analysis.

Broad implementation of on-line monitoring has been slow, however, because it requires a significant effort to define, select, install and integrate systems and must be carried out by an interdisciplinary group of people working cooperatively together. At the same time, some of the people involved may lack confidence in on-line analysis and no one can guarantee them an adequate return for their invested effort.

Thus, the biggest advance in the near future could be the widespread conviction that the reward is worth the effort. The recognition of the necessity and value of on-line water quality monitoring will spread into many areas in a ripple effect from the acceptance of process analytical chemistry as a valid and separate discipline. This acceptance will result in more respect, more effort and more willingness to try new applications. The success of these new applications will show the value of process monitoring and control.

One of the rocks that has been thrown in the pool to create these ripples is the formation of academic institutes such as the Center for Process Analytical Chemistry (CPAC) at the University of Washington (Callis et al. 1987). These centers are bringing a new perspective and fresh look at the basic principles of on-line analysis. One of CPAC's key contributions is in clearly communicating the idea that on-line analysis is not optimized by retroactively fitting lab instruments into the process environment. What is required is looking at user needs to define the ideal analyzer for a given application and establishing principles of design for that situation.

In general terms, however, some of the critical needs for an on-line analyzer are to be: (1)reliable, to keep working and giving accurate answers; (2)simple, to avoid the "black box" syndrome which creates distrust of results due to lack of understanding of the method involved in obtaining them; (3) low maintenance, to be reliable without effort and upkeep; and (4)easy to use with little training or set-up. Meeting all these criteria generally requires an instrument dedicated to a particular analysis in a particular application.

ANALYZERS USING SENSORS

In looking at what is currently available for on-line water quality monitoring that meets these criteria, it is generally conceded that the mature technologies consist of monitors for pH, dissolved oxygen and conductivity. These instruments consist of a very simple system with sensor in the sample and a readout of results (Fig. 3). In general, the two parts involved are fairly well developed and proven.

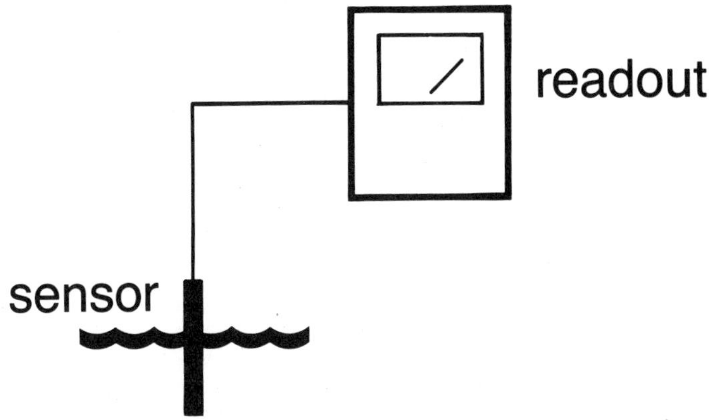

Figure 3: Electrochemical analyzers are the simplest type of system consisting of a sensor immersed in the sample and a readout of the signal generated.

There will be some improvements in this group of instruments in the near future such as broad application of microprocessors for self-diagnostics and calibrations; better reliability particularly in the area of sensors; and non-clogging reference junctions for pH.

Currently, as high as 90% of pH electrode failures (excluding breakage) on process lines in power plants (with very clean water) are due to clogged junctions. Several approaches are being used in the laboratory to avoid this problem and they should soon be available. For example, the Hach One Junction uses an open tube with a positive pumping of reference solutions to continuously refresh the junction. This is contrasted with the conventional frit junction (Fig. 4). This open junction has increased useful lifetime under difficult conditions by over 5 times.

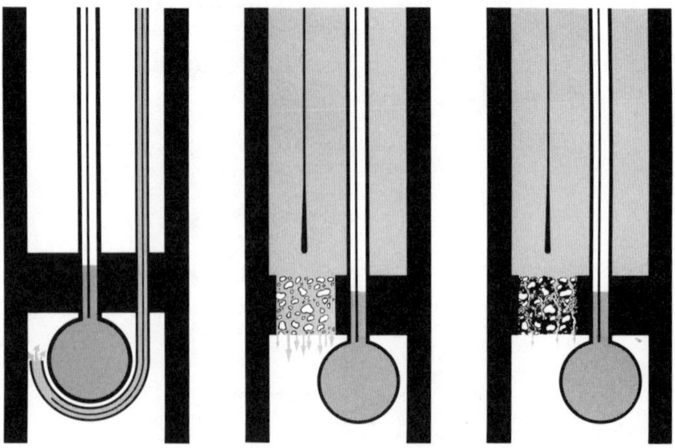

Figure 4: The reference portion of a conventional pH electrode (center) works by leaking electrolyte through a porous frit. Eventually, the frit clogs (right) and the electrode must be replaced. The Hach One electrode (left) is an example of a free-flowing junction with the electrolyte pumped out through an open tube which does not clog thus producing longer useful life.

It would be particularly useful if other parameters could be monitored in a similar fashion to pH by just substituting other sensors in this type of analytical system. For this reason, a great deal of research is under way in the areas of solid state and fiber optic sensors and biosensors.

All these types of sensors require a two-stage process: (1) selectivity to isolate the parameter of interest by chemical reaction, adsorption, or membrane transport, or other process; and (2) measurement of response resulting in an electrical signal. Work is occurring in both aspects but is concentrated in development of the selectivity step since that offers the potential for the most significant advances.

Solid state sensors include devices like Field Effect Transistors (FETs) and Piezoelectric and Surface Acoustical Wave (SAW) sensors (Harmer

1987; Goldman 1988; Bergveld and van der Schoot 1988). They all consist of a coating applied to an electronic device so that as the coating changes in response to a change in the concentration of the analyte, the output of the device changes (Fig. 5). For example, the vibrational frequency of a piezoelectric crystal might change as water is absorbed by a surface coating. This has been used in monitoring drying ovens. Solid state sensors are in the early stages of work in selectivity and improvements in ruggedness of coatings and speed, range and isolation of response.

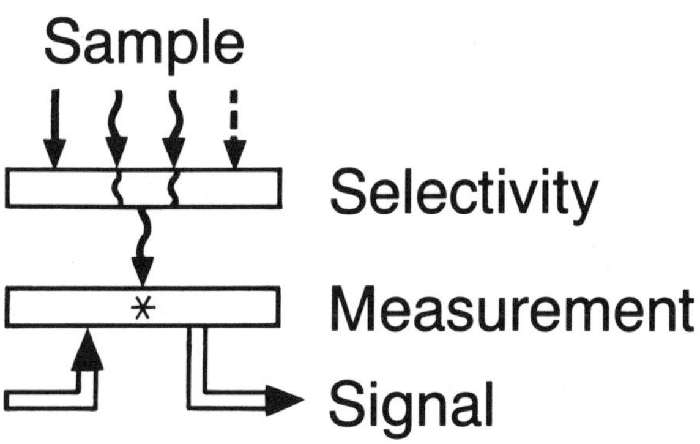

Figure 5: Most sensors require both a method of selecting the species of interest from the sample and a means of generating a signal proportional to the species to complete the measurement.

These types of sensors do offer real opportunities because of their potential for low cost due to mass production techniques. An additional possibility lies in the use of arrays of semi-selective sensors combined with chemometric computational techniques to separate overlapping signals. Their first successes will likely be in applications in defined atmosphere gaseous processes in the chemical industry but there is no reason their application should be limited.

Optrodes (optical fiber sensors) measure an optical change in a chemical indicating substance caused by the parameter being monitored (Miller 1987; Nayanaswamy 1987). Typical techniques include use of single and bifurcated fiber reflective sensors and evanescent sensors which measure a change in transmission as the coating on the fiber changes (Fig. 6).

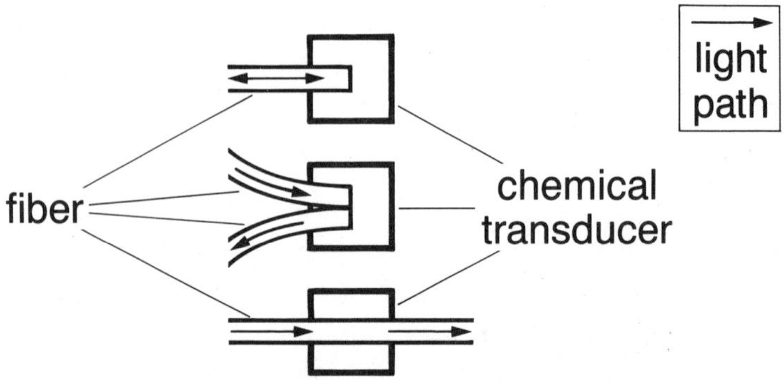

Figure 6: Three typical configurations of sensors using optical fibers.

Examples of analyses which have been demonstrated using optrodes include changes in color or fluorescence of bonded indicators and changes in the color of dissolved indicators caused by diffusion of a gas through the membrane retaining the solution. Immunoreagents have also found use in optrodes.

While optrodes are probably closer to commercial availability than solid state sensors, work is needed in many areas. Selectivity of membranes or indicator reactions and efficiency of mass transference steps must be improved. Stability of bound reagents against the effects of bleaching or leaching by samples must be lengthened for longer sensor life. The problem of limited range of response due to transport equilibrium effects needs to be solved. In addition, reduction of environmental effects like temperature or ionic strength on response will facilitate application.

Biosensors are merely sensors using biological-based chemistry. They are commercially available in the laboratory for glucose measurements (Best 1987) and appear to be particularly applicable to biomolecules and organic compounds such as pesticides and mycotoxins. The strength of biosensors is that they possess, in theory, the ideal selectivity characteristics of enzymes or antibodies. A great deal of basic research has been done of this technology in the medical and biochemistry area which can be applied to analytical applications (Coulet 1988; Turner et al 1987). Their weakness is the difficulty of transducing the reaction to an output signal in a simple and reliable manner.

A great deal of active applied analytical research on biosensors is being carried out primarily in the areas of: (1) monoclonal antibodies which can improve selectivity and uniformity while reducing costs; (2) combined systems of reactions and improvement of their efficiency; and (3) applying the advantages of fiber optics to enzyme-indicator systems. Biosensors will be a major force in laboratory analysis in the future because the methods they replace are often clumsy, expensive chromatographic techniques (Rechnitz 1988). From the lab they should move swiftly to the field as stability and system integration challenges are met.

WET-CHEMISTRY ANALYZERS

Analytical systems which use reagents, such as Ion Selective Electrodes (ISE), or Colorimetric and Titrimetric methods, are another major approach to on-line monitoring and will remain a mainstay in the future. A major disadvantage of this system (Fig. 7) is the requirement for transport of reagent and sample. However, the vast amount of chemical knowledge that can be applied to these systems is a big advantage. Many problems of selectivity, response and stability have been solved.

Colorimetric methods remain the main testing approach to water quality. They have been applied to on-line analysis in two ways: by conversion of laboratory analyzers designed for multiple samples; and by design of new instruments to sample one or more streams continuously or periodically. In fact, colorimetric on-line water quality monitoring instruments of the latter type have been available for nearly 20 years but it can not yet be considered a mature technology with the exception of certain applications.

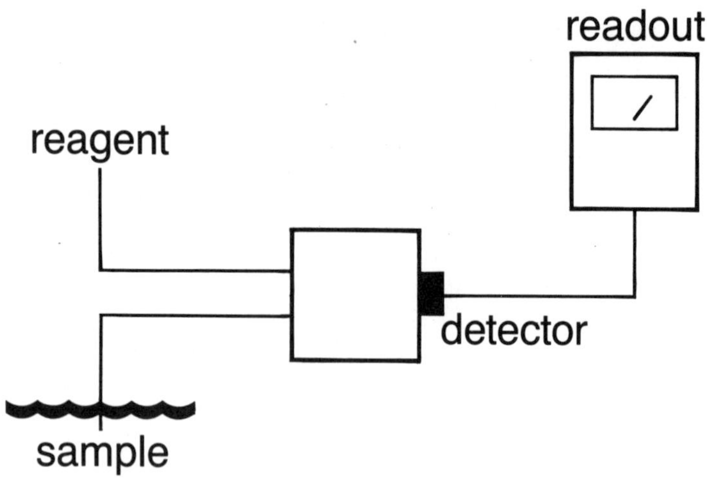

Figure 7: A reagent-using analyzer must transport the sample and reagent to the detector for measurement

Similar approaches to those just mentioned are being used for automating ISE methods. This technique could see more activity in the future because it is a good method for continuous monitoring. Work is continuing in solving problems in the areas of: (1) stability—to reduce noise, drift, reference junction clogging and response time while improving sensitivity; (2) reliability—to prolong electrode life and improve solution transport mechanisms, (3) chemistry—to improve selectivity and remove interferences; and (4) mechanical components—to simplify systems and reduce costs. Nevertheless, in the next 5 years you could see a variety of on-line ISE analyzers available at costs below $3000 that will be very useful for water quality applications.

Flow Injection Analysis (FIA) is an approach to instrument automation that is producing a great deal of activity in the area of colorimetric analysis (Ruzicka and Hansen 1981; Valcarcel and Luque de Castro 1987). At this point FIA is mainly used in the laboratory and for research. Many variations are possible besides the concept shown in Figure 8, but all FIA schemes are dependent upon non-steady state measurements resulting from controlled dispersion thru precise timing of pumps and valves. A sample is injected into

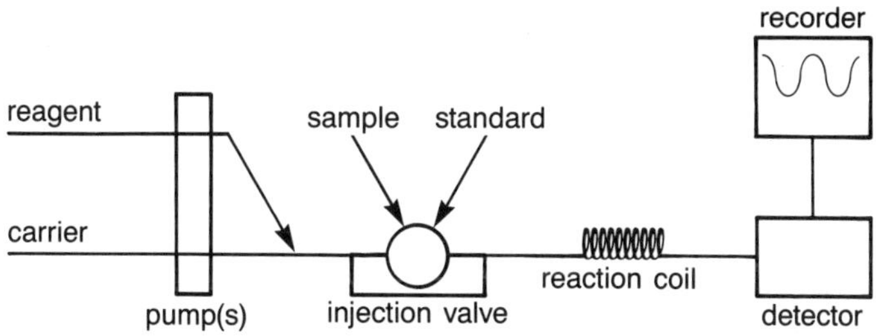

Figure 8: Typical schematic of a flow injection analyzer.

a continuously moving carrier/reagent. As it moves along it diffuses into the stream and reacts to form a color which is measured in a flow-thru cell and the peaks of absorbance are recorded and quantified.

FIA is derived from segmented flow analyzers such as the famous Technicon Autoanalyzer and has similar strengths and weaknesses: high throughput of hundreds of samples/hr; highly repeatable precision and timing; use of small sample volume (significant if sample is limited or valuable); ability to be switched between chemistries with a reasonable effort; and high cost and complexity due to the features listed above. In addition, start-up and operating expenses are significant and a trained technician is required.

Some commercial FIA on-line models have recently become available. The emerging patterns in research (Hansen 1988; Frenzel 1988) seem to include: (1) more variety and versatility in the methods and pretreatments possible; (2) miniaturization including dedicated microchannel manifold and packed minireactors; (3) exploitation of controlled dispersion and its capabilities for providing more information than merely peak height; and (4) new flow through-detectors.

This type of research is moving FIA into a viable position for on-line monitoring applications (Christian and Ruzicka 1988). What remains to be done for FIA to achieve wide spread on-line use includes the development and availability of off-the-shelf dedicated models with lower instrument cost and reduced reagent consumption. In addition, improved reliability because of the critical nature of flow and timing is required along with ease of use so that less-trained personnel can just plug the instrument in and keep it going. Some of the problems with FIA as an on-line monitoring approach are inherent in the concept. Because it is not a steady-state measurement, FIA requires frequent standardization and maintenance of precision. Pumps are always a weak point as is the need for flow recalibrations after tubing changes. There is limited flexibility for multiple reagents and variable sampling. Finally, the use of small samples and a throughput of hundreds of samples per hour are definitely more than necessary for on-line water quality monitoring.

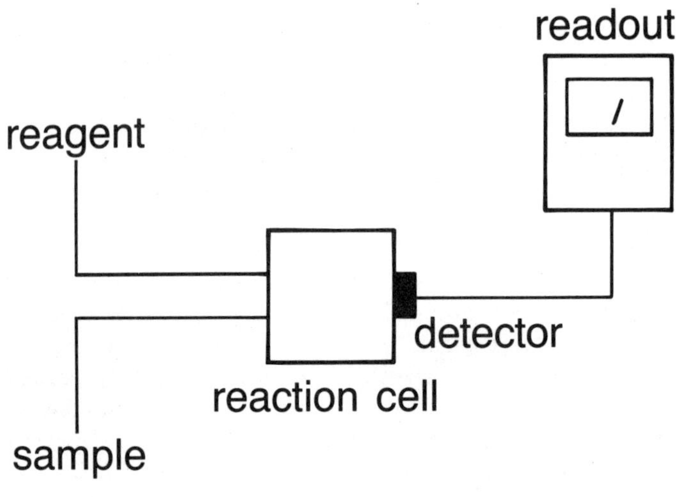

Figure 9: Stopped flow or batch mode analyzers simply bring the reagent and sample to a reaction cell where the mixture resides for the appropriate length of time until the analytical reaction is completed and it is ready for measurement.

An alternate approach, Stopped-Flow or Batch Mode Analyzers, is a simpler concept, more applicable to on-line analysis (Fig. 9). The carrier stream is eliminated. A steady state is easily achieved, and it has a well developed history for on-line water quality monitoring.

For example, the chlorine analyzer shown in Figure 10 costs less than $2000 (one-tenth an on-line FIA) and has been installed in over 700 water and wastewater plants during the last two years. Its operational features are typical of those that a good, practical monitor should have: (1) one analysis is carried out on a regular basis (every 2 min in this case); (2) zero is

Figure 10: This chlorine analyzer (Model CL-17, Hach Company) is a typical batch mode instrument using a colorimetric method for water quality monitoring. Reagents are at left, reagent pump/sample controller at center-right, colorimeter cell at bottom-right and readout at top.

automatically set each time; 3)reagent use is limited (0.5 liter each of buffer and indicator every 4 weeks); (4) maintenance is minimal (15 minutes every 4 weeks); and (5) alarm relays and recorder outputs are built in. It is a good analyzer for its particular applications, yet not perfect for all applications, so research continues to be needed.

While Stopped-Flow analyzers have not been as much a focus of interest as continuous flow analyzers, several independent groups have recently reported work that is revitalizing the concept. Their work has resulted in the schematic shown here in Figure 11 which has then been applied in various ways. The general concept consists of pressure-driven reagents (no pumps involved) with electrically actuated valves controlling the flow. Because pressure and tubing size are fixed the volumes dispensed are controlled by the time the valve is open.

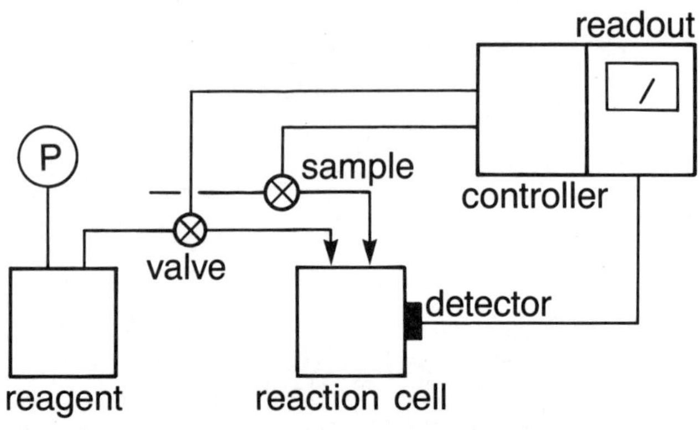

Figure 11: Schematic of a batch mode analyzer using gas pressure to pump reagents. With minimal moving parts—the solenoid valves which control flows—this type of analyzer offers the potential for reliability and low maintenance.

This scheme has appeared recently in a commercial form in the power industry (Fig. 12) (Wasson and Hach 1987). One can see the inherent simplicity and obviousness with the pressurized reagents, valves, reactor

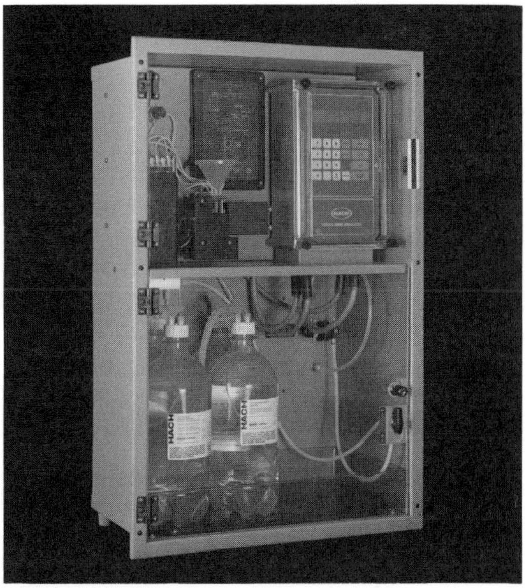

Figure 12: This silica analyzer used in the power and electronics industries (Series 5000 Analyzer, Hach Company) is an example of a pressure driven analyzer. Reagents are at bottom-left, valves at top-left, colorimeter/reaction cell at top-center and controller/readout top-right.

block and detector, and readout/controller as the only components involved. This instrument, or any others using this approach, will have a number of benefits—which are those needed by a practical on-line monitor: (1) minimum moving parts—basically only the valves which are designed for repeated opening and closing; (2) flexibility for application development; (3) dedicated and easy to use; (4) microprocessor control allows for self-calibrating, zeroing, alarms, trends, etc.; (5) appropriate sample frequency and reagent use; (6) lower cost due to simplicity and macro approach which loosens precision requirements.

A group at Texas Technological University (Sweileh et al. 1988) independently has reported a 200 sample per hour microscale lab analyzer using a very similar batch concept with pressurized reagents and valve flow control. They used a sample loop for injection and measurement in a flow-after color formation—both techniques similar those used in FIA. They reported a number of advantages of a Batch Mode Analyzer over FIA. There is no tubing replacement with subsequent flow recalibration and it is functionally simpler resulting in less problems and lower cost. There is deliberate control of mixing as opposed to control by timing of a diffusion process. All inert materials could be used. Finally, the Batch Mode is discretely flexible in timing and sequence of individual reagent addition for easier method development.

Interestingly, when one looks at future developments in reagent-using analyzers, one finds an English group (Malcolme-Lawes et al. 1987; Malcolme-Lawes and Pasquini 1988) has recently reported a pressure-driven flow injection analyzer which, they report, offers significant benefits over peristaltic pumps as a carrier and reagent transport mechanism. This is an indication that the pneumatic concept will find broad application. This report is also an indication that one will probably see a blurring of the line between the batch and flow approaches with the best features of each being applied as appropriate. It reinforces the value of the goals of simplicity and reliability for on- line monitors. In addition, many more applications and commercial models of batch-mode analyzers, including on-line titrators, as well as additional products and applications of continuous flow analyzers will be seen over the next few years. As the field continues to progress, reagent-using analyzers should approach the current state of the sensor systems discussed earlier in terms of becoming a mature, cost-effective technology.

CONCLUSION

Looking at the future of on-line water quality monitoring as a whole, a number of major trends are obvious. Process control testing is becoming accepted, creating a market demand for many general water quality monitoring applications. This will yield the increased volume needed for commercial interest which in turn will result in tools to solve new problems. The field could enter a growth cycle which will benefit the user with better analyzers at lower cost. These new applications will start with manual testing and then move on-line as the need for such monitoring is demonstrated. The overall result will be the wide-spread use of low-cost dedicated instruments with the characteristics needed for successful system integration to yield a more efficient aquaculture industry producing high quality food for the world market.

LITERATURE CITED

Bergveld, P. and B.H van der Schoot, 1988. A classification of chemical sensors as electronic components. Selective Electrode Review 10: 5-31.

Best, D. 1987. Biosensors revolutionize quality control. Prepared Foods October: 182-186.

Borman, S. 1987. Progress in process instrumentation. Analytical Chemistry 59:901A-903A.

Callis, J. B., D.L. Illman, and B.R. Kowalski, 1987. Process analytical chemistry. Analytical Chemistry 59: 624A-637A.

Christian, G. D. and J. Ruzicka, 1988. Analysis technique allows online control. Chemical Engineering March 28: 57-60.

Coulet, P. R. 1988. Biosensor-based analyzers: from design to instrument. Pages 319-327 in Guilbault, G. G. and Mascini, M. (eds.) Analytical uses of immobilized biological compounds for detection, medical and industrial uses. D. Reidel Publishing. Dordrecht, The Netherlands.

Frenzel, W. 1988. Potential of modified reverse flow injection analysis for continuous monitoring and process control. Fresenius Zeitschrift fur Analytische Chemie 329: 668-674.

Goldman, D. S. 1988. Chemical sensor development for process and environmental monitoring. American Laboratory November: 102-109.

Hansen, E. H. 1988. Flow injection analysis - recent developments and future trends. Pages 291-308 in Guibault, G. G. and Mascini, M. (eds.) Analytical uses of immobilized biological compounds for detection, medical and industrial uses. D. Reidel Publishing. Dordrecht, The Netherlands.

Harmer, A. L. 1987. Guided-wave chemical sensors. Pages 409-427 *In* D.R. Turner (ed.), Proceedings of the Electrochemical Society Symposium on Chemical Sensors. Electrochemical Society, Inc. Pennington, NJ..

Malcolme-Lawes, D. J., G.A. Mulligan, and R. Newton, 1987. A novel approach to non-segmented flow analysis—part 1: an experimental system. Journal of Automatic Chemistry 9: 179-183.

Malcolme-Lawes, D. J. and C. Pasquini, 1988. Part 2: a prototype high-performance analyzer. Journal of Automatic Chemistry 10: 25-30.

Miller, R. M. 1987. Factors affecting the performance of fiber optic probes. Journal of Automatic Chemistry: 449-455.

Narayanaswamy, R. 1987. Optical fiber chemical sensors. Proceedings of the Electrochemical Society 87: 428-437

Rechnitz, G. A. 1988. Biosensors. Chemical and Engineering News September 5: 24-36.

Ruzicka, J. and E.H. Hansen, 1981. Flow injection analysis. John Wiley & Sons, Inc. New York, New York USA.

Sherman, R. E. 1988. Process analyzers: vital control system components. Chemical Processing June: 70-78.

Sweileh, J. A., J.L. Lopez, and P.K. Dasgupta, 1988. Novel automated micro batch analyzer. Review of Scientific Instrumentation 59: 2609-2615.

Turner, A. P. F., I. Karube, and G.S. Wilson, 1987. Biosensors: fundamentals and applications. Oxford Press. Oxford, England

Valcarcel, M. and M.D. Luque de Castro, 1987. Flow injection analysis: principles and applications. John Wiley & Sons, Inc. New York, New York USA.

Wasson, J. and C. Hach, 1987. A pressure driven titrimetric/colorimetric analyzer for process applications. Pittsburgh Conference on Analytical Chemistry. Society for Analytical Chemistry and Applied Spectroscopy, Pittsburg. Abstract No. 288.

Yeh, J. T. Y. 1986. Online composition analyzers. Chemical Engineering January 20: 55-68.

MODELING WATER QUALITY IN AQUACULTURE ECOSYSTEMS

Raul H. Piedrahita

ABSTRACT

The techniques and uses of computer models, future developments and applications of computer modeling in aquaculture are reviewed. The general characteristics of aquaculture ecosystems are summarized from a computer modeling point of view. Various types of models of aquaculture systems are presented, including models that simulate dissolved oxygen concentration, water temperature, and other water quality variables. The models presented are reviewed from perspectives of modeling technique, model uses, data requirements, and predictive ability. Future applications of computer models in aquaculture research and commercial practice are proposed. These applications include the development of models integrated with automated data acquisition and control systems, models that simulate water temperature and water quality parameters simultaneously, and models for use in stratified ponds.

INTRODUCTION

The complexity of aquaculture pond ecosystems is well documented, and computer models provide an extremely useful tool for organizing information, reviewing existing knowledge about the system, improving our understanding of the ponds, and developing new and improved management methods. Models of aquaculture ponds have been developed following the traditional methods and techniques used for the simulation of other ecosystems. Straskraba and Gnauck (1985) have recognized four major areas for model use: "pedagogical, understanding theory, prediction, and management." There are examples of all four areas in aquaculture, and the choice of model type, computer language used, model complexity are determined, in large part, by the intended use of the model. The two common model types, empirical and mechanistic, have possible applications in the four areas of model use defined above, but the characteristic of each model type makes it more suitable for specific applications. Empirical models have been developed based on data collected, and these models have been used primarily for predictive and management purposes. Mechanistic, or analytical models have been developed primarily for research and teaching purposes, and these models have tended to be more complex in structure, and less accurate predictors of specific field conditions.

The choice of language for computer implementation of a model has important consequences on the resulting model structure, and ease of use (Kreutzer, 1986). Computer languages and systems used for the models described here range from BASIC to ALGOL, with most of the models developed in FORTRAN. A recent trend (Wolfe et al. 1986; Losordo 1988) has been the use of computer languages specifically designed for dynamic simulations. These languages (eg. DYNAMO, and STELLA™ for Wolfe's and Losordo's models respectively) simplify the process of computer implementation of a model, and reduce the effort required for coding procedures for data input and output.

With the advent of improved data acquisition and management systems, there are new developments in pond modeling strategies taking place. Whereas the traditional methods relied on limited data bases for model development (for empirical models primarily) or for model calibration and validation (for mechanistic models primarily), some concerted efforts at organizing large databases containing measurements of water quality, environmental conditions, and fish production, are making possible the development of new models (Hopkins et al. 1987; Losordo 1988). Modeling is also entering a new stage, where the criteria for judgment and success of models will probably continue to get more stringent as the availability of data increases and model calibration and validation become more common place.

General principles of computer modeling as they apply to the modeling and simulation of aquaculture systems have been reviewed recently (Bolte et al. 1986; Piedrahita 1988), and another review of modeling in aquaculture is in progress (M.L. Cuenco 1988, personal communication). In this paper, specific models developed for aquaculture ponds are presented and some of their characteristics are summarized. The models presented here are grouped according to the variables simulated: dissolved oxygen, temperature, and multiple variables or "ecosystem" models. Models that do not explicitly simulate water quality parameters are not included in this review. There are a multitude of fish and shrimp growth models available, some of which simulate some of the bioenergetics of food conversion in ponds (eg. Ma 1979; Cuenco et al. 1985 a,b,c), but those are not considered here. Conceptual models that have not been implemented as computer models are not included in this review (eg. Farnsworth 1979). There are a considerable number of models that have been developed for various types of aquatic environments: wastewater treatment ponds, reservoirs, lakes, rivers, estuaries, the oceans. These models, in many cases, have provided essential components and information for the development of models specifically for aquaculture systems. In this paper, however, only models developed explicitly for aquaculture ponds will be reviewed. A brief overview of pond ecosystem characteristics is presented to set the stage for the discussion of the models.

GENERAL CHARACTERISTICS OF AQUACULTURE POND ECOSYSTEMS

The ecology of warm water aquaculture ponds has been reviewed in the past (Lannan et al. 1986; Moriarty and Pullin 1987). Aquaculture ponds in general, and under normal management practices can be considered as "internally driven systems" (George Tchobanoglous, personal communication), in which processes taking place within the system are the primary determinants of water quality. In contrast, a raceway would be considered "externally driven," since water quality is primarily determined by "external" factors which can be managed by adjustments of water flow and aeration.

In an internally driven system, the rates of biological and chemical processes determine the conditions for culture of the target species. These processes are affected by conditions inside the ponds, as well as by external factors such as solar radiation, air temperature, wind velocity and direction, and by water exchange rates.

Previous models, and pond water quality characterization studies have invariably led to the conclusion that phytoplankton related processes are the main determinants of water quality in a pond. This is evident not only from detailed studies, but also from simple observations of dissolved oxygen fluctuations in ponds. The diurnal cycle of oxygen concentration in a well mixed pond is well known, and goes through a maximum sometime during the afternoon, and through a minimum shortly before sunrise. The increase in oxygen concentration during the day is due to photosynthetic oxygen production, and the decline at night is due to the combined respiration by organisms in the pond. The relative magnitude of the various oxygen consumers has been estimated by various authors. Boyd (1982) estimated a dissolved oxygen budget for a catfish pond. In it, "plankton" account for 52 % of the dissolved oxygen consumption, while diffusion to the atmosphere accounts for 26 %, fish respiration 13 % and benthic respiration 9 %. In Boyd's oxygen budget, the "plankton" oxygen demand includes the use by phytoplankton, zooplankton, and the decomposition of dissolved and particulate organics by bacterial action. Losordo (1988) carried out a similar analysis of some experimental results, and came up with values ranging between 40 and 73 % for what he called "water column respiration," and is equivalent to Boyd's "plankton" term. In Losordo's ponds, fish accounted for 14 to 31 % of oxygen consumption, reaeration between 0 and 35 %, and sediment respiration between 5 and 14 %. It may come as a surprise to some that fish constitute such a small proportion of the oxygen consumers in the pond. This, of course can vary widely depending on stocking densities and other operational characteristics of ponds. The highest proportion of oxygen use by fish in Losordo's experiments occurred in a pond stocked with approximately 5300 kg/ha of catfish. Fish, on the other hand, can have a profound impact on the ecology of the pond, and therefore ultimately on the dissolved oxygen cycle, by their feeding habits. In particular, phytophagous fish, as well as zooplanktivorous fish can affect the type, cell size and numbers of phytoplankton present in a pond (Opuszinski 1978; Colman and Edwards 1987; Smith 1987).

Some of the critical groups of processes affecting pond conditions are reviewed briefly below. It is no accident that phytoplankton processes are treated with somewhat greater depth than the others. This is in part due to the relative importance that has been traditionally given to the various processes, and also due in part to our level of knowledge and information on them. In relative terms, a great deal more is known about phytoplankton than about any of the other groups of processes, and that information has been incorporated into models to a greater extent. Research on the other groups of processes is, in some cases, just starting in aquaculture systems. Results from field and laboratory research will be incorporated into models as the information becomes available. The structure of a model will have a marked effect on the easiness with which new information is incorporated. In general, mechanistic models may be modified to include new processes or rates easier than empirical models.

Phytoplankton Processes

Photosynthetic oxygen production by phytoplankton is normally the principal source of dissolved oxygen for nonaerated ponds. In some intensive pond culture systems, mechanical aeration may provide a larger proportion of the oxygen used in the pond than phytoplankton.

Phytoplankton growth can be represented in chemical terms by an equation as proposed by Redfield (1958), Stumm and Morgan (1981), and others.

$$106\ CO_2 + 16\ NO_3^- + HPO_4^{2-} + 122\ H_2O + 18\ H^+ \\ (+\text{trace elements} + \text{energy}) \Leftrightarrow \{C_{106}H_{263}O_{110}N_{16}P_1\} \\ + 138\ O_2 \qquad (1)$$

The products in this equation represent phytoplankton biomass and oxygen produced by photosynthesis. During respiration, phytoplankton biomass and oxygen are consumed and inorganic nutrients are released, and the equation proceeds to the left as written. Equation (1) may be used to develop stoichiometric ratios and determine relationships between the consumption and production rates of the various terms in the equation based on the measurement of any one of the other terms. For instance, the rate of oxygen production by photosynthesis can be used to estimate the rate of phytoplankton biomass produced, or the rate of carbon dioxide consumption. The stoichiometric ratios for equation (1) represent only those changes in the parameters that are due to phytoplankton photosynthesis and respiration. Other processes are likely to affect the concentrations of the various components, and these processes would have to be quantified separately.

Phytoplankton growth rate is affected by nutrient concentration, light intensity, water temperature, and the intrinsic characteristics of the phytoplankton species. These intrinsic characteristics determine the maximum attainable photosynthetic production rate for a particular species, and the response functions to the environmental and water conditions mentioned above.

Traditionally, primary nutrients considered as possible limiting factors for phytoplankton growth have been phosphorus and nitrogen (eg. Wetzel, 1983). In highly eutrophic environments, of which aquaculture ponds are examples, inorganic carbon availability can also constitute a limiting factor (Brune, 1978; Fritz et al., 1979).

The effect of light intensity on photosynthesis rate has been the subject of numerous publications and reviews. At relatively low light intensities, photosynthesis rates are directly proportional to light intensity. As the intensity of the light increases, the "quantum yield" (Smith 1980), the photosynthesis rate per unit of light, decreases. The rate of photosynthesis goes through a maximum after which any increases in light intensity will actually cause a reduction in photosynthetic rate. The light intensity at which the rate of photosynthesis starts to decline is normally referred to as the light saturation level. Phytoplankton exposed to a light intensity lower than light saturation are considered to be "light limited," while those exposed to high light intensities are "light inhibited" (Groden, 1977). The precise form of the photosynthesis light relationship is not well defined, as exemplified by the number of formulations proposed (Field and Effler 1982). The differences between the various formulations are primarily for high light intensities.

Estimation of the overall rate of photosynthesis requires that the functions relating light intensity and photosynthesis rate be integrated over the depth of the water column. Light intensity decays exponentially with depth (Wetzel 1983). When the rate functions are integrated, it becomes clear that the saturation light intensity for the water column can be substantially higher than for the surface layer (Giovannini 1989), and it is unlikely that phytoplankton growth will be limited by excessive light unless other types of stresses (eg. nutrient limitations) are compounded on the plankton cells (Smith 1987).

Heterotrophic Processes

Much of the traditional aquaculture practiced around the world is based on the transformation and utilization of nutrient inputs from plant and animal waste products, in what are called detritus driven systems (Moriarty and Pullin 1987). In these systems, the degradation and transformation of organic matter by bacteria and other pond organisms may be responsible for the production of a large proportion of the "food" available for fish in the ponds (Schroeder 1987). End products of these transformations include primary nutrients for phytoplankton and organic matter that is useable as a food by fish. Questions about the relative importance of the various possible paths of transformation of nutrients are still unresolved. Traditional thinking has been that detritus added to water would be decomposed by bacterial action and primary nutrients would be released into the water to be picked up by phytoplankton. Phytoplankton would in turn utilize these nutrients and solar energy to produce new cell biomass that would be the base of a food chain that would eventually result in fish growth by direct consumption of phytoplankton, or by consumption of zooplankton or other phytoplankton grazers. An alternate pathway is by direct consumption of detritus particles by the fish, but the digestibility of these particles is questionable (Lu and Kevern 1975), and the nutrient composition is less than optimal for fish growth (Wohlfarth and Schroeder, 1979; Wholfarth and Hulata 1987).

Some workers (Little and Muir 1987; Kerns and Roelofs 1977) suggest that the microorganisms attached to the particles are digested but that the particles themselves are not digested by most fish. Ongoing research into this topic has important implications to the management of ponds for optimum resource use and maximum fish production.

Quantified information on the various detritus-related processes is not widely available. The structure of some of the existing dissolved oxygen and ecosystem type models will permit updating to include the new information as it becomes available. Some of these models have included the basic structures and processes of a heterotrophic based food chain, but suffer from a lack of data on the rates at which the processes take place. This information needs to be developed in ponds, or similar aquatic systems, for use in pond management and modeling.

Sediment-water interactions

The importance of sediment-water interactions in aquaculture ponds has been known for a long time, and yet information on the topic has been scarce.

Biological, physical and chemical processes at the sediment-water interface account for the rate of recycling, resuspension, and availability of some critical nutrients in ponds. The availability of phosphorus has long been known to depend to a large extent on dissolved oxygen concentration at the sediment-water interface (Welch 1980; Boyd 1982). It is also known that the rate at which nutrients are released from sediments is highly dependent on the degree of turbulence over the sediments (Welch 1980).

Anaerobic processes in pond sediments have been reviewed recently by Blackburn (1987), who noted that oxygen penetration in sediments is limited to a few millimeters even under light organic loads. Blackburn also summarized the various groups of microorganisms acting in anaerobic sediments, and their effect on the cycling of primary nutrients. Reduced conditions created by the lack of oxygen in the sediments may also cause the formation of toxic substances such as hydrogen sulfide. This is especially problematic in sea waters due to the high sulfur content of sea water.

Hydrodynamics

As will be seen in the models reviewed below, conditions in ponds were assumed to be uniform in all but two models (Cathcart 1987; Losordo 1988), even though most modelers and other researchers acknowledge the lack of uniformity in ponds. Stratification in shallow (less than 1 m deep) ponds is a common phenomenon. Horizontal variations in water quality may not be as marked as vertical variations, but may be significant, especially in ponds receiving substantial flows, or exposed to moderate winds. The effect of turbulence on sediment-water interactions has already been mentioned. And yet, with all of these indications of the importance of pond hydrodynamics, this has been a topic ignored in aquaculture research and modeling. Only now is information generated with other types of water bodies being extended to aquaculture ponds (Losordo 1988). Incorporation of some treatment of stratification will be essential for future models and improved management of ponds used for benthic animals.

DISSOLVED OXYGEN MODELS

The simulation of dissolved oxygen concentration in ponds has been attempted by several authors, but with different purposes, methods, and results. The models are described below in chronological order of publication, followed by some comparisons of their general characteristics, data requirements and quality of results obtained.

DO Model 1

The model developed by Busch and coworkers (1977) was designed to simulate dissolved oxygen concentration changes during the night in a catfish pond. The model uses a very general mass balance equation that includes aeration from paddle wheels as an oxygen source. The mass balance equation used is:

$$DO_{t+\Delta t} = DO_t - ((DO_f + DO_w) + DO_a) \cdot \Delta t \qquad (2)$$

Model execution was started with dissolved oxygen concentration at dusk. Fish and water column oxygen demands were estimated as functions of temperature from relationships proposed by Boyd (1973). Water column dissolved oxygen consumption rate was based on measurements for a single pond, considered to be the most "likely to produce critical conditions under intensive fish production" (Busch et al. 1977). Aeration rates with the paddle wheel were estimated based on transfer coefficients developed by Knight (1965).

The model was tested with data obtained from two ponds at Auburn and predictions for three days during the growing season were reported. Agreement between the measurements and simulation is generally good (within 10 % after 8 hours of simulation) for a small (0.1 acre) aerated pond, but poor (within 110% after 9 hours of simulation) for a larger (1.5 acre) pond. Deviations are attributed primarily to the mechanical aeration estimates, as measured and simulated water column oxygen consumptions agree within 18 % for both ponds.

No further developments or applications of this model have been reported.

DO Model 2

The dissolved oxygen model developed by Romaire and coworkers (1978) was designed to be used to predict early morning dissolved oxygen concentration based on measurements of dissolved oxygen concentration at dusk, and on estimates of oxygen consumption by various physical (diffusion) and biological (respiration) processes.The model was first proposed for use in catfish (Boyd et al. 1978), and tilapia ponds (Romaire et al. 1978). The adaptation of the model to trout ponds was also mentioned in a later publication (Boyd 1982). The model is based on a mass balance for dissolved oxygen, and is used to estimate dissolved oxygen decline at night. The mass balance, as proposed by Boyd and coworkers is:

$$DO_{Dt} = DO_{dusk1} + (\pm DO_{rea} - DO_f - DO_b - DO_{pr} - DO_{om}) \bullet \Delta t_d \qquad (3)$$

Because the model is designed to calculate dissolved oxygen concentrations changes during the night, there is no need for inclusion of a photosynthetic oxygen production term. Estimates of diffusion are obtained from relationships developed after Schroeder (1975). These relationships, as used by Boyd and coworkers (1978) are based solely on the dissolved oxygen concentration with respect to saturation at dusk. The relationship ignores wind, an important factor affecting the gas exchange rate (Banks and Herrera 1977). Fish respiration is estimated with regression equations developed after Andrews and Matsuda (1975) for catfish, after Schroeder (1975) for tilapia, and after Muller-Feuga and coworkers (1978) for trout. These relationships are used to estimate oxygen consumption by fish as a function of fish size (individual mass), and temperature in the case of catfish and trout. "Mud" respiration includes oxygen consumed by benthic organisms and is assigned a constant value of 0.61 mg m^{-2} h^{-1} , based on work by Mezainis (1977) and Schroeder (1975). The mud respiration term is calculated based on an estimate of the area of the pond bottom covered by oxygenated water, given the assumption that "muds beneath anaerobic waters do not contribute directly to oxygen demand" (Boyd et al. 1978). Oxygen consumed by plankton constitutes the largest term in the mass balance equation, and is estimated from either COD or Secchi disk measurements using relationships developed by Boyd and coworkers (1978). The planktonic oxygen consumption equations were developed from samples collected and analyzed at the Auburn University Fisheries Research Unit. Oxygen consumption by organic matter was included in the model for calculations of dissolved oxygen changes in tilapia ponds (Romaire et al. 1978), and was estimated based on data from Schroeder (1975) adjusted for temperature. Romaire and Boyd (1979) later expanded the model by including oxygen produced by photosynthesis. The equation used to estimate the oxygen produced was based on measurements of dissolved oxygen concentration change during the daylight hours, adjusted by adding the oxygen consumed by fish. Statistical analysis of these measurements was carried out with respect to solar radiation, and either chlorophyll concentration or Secchi disk depth. Data used to develop the regression equations were collected at the Auburn University Fisheries Research Unit.

The model, as shown in simulations by Boyd and coworkers (1978, Romaire et al. 1978; Romaire and Boyd 1979; Boyd, 1982) provides good estimates of dissolved oxygen changes in Auburn ponds. The empirical nature of the relationships included in the model limits the application of the model to conditions that must be very similar to those found in the ponds used to develop the empirical relationships. In particular, the nature of the phytoplankton population, the non-algal turbidity, and plankton growth limiting factors other than light, are factors that will affect how applicable the model results may be at different sites (Smith and Piedrahita 1987).

DO Model 3

The objectives of the model developed by Cassinelli and coworkers (1978) included the identification of oxygen sources and sinks, and of environmental parameters affecting dissolved oxygen concentration in marine aquaculture ponds. Of the models reviewed, this is the only one developed specifically for salt or brackishwater ponds. The model is based on existing expressions for the rate at which the various processes occur in marine systems. The mass balance equation used as the basis of the model is:

$$DO / dt = DO_s \pm DO_{rea} - DO_{om} + DO_{gpp} - DO_{pr} + DO_{wto} \quad (4)$$

where: $$DO_{wto} = (Q / V) * (DO_{supply} - DO_{pond}) \quad (5)$$

The introduction of the water turnover term is a significant variation from the "freshwater" models described above. This is necessary due to the levels of water exchange normally used in shrimp ponds, ranging up to 50 % per day . Rate expressions for the various producers and consumers of oxygen are based on published reports (eg. Steele 1962 for photosynthesis as a function of light; Kester 1975 for reaeration rates). Sensitivity analysis of the model indicated that major factors affecting dissolved oxygen concentration in the ponds included the decomposition of organic matter, the concentration of phytoplankton, and light intensity. Factors that showed little effect on dissolved oxygen concentration included shrimp respiration and exchange with the atmosphere. Casinnelli and coworkers (1978) postulated that the effect of wind on water quality may be by the prevention of stratification.

DO Model 4

The model developed by Meyer and Brune (1982; Meyer 1980) was based on a strictly mechanistic approach, and no specific pond data were used to estimate parameters or to develop regression equations. The model was developed to provide a tool for improved understanding of the dissolved oxygen cycle in ponds, and to identify areas for possible field research (Meyer 1980). A mass balance equation was proposed for dissolved oxygen in a pond as (Meyer 1980):

$$DO_{t+\Delta t} = DO_t + (DO_{npp} - DO_f - DO_{BOD} + DO_{rea}) \bullet \Delta t \qquad (6)$$

Calculation of the various terms in equation (6) was based on relationships in the published literature. Algal oxygen production was estimated using equations originally proposed by Smith (1936) for light limited growth, and by Steele (1962) for light inhibited growth, following the procedure of Groden (1977). The equation for algal growth included provisions for accounting for temperature effects, and for nutrient limitations, although the model was only run with assumed nutrient saturation. Phytoplankton respiration was estimated as a function of temperature (Riley 1946, Jewell and McCarty 1971), and used to adjust the gross photosynthesis values and arrive at a net oxygen production term for phytoplankton. Fish oxygen demand was estimated as in Boyd and coworkers' model (1978).

The other respiration term included in the model accounts for the oxygen consumed by "decaying phytoplankton, unconsumed fish feed and fish waste products" (Meyer and Brune 1982), and uses an expression based on an estimate of the ultimate BOD of the material, and a temperature dependent decay rate. The sources of detritus were estimated from equations describing the sinking and death rate of phytoplankton, the fish feeding rate and the amount of uneaten feed (Meyer 1980). Reaeration was estimated using a wind-dependent oxygen transfer rate (Banks and Herrera 1977), and the difference between dissolved oxygen saturation and the calculated concentration at each time step.

Model testing and calibration were based on sensitivity analysis to changes in algal concentration, BOD, wind speed, and cloud cover (Meyer 1980; Meyer and Brune 1982). The model was run with data corresponding to diurnal oxygen and chlorophyll measurements from Auburn, Alabama (Augusto 1975; Mezainis 1977), and to measurements obtained from a small pond in Davis, California (Brune et al. 1983). Simulation results show good agreement with field measurements, with the exception of bottom samples in the Davis pond. Discrepancies with some of the Auburn data are also attributed to the existence of non-homogeneous conditions in the ponds (Meyer and Brune, 1982).

DO Model 5

A nutrient saturated phytoplankton growth model developed by Bannister (1979), was adapted by Laws and Malecha (1981) to freshwater prawn (*Machrobrachium rosenbergii*) ponds in Hawaii. The model was used to study the effect of phytoplankton density (as measured by chlorophyll-a content) on dissolved oxygen concentration. Optimum phytoplankton concentrations for dissolved oxygen maintenance in prawn ponds were recommended to be between 150 and 400 µg/L. Laws and Malecha (1981) found that the model gave an accurate description of gross primary productivity after adjusting some of the parameters originally used by Bannister (1979).

DO Model 6

The model developed by Madenjian and coworkers (1987a, b) was based on data collected in prawn (*Machrobrachium rosenbergii*) ponds in Hawaii. The purpose of the model was the prediction of early morning dissolved oxygen concentration based on measurements of DO and water temperature at dusk and several hours after dusk, and estimates of wind speed. In this model, the mass balance equation for DO includes only two terms, whole pond respiration and transfer with the atmosphere:

$$DO_{t+\Delta t} = DO_t + (DO_{wpr} \pm DO_{rea}) \cdot \Delta t \tag{7}$$

Whole pond respiration estimates can be calculated as a function of temperature using an equation developed from measurements at two Hawaiian sites. The whole pond respiration coefficient for a particular run of the model is estimated from the two evening dissolved oxygen measurements (dusk and several hours after dusk), and from estimates of dissolved oxygen exchange with the atmosphere during the time interval between measurements. In other words, the two dissolved oxygen measurements are used to estimate the whole pond respiration rate. This rate is then used to project dissolved oxygen concentration at dawn the next morning, assuming the whole pond respiration is a function of temperature only and that exchange with the atmosphere is a function of wind speed (Banks and Herrera 1977) and the difference between DO and saturation DO. The model was run for ponds at two Hawaiian sites. As a result of simulations, and for practical reasons, Madenjian and coworkers (1987b) recommend that the interval between the two measurements be at least 3 hours. Dissolved oxygen prediction from the model underpredicted the measured values by as much as 1.7 mg/L, and overpredicted by as much as 0.5 mg/L for a windy site, and underpredicted by as much as 0.8 mg/L, and overpredicted by as much as 0.4 mg/L in a relatively windless site. Discrepancies between simulation and measured values were attributed in part to the modeling of respiration rates as functions of temperature only, the lack of homogeneity in water quality in the ponds, and to inaccuracies in the estimation of gas transfer rates using the expression developed by Banks and Herrera (1977).

DO Model 7

In a recent paper, Smith and Piedrahita (1988) used the model developed by Boyd and coworkers (1978), and a mechanistic model based on a mass balance equation to study the relationship between phytoplankton concentration and the dissolved oxygen cycle. The mechanistic model was also used to determine possible growth limiting factors not specifically accounted for by Boyd and his coworkers (1978).

The mass balance equation used, by Smith and Piedrahita is similar to that used in previous models, but it aggregates oxygen consumers and producers slightly differently:

$$DO_{dawn2} = DO_{dawn1} + DO_{gpp} - \Sigma((DO_{pr} + DO_{f} \pm DO_{rea} + DO_{b}) \cdot \Delta t) \qquad (8)$$

In the mechanistic model proposed by Smith and Piedrahita (1988), gross primary production is estimated from an expression proposed by Baumert and Uhlmann (1983) for calculating growth at optimum conditions (nutrient saturation, no non-algal turbidity, and light saturation), as a function of incident light, light extinction and maximum photosynthetic rate. Non-optimal conditions are simulated by adjusting the light extinction coefficient, and by introducing a nutrient limitation term.

Results from simulations run by Smith and Piedrahita (1988) indicated that phytoplankton in the catfish ponds used for development of the empirical relationships used by Boyd et al. (1978) were probably nutrient limited, that there was a substantial amount of non-algal turbidity, and that there was a substantial amount of non-phytoplankton respiration in the water column. Comparison of the two models run by Smith and Piedrahita (1988) showed that optimum dissolved oxygen concentrations after 24 hours (DO dawn2) were obtained for intermediate concentrations of phytoplankton (between 200 and 300 μg/L chlorophyll-a).

DO Model 8

The model developed by Losordo represents the first attempt at simulating dissolved oxygen (and temperature) in a stratified pond (Losordo 1988; Losordo and Piedrahita 1988a). Losordo's model is coupled to a model used to calculate water temperature needed for the estimation of reaction rates in the dissolved oxygen mass balance. The model is based on a mass balance for dissolved oxygen in each of the volume elements (horizontal layers) in which the pond is divided. Thickness of the volume elements is 10 cm for the surface element, and 20 cm for all other volume elements. Transfer of dissolved oxygen between the volume elements is accounted for with diffusion equations that include a mass transfer coefficient and the concentration difference between adjacent volume elements. The mass balance equation for the surface volume element is:

$$d\,DO_{sve} / dt = DO_{fss} \quad DO_{rea} - DO_{ed,s} + DO_{gpp,s} - DO_{pr,s} - DO_{wc,s} \qquad (9)$$

The subscript s (,s) represents the dissolved oxygen production or consumption exerted in the surface volume element by the various processes.

The mass balance equation for a mid-water volume element is similar to that presented in equation (9), with the exclusion of the reaeration term, and with transfer by effective diffusion with the adjacent volume elements above and below the volume element in question. The mass balance for the bottom volume element includes a sediment oxygen demand term (DO_b), transfer by effective diffusion with the volume element directly over it, and does not include a reaeration term.

Production and consumption rates are calculated using expressions from the published literature, with parameters adjusted based on field measurements in catfish ponds in California. Dissolved oxygen calculations are carried out for 24 hour simulation runs. The dissolved oxygen model is coupled to a temperature model described below. Furthermore, the effective diffusion coefficient used for heat (in the temperature model) is also used for dissolved oxygen and is estimated as a function of wind speed, water and air density, and temperature gradient after an expression developed by Sunderam and Rehm (1973).

The model was used to simulate diurnal dissolved oxygen cycles in ponds at four sites in California and one in Hawaii. Good agreement between simulated and measured oxygen decline rates at night was obtained. Oxygen production calculations during the day (using Steel's equation (Steel 1962)) tended to underestimate the measured dissolved oxygen increases. The model was useful in predicting the occurrence of dissolved oxygen stratification events. Sensitivity analysis of the model indicated that the greatest effect on dissolved oxygen concentration was due to changes in the phytoplankton production and reaeration terms.

General Comments on Dissolved Oxygen Models

The models presented followed surprising similar approaches to the simulation of dissolved oxygen concentration in ponds. In all cases, some form of mass balance equation was developed, and rate expressions were obtained either from the published literature, or by analysis of experimental data for a specific pond situation. The level of aggregation in the mass balance equation varied considerably depending on the information available and on the intended use for the models (Table 1). Models that relied more

Table 1. Aggregation of dissolved oxygen production and consumption processes in the mass balance equations.

DO Model#	1	2	3	4	6	7	8	8	8
eqtn (#)	(2)	(3)	(4)	(6)	(7)	(8)	(9)	Mid	Bot
Oxygen Sources									
Photosynthesis			d	b		b	c	b	b
Reaeration		b	b	d	b	d	b		
Inflow			f						
Mechanical Aeration	c*								
Oxygen Sinks									
Phytoplankton Resp.	b	d	e	b	a	c	d	c	c
Zooplankton Resp.	b				a		e	d	d
Fish/Shrimp Resp.	a	a	a	a	a	a	a	a	a
Benthic Animal Resp.		c			a	e			e
Suspended Bacteria Resp.	b	e	c	c	a		e	d	d
Settled Bacteria Resp.		c		c	a	e			e
Degassing		b	b	d	b	d	b		
Outflow			f						

* Processes with the same letter were included as a single term in the mass balance equation.

on experimental data tended to have more aggregated terms due to the difficulties in separating out the various oxygen consumption processes from dissolved oxygen measurements in ponds. More theoretical, mechanistic models, on the other hand, tended to separate out the various consuming and producing processes, and use published equations to calculate the rates of oxygen change. Sensitivity analyses of the models indicated invariably that phytoplankton related processes (photosynthesis and respiration) are the primary determinants of dissolved oxygen concentration in ponds. The models also indicated that fish (or shrimp) biomass has a relatively small direct effect on dissolved oxygen concentration. There was considerable discrepancy, however, on the relative importance of wind. Some modelers (eg. Losordo 1988; Madenjian et al. 1987b) suggest that reaeration across the air-water interface plays a significant role in the overall oxygen balance, while others (eg. Cassinelli et al. 1978), indicate that the direct effect is minimal.

In general, the trend over the years has been away from empirical, site specific models to more general models that can be adapted to specific sites by use of parameter values. The models have been useful in the study of dissolved oxygen-related processes, but have not reached the point where they can be used reliably for management purposes. Some of the limitations for the use for management purposes are in the difficulty in collecting the data required for the routine execution of the models. Routine model use for water management purposes will not become a reality until data acquisition has been automated and simplified enough to be commercially feasible for a farmer.

The time scale used for simulations was almost always limited to short term, diurnal fluctuations in dissolved oxygen concentration. This reflects the frequency of fluctuations observed in ponds, and the importance of unpredictable factors (i.e., weather) in determining the various rates affecting dissolved oxygen. The time scale used also reflects the management needs for farmers in dealing with the daily cycles and emergencies. Some inferences are made in a few of the models about long term management goals that could result in more stable dissolved oxygen environments, but no practical management practices are proposed (Laws and Malecha 1981; Smith and Piedrahita 1988a).

With one exception (Losordo, 1988), conditions in the ponds were assumed to be uniform for modelling purposes. Most authors, however, acknowledged the fact that the ponds were in fact not homogeneous, and that stratification might play a significant role in determining the oxygen concentration. The magnitude of oxygen stratification noted by Losordo (1988; Losordo and Piedrahita, 1988), suggests that this effect must be explicitly accounted for to obtain accurate predictions of dissolved oxygen concentration. Stratification becomes even more important when the ponds modeled are used for the culture of benthic animals (eg. shrimp). Simulations with fully-mixed models do not provide accurate predictions of the dissolved oxygen concentrations to which the animals are exposed.

TEMPERATURE MODELS

The simulation of water temperature in ponds has not been pursued with the same intensity as dissolved oxygen. With one exception (Losordo 1988), dissolved oxygen models (and ecosystem models described below) require water temperature to be input for model execution, even though temperature is universally recognized as a critical variable determining the rates of processes taking place in the ponds. Models developed to simulate pond water temperature are presented below.

Temperature Model 1

The model developed by Krant et al. (1982) was based on an energy balance for a "mixed pond of constant volume":

$$c_o \cdot (V \cdot dT / dt) = q_{ad} + q_{nr} - q_e - q_c \quad (10)$$

In this model, the heat capacity of the water is assumed to be constant, and the temperature in the pond is equal to the temperature of the water leaving the pond (i.e., the pond is fully mixed). Since the model was developed for saltwater ponds which normally have substantial water replenishment rates, heat transfer by the water flow through is considered significant and is included in the model. The net radiation is the resultant of solar and atmospheric radiation, and backradiation from the pond:

$$q_{nr} = q_r - q_L \quad (11)$$

The net solar and atmospheric radiation is the resultant of long and short wave radiation on the pond, adjusted for reflection:

$$q_r = (q_s - q_{sr}) + (q_a - q_{ar}) \qquad (12)$$

Heat transfer between the pond water and the soil was neglected based on work by Keijman (1984) on a lake. Expressions used for the calculation of the various heat flux terms were derived from the published literature (eg. TVA, 1972). The model was tested by simulating water temperature in small ponds (200 m^2) in Israel for 1 and 5 day periods. Heat transfer modes found to have the greatest effect on temperature were solar radiation, and evaporative heat loss. The relative importance of heat transfer due to water flow was found to be highly variable, and depended not only on the residence time for the ponds (20-175 hours for the ponds tested), but also on the difference between the incoming and pond waters. The limited simulations shown for the model indicate good agreement between simulated and measured temperatures. The model was later expanded (Krant et al. 1985) to include pond heating with a heat exchanger for simulation of low grade heat utilization. No data were used to validate the model, and simulations were used to compare the hypothetical temperatures in heated and unheated ponds and estimate the efficiency of heat transfer and heat loss by evaporation. Results were compared with alternative uses of low grade heat, namely warming of soil for plant culture.

Temperature Model 2

The model developed by de Jaeger and Walmsley (1984) was designed to simulate the temperature in a small (100 m^2), shallow (0.3 m) pond covered with plastic and used for the culture of algae. The model was used to simulate the maximum and minimum water temperatures in the pond. Theoretical energy transfer equations were used, and the energy balance used for the pond water included heat transfer by conduction with the sides and bottom of the pond. The plastic cover acted as an effective barrier to prevent evaporation, and evaporative heat loss was substantially reduced since the air under the plastic cover was assumed to be saturated with water. Furthermore, heat generated by condensation on the plastic was assumed to be lost by conduction across the plastic film and convection to the atmosphere. Data requirements for model execution were modest and included the initial water temperature, temperature and amount of water added to the system, date, latitude and longitude of the site, and pond dimensions. The model was validated against data from a pond in Bloemfontein (South Africa), and simulation results tended to underestimate the maximum water temperature (by an average of 2.4 C) and overestimate the minimum temperature by an average of 2.8 C.

Temperature Model 3

The model by Klemetson and Rogers (1985) was developed to estimate monthly average water temperature in a pond. The proposed use for the model is the analysis of expected temperatures in proposed aquaculture sites, and preparation of production estimates for these temperatures. The model is designed to account for possible heat inputs from geothermal or power plant cooling waters. In the model, diurnal temperature fluctuations are ignored, and a heat balance is carried out that accounts for average inputs and outputs of energy, following the technique normally used for large water bodies. The mass balance equation used to estimate water temperature is:

$$\Delta \text{Heat} = q_{ad} + q_s - q_L - q_e - q_c \qquad (13)$$

Expressions for the various heat transfer rates are primarily based on Velz (1970). Data required for model execution are limited to incoming water flow rate and temperature, pond morphology (depth and area), initial pond temperature, average wind speed, air temperature, relative humidity and solar radiation. The model was used to simulate water temperature for ponds at a location in Colorado, but the results were not compared to measured values to establish the validity of the simulations. Results of the simulations were presented in terms of average monthly temperatures, pond productivity, and pond area required to achieve a selected productivity level. The effect of several methods of heat loss reduction by evaporation were investigated with the model by adjusting evaporation rates.

Temperature Model 4

The model developed by Cathcart and Wheaton (1986, 1987; Cathcart 1987) presents a new approach to the modelling of temperature in ponds (later followed by Losordo 1988), in that vertical variations in pond temperature are specifically accounted for, and the ponds are not assumed to be completely mixed. An assumption that remains in Cathcart's and Losordo's models is that conditions in the ponds are uniform on a horizontal plane; i.e., the pond is composed of horizontal volume elements, each of which is essentially completely mixed. Cathcart's model was developed with the objective of making it usable to farmers; therefore, an attempt was made at minimizing the amount of data required for model execution. As a result, surface water temperature was used as a model input, eliminating the need for calculating surface heat flux, and for the weather data normally required for such calculations. The resulting model is designed to estimate the temperature distribution in a pond when the surface temperature is known. Surface water temperature is estimated using a third degree polynomial function of time of day, and using daily readings of maximum and minimum temperature 10-15 cm below the water surface. Maximum and minimum temperatures are assumed to occur at 4 p.m. and 8 a.m. respectively.

Heat transfer modes between adjacent layers are due to molecular diffusion, absorption of incoming solar radiation, buoyancy instability, and turbulent diffusion due to wind action. The first three heat transfer modes are quantified using conventional expressions from the published literature. The estimation of heat energy mixed into the water column by turbulent diffusion is based on heat balances for surface and mid-water volume elements, and on estimates of a "reservoir of energy which cannot be accounted for in the pond heat balance" (Cathcart and Wheaton 1987). The approach is then one of trying to account for how the heat is distributed in the water column, rather than to analyze and quantify the specific mechanisms of wind induced heat transfer.

The model was validated against data collected from a 0.17 ha pond in Maryland with a maximum depth of 2.2 m . Temperature was measured using thermocouples at 30 cm depth intervals. Initial validation runs were carried out using the measured surface temperatures as initial conditions. Predictions from the model tended to underestimate the measured temperatures, and at least 70 % of predictions were within 1 C of observed temperatures for the first midwater volume element (volume element number two, with the surface element being number one). Agreement between predicted and measured temperatures improved for the deeper volume elements, and was best (over 80 % within 1 C of measured) for the sixth and seventh elements (180 and 210 cm deep). Use of the predicted surface temperatures for the water column temperature simulations caused a decrease in accuracy of about 7 % in the number of predictions within 1 C of measured values. Measured and predicted temperatures in the analysis presented by Cathcart and Wheaton (1987) were corrected for seasonal changes observed during the period of data collection (August to October). As a result, temperatures, called "seasonally adjusted temperatures", are presented with respect to a predicted temperature at 4 p.m for a given date. The 4 p.m. prediction is obtained from a regression analysis of 4 p.m. temperatures over the sampling period. The adjustment of temperatures was justified by the seasonal decline in temperatures over the sampling period of approximately 15 C.

Temperature Model 5

The model developed by Losordo (1988) was designed to simulate diurnal temperature cycles in a stratified aquaculture pond and to provide information on the processes of heat transfer in ponds. Water temperatures were estimated for 24 hour periods, starting at 6 a.m. The procedure used for estimating temperatures was based on heat energy balances for horizontal volume elements, as was the case in Cathcart and Wheaton's (1987) model. The primary difference between Losordo's and Cathcart and Wheaton's models is that Losordo carried out a complete heat balance for the surface volume element, and used it, in conjunction with the energy balances for the other volume elements, to calculate the temperature in all volume elements. Data requirements for Losordo's model were, therefore, substantially greater than for Cathcart and Wheaton's model, and included solar radiation, air temperature and relative humidity, wind direction and magnitude. In addition, underwater radiation was measured and used to obtain estimates of the light extinction coefficient in the pond. A specially designed data acquisition system was used to collect the weather data required for model execution, as well as the data required for model calibration and validation (Losordo et al. 1986, 1988; Piedrahita et al. 1986). The depth of the surface volume element used in Losordo's model was 10 cm, while the subsurface elements were 20 cm deep. The form of the heat balance equations was slightly different for the surface, midwater and bottom volume elements. The surface heat balance was similar to that used by Krant and coworkers (1982) (shown in equation 10), and included terms for long and short wave radiation, evaporative cooling, sensible heat transfer with the atmosphere, as well as heat transfer with the underlying volume element. Radiation calculations were carried out using published relationships that included evaporation rate estimates, estimates of the light extinction coefficient, reflected radiation, and others.

Heat transfer by diffusion between two adjacent volume elements was calculated using an effective diffusion coefficient that combines the effect of molecular diffusivity and turbulent diffusivity (Water Resources Engineering (WRE) 1968; Rossi et al. 1982; Losordo and Piedrahita 1988b). The effective diffusion coefficient is calculated as a function of wind speed, pond depth, and temperature gradient following procedures outlined by Henderson-Sellers (1984) and others (see Losordo 1988; and Losordo and Piedrahita 1988a,b). Buoyant instabilities were simulated by the use of a maximum diffusion coefficient (Sundaram and Rehm 1973), rather than by "numerically mixing" or averaging the temperature in adjacent volume elements when the top element was cooler than the one underneath (Losordo and Piedrahita 1988b).The heat balance for the bottom element included a term for conductive heat transfer with the sediments.

Calibration of the model was carried out by adjusting a minimum of parameters (the fraction of solar irradiance absorbed in the first cm of the water column, the maximum effective fetch length, and a constant used for the calculation of the effective diffusion coefficient) for a specific pond. All other expressions were used in the form in which they were developed originally for reservoirs, lakes, or waste treatment ponds (Losordo 1988). The model was thus kept as a general, mechanistic model that could be applied to different sites. Model execution was done by inputting climatic data, pond morphometry and light transmission properties of the water for a particular site. Sensitivity analysis and model validation was done for dates of completely mixed conditions, as well as severe stratification. Agreement between simulated and measured temperatures in fully mixed ponds ranged between 0.25 and 1 C (or an average relative error of 0.006 to 0.024). For stratified ponds, the agreement between measured and simulated values ranged from a relative error of 0.006 to 0.070. The model was remarkably accurate in predicting the onset of stratification, the time of maximum stratification, and the time of destratification and pond mixing (within less than one hour in all cases). Sensitivity analysis showed that the parameters having the greatest effect on the model were (in order of importance) wind speed, solar radiation, and air temperature. Analysis of simulation results revealed that during the daylight hours, the dominant heat flux was by short wave solar radiation. Surprisingly, though, over a full 24 hour cycle, the primary form of heat transfer was long wave radiation from the water surface (Losordo, 1988). Sensible heat flux and heat transfer with the sediments appeared to be the least important in the overall heat balance for the pond.

Losordo's model succeeded in creating a mechanistic structure for simulating temperature in stratified ponds. Data requirements, while considerable, can be easily automated with the use of computerized weather station and data acquisition system.

General Comments on Temperature Models

Whereas dissolved oxygen models were based on mass balances, temperature models are based on energy balances. Data requirements for temperature simulations are considerable, but collection of the information is easily automated using existing, field proven technology, such as weather stations and thermocouples. Portable, microprocessor controlled weather stations are readily available, and they can be easily interfaced with models for real time simulations, projections, and management decisions. Thermocouples are inexpensive and easy to use. As was the case with dissolved oxygen models, stratification is a phenomenon that cannot be neglected in most culture situations, especially in ponds used for benthic animals.

ECOSYSTEM MODELS

There are a small number of models that have been developed to simulate a number of variables in addition to dissolved oxygen and temperature. In these models there has been an attempt at providing quantified descriptions of the behavior of pond ecosystems. The models have been developed for research purposes, and have been based on an assumption of completely mixed pond conditions. Difficulties in collecting comprehensive data sets have limited the degree of model validation and use. The models currently available are described below.

Ecosystem Model 1

The model developed by Piedrahita et al. (1983; 1984; Piedrahita 1984; 1986) was designed to simulate up to 22 state variables in a pond. The variables included biological (fish, zooplankton, phytoplankton), chemical (inorganic nitrogen (ammonia, nitrate, nitrite), inorganic phosphorus, inorganic carbon, alkalinity, pH, dissolved organic matter), and mixed chemical and biological (particulate organic matter and sediments) terms. The mixed terms included both a living component such as bacteria, and a non-living component usually composed of particulate organic matter that may include added feeds. Concentrations of the various state variables were calculated with mass balance equations that included sources, sinks, and advection terms as in the dissolved oxygen models mentioned above. Rate equations were based on published relationships. The model was designed for short term (24 to 48 hour) simulations, with values calculated at hourly intervals.

Sensitivity analysis of the model has been carried out, particularly for changes in environmental conditions, and in rate coefficients. Changes in wind velocity were shown to have a marked effect on dissolved oxygen and pH, while cloud cover had a minor effect on the same parameters (Piedrahita et al. 1984; Piedrahita 1984). Rate coefficients having the greatest effect on state variables were phytoplankton growth rate (rate of photosynthesis), and particulate organic matter decomposition rate (Piedrahita 1986). The variable showing the greater sensitivity to changes in environmental conditions and rate coefficients was dissolved oxygen. Sensitivity analysis also corroborated what has already been mentioned in the discussion on oxygen models, that fish are relatively unimportant as direct contributors to the dissolved oxygen balance in the pond. The influence of fish on water quality in ponds tends to be more through the production of metabolites (ammonia), and through the grazing pressure exerted on the various foods present in the pond. The model required a comprehensive data base for calibration and validation. Original calibration and sensitivity analysis was carried out with hypothetical data sets assembled for various sources. Only recently, has the model been tested against field data (Piedrahita 1988, Unpublished Report).

Ecosystem Model 2

The model developed by Svireshev and coworkers (1984) was designed specifically to simulate a polyculture pond that included common carp, bighead carp, and silver carp. Other state variables included in the model are phytoplankton, zooplankton, benthic organisms, dissolved inorganic phosphorus, dissolved inorganic nitrogen, dissolved oxygen, feed, and detritus combined with bacteria. Transformations between the various state variables are simulated using mass balance equations. The time scale used for the simulations is different for dissolved oxygen ("hours") and the other variables ("days"). External variables affecting the pond processes include water temperature and total solar radiation, artificial feeding, fertilization, and mechanical aeration. Expressions and parameter values used for the various rate functions were derived from the published literature. The model was calibrated with data from Polish ponds (Opuszynski 1978, and others), and limited sensitivity to inorganic nutrients and temperature was tested. The model can be used to simulate the seasonal changes in water quality as well as diurnal cycles of dissolved oxygen. Simulation of fish growth over a season provides a useful structure for analysis of alternative production and management strategies.

Ecosystem Model 3

The model developed by Wolfe et al. (1986) was designed to simulate water quality in translucent 2300 L cylinders ("solar-algae ponds") used for the culture of tilapia (*Oreochromis aureus*). These systems were operated with a minimum of make up water (approximately 20 % per week), and received pelleted feeds as the nutrient input. The characteristics of the ecosystem, and the processes taking place in the cylinders are very similar to those in conventional aquaculture ponds. The similarities make the procedure used, and the basic structure developed applicable to ponds, even though some specific details and rate coefficients may not be directly applicable.

The procedure used for development of the model is similar to that used for the other ecosystem models mentioned, and mass balance equations are used to account for the changes in the state variables. The variables included in the model are: fish, phytoplankton, suspended particular organics, inorganic nitrogen (ammonia, nitrite, and nitrate), dissolved oxygen, carbon dioxide, alkalinity, pH, and bacteria. The explicit simulation of bacteria is unique to this model. Furthermore, several groups of bacteria are included, and each is simulated as a separate state variable: decomposing bacteria, ammonia oxidizing bacteria, nitrite oxidizing bacteria, nitrate reducing bacteria, and nitrite reducing bacteria. The data required to run a simulation are considerable, as is the number of rate coefficients that need to be estimated from measurements or from literature values. The time step used for the simulations is 0.05 d, although all calculations are based on daily reaction rates, and daily inputs and outputs.

The model was calibrated against data collected from three cylinders monitored during 100 days. Simulations show good agreement with the data.

General Comments on Ecosystems Models

The greatest limitation on the continued development of ecosystem type models has been the lack of adequate data sets for model validation and calibration. There has been tremendous progress, however, due to the availability of data acquisition systems and sampling probes and equipment. The limitations remain on the number of parameters that can be monitored practically. Even critical water quality parameters such as ammonia concentration cannot be monitored automatically in a field situation. Equipment is being developed for this purpose, but severe limitations still exist (Kaiser and Wheaton 1988; Fuss 1988).

The level of resolution possible with ecosystems models makes them extremely useful tools for the analysis of complex systems such as aquaculture ponds. By analyzing complete systems, it is possible to quantify complex interactions that may not be evident to researchers and managers. An example is the effect of aeration on variables other than dissolved oxygen, by affecting temperature, mixing, and the transfer rate of other gases between the atmosphere and the pond.

FUTURE DEVELOPMENTS AND CONCLUSIONS

The modeling of pond ecosystems has seen remarkable progress in the 12 years since the appearance of the first dissolved oxygen model for aquaculture ponds. Models have evolved as the quality of data available and the understanding and knowledge base have improved. This parallel evolution is likely to continue at an accelerated pace. Improvements in computer hardware, the availability of modeling and statistical software in powerful microcomputers, and the use of automated data acquisition systems will result in improvements in models, and expanded model use. Currently models are used primarily for research purposes, but the natural trend should be towards the application of models in planning and management.

Some future developments in pond ecosystem modeling include:

1. One dimensional models for vertically stratified ponds. The first attempts at quantifying and modeling stratification in ponds have been quite successful for temperature predictions. The simulation of other variables, especially dissolved oxygen, is still in need of further development. Extension of the work to saltwater ponds, where residence times are likely to be considerably shorter than in the ponds used for the current temperature models, is important, given the size of the shrimp industry.

2. Integrated temperature-ecosystem models. All ecosystems models developed to date have used temperature as a model input. Efforts in temperature modeling have shown the feasibility of calculating temperature in ponds. At least one model of dissolved oxygen has been developed that is coupled to a temperature model. Given the importance of temperature values in determining the rates of reactions, temperature models should be integrated with dissolved oxygen and more general ecosystem models.

3. A model is only as good as the information used to develop it. As the quality of the information improves, so will the quality of the models. Areas in particular need of further research include sediment-water interactions, heterotrophic processes, pond hydromechanics, and phytoplankton dynamics. A great deal of work has been done on phytoplankton dynamics, but we still lack enough understanding to be able to predict dissolved oxygen accurately.

4. Simplification of data collection for model execution is important. As an example, dissolved oxygen production rates depend on phytoplankton

concentration, light intensity, temperature, nutrient concentrations, and the particular physiological state of the algal cells that determine how they react to their environment. In modeling, measurements of each of these factors are needed, or in lieu of them, a gross indication of algal physiological state.

5. Models integrated with automated data acquisition and control systems. This has been proposed before, and the first stages are being implemented in research and production facilities. Initially, the data acquisition and controls are being used for monitoring and alarms. The next step in development is the use of models to forecast water quality and to control water flows, aeration and feeding systems. Forecasts can be updated continuously with field data on pond water quality and climate.

6. The use of computer models to develop and evaluate pond management practices is already being attempted and is likely to continue. Information generated from models can be presented to pond managers and other users in the form of guidelines, and possibly with the use of expert computer systems (J. Lannan 1988, personal communication).

ACKNOWLEDGEMENT

This is contribution No. 89-29 of the Pond Dynamics/Aquaculture Collaborative Research Support Program (CRSP) supported by the Agency for International Developments Grant No. DAN-4023-G-SS-7066-00.

NOMENCLATURE

c_o	=	Volumetric heat capacity of water
DO_a	=	Dissolved oxygen addition rate by mechanical aeration
DO_b	=	Rate of oxygen use by the "mud" or sediment, including benthic organisms
DO_{BOD}	=	Total oxygen consumption rate due to detritus
DOdawn2	=	Calculated dissolved oxygen concentration at dawn after 24 hours of simulation
DO_{Dt}	=	DO concentration after t hours of darkness
DO_{dusk1}	=	Initial DO concentration at dusk
$DO_{ed,s}$	=	Dissolved oxygen transfer by effective diffusion between the surface volume element and the adjacent volume element.

DO_f	=	Dissolved oxygen consumption rate by fish
DO_{gpp}	=	Gross oxygen production rate by phytoplankton
DO_{Gpp}	=	Gross primary production by phytoplankton during the day
DO_{npp}	=	Net oxygen production rate due to photosynthesis
DO_{om}	=	Oxygen consumption rate by organic matter
DO_{pond}	=	Dissolved oxygen concentration in the water exiting the pond, assumed equal to the concentration in the pond (i.e., fully mixed conditions)
DO_{pr}	=	Oxygen consumption by plankton respiration
DO_{rea}	=	Rate of oxygen transfer between the atmosphere and the pond
DO_s	=	Dissolved oxygen consumption by shrimp
DO_{supply}	=	Dissolved oxygen concentration in the water coming into the pond
DO_{sve}	=	Dissolved oxygen concentration in the volume element
DO_t	=	Dissolved oxygen concentration at time t
$DO_{t+\Delta t}$	=	Dissolved oxygen concentration at time t+Δt
DO_w	=	Dissolved oxygen consumption rate in the water column
$DO_{wc,s}$	=	Dissolved oxygen consumption by organisms and chemicals suspended in the water column, other than fish and phytoplankton.
DO_{wpr}	=	DO consumption rate by all processes in a pond ("whole pond respiration")
DO_{wto}	=	Dissolved oxygen transfer due to water turnover, and assuming that the rate of water flow into and out of a pond are equal (i.e. pond volume remains constant):
E	=	Evaporation rate
L	=	Latent heat of vaporization of water
Q	=	Water flow rate through the pond
q_a	=	Long wave atmospheric radiation
q_{ad}	=	Advective heat transfer = $c_o \cdot Q \cdot (T - T_i)$
q_{ar}	=	Reflected long wave radiation

q_c	=	Sensible heat flux
q_e	=	Evaporative heat loss = L· E
q_L	=	Long wave radiation from the water surface
q_{nr}	=	Net radiative heat flux
q_r	=	Net solar and atmospheric radiation
q_s	=	Incoming short wave solar radiation
q_{sr}	=	Reflected short wave radiation
T	=	Water temperature in the pond
T_i	=	Temperature of the water entering the pond
V	=	Volume of water in the pond
Δt	=	Duration of time interval between calculations
Δt_d	=	Hours after the measurement of DO_{dusk1}

LITERATURE CITED

Andrews, J.W. and Y. Matsuda 1975. The influence of various culture conditions on the oxygen consumption of channel catfish. Trans.Am. Fish. Soc. 104:322-327.

Augusto, J.A.F. 1975. Effects of inorganic fertilization on phytoplankton in fed catfish pond. M.S. thesis, Auburn University, Alabama.

Banks, R.B. and H. Herrera 1977. Effect of wind and rain on surface reaeration. J. of Env.Eng. Div., ASCE 103(EE3):489-504.

Bannister, T.T. 1979. Quantitative description of steady state, nutrient-saturated algal growth, including adaptations. Limnol. Oceanogr., 24:76-96.

Baumert, H. and D. Uhlmann 1983. Theory of the upper limit to phytoplankton production per unit area in natural waters. Int. Rev. Gesamten Hydrobiol., 68:753-783.

Blackburn, T.H. 1987. Role and impact of anaerobic microbial processes in aquatic systems, p. 32-53. In D.J.W. Moriarty and R.S.V.Pullin (eds.) Detritus and microbial ecology in aquaculture. ICLARM Conference Proceedings 14, 420 p. International Center for Living Aquatic Resources Management, Manila, Philippines.

Brune, D.E. 1978. The growth kinetics of freshwater algae. Ph.D. dissertation. University of Missouri, Columbia.

Brune, D.E., D.I. Meyer, and B. Haight 1983. Mathematical and computer modeling of aquatic culture systems. Presented at the Annual Meeting of the World Mariculture Society, Washington, D.C. 17 pp.

Bolte, J. P., R.D. Smitherman, and D.T. Hill 1986. Applications of mathematical modeling to pond aquaculture: A state-of-the-art review. Am Soc. Agric. Eng. Paper No 86-5049. St. Joseph, Michigan, 15 pp.

Boyd, C.E. 1973. The chemical oxygen demand of waters and biological materials from ponds. Trans. Am. Fish. Soc., 102:606-611.

Boyd, C.E. 1982. Water Quality Management for Pond Fish Culture. Elsevier, Amsterdam. 318 pp.

Boyd, C.E., R.P. Romaire, and E. Johnston 1978. Predicting early morning dissolved oxygen concentrations in channel catfish ponds. Trans. Am.Fish.Soc. 107:484-492.

Busch, C.D., C.A. Flood, J.R. Koon, and R. Allison 1977. Modeling catfish pond nighttime dissolved oxygen levels. Trans. Amer. Soc. Agr. Eng., 20:394-396.

Cassinelli, R.D., J.T. Farnsworth, V.E. Sweat, and D.L. Stoner 1978. Variations in dissolved oxygen concentration in mariculture ponds - a preliminary model, p. 310-329. In Proceedings of the Third Annual Tropical and Subtropical Fisheries technological Conference of the Americas. New Orleans, Louisiana. Texas A & M Sea Grant College Program, TAMU-SG-79-101.

Cathcart, T.P. and F.W. Wheaton 1986. Modeling temperature distribution in freshwater ponds. Am.Soc. Agric. Eng. Paper No. 86-5050. Saint Joseph, Michigan. 30 pp.

Cathcart, T.P, and F.W. Wheaton 1987. Modeling temperature distribution in freshwater ponds. Aquacultural Engineering, 6:237-257.

Cathcart, T.P. 1987. Heat transfer and temperature prediction in small freshwater ponds. Ph.D. Dissertation, University of Maryland, College Park, Maryland.

Colman, J.A. and P. Edwards 1987. Feeding pathways and environmental constraints in waste-fed aquaculture: balance and optimization, p 240-281. In D.J.W. Moriarty and R.S.V.Pullin (eds.) Detritus and microbial ecology in aquaculture. ICLARM Conference Proceedings 14, 420 p. International Center for Living Aquatic Resources Management, Manila, Philippines.

Cuenco, M. L., R.R. Stickney, and W.E. Grant 1985a. Fish bioenergetics and growth in aquaculture ponds: I. Individual fish model development. Ecological Modelling, 27:169-190.

Cuenco, M. L., R.R. Stickney, and W.E. Grant 1985b. Fish bioenergetics and growth in aquaculture ponds: II. Effects of interactions among size, temperature, dissolved oxygen, unionized ammonia and food on growth of individual fish. Ecological Modelling, 27:191-206.

Cuenco, M.L., R.R. Stickney, and W.E. Grant 1985c. Fish bioenergetics and growth in aquaculture ponds: III. Effects of intraspecific competition, stocking rate, stocking size and feeding rate on fish productivity. Ecological Modelling 28:73-95.

de Jaeger, J.M. and R.D. Walmsley 1984. A model to predict water temperature in plastic-covered outdoor mass algal culture systems. Aquacultural Engineering, 3:191-206.

Farnsworth, J.T. 1979. Systems analysis of a shrimp mariculture system. M.S. Thesis. Texas A & M University.

Field, S.D. and S.W. Effler 1982. Photosynthesis-light mathematical formulations. Jour. of the Env. Eng. Div., ASCE, 108(EE1):199-203.

Fritz, J.J., A.C. Middleton and D.D. Meredith 1979. Dynamic process modeling of wastewater stabilization ponds. Jour. Water Poll. Control Fed. 51:2724-2743.

Fuss, J. 1988. Device for the continuous monitoring of ammonia levels. Presented at the Fisheries Bioengineering Symposium, Am. Fish. Soc., Portland, Oregon.

Giovannini, P. 1989. Analysis and modeling of dissolved oxygen concentration in warm water aquaculture ponds with special regard to photosynthesis dynamics. M.S. Thesis. University of California, Davis.

Groden, T.W. 1977. Modeling temperature and light adaptation of phytoplankton. Report No. 2 . Center for Ecological Modeling. Rensselaer Polytechnical Institute, Troy, New York. 17 pp.

Henderson-Sellers, B. 1984. Engineering limnology. Pitman Advanced Publishing Program, Boston. 356 pp.

Hopkins K.D., J.E. Lannan, and J. R. Bowman. 1987. A data base management system for research in pond dynamics. CRSP Research Report No. 87-1. Pond Dynamics/Aquaculture CRSP, Oregon State University, Corvallis, Oregon.

Jewell, W.J. and P.L. McCarty. 1971. Aerobic decomposition of algae. Environ. Sci.Tech., 5:1023-1031.

Kaiser, E., and F. Wheaton. 1988. Automated water quality analysis and control. Presented at the Fisheries Bioengineering Symposium, Am. Fish. Soc., Portland, Oregon.

Keijman, J.Q. 1974. The estimation of the energy balance of a lake from simple weather data. Boundary-Layer Meteorol., 7:389-407.

Kerns, C.L. and E.W. Roelofs. 1977. Poultry wastes in the diet of Israeli carp. Bamidgeh, 29(4): 125-135.

Kester, D.R. 1975. Dissolved gases other than CO_2, p. 2:497-589 In J.P. Riley and G. Skirrow (eds.) Chemical Oceanography. John Wiley and Sons, New York.

Klemetson, S.L. and G.L. Rogers 1985. Aquaculture pond temperature modelling. Aquacultural Engineering, 4:191-208.

Knight, R.S. 1965. Performance of a cage rotor in an oxidation ditch. M.S. thesis, Iowa State University, Ames.

Krant, J., F. Motzkin, and H. Gordin 1982. Modelling temperatures and salinities of mixed seawater fish ponds. Aquaculture, 27:377-388.

Krant, J., I. Segal, and A. Dayan 1985. A simple model for comparing energy exchange in warmed soils and fish ponds. Energy Agric., 4:25-39.

Kreutzer, W. 1986. System simulation programming styles and languages. Addison-Wesley, Sydney.

Lannan, J. E., R.O. Smitherman, and G. Tchobanoglous (eds.) 1986. Principles and practices of pond aquaculture. Oregon State University Press, Corvallis, Oregon.

Laws, E.A. and S.R. Malecha 1981. Application of a nutrient saturated growth model to phytoplankton management in freshwater prawn (*Machrobrachium rosenbergii*) ponds in Hawaii. Aquaculture, 24:91-101.

Little, D. and J. Muir 1987. A guide to integrated warm water aquaculture. Institute of Aquaculture, Univ. of Stirling, Stirling, Scotland. 238 pp.

Losordo, T.M. 1988. Characterization and modeling of thermal and oxygen stratification in aquaculture ponds. Ph.D. Dissertation. University of California, Davis.

Losordo, T.M., J.M. Ebeling, and R.H. Piedrahita 1986. Stratification measurement techniques in aquaculture ponds. Am.Soc. Agr. Eng. Paper No. 86-5047. Saint Joseph, Michigan. 14 pp.

Losordo, T.M. and R.H. Piedrahita 1988a. Simulating aquacultural pond thermal stratification with a spreadsheet model. Am. Soc. Agric. Eng. Paper No. 88-5003. Saint Joseph, Michigan. 37 pp.

Losordo, T.M. and R.H. Piedrahita 1988b. Modeling vertical water quality profiles in aquaculture ponds, review and evaluation. Presented at the International Conference on Aquaculture Engineering: Technologies for the Future, The Institution of Chemical Engineers. University of Stirling, Scotland. 15 pp.

Losordo, T.M., R.H. Piedrahita, and J.M. Ebeling 1988. An automated water quality data acquisition system for use in aquaculture ponds. Aquacultural Engineering, 7:265-278.

Lu, J. and N. Kevern 1975. The feasibility of using waste materials as supplemental fish feed. Prog. Fish-Cult., 37:241-244.

Ma, J.C. 1979. A systems analysis approach to the study of the energetics of shrimp production in the mariculture ponds near Angleton, Texas. M.S. Thesis. Texas A & M University.

Madenjian, C.P., G.L. Rogers, and A.W. Fast 1987a. Predicting night time dissolved oxygen loss in prawn ponds of Hawaii: Part I. Evaluation of traditional methods. Aquacultural Engineering, 6:191-208.

Madenjian, C.P., G.L. Rogers, and A.W. Fast 1987b. Predicting night time dissolved oxygen loss in prawn ponds of Hawaii: Part II. A new method. Aquacultural Engineering, 6:209-225.

Meyer, D.I. 1980. Modelling diel oxygen flux in a simulated catfish pond. M.S. thesis. University of California, Davis.

Meyer, D.I. and D.E. Brune 1982. Computer modeling of the diurnal oxygen levels in a stillwater aquaculture pond. Aquacultural Engineering, 1:245-261.

Mezainis, V.E. 1977. Metabolic rates of pond ecosystems under intensive catfish cultivation. M.S. Thesis. Auburn University, Auburn, Alabama. 107 pp.

Moriarty, D.J.W. and R.S.V. Pullin (eds.) 1987. Detritus and microbial ecology in aquaculture. ICLARM Conference Proceedings 14. International Center for Living Aquatic Resources Management, Manila, Philippines. 420 pp.

Muller-Feuga, A.,A. Petit, and J.J. Sabbaut 1978. The influence of temperature and wet weight on the oxygen demand of rainbow trout (Salmo gairdneri R.) in fresh water. Aquaculture, 14:355-363.

Opuszinski, 1978. The influence of silver carp on eutrophication of carp ponds. III. Phytoplankton. Rocz. Nauk. rol., H, 99:127-151.

Piedrahita, R.H. 1984. Development of a computer model of the aquaculture pond ecosystem. Ph.D. Dissertation. University of California, Davis.

Piedrahita, R.H. 1986. Sensitivity analysis for an aquaculture pond model, p. 119-123. In J.G. Balchen (ed.) Automation and data processing in aquaculture. International Federation of Automatic Control. Pergamon Press, Oxford.

Piedrahita, R.H. 1988. Introduction to computer modelling of aquaculture pond ecosystems. Aqua. Fisheries Mgmt., 19:1-12.

Piedrahita, R.H., D.E. Brune, G.T. Orlob, and G. Tchobanoglous 1983. Ecologic simulation of warm water aquaculture ponds. Am. Soc. Agr. Eng. Paper No. 83-5041. Saint Joseph, Michigan. 14 pp.

Piedrahita, R.H., D.E. Brune, G. Tchobanoglous, and G.T. Orlob 1984. A general model of the aquaculture pond ecosystem. J. World Maricul. Soc. 15:355-366.

Piedrahita, R.H., J.M. Ebeling, and T.M. Losordo 1986. Use of data acquisition systems in aquaculture, p. 259-262 . In J.G. Balchen (ed.) Automation and data processing in aquaculture. International Federation of Automatic Control. Pergamon Press, Oxford.

Redfield, A.C. 1958. The biological control of chemical factors in the environment. Amer. Scient. 46:205-221.

Riley, G.A. 1946. Factors controlling phytoplankton populations on Georges Banks. J. Mar. Res., 6:54-73.

Romaire, R.P. and C.E. Boyd 1979. Effects of solar radiation on the dynamics of dissolved oxygen in channel catfish ponds. Trans. Am. Fish. Soc. 108:473-478.

Romaire, R.P., C.E. Boyd, and W.J. Collis 1978. Predicting nighttime dissolved oxygen decline in ponds used for tilapia culture. Trans. Am. Fish. Soc., 107:804-808.

Rossi, G., F. Beonio-Brocchieri, and J. Sangregorio 1982. On the evaluation of the vertical eddy coefficient in lakes by means of measured temperature profiles. Presented at the Third International Conference on State of the Art in Ecological Modelling. Colorado State University.

Schroeder, G.L. 1975. Nighttime material balance for oxygen in fish ponds receiving organic wastes. Bamidgeh, 27:65-74.

Schroeder, G.L. 1987. Carbon pathways in aquatic detrital systems, p. 217-236. In D.J.W. Moriarty and R.S.V.Pullin (eds.) Detritus and microbial ecology in aquaculture. ICLARM Conference Proceedings 14, 420 p. International Center for Living Aquatic Resources Management, Manila, Philippines.

Smith, D.W. 1987. Biological control of excessive phytoplankton and enhancement of aquacultural production. Ph.D. Dissertation. University of California, Santa Barbara.

Smith, D.W. and R.H. Piedrahita 1988. The relation between phytoplankton and dissolved oxygen in fish ponds. Aquaculture, 68:249-265.

Smith, E.L. 1936. Photosynthesis in relation to light and carbon dioxide. Natl. Ac. of Sci. Proc., 22:504-511.

Smith, R.A. 1980. The theoretical basis for estimating phytoplankton production and specific growth rate from chlorophyll, light and temperature data. Ecological Modelling, 10:243-264.

Steele, J.H. 1962. Environmental control of photosynthesis in the sea. Limnol. Oceanogr. 7:137-150.

Straskraba, M. and A.H. Gnauck 1985. Freshwater ecosystems modelling and simulation. Elsevier, Amsterdam. 309 pp.

Stumm, W. and J.J. Morgan 1981. Aquatic chemistry, 2nd ed. Wiley, New York. 780 pp.

Sunderam, T.R. and R.G. Rehm 1973. The seasonal thermal structure of deep temperate lakes. Tellus, 25:157-167.

Svireshev, Yu.M., V.P. Krysanova, and A.A. Voinov 1984. Mathematical modelling of a fish pond ecosystem. Ecological Modelling, 21:315-337.

Tennessee Valley Authority. 1972. Heat and mass transfer between a water surface and the atmosphere. Water Resour. Res. Lab. Report 14. Norris, Tennessee. 169 pp.

Velz, C.T. 1970. Applied stream sanitation. John Wiley and Sons, New York. 619 pp.

Water Resources Engineers, Inc. 1968. Report prepared for the Department of Fish and Game, State of California. 90 pp.

Welch, E.B. 1980. Ecological effects of wastewater. Cambridge Univ. Press. Cambridge. 337 pp.

Wetzel, R.G. 1983. Limnology, 2nd ed. Saunders, Philadelphia. 767 pp.

Wholfarth, G.W., and G. Hulata 1987. Use of manures in aquaculture, p 353-367. In D.J.W. Moriarty and R.S.V.Pullin (eds.) Detritus and microbial ecology in aquaculture. ICLARM Conference Proceedings 14, 420 p. International Center for Living Aquatic Resources Management, Manila, Philippines.

Wholfarth, G.W., and G.L. Schroeder 1979. Use of manure in fish farming - a review. Agricultural Wastes, 1:279-299.

Wolfe, J.R., R.D. Zweig, and D.G. Engstrom 1986. A computer simulation model of the solar-algae pond ecosystem. Ecol. Modelling, 34:1-59.

EMPIRICAL MODELING OF PHYTOPLANKTON GROWTH AND OXYGEN PRODUCTION IN AQUACULTURE PONDS

Claude E. Boyd

ABSTRACT

Phytoplankton growth in aquaculture ponds is regulated in freshwater primarily by phosphorus concentration, but in brackishwater both nitrogen and phosphorus concentrations often limit abundance of phytoplankton. Abundance of phytoplankton and production of aquacultural species can be enhanced by fertilization or manuring. Even greater production of aquacultural species can be achieved through feed applications. Phytoplankton abundance in ponds with feeding increases as feeding rates are raised. Dissolved oxygen (DO) dynamics in ponds are regulated primarily by phytoplankton abundance. As phytoplankton density increases, a pronounced vertical stratification of DO concentration develops and DO exhibits a strong daily cycle with the highest concentration in the afternoon and the lowest concentration around dawn. Problems with low DO during the night increase in frequency and severity as feeding rates are elevated. Cloudy weather and phytoplankton die-offs also may cause DO depletion in ponds. Computer simulation techniques for predicting DO concentrations in ponds are available, but these methods require data that often are not easily available to pond managers. A simple projection technique for extrapolating DO concentrations measured early in the night to provide a forecast of DO concentration later in the night is popular with pond managers. Aeration is the only reliable means of combating DO depletion.

INTRODUCTION

Phytoplankton photosynthesis and respiration dominate the dynamics of dissolved oxygen (DO) in aquaculture ponds. Phytoplankton abundance and photosynthetic rate are determined by nutrient inputs into ponds in chemical fertilizers, manures, and feed. As the level of aquacultural production increases, fluctuations in DO concentrations are accentuated and the frequency and severity of DO problems increased. Many researchers are interested in preparing DO models for ponds. Although elaborate models are useful exercises for understanding the dynamics of DO in ponds, these models are of little utility to pond managers. Practical aquaculturists need to understand basic relationships among environmental factors and DO concentrations and to have means of making rapid and approximate forecasts of DO concentrations. Here I attempt to provide a practical discussion of oxygen dynamics and prediction in ponds.

PHYTOPLANKTON GROWTH AND OXYGEN PRODUCTION

Growth Factors

Phytoplankton propagules are practically everywhere, so it is not necessary to seed ponds with phytoplankton. If water contains adequate concentrations of inorganic nutrients and there is sufficient sunlight, phytoplankton usually will flourish. Of course, warmth favors phytoplankton growth, but phytoplankton will grow in cold water.

Pond aquaculture is most commonly practiced in temperate, subtropical, or tropical climates where there is plenty of sunlight. Therefore, the growth of phytoplankton usually is limited by nutrient concentrations. Nutritive elements required for phytoplankton growth are listed in Table 1 along with the approximate average concentrations of these elements in freshwater, seawater, and phytoplankton (Boyd 1988). Concentration factors (concentration of element in phytoplankton/concentration in water) are greatest for nitrogen and phosphorus. Therefore, nitrogen and phosphorus are the two essential elements present in the shortest supply, and they are most likely to regulate the amount of phytoplankton growth in an aquaculture pond. Of course, phytoplankton also require oxygen and hydrogen, but oxygen can be produced in photosynthesis and hydrogen obtained from metabolic pathways.

Table 1. Concentrations of elements in seawater, freshwater and phytoplankton.

Element	mg/liter			Concentration factors	
	Seawater	Freshwater	Phytoplankton (ppm)[a]	Seawater	Freshwater
Phosphorus	0.07	0.03	230	3,286	7,667
Nitrogen	0.5	0.3	1,800	3,600	6,000
Iron	0.01	0.2	25	2,500	125
Manganese	0.002	0.03	4	2,000	133
Copper	0.003	0.03	2	667	100
Silicon	3	2	250[b]	83	125
Zinc	0.01	0.07	1.6	160	23
Carbon	28	20	12,000	429	600
Potassium	380	2	190	0.5	95
Calcium	400	20	220	0.55	11
Sulfur	900	5	160	0.18	32
Boron	4.6	0.02	0.1	0.02	5
Magnesium	1,350	4	90	0.07	22.5
Sodium	10,500	5	1,520	0.14	304

[a]Wet weight basis. Assumed 2.5% dry weight.
[b]Concentration is much greater for diatoms.

Blue-green algae, which often abound in freshwater aquaculture ponds can fix nitrogen, so phosphorus is more likely to limit phytoplankton growth than is nitrogen in freshwater ponds. In brackishwater and seawater, blue-green algae usually are less abundant, and both nitrogen and phosphorus are important limiting factors.

The availability of phosphorus is regulated by reactions between dissolved phosphate and phosphate compounds in bottom muds (Boyd 1982). Acid muds bind phosphate tightly and restrict its solubility. Also, ponds with acid muds usually have waters with low concentration of total alkalinity and free carbon dioxide. Liming of ponds will raise mud pH and total alkalinity. These changes improve phosphate availability and the concentration of free carbon dioxide. Problems of low pH and alkalinity are more common in freshwater ponds than in brackishwater or seawater ponds. However, brackishwater ponds constructed in areas with potential acid-sulfate soils may be extremely acidic (Dent 1986).

Color in water resulting from high concentrations of dissolved humic substances and turbidity resulting from suspended, colloidal, soil particles restrict light penetration into water. Ponds receiving water from acidic forest or jungle watersheds often have waters highly stained with humates, while ponds with denuded watershed may have extremely turbid waters. Excessive color or turbidity can restrict phytoplankton growth.

Food Webs

One popular means of increasing aquacultural production is to add manures, chemical fertilizers, or both to ponds. Inorganic nutrients from these substances foster greater phytoplankton productivity, which serves as the base for increased production of shrimp, fish, or other species. Representative food webs are presented in Figures 1 and 2. An alternative for

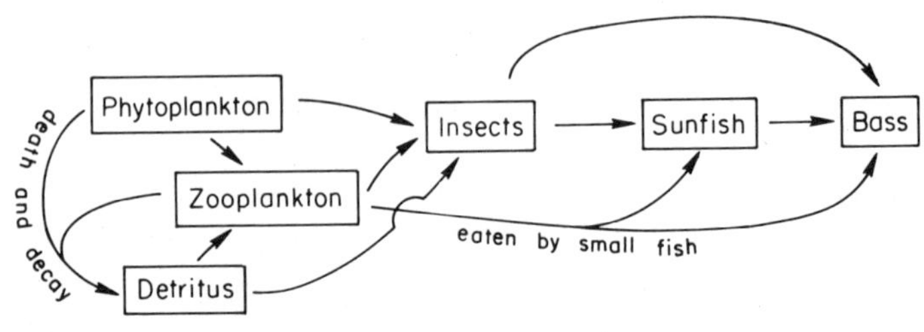

Figure 1. A representative food web in a sunfish-bass pond.

enhancing production is to add feed for shrimp or fish. Depending on the amount and quality of feed, the feed supplements natural food organisms or serves as the primary source of nutrition. Metabolic wastes from the feed increase nutrient concentrations in water, and phytoplankton growth increases in direct proportion to feeding rate.

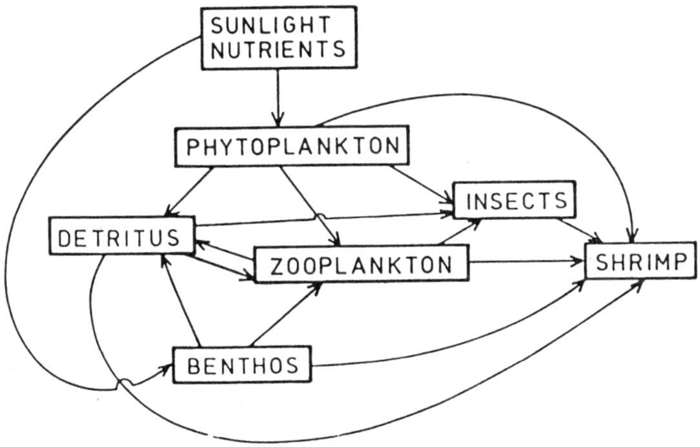

Figure 2. A food web in a shrimp pond. Feed may be applied to supplement natural food.

Weeds

The problems caused by aquatic weeds in aquaculture ponds are well known. Many methods of control have been attempted, but in general, biological control techniques are superior to chemical or mechanical means of aquatic weed control. Many aquatic weeds begin their growth at the pond bottom, and infestation of these weeds depend upon light penetrating to pond bottoms. If phytoplankton blooms are initiated before aquatic weed infestations begin,the phytoplankton will create turbidity, reduce light penetration, and shade the pond bottom to prevent aquatic weed growth. Conversely, if a weed infestation is present in a pond, the weeds will compete with phytoplankton for nutrients, and phytoplankton growth often is inhibited (Boyd 1982).

PHYTOPLANKTON AND DISSOLVED OXYGEN

The summary reaction for photosynthesis is:

$$6CO_2 + 6H_2O \xrightarrow[\text{nutrients}]{\text{sunlight}} C_6H_{12}O_6 + 6O_2$$

In this reaction, organic matter is formed, energy is stored in the organic matter, and oxygen is released to the water. This reaction is the primary source of DO in most aquaculture ponds. Respiration is the process whereby energy in organic matter is used by living organisms. Ecologically, respiration is simply the reverse of photosynthesis:

$$6CO_2 + 6H_2O \xrightarrow[\text{nutrients}]{\text{sunlight}} C_6H_{12}O_6 + 6O_2$$

In respiration, oxygen is consumed, and the chemical energy stored in organic matter is used to do biological work or is dissipated as heat.

All living organisms carry out respiration, but only plants and a few bacteria are capable of photosynthesis. In ponds, phytoplankton usually are abundant. They not only produce most of the oxygen, but they also usually use more oxygen in respiration than all other organisms combined (Boyd 1985).

At first, it seems that increasing phytoplankton abundance would increase oxygen production by photosynthesis with an increase in DO supplies for ponds. Actually, increasing phytoplankton abundance leads to the production of more organic matter and greater respiration rates. High densities of phytoplankton restrict light penetration so that photosynthetic oxygen production is restricted to shallow depths (Hepher 1962). Oxygen production will be high near the surface, but when oxygen production is considered for the entire column of water, a pond with high density of phytoplankton may not produce any more oxygen than a pond with a moderate density of phytoplankton (Boyd 1982). The stratification of DO concentration with depth is much more pronounced in a pond with a heavy phytoplankton bloom than in ponds with less phytoplankton (Fig. 3).

There also is a daily fluctuation of DO in ponds. During daylight, photosynthesis releases oxygen faster than oxygen is used in respiration. Concentrations of DO increase during the morning and usually reach a peak in the mid afternoon. Photosynthesis stops at night, but respiratory processes continue to use oxygen. Thus, DO concentrations decline during the night and reach their lowest values about dawn. The difference between daily maximum and minimum DO concentrations in a pond tends to become greater as phytoplankton density increases (Fig. 4).

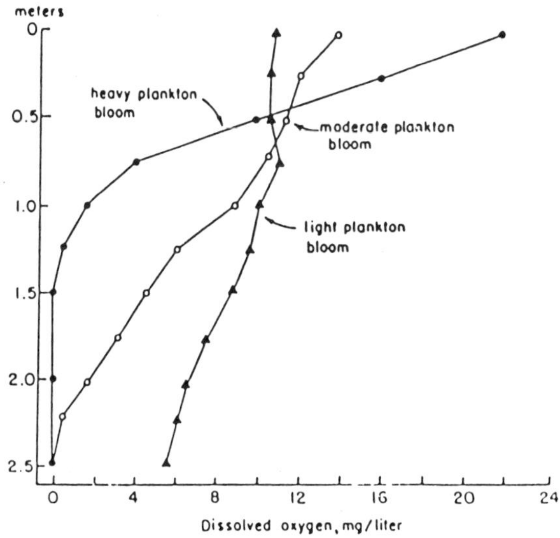

FIGURE 3. Influence of depth on dissolved oxygen concentrations in ponds with different amounts of plankton.

From the standpoint of fish and shrimp physiology, ponds with low abundances of phytoplankton are better because DO concentrations almost always are high enough to satisfy respiratory requirements. However, in order to produce high yields of aquaculture species, fertilizers, manures, or feeds are applied to ponds. The resulting phytoplankton blooms often cause shallow DO stratification and daily fluctuations of DO concentrations.

PREDICTING AND MODIFYING PHYTOPLANKTON GROWTH

Fertilizers and Manures

In ponds where no nutrients are applied directly in manures, fertilizers, or feeds, nutrient concentrations in ponds depend upon the fertility of watershed soils and watershed management practices. Phytoplankton growth

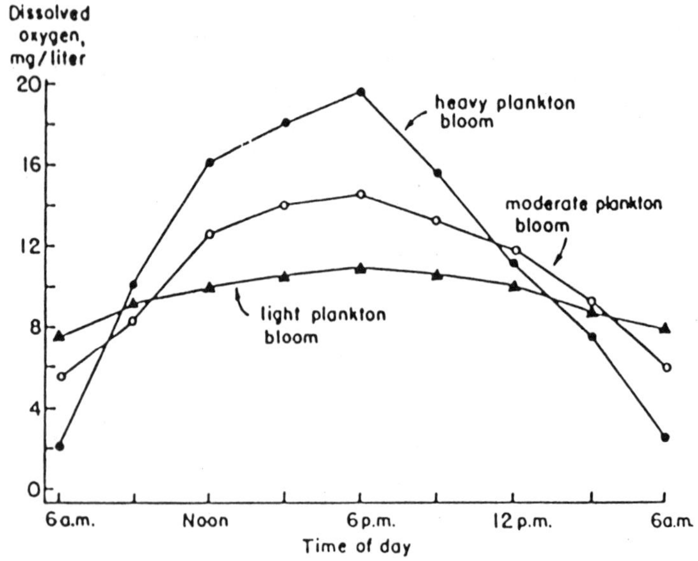

FIGURE 4. Effect of time of day and plankton density on concentrations of dissolved oxygen in surface water.

in ponds is greater in ponds on fertile soils than in ponds on infertile soils. In the United States, ponds usually are constructed on one of three types of watersheds: woodland, pasture, or cropland. Ponds on wooded watersheds usually have low concentrations of nutrients and phytoplankton abundance is low (Boyd 1976). Ponds on row-cropped watersheds often have high concentrations of nutrients because of runoff from plowed and fertilized fields, but suspended soil particles in runoff makes ponds turbid and phytoplankton growth often is slow. Hay fields do not yield turbid runoff, but neither do they yield runoff enriched with nutrients. Therefore, ponds which have ungrazed grassland for watersheds normally have low phytoplankton abundance. Ponds with pastures for watersheds usually have high concentrations of nutrients and an abundance of phytoplankton. Livestock on watersheds eat the grass and deposit large amounts of manure. Runoff delivers much of the manure to ponds where it serves as fertilizer.

In many tropical regions, water from the estuarine reaches of rivers is used in brackishwater ponds. The water in rivers often is polluted with agricultural and domestic wastes which contain nutrients. When held in ponds, nutrient concentrations in such waters often are high enough to foster phytoplankton blooms. For example, in the shrimp-farming region around Bangkok, Thailand, river waters used in brackishwater ponds contain sufficient nitrogen and phosphorus to cause dense phytoplankton blooms even without applications of fertilizers or manures (Ruttanagosrigit et al. 1980).

In a few places, ground water may have high concentrations of phosphate. In west-central Alabama, water from some wells contain from 0.2 to 0.5 mg/L soluble orthophosphate. When this water is put in ponds, no additional phosphate is required to promote phytoplankton growth.

Studies have not been conducted to determine the concentrations of nitrogen, phosphorus, and other nutrients necessary to cause phytoplankton blooms in ponds, but fertilization and manuring rates necessary to promote phytoplankton blooms have been determined. These rates seldom have been determined as the amount of fertilizer or manure necessary to provide particular concentrations of nutrients. Instead, fertilizer or manure rates have been correlated with aquacultural production (Swingle 1947; Hickling 1962; Boyd 1982). Many different fertilization programs have been used, but most programs consist of applications at 2- to 4-week intervals of 2 to 8 kg/ha of N and P_2O_5 or P_2O_5 alone (Boyd 1982; Murad and Boyd 1987). Many different sources of N and P_2O_5 have been used, but the most common are urea, ammonium nitrate, ammonium sulfate, diammonium phosphate, superphosphate, triple superphosphate, and the liquid fertilizer, ammonium polyphosphate. Liquid fertilizers are much more efficient than granular fertilizers (Murad and Boyd 1987). In brackishwater, nitrogen fertilizer is more important than in freshwater, because nitrogen-fixing blue-green algae are usually not as abundant in brackishwater as in freshwater. Fertilization with chemical fertilizers can result in 2- to 5-fold increases in fish or shrimp production.

A variety of types of manures also have been used in ponds. Concentrations of nitrogen and phosphorus are much lower in manures than in chemical fertilizers, and application rates for manures are much higher than rates for fertilizers. Manures often are applied at rates of 200 to 500 kg/ha at 1- to 2-week intervals. Manures can be as effective as fertilizers in increasing fish production. However, because manures consist primarily of organic matter, they exert an oxygen demand, and excessive use of manures can result in DO depletion (Schroeder 1975; Schroeder 1978; Wohlfarth and Schroeder 1979).

Phytoplankton abundance in ponds fluctuates considerably over brief periods in response to a myriad of variables, but in general, the average abundance of phytoplankton in ponds increases in response to greater concentrations of phosphorus and nitrogen in water which results when fertilization, manuring, or feeding rates are increased. The optimum abundance of phytoplankton is not known, but chlorophyll *a* concentrations of 60 to 150 µg/L are normally encountered in productive fish and shrimp ponds. Where waters are not extremely turbid with suspended clay particles or are not highly colored with humates, the Secchi disk visibility is a good index of phytoplankton abundance (Almazan and Boyd 1978). Most productive shrimp and fish ponds will have a Secchi disk visibility of 30 to 60 cm. When Secchi disk visibility is less than 20 to 25 cm and chlorophyll a concentrations exceed 150 or 200 µg/L, nighttime depletion of DO is highly probable in ponds.

Sometimes, fertilization or manuring will not evoke phytoplankton blooms. The major reasons for failure are as follows: low alkalinity waters, aquatic weed infestations, excess water flow through ponds, and turbidity from suspended soil particles. Alkalinity can be increased by liming. Water with less than 20 mg/L total alkalinity should be limed (Boyd 1982). Methods for calculating lime requirements for fish ponds are available (Boyd 1982; Pillai and Boyd 1985). Agricultural limestone, finely pulverized limestone, is the best liming material. Aquatic weeds compete with phytoplankton for nutrients. Weeds usually cannot be controlled successfully with chemicals, as mentioned earlier. Biological control methods such as fertilization to produce turbidity and grass carp to remove weeds by grazing are more reliable. Of course, when a production pond is infested with weeds, it may be drained and the bottom dried, treated with calcium hydroxide or calcium oxide, and tilled (Snow and Jones 1959). This procedure often will eliminate weeds, but weeds will return unless the pond is managed properly. If the water retention time in a pond is less than 3 or 4 weeks, nutrients are flushed from ponds so fast that it is difficult to maintain adequate phytoplankton blooms. The problem of excess flow usually is limited to watershed ponds which have a large watershed area to pond volume ratio or which have a large stream flowing through the pond. The best solution is to divert a part of the flow to prevent it from entering the pond. Turbidity from suspended clay particles often results from denuded areas on watersheds which yield turbid runoff to ponds. High rates of aeration, wind action, or high densities of bottom-feeding fish also can cause clay turbidity. Denuded areas on watersheds should be revegetated. Trees or shrubs may be planted around ponds to provide windbreaks. In most ponds it is not necessary to operate aerators in the daytime. Ponds with continuous aeration usually contain more clay turbidity than ponds aerated only during the night.

In brackishwater ponds for shrimp, water from estuaries often are heavily laden with settleable soil material. When this water is pumped into ponds the particles quickly settle to the bottom for sedimentation is favored by high concentrations of ions in water (Boyd 1988). Because sediment fills ponds and exerts an oxygen demand, settling ponds should be used to presettle soil particles from water for shrimp ponds (Boyd 1988).

Feeding

High quality feeds may be used to increase aquacultural production above that possible with fertilizers or manures. Feeds are used in two ways. Sometimes feeding rates are low and the feed merely supplements natural productivity of food organisms. In other situations, high feeding rates are used an fish or shrimp depend almost entirely on the feed for nutriment.

The feed cannot be assimilated and converted to fish or shrimp flesh with 100% efficiency. To illustrate, consider channel catfish production in a pond with feeding. A good manager might obtain a net production of 4,000 kg/ha of fish by applying 7,000 kg/ha of feed. This is a feed conversion ratio (feed applied ÷ net fish production) of 1.75. In other words, 1.75 kg feed was required to produce 1.00 kg of fish. From an economic standpoint, this conversion ratio is logical, for one buys feed and sells live fish. However, the conversion ratio is not sensible with regards to water quality considerations, for the feed is about 90% dry matter and the fish are about 25% dry matter.

In this example, 6,300 kg dry substance in feed (7,000 kg feed x 0.9) yielded 1,000 kg dry substance in fish. The difference, 5,300 kg, represents uneaten feed and metabolic wastes which contaminate the pond water. The organic portion of uneaten feed and metabolic waste exert an oxygen demand, but more importantly, these substances contain nitrogen and phosphorus which stimulate phytoplankton production. The amount of organic matter and oxygen demand resulting from phytoplankton production in ponds receiving feed greatly exceeds that of uneaten feed and metabolic wastes (Boyd 1985).

As feeding rates in ponds increase, concentrations of phosphorus and nitrogen rise with a simultaneous enhancement of phytoplankton productivity. Finally, a feeding rate is reached where the nighttime oxygen demand is so great that the risk of DO depletion is unacceptable (Cole and Boyd 1986). The relationships among feeding rate, phytoplankton abundance as measured by Secchi disk visibility, and early morning DO concentration are illustrated in Figure 5.

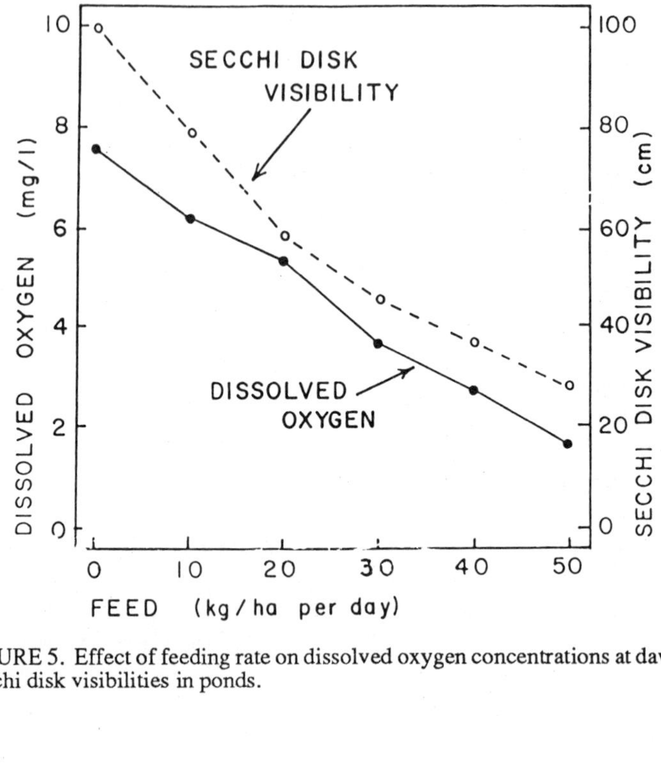

FIGURE 5. Effect of feeding rate on dissolved oxygen concentrations at dawn and Secchi disk visibilities in ponds.

Modifying Phytoplankton Growth

Of course, phytoplankton growth can be enhanced by adding fertilizers or manures as indicated above. Excessive blooms of phytoplankton in fertilized or manured ponds indicate that application rates of fertilizers or manures should be reduced. In ponds receiving feed, phytoplankton growth usually can be maintained within a desirable range by prudent feeding schedules.

Nevertheless, phytoplankton blooms sometimes become excessive despite the efforts of pond managers. Control of phytoplankton has been attempted by application of copper sulfate, simazine, and other algicides, but these efforts have been largely unsuccessful in aquaculture ponds. When phytoplankton is killed, it decays and exerts an oxygen demand. Because the phytoplankton has been killed, DO supplies are not replaced by photosynthesis and DO depletion may occur (Tucker and Boyd 1977). Also, once the toxicity of the algicide has dissipated, nutrients are still present in ponds to promote rapid growth of a new phytoplankton bloom. Dyes which color water and restrict light penetration have not proven successful for reducing phytoplankton blooms in aquaculture ponds (Boyd and Noor 1982).

There have been studies of the utility of cultivating water hyacinths (*Eichhornia crassipes*) in enclosures in ponds to remove nutrients and reduce phytoplankton growth (McVea and Boyd 1975). Although phytoplankton growth can be controlled in this manner, the amount of water hyacinths necessary to accomplish the task is great and the practicality of this method is highly questionable.

In brackishwater ponds, water often is obtained from estuaries and it is possible to flush water through ponds. Excess nutrients and phytoplankton can be removed from ponds by water exchange. Also, if water from estuaries contains more DO than pond water, some improvement in DO concentrations may be affected through water exchange. Unfortunately, many shrimp farmers do not understand how to use water exchange. There is a common belief that water must be flushed through ponds on a daily basis - exchange rates of 10 to 15% of pond volume per day often are used. This practice makes no sense as long as water quality in ponds is adequate for good shrimp growth. Nutrients and phytoplankton are flushed from ponds, often necessitating additional fertilization, and the added expense of unnecessary pump operation.

Algal Die-Offs

Phytoplankton blooms normally increase and decrease in abundance, but under certain circumstances phytoplankton blooms may suddenly die. There are probably several causes of phytoplankton die-offs, but one of the most common causes is light injury. In ponds with heavy blooms of certain blue-green algae, scums of these algae may form at pond surfaces during calm weather. On bright days, intense light at the surface results in high rates of photosynthesis; DO concentrations and pH rise, and carbon dioxide concentrations fall. Under these conditions, blue-green algae cells may die because of a metabolic malfunction known as photooxidation (Aveliovich and Shilo 1972). If a large quantity of blue-green algae die, their rapid decomposition can cause DO depletion. Changes in phytoplankton abundance and DO concentration during a massive phytoplankton die-off (Boyd et al. 1975) are illustrated in Figures 6 and 7. Sudden die-offs of mats of filamentous algae also have been reported (Dobbins and Boyd 1976), but the causes of such die-offs have not been described.

Feed-Oxygen Relationships in Raceways

In raceways, water is exchanged rapidly in production units. Many times the exchange rate may be as rapid as 3 to 6 times per hour. Phytoplankton do not grow in raceways because of the short water retention time. Thus, DO dynamics are simple; almost all DO is used in fish respiration. About 0.2 kg of oxygen is used in fish respiration for each kilogram of feed applied in a raceway (Ray 1981; Westers 1981). This means that the maximum permissible feeding rate may be estimated from the quantity of DO available. For example, suppose that a raceway segment has a water flow rate of 2 m^3/min, the water enters at 12 mg/L DO, and it is desired to have a DO concentration of 6 mg/L at the lower end of the raceway segment. In 1 day, the flow is 2,880 m^3 (1,440 min/day x 2 m^3/min). The available DO is 6 mg/L (12 mg/L in inflow - 6 mg/L in outflow), so the total quantity of available DO is 17.3 kg (2,880 m^3 x 6 g/m^3 x 10^{-3} kg/g). The maximum permissible feeding rate is

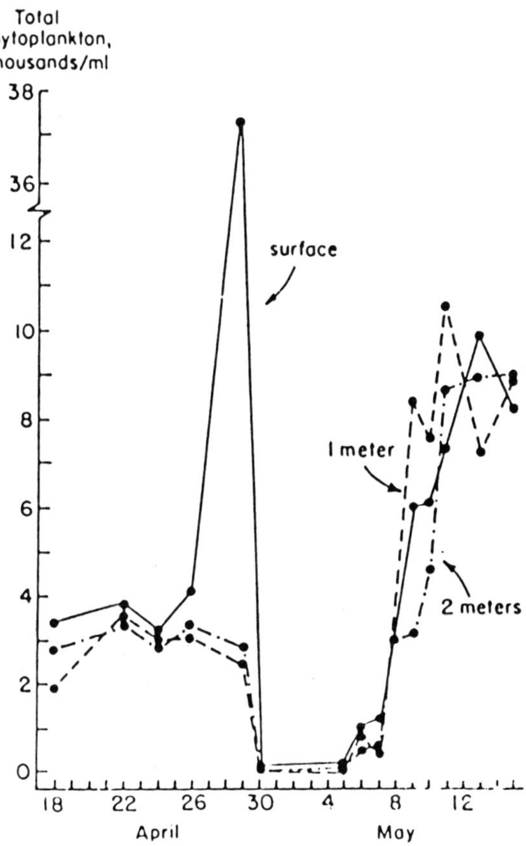

FIGURE 6. Changes in phytoplankton abundance before, during, and after a massive phytoplankton die-off in a pond.

86.5 kg/day (17.3 kg oxygen ÷ 0.2 kg O_2/kg feed). This calculation assumes that solid wastes are removed from the raceway and do not contribute to oxygen demand.

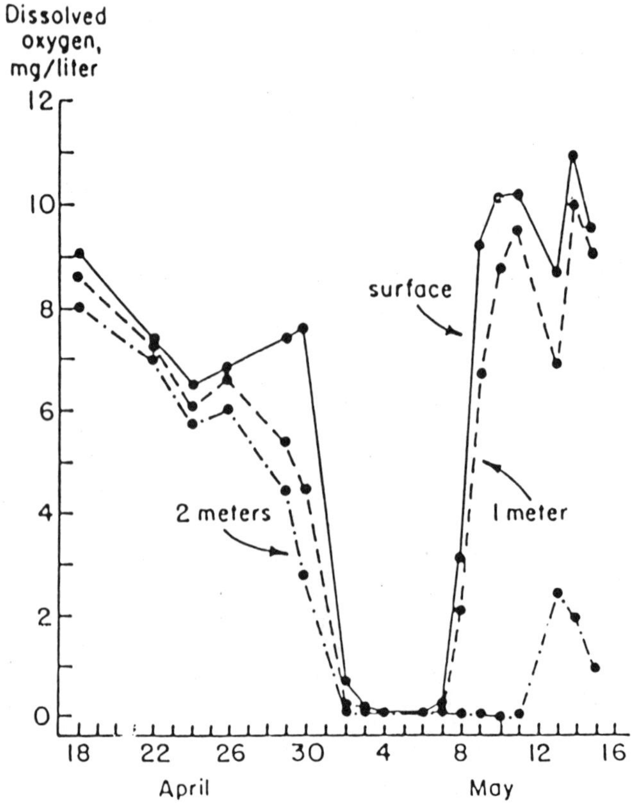

FIGURE 7. Dissolved oxygen concentrations before, during and after a massive phytoplankton die-off in a pond.

PREDICTING OXYGEN CONCENTRATIONS

Projection Technique

A simple procedure for estimating the nighttime decline in DO concentration is illustrated in Figure 8. The DO concentration is measured early in the night and again after a few hours. A line projected through the two points and extrapolated predicts DO concentrations at later times during the night.

This method often predicts a lower DO concentration than actually occurs, because the rate of respiration by phytoplankton tends to decline with time during darkness (Boyd et al. 1978). Also, as the DO concentration falls, the potential for diffusion of oxygen into the water from the air increases. Thus, the contribution of DO from diffusion increases as the night progresses, because DO concentrations continually decline during darkness. The projection technique illustrated in Figure 8, or some modification of it, is widely and successfully used in commercial aquaculture ponds.

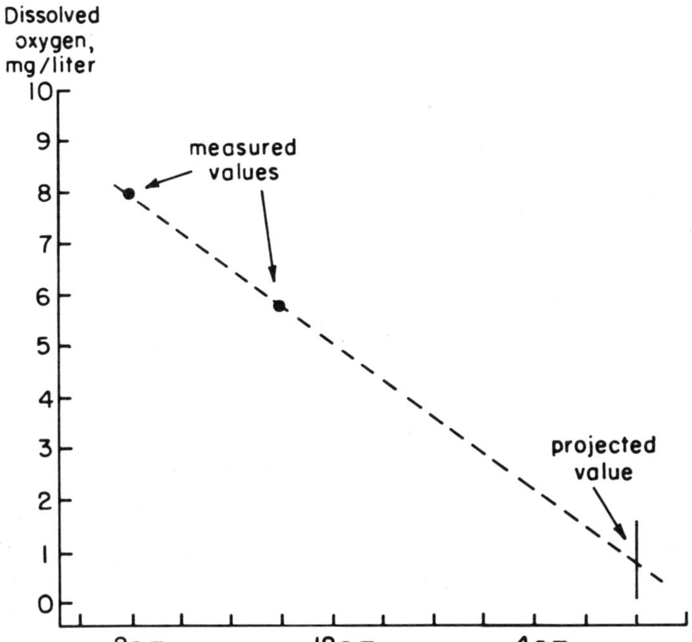

FIGURE 8. A graphical method for predicting the nighttime decline in dissolved oxygen concentration in fish ponds.

Empirical Models

The dynamics of DO concentrations in aquaculture ponds are extremely complex. Considerable research has been conducted on DO dynamics in channel catfish ponds, and the findings are indicative of DO dynamics for other aquaculture species.

The DO concentration in a catfish pond may be computed from the following equation (Boyd 1982) in which all variables are expressed in mg/L:

$$DO_2 = DO_1 \pm DO_{df} + DO_{ps} - DO_{pr} - DO_{fr} - DO_{br}$$

where DO_2 = DO at time 2
DO_1 = Initial DO
DO_{df} = DO gained or lost by diffusion
DO_{ps} = DO gained from photosynthesis
DO_{pr} = DO used by respiration of planktonic community
DO_{fr} = DO used in fish respiration
DO_{br} = DO used in benthic respiration.

Procedures were developed for predicting DO concentrations in channel catfish ponds. The daytime equation permits estimation of the daytime change in DO concentration adjusted to remove the influence of fish respiration (Romaire and Boyd 1979). The equation is:

$$\Delta DO_{day} = -3.240 + 0.0124\ Rad - 0.000012\ Rad^2 + 0.0582\ Cha - 0.000144\ Cha^2 + 0.0858\ Sat + 0.000651\ Sat^2 + 0.0000520\ Rad \cdot Cha - 0.000355\ Cha \cdot Sat$$

where ΔDO_{day} = adjusted daytime DO change (mg/L)
Rad = solar radiation in langleys per day (range = 60 - 600 langleys per day)
Cha = chlorophyll *a* in µg/l (range = 8 - 280 µg/L)
Sat = percentage DO saturation at dawn (range = 10 - 115%)

The correlation coefficient for the equation was 0.91. In order to calculate the DO concentration at dusk, one must add ΔDO_{day} to the DO concentration at dawn and from the sum subtract the estimated shrimp or fish respiration. Equations for the respiration of various species of shrimp and fish are available in the literature. Because the daytime DO equation is based on DO changes between dawn and dusk, DO concentrations at other times during the photoperiod cannot be predicted. Of course the daytime equation has little practical value, because it is impossible to know solar radiation in advance.

However, the equation permits one to determine relationships among chlorophyll *a* concentrations, daily solar radiation, and DO changes during the day.

The nighttime equation (Boyd et al. 1978) has the form:

$$DO_t = DO_{dusk} \pm DO_{df} - DO_{pr} - DO_{fr} - DO_{br}$$

where DO_t = DO concentrations after t hours of darkness
DO_{dusk} = DO concentrations at dusk.

The gain or loss of DO by diffusion may be estimated from the DO concentration at dusk and Table 2. Respiration by the planktonic community may be determined from the equation of Boyd et al. (1978) as follows:

$$\text{Oxygen consumption (mg/L)} = 1.006 - 0.00148\ C - 0.0000125\ C^2 + 0.0766\ T - 0.00144\ T^2 + 0.000253\ CT$$

where C = chemical oxygen demand (mg/L)
T = water temperature (C).

Data used in preparing the equation ranged from 20 to 160 mg/L chemical oxygen demand and from 20 to 32 C. The correlation coefficient was 0.92. The equation for channel catfish respiration has already been presented. Benthic respiration was taken as a constant value of 61 mg O^2/m^2 per hour. An oxygen consumption rate of 61 mg O^2/m^2 per hour represents a DO reduction of 0.72 mg/L during 12 hours of darkness in a pond with an average depth of 1 m provided the entire bottom is aerobic.

The nighttime equation provided highly reliable estimates of DO concentrations at dawn. The procedure is probably too complex for use on commercial shrimp and fish farms. However, chemical oxygen demand may be estimated from Secchi disk visibility, and there is usually a close relationship between average fish weight and total weight of channel catfish in ponds. These relationships permitted preparation of a series of tables for

Table 2. Gains (+) and losses (-) of DO because of diffusion during the night (approximately 12 hours of darkness) in ponds with different dissolved oxygen saturation values at dusk. The values represent the average change in the water column for a pond of 1-meter depth. Modified after Schroeder (1975).

DO concentrations at dusk (% of air saturation)	Gain or Loss of DO during the night (mg/liter)	DO concentrations at dusk (% of air saturation)	Gain or Loss of DO during the night (mg/liter)
50	+1.69	160	-1.64
60	+1.49	170	-1.82
70	+1.18	180	-1.98
80	+1.00	190	-2.11
90	+0.77	200	-2.37
100	+0.44	210	-2.42
110	+0.16	220	-2.54
120	-0.18	230	-2.67
130	-0.55	240	-2.76
140	-0.94	250	-2.91
150	-1.48		

use by fish farmers in predicting the DO concentration at dawn from water temperature, DO concentration at dusk, total weight of fish in a pond, and Secchi disk visibility (Romaire and Boyd 1980). An example of one of the tables is provided (Table 3). Even though the tables permit accurate estimates of DO concentrations at dawn, they have not been used widely for they require several pieces of information.

The success of the nighttime equation in predicting DO decline results primarily from the large contribution of plankton respiration to DO loss and to a reliable procedure for calculating plankton respiration. The partitioning of the nighttime DO loss for a representative channel catfish pond was: plankton respiration, 82%; fish respiration, 9%; benthic respiration, 5%; diffusion, 4% (Boyd 1982). The daytime equation showed that the daily change in DO was related to phytoplankton abundance and solar radiation.

Table 3. Minimum Secchi disk visibility (cm) to maintain a dissolved oxygen concentration above 2 mg/l at dawn in a 1-m deep pond containing 1,000 kg of channel catfish per hectare.

Temperature	Dissolved oxygen concentration at dusk (mg/l)									
(C)	2	3	4	5	6	7	8	9	10	11
20	37									
21	58	26								
22	79	42	21							
23	90	58	32	16						
24	[a]	69	53	37	21					
25	[a]	79	53	37	21					
26	[a]	85	63	48	32	16				
27	[a]	90	69	53	37	26				
28	[a]	95	74	58	45	32	21			
29	[a]	95	79	63	53	40	29	18		
30	[a]	[a]	85	69	58	45	34	26	16	
31	[a]	[a]	87	74	63	50	40	32	21	
32	[a]	[a]	90	79	66	55	45	37	29	18

[a]Concentration will fall below 2 mg/l.

Clearly, oxygen dynamics in a channel catfish pond are dominated by the planktonic community and the influence of solar radiation on photosynthesis by phytoplankton. The abundance of plankton, especially phytoplankton, is regulated primarily by feeding rate - the greater the feeding rate, the more plankton (Boyd 1982).

Madenjian et al. (1987a, b) made an extensive study of methods for predicting DO loss from prawn ponds in Hawaii. These workers stated that the DO prediction methods developed at Auburn University (Boyd et al. 1978; Romaire and Boyd 1979) were inadequate for estimating DO loss from Hawaiian prawn ponds because of environmental differences between the two locations. In particular, they claimed that benthic respiration was comparatively greater in prawn ponds than in catfish ponds and that wind was not taken into account in the methods developed at Auburn University. Of course, wind has an influence on diffusion of oxygen into and out of ponds, and in areas with appreciable wind, methods for predicting DO loss should have a wind component. Wind usually is not an important factor in DO budgets of catfish ponds in the southeastern United States. Most problems with dissolved oxygen depletion occur in summer when feeding rates are high. At this time, wind velocities normally are very low, and little reaeration is affected by wind action (Romaire and Boyd 1979). Furthermore, by not accounting for wind, any wind that occurs on a particular night provides a safety factor. For this reason, leaving the wind component out of a DO loss procedure is not especially critical to practical applications.

The methods developed by Madenjian et al. (1987a, b) are probably more accurate than those developed at Auburn University. However, the DO loss method developed by Madenjian et al. is quite complex, and it is doubtful that it will be useful in practical situations. Nevertheless, the studies show that the factors affecting DO concentrations in ponds will differ geographically, and methods for predicting DO loss should be developed for specific localities and specific types of ponds. Of course, the more complex models have more variables incorporated in an attempt to make them generally applicable. This feature of complex models makes them more accurate, but more difficult to use in practical aquaculture.

Busch et al. (1977) developed a computer simulation model which provided reliable estimates of DO changes during the nighttime in channel catfish ponds. This model was similar to the one described by Boyd et al. (1978). Meyer and Brune (1982) also present a dynamic DO model for stillwater aquaculture ponds which included the following components: phytoplankton, fish, detritus, and diffusion. All of the other simulation models require several pieces of information often not easily available to practical pond managers.

Weather

Major meteorologic factors affecting DO dynamics in ponds are wind, temperature, solar radiation, and cloud cover. Wind provides reaeration by creating turbulence in surface water and providing a greater area of contact between pond water and air to facilitate diffusion. Wind also mixes pond water so that oxygen entering the surface water is mixed throughout the pond volume. Wind action becomes a more important factor as ponds become larger and the fetch increases. The fetch can be increased by orienting ponds with their long axis parallel to the direction of prevailing winds. Wind will be more important in some areas than others. It was pointed out earlier that wind action was important in reaeration of shrimp ponds in Hawaii, but often a minor factor in reaeration of channel catfish ponds in the southeastern United States during summer.

Water temperature is affected primarily by season, for daily cycles and normal day to day fluctuation in air temperature do not cause drastic changes in the temperature of pond water. In tropical climates, water temperatures do not change drastically over time, but in some places there may be a few degrees difference in water temperatures between wet and dry seasons. In temperate climates, pond water temperatures change drastically with season. Cool water holds more DO at saturation than warmer water, but the demand for oxygen by organisms is much greater in warm water. Therefore, problems with DO depletion in ponds is more likely when water temperatures exceed 20 to 25 C.

Solar radiation, temperature, and cloud cover are closely related. Water temperatures increase in response to greater solar radiation, so the comments about water temperature apply equally to solar radiation. Cloud cover can quickly and drastically reduce the amount of solar radiation reaching a pond surface. Phytoplankton photosynthesis is positively correlated with increasing solar radiation. In temperate climates, the potential rates of photosynthesis and oxygen production vary with season, but in the tropics the potential for photosynthesis is high year around. However, cloudy days may drastically affect photosynthesis and DO concentrations in ponds. On a cloudy day, there is less light, so less oxygen is produced in photosynthesis and DO concentrations in ponds are lower following a cloudy day than after a clear day. For this reason, DO concentrations during the night decline more during cloudy weather than in clear weather. Romaire and Boyd (1979) developed a computer simulation technique for predicting the influence of different amounts of daily solar radiation on the DO dynamics in ponds. This procedure is of limited value to pond managers because forecasts of cloudy weather are not highly accurate or even available in some places. A method for using local weather data to predict the probability of cloudy days at different times during the year was described by Romaire and Boyd (1979).

Feeding Rate and Dissolved Oxygen Concentration

Experience and studies of catfish ponds (Tucker et al. 1979; Boyd et al. 1979) and shrimp ponds (Daniels and Boyd, unpublished) suggest that problems with low DO concentrations in aquaculture ponds are unlikely if feeding rates do not exceed 30 kg/ha per day. At feeding rates between 30 and 50 kg/ha per day, DO depletion may occur in a few ponds and all ponds will probably experience undesirable low DO one or more times during a production cycle. When feeding rates exceed 50 kg/ha per day, low DO concentrations will be common in all ponds and DO depletion can be expected in most ponds.

MODIFYING DISSOLVED OXYGEN CONCENTRATIONS

Aeration

There are three basic ways of applying aeration in aquaculture: emergency aeration, nightly aeration, and continuous aeration. If we consider pond aquaculture in general, most ponds are not aerated. This is because the level of production is low enough in the majority of ponds that natural reaeration is adequate to offset the oxygen demand created by nutrient additions. When production levels are raised, problems with insufficient DO becomes more common and it is necessary to use aeration. Emergency aeration is probably employed more widely today than other methods of aeration. Pond managers check DO concentrations during the night, and when DO concentrations are expected to fall below what is considered a critical level by a manager, aeration is applied until DO concentrations improve. Emergency aeration is a proven management technique. In the United States, emergency aeration often is applied with large tractor-powered paddle wheel aerators (Fig. 9) or pump-sprayer aerators (Fig. 10). However, there is a growing tendency to use floating electric aerators instead of tractor-powered ones.

Figure 9. A tractor powered paddle wheel aerator.

Figure 10. A tractor powered pump-sprayer aerator.

A better way to aerate ponds is to install an electric aerator in each pond and operate the aerator every night during the period when low DO concentrations are likely. If ponds are stocked and fish fed at reasonable rates, nightly aeration will prevent low concentrations of DO (Fig. 11). Lai-fa and Boyd (1988) compared emergency and nightly aeration in ponds of channel catfish stocked and fed at identical rates (Table 4). Concentrations of DO were higher at night in nightly-aerated ponds than in emergency-aerated ponds, and because of this fish converted feed to flesh more efficiently in the nightly-aerated ponds. Production was about 1,000 kg/ha greater in the nightly-aerated ponds.

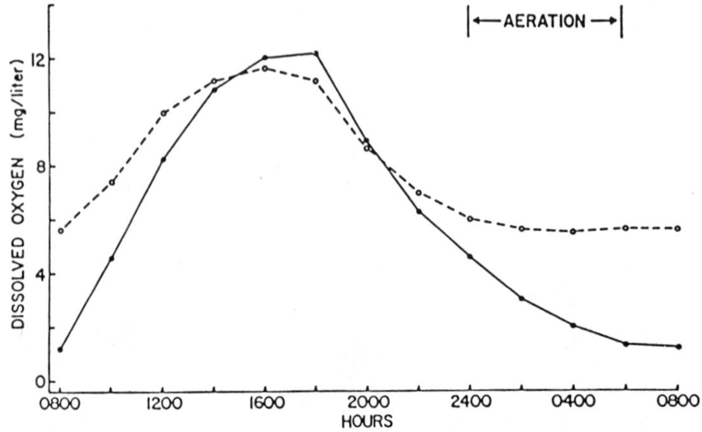

FIGURE 11. Effect of nighttime aeration on dissolved oxygen concentrations in ponds: aerated pond, dashed line; un-aerated pond, solid line.

Concentrations of DO usually are high during the day in ponds, so aeration often is counterproductive. When waters are supersaturated with DO, aerators actually increase the rate at which oxygen diffuses from pond water to air; aerators become de-aerators. Thomforde and Boyd (unpublished data) obtained no greater production of channel catfish in ponds with continuous aeration than in ponds with nightly aeration. Of course, in extremely intensive aquaculture systems it may be necessary to use continuous aeration.

Table 4. Comparison of channel catfish production characteristics in ponds aerated every night from midnight to dawn and in ponds aerated only in emergencies resulting from low dissolved oxygen concentrations. Treatments were replicated three times.

Variable	Nightly aeration	Emergency aeration	t-value
Fish stocked (no./ha)	10,000	10,000	——
Fish harvested (no./ha)	9,708	9,410	0.96
Harvest weight (kg/ha)	4,813	3,657	4.23*
Avg. weight individual fish (g)	496	389	3.25*
Feed applied (kg/ha)	6,229	6,229	——
Feed conversion ratio[a]	1.32	1.75	2.99*

[a]Weight feed applied/net production.
*Significant at P = 0.05.

Many types and brands of aerators are available for aquaculture. Studies at Auburn University (Boyd and Ahmad 1987) suggest that large (3.75 to 7.5-kw), floating, electric paddle wheel aerators (Fig. 12) are the most efficient type of aerator for large ponds (2 or 3 ha and larger). Small paddle wheel aerators also are suitable for smaller ponds, but propeller-aspirator pump aerators, vertical pump aerators, and diffused air aeration systems also can be effective in small ponds.

Water Exchange

If water of high DO concentration and sufficient volume is available, it may be discharged into DO-depleted ponds to improve DO concentrations. Flushing oxygen-enriched water through ponds is not likely to improve DO concentrations appreciably unless 25 to 50% of pond volume can be exchanged in a single day. Also, water exchange is more efficient if the volume to be exchanged can be drained quickly from the pond before the oxygen-enriched water is added. Well water often is depleted of DO, so it should be passed over a gravity aerator before it enters the oxygen-depleted pond. Water exchange is almost never as effective as aeration for combating DO depletion.

FIGURE 12. A floating electric paddle wheel aerator.

Other Techniques

Some pond managers stop applying feed to ponds when DO is low. Actually, this practice probably does little to improve DO concentration, but if DO concentration is extremely low fish or shrimp will not eat well. Anytime that fish or shrimp will not eat, food should not be offered. The problem of low DO concentration on a particular night is not a result of the feed applied on that day; rather, the problem will have developed over time because of feeding for many days or weeks.

Applications of potassium permanganate ($KMnO_4$) and bacterial suspensions have been recommended to improve DO concentrations. Research has demonstrated that neither treatment is effective for this application (Tucker and Boyd 1977; Boyd et al. 1984).

ACKNOWLEDGMENTS

Manuscript preparation was supported by funds from BARD Project Number US-1204-86 and a project from the USDA - Southern Regional Aquaculture Center.

LITERATURE CITED

Almazan, G. and C. E. Boyd. 1978. An evaluation of Secchi disk visibility for estimating plankton density in fish ponds. Hydrobiologia 65: 601-608.

Aveliovich, A. and M. Shilo. 1972. Photo-oxidative death in blue-green algae. Journal of Bacteriology 111: 682-689.

Boyd, C. E. 1976. Water chemistry and plankton in unfertilized ponds in pastures and woods. Transactions of the American Fisheries Society 105: 634-636.

Boyd, C. E. 1982. Water quality management for pond fish culture. Elsevier Scientific Publishing Company, Amsterdam, The Netherlands.

Boyd, C. E. 1985. Chemical budgets for channel catfish ponds. Transactions of the American Fisheries Society 114: 291-298.

Boyd, C. E. 1988. Water quality and aeration in shrimp farming. Alabama Agricultural Experiment Station, Auburn University, Fisheries and Applied Aquaculture's Departmental Series No 2.

Boyd, C. E. and M. H. M. Noor. 1982. Aquashade treatment of channel catfish ponds. North American Journal of Fish Management 2: 193-196.

Boyd, C. E. and T. Ahmad. 1987. Evaluation of aerators for channel catfish farming. Alabama Agricultural Experiment Station, Auburn University, Alabama, Bulletin 584.

Boyd, C. E., E. E. Prather and R. W. Parks. 1975. Sudden mortality of massive phytoplankton bloom. Weed Science 23: 61-67.

Boyd, C. E., R. P. Romaire and E. Johnston. 1978. Predicting early morning dissolved oxygen concentrations in channel catfish ponds. Transactions of the American Fisheries Society 107: 484-492.

Boyd, C. E., R. P. Romaire and E. Johnston. 1979. Water quality in channel catfish production ponds. Journal of Environmental Quality 8: 423-429.

Boyd, C. E., W. D. Hollerman, J. A. Plumb and M. Saeed. 1984. Effect of treatment with a commercial bacterial suspension on water quality in channel catfish ponds. Progressive Fish-Culturist 46: 36-40.

Busch, C. D., C. A. Flood, Jr., J. L. Koon, and R. Allison. 1977. Modeling catfish pond nighttime dissolved oxygen levels. Transactions of the American Society of Agricultural Engineers 20: 394-396.

Cole, B. A. and C. E. Boyd. 1986. Feeding rate, water quality, and channel catfish production in ponds. Progressive Fish-Culturist 81: 25-29.

Dent, D. 1986. Acid Sulfate Soils: A Baseline for Research and Development. International Institute Land Reclamation and Improvement, Publication 39, Wageningen, The Netherlands.

Dobbins, D. A. and C. E. Boyd. 1976. Phosphorus and potassium fertilization of sunfish ponds. Transactions of the American Fisheries Society 105: 536-540.

Hepher, B. 1962. Primary production in fishponds and its application to fertilization experiments. Limnology and Oceanography 7: 131-135.

Hickling, C. F. 1962. Fish culture. Faber and Faber, London.

Lai-fa, Z. and C. E. Boyd. 1988. Nightly aeration to increase the efficiency of channel catfish production. Progressive Fish-Culturist 50: 237-242.

Madenjian, C. P., G. L. Rogers and A. W. Fast. 1987a. Predicting night time dissolved oxygen loss in prawn ponds of Hawaii: Part I, Evaluation of traditional methods. Aquacultural Engineering 6: 191-208.

Madenjian, C. P., G. L. Rogers and A. W. Fast. 1987b. Predicting night time dissolved oxygen loss in prawn ponds in Hawaii: Part II, A new method. Aquacultural Engineering 6: 209-225.

McVea, C. and C. E. Boyd. 1975. Effects of water hyacinths cover on water chemistry, phytoplankton, and fish in ponds. Journal of Environmental Quality 4: 375-378.

Meyer, D. I. and D. E. Brune. 1982. Computer modeling of the diurnal oxygen levels in a stillwater aquaculture pond. Aquacultural Engineering 1: 245-261.

Murad, A. and C. E. Boyd. 1987. Experiments on fertilization of sportfish ponds. Progressive Fish-Culturist 49: 100-107.

Pillai, V. K. and C. E. Boyd. 1985. A simple method for calculating liming rates for fish ponds. Aquaculture 46: 157-162.

Ray, L. 1981. Channel catfish production in geothermal water. Pages 192-195 in L. J. Allen and E. C. Kinney, editors. Proceedings of the bioengineering symposium for fish culture. Fish Culture Section of the American Fisheries Society, Bethesda, Maryland, USA.

Romaire, R. P. and C. E. Boyd. 1979. Effects of solar ratiation on the dynamics of dissolved oxygen in channel catfish ponds. Transactions of the American Fisheries Society 108: 473-478.

Romaire, R. P. and C. E. Boyd. 1980. Predicting oxygen depletion in catfish ponds. Alabama Agricultural Experiment Station, Auburn University, Alabama, Bulletin 505.

Ruttanagosrigit, W., Y Musig and T. Buarark. 1980. The study of nutrient element in the shrimp farm area of Samut Sakorn and Samut Prakarn. First Seminar on Brackishwater Fisheries, Thailand Department of Fisheries, Bangkok, Thailand.

Schroeder, G. L. 1978. Autotrophic and heterotrophic production of microorganisms in intensely-manured fish ponds, and related fish yields. Aquaculture 14: 303-325.

Schroeder, G. L. 1975. Nighttime material balance for oxygen in fish ponds receiving organic wastes. Bamidgeh 27: 65-74.

Snow, J. R. and R. O. Jones. 1959. Some effects of lime applications to warmwater hatchery ponds. Proceedings Annual Conference of Southeastern Association of Game and Fish Commissioners 13: 95-101.

Swingle, H. S. 1947. Experiments on pond fertilization. Alabama Polytechnic Institute Agricultural Experiment Station, Auburn, Alabama, Bulletin 264.

Tucker, C. S. and C. E. Boyd. 1977. Relationships between potassium permanganate treatment and water quality. Transactions of the American Fisheries Society 106: 481-488.

Tucker, C. S. and C. E. Boyd. 1978. Consequences of periodic application of copper sulfate and simazine for phytoplankton control in catfish ponds. Transactions of the American Fisheries 107: 316-320.

Tucker, L., C. E. Boyd and E. W. McCoy. 1979. Effects of feeding rate on water quality, production of channel catfish, and economic returns. Transactions of the American Fisheries Society 108: 389-396.

Westers, H. 1981. Fish culture manual for the state of Michigan, Michigan Department of Natural Resources, Lansing, Michigan.

Wohlfarth, G. W. and G. L. Schroeder. 1979. Use of manure in fish farming - a review. Agricultural Wastes 1: 279-299.

STRATIFICATION AND MIXING IN SHALLOW AQUACULTURE PONDS

Thomas M. Losordo

ABSTRACT

Shallow aquaculture ponds are known to frequently become thermally and chemically stratified during hot, quiescent weather conditions. Vertical temperature and dissolved oxygen concentration gradients of up to 12 °C/m and 21 mg/L/m, respectively, have been recorded in aquaculture production ponds. Thermal and chemical stratification in aquaculture ponds produces areas of poor environmental quality in ponds that can effectively limit the usable culture volume. Poor environmental quality and reduced culture volume can significantly effect growth rates and reduce pond crop yield. To date, most water quality simulation models are based on the assumption of completely mixed pond conditions. Models based on the assumption of completely mixed conditions cannot adequately simulate temperature and water quality in stratified ponds.

Thermal and dissolved oxygen stratification in commercial aquaculture production ponds have been characterized and modeled in recent studies by the author. The processes of thermal and dissolved oxygen stratification in shallow aquaculture ponds are quantified and described. Modeling techniques for simulating vertical stratification and mixing in water bodies are reviewed and special considerations in modeling shallow aquaculture ponds are discussed. Areas in need of further research in the modeling of stratification and mixing in shallow aquaculture ponds are noted.

INTRODUCTION

Shallow earthen ponds are used throughout the world for the culture of marine and freshwater species of fish and crustaceans. Although considerable research has been conducted in the areas of pond water quality dynamics, simulation, and control, water quality management by commercial aquacultists continues to be an art rather than a science. Commonly, the farmer's water quality management procedures are limited to short-term reactions to adverse pond water quality conditions by increasing pond water flushing rates and/or application of emergency mechanical aeration.

Most of the models that have been developed by researchers to simulate water quality in aquaculture ponds are based on the assumption of completely mixed conditions. However, shallow aquaculture ponds have been known to frequently become thermally and chemically stratified during hot, quiescent weather conditions. Vertical temperature and dissolved oxygen concentration gradients of up to 12 C/m and 21 mg/L/m, respectively, have been recorded in aquaculture production ponds in California. Thermal and dissolved oxygen stratification in commercial aquaculture production ponds have been characterized and modeled in recent studies by the author. This paper will quantify and describe the processes of thermal and dissolved oxygen stratification in shallow aquaculture ponds, review modeling techniques that have been used for simulating vertical stratification and mixing in deeper water bodies, and discuss special considerations in modeling shallow aquaculture ponds. Topics in need of further research are also noted.

Pond Thermal Stratification

Water column stratification is most often a seasonal phenomenon in deep lakes and reservoirs (Welch 1980; Wetzel 1983). In shallow water bodies, however, stratification can be both a seasonal and diurnal process (Boyd 1979; Boyd et al. 1978; Busch et al. 1978; Ganf 1974; Kersting 1983; Young 1975; Veeningen 1982).

Stratification takes place as a result of density gradients caused by temperature differences in the water column. The thermal gradients are the result of the differential heating and cooling of the water column caused in part by the attenuation of shortwave solar radiation in the water column. Light attenuation is caused by the scattering and absorption of solar radiation as it passes through the water column. The attenuation of light limits the availability of energy for the heating of the deeper waters. Light attenuation is a function of water turbidity. Turbidity in aquaculture ponds is often caused by phytoplankton blooms and/or suspended clay particles (Almazan and Boyd 1978; Boyd 1979).

During windless weather conditions, thermal stratification is increased by the reduction of the turbulent diffusion of heat through the water column. Additionally, the transport of energy and mass by advection and turbulent diffusion becomes even more limited as shallow water bodies become stratified. Convective mixing occurs and the pond becomes completely mixed and isothermal, under low or no wind conditions, only after the density gradient in the water column breaks down, usually due to surface cooling after sunset.

Under variable wind conditions, the stratification and destratification processes within shallow water bodies are driven by heat flux and mechanical energy flux. The changing vertical density gradients in the water column are equivalent to changes in the potential energy state of the water column (Harleman 1982). Concurrently, the input of mechanical energy from flowing water (advective forces), convective forces due to unstable buoyancy gradients (Stefan and Ford 1975), or wind shear at the surface (Walters 1980) impart turbulent kinetic energy which tends to destroy the density gradients by turbulent mass exchange (Imboden 1978). The degree of stratification of a water body is the result of a dynamic equilibrium between the heat and mechanical energy fluxes. While these basic mechanisms have been qualitatively described, quantification is restricted to calculations based on the thermal structure of water bodies and accompanying input (forcing) parameters such as wind action and heat flux (Laska 1981).

Dissolved Oxygen Stratification

In stratified ponds, heat and oxygen in the surface strata of the pond are effectively segregated from the lower pond water column by the density gradient acting as a "barrier." The development of temperature and density gradients in the water column gives rise to other stratified water quality conditions. Variation of the physical environment of the various strata of the water column produces differing controls for the chemical and biochemical processes occurring at each depth (Markofsky and Harleman 1973).

Under thermally stratified conditions, the variation of light intensity and temperature with depth gives rise to DO stratification in the water column. Due to high turbidity, the compensation point (the depth at which water column respiration and photosynthetic oxygen production are balanced) is usually shallow in aquaculture ponds. As a result, a large portion of the deeper pond volumes in aquaculture ponds are net oxygen consumers. Near the surface, during periods of high light intensity, photosynthesis by microalgae blooms produce supersaturated DO concentrations. In the absence of oxygen transfer by vertical mixing in the water column, the dissolved oxygen concentration in the deeper regions of these ponds can become significantly reduced. The inhibition of vertical mixing can also lead to stratification in other water quality parameters such as pH, alkalinity, and in some cases conductivity (Stahl and May 1967).

Adverse Consequences of Stratification

The maintenance of optimum water quality conditions in ponds is of vital concern to aquaculturists. Water temperature directly affects the growth rate of all cultured aquatic species. The dissolved oxygen concentration of water in aquaculture ponds is perhaps the single most important factor involved in the survival and proper growth of cultured aquatic species (Boyd et al. 1978; Hepher and Pruginin 1981).

Shallow aquaculture ponds frequently undergo daily cycles of thermal and chemical vertical stratification and destratification during periods of hot and windless weather. Dissolved oxygen problems and "oxygen kills" most frequently occur during these hot, windless periods. Methods have been developed for the short-term (12 hours or less) prediction of nighttime DO decline in ponds (Boyd et al. 1978; Madenjian et al. 1987). However, these methods are invariably based on the assumption that the pond is a completely mixed, homogeneous water body and are not appropriate for use with stratified pond conditions. In addition, all previous methods require an a priori knowledge of the pond water column DO concentration at sunset. In stratified ponds, it is difficult to estimate the true integrated average DO concentration of the water column during the late afternoon. Thus, it is very difficult to predict the oxygen decline rate and early morning DO concentrations in the continuously changing thermal and DO profile of the stratified water column.

Thermal and oxygen stratification in aquaculture production ponds can also have more subtle, chronic effects on pond production. Pond stratification can result in areas (volumes) of poor environmental quality that can effectively limit the usable pond culture volume. Krom and co-workers (1985) documented mortality and stress in aquacultured fish due to chronically low DO and the combined effect of super saturated levels of oxygen and high water temperatures. Both of these conditions can occur simultaneously at the bottom and surface of stratified aquaculture ponds, respectively. Gebhart and Summerfelt (1976) indicated that fish tend to avoid areas of environmental extremes such as excessively low oxygen and high or low temperatures in stratified lakes and ponds. This behavior could effectively limit the movement of fish to the mid-depth region of the water column in aquaculture ponds. Given that farmers use either floating or sinking artificial feeds, production could be significantly affected by prolonged or frequent periods of stratified thermal and water quality conditions.

Pond stratification may have even more dire implications for benthic (bottom dwelling) species such as freshwater prawns and marine shrimp. Stratified conditions can create areas of low temperatures and low or no DO concentrations on or near the bottom of culture ponds. These conditions can severely impact animal growth and survival. In some cases, stratification has been shown to restrict a benthic population's movement to the shallow areas of the pond (Fast et al. 1983). Such conditions could further reduce shrimp survival due to over crowding which often leads to aggressive behavior (Fast et al. 1983).

POND STRATIFICATION CHARACTERIZATION

Thermal and chemical stratification processes in aquaculture ponds are not well documented phenomena. In an extensive review of the literature, Marjanovic and Orlob (1986), concluded that published works relevant to the hydromechanics of aquaculture ponds were "virtually nonexistent."

To more fully understand the nature of aquaculture pond stratification and mixing, the author monitored temperature and dissolved oxygen conditions in 5 catfish production ponds in California and 1 freshwater aquaculture research pond in Hawaii. The pond studies were implemented utilizing an automated pond water quality and temperature data acquisition system described by Losordo et al. (1988). In total, 55 days of continuous hourly pond water temperature, DO, and associated atmospheric data were compiled for the 6 aquaculture ponds in this study. The information presented below describes typical patterns of pond stratification and the timing, magnitude and frequency of the thermal and DO stratification that were monitored.

Typical Patterns of Pond Stratification

The time series data presented in Figures 1 and 2 are typical of the diel cycles of thermal and DO stratification and destratification, respectively, that were observed in the ponds during the study. For a given pond, the diel patterns of thermal and DO stratification tended to be similar. Given that the turbulent diffusion of heat and mass are controlled by many of the same external forces, the similarity of the patterns is not surprising.

Thermal stratification typically began in the morning, between 9:00 and

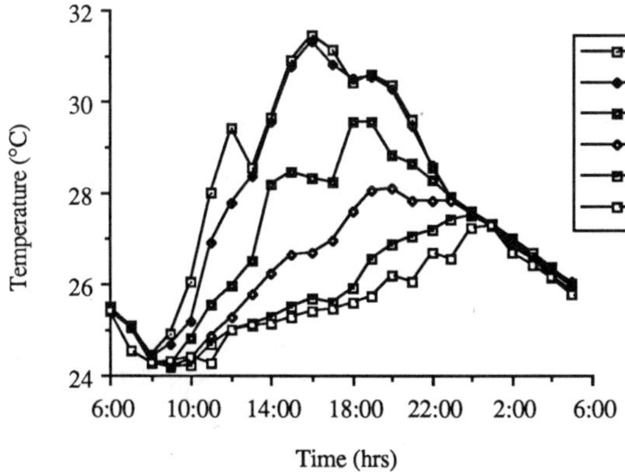

Figure 1. Typical pattern of thermal stratification in shallow aquaculture ponds.

11:00 am, as the solar irradiance, impinging on the pond, heated the surface (Fig. 1). Quite often, the water at the 5 cm and 20 cm depths remain completely mixed throughout the day as indicated by the overlapping temperature lines in Figure 1. On this particular day (Fig. 1) the water column

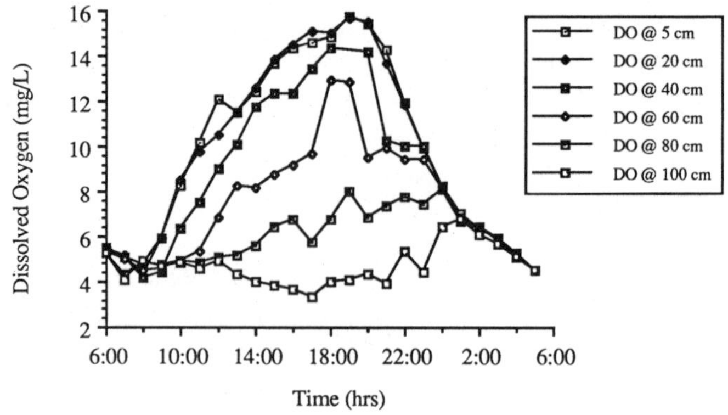

Figure 2. Typical pattern of dissolved oxygen stratification in shallow aquaculture ponds.

from 20 to 80 cm depth was thermally stratified as indicated by distinctively separate temperature lines. The water column below the 80 cm depth remained nearly isothermal during the day. The temperature graph also shows another common feature of the stratification process observed in most of the ponds during periods of quiescent weather conditions. Under quiescent conditions, the ponds mix vertically in the evening, starting at the surface and entraining the deeper waters into the completely mixed upper layer as the evening progressed. This process occurs as the surface waters cools to the temperature of the underlying water and density gradient breaks down. The speed at which this process occurrs is a function of the degree of pond stratification at sunset, the rate of cooling of the mixed surface layer, and wind conditions at the surface of the pond. In this example (Fig. 1) the pond became isothermal at 1:00 hrs in the morning at a temperature of 27 C. The water columns of most of the ponds in this study were isothermal by 6:00 hrs in the morning.

In each pond on any given day, the exact shapes of the thermal stratification patterns were uniquely determined by the turbidity of the pond and interaction of the solar irradiance and wind speed at the surface of the pond. The effect of each of these parameters (turbidity, wind speed, and solar irradiance) are discussed below.

Effect of Pond Turbidity on Stratification

The turbidity of the water column is an important factor in determining if the pond will become thermally stratified. In windless weather conditions, direct solar irradiance penetrating the ponds surface is the only source of heat for the pond's lower strata. Direct solar irradiance can be completely attenuated within the first 20 cm of the water column in very turbid ponds. A highly turbid pond will stratify more intensely and more often than a pond of lesser turbidity given that all other pond and weather parameters are equal.

An example of this can be observed in comparing data displayed in Figures 3 and 4 from two 4.45 ha catfish production ponds located on the same site in southern California (the pond corresponding to data in Fig. 3 will be referred to as pond A and the pond corresponding to data in Fig. 4 will be refered to as pond B). The average Secchi disk depth for ponds A and B were 0.13 and 0.28 m, respectively, during the period of the study. The intensities of underwater photosynthetically active radiation (PAR) at the 20 cm pond depth were approximately 5% and 34% of the surface intensity, respectively. Pond B did not become significantly stratified (maximum stratification was 1.7 °C) during the monitoring period (Figure 4). This was as a result of solar irradiance reaching and heating the lower strata of the water column. In contrast, the more turbid Pond A became thermally stratified by up to 6 °C on Julian Day (JD) 245 (Fig. 3). A review of the wind speed and solar irradiance data for the monitoring periods indicates that nearly identical conditions occurred on both days. Clearly pond turbidity played a major role in the degree to which thermal stratification occurred in these two ponds.

To avoid the adverse consequences of thermal stratification described previously, it is desirable to keep thermal stratification to a minimum. Turbidity should be kept to a minimum in aquaculture ponds from this point of view. From a practical standpoint, however, pond turbidity must be maintained at a level high enough to "shade" the pond bottom and inhibit the growth of benthic plants (i.e., aquatic weeds). Excessive bottom growth can severely inhibit harvesting activities in ponds. Microscopic algae in the water column are both a major source of DO and turbidity in ponds. The aquaculturist must consider the beneficial effects of algae and maintain the algal population at some optimum level. The maximum level of turbidity at which a pond would remain unstratified is determined by a dynamic interaction between the wind speed and solar irradiance.

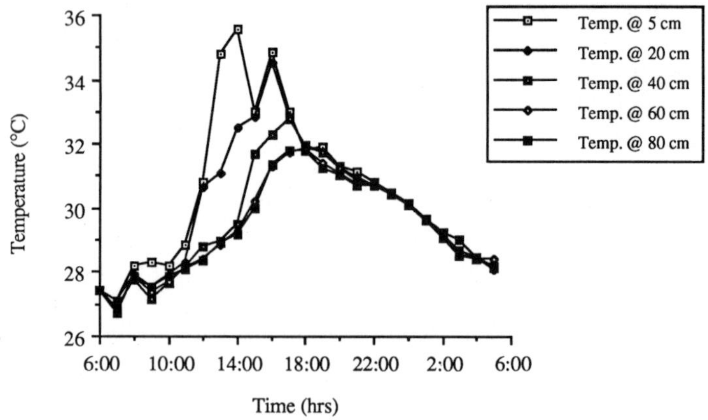

Figure 3. Diel temperature profile of a 4.45 ha catfish production pond (A) in Southern California with a secchi disk depth of 0.13 m.

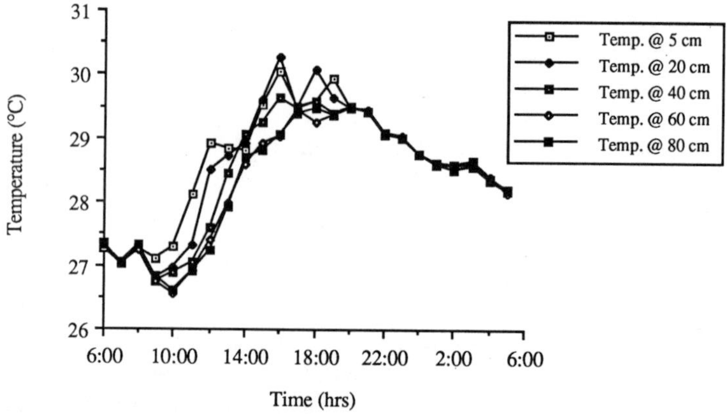

Figure 4. Diel temperature profile of a 4.45 ha catfish production pond (B) in Southern California with a secchi disk depth of 0.28 m.

Effect of Solar Irradiance on Stratification

The importance of the intensity of solar irradiance on thermal stratification formation can be illustrated in reviewing and comparing the data in Figures 5 and 6 taken from the same small (0.30 ha) research pond in Hawaii on two different days. On Julian Day (JD) 231 (Fig. 5), thermal stratification in the water column of the pond was at a maximum of approximately 1 °C at noon-time. The pond remained slightly stratified until 21:00 hrs. In contrast, the same pond reached a maximum stratification of approximately 4 °C at 14:00 hrs on JD 239 (Fig. 6) and remained stratified until nearly midnight. The level of total solar irradiance for JD 231 was only approximately 36% of that measured on JD 239. An on-site manual weather observation note for JD 231 indicated that the sky was completely overcast for most of the day with occasional morning rain. Analysis of the wind data from both days indicates that the wind speed on JD 239 (Fig. 6, the day on which the pond stratified) was in fact greater than on JD 231. The average wind speed on JD 239 was 1.84 m/sec (st.d. = 0.40) in comparison to only 0.84 m/sec (st.d. = 0.66) on JD 231. The turbidity of the pond was similar on both days.

The important point demonstrated by the failure of the pond to stratify on JD 231 (Fig. 5) is that even under quiescent conditions, there is a minimum amount of solar irradiance required to provide the energy (buoyant potential energy) for stratification formation.

Effect of Wind Speed on Stratification

The importance of the winds' effect on the thermal stratification process can be illustrated in reviewing and comparing temperature data from a 4 ha catfish production pond located in northern California on two different days, JD 170-171 (Fig. 7) and JD 173-174 (Fig. 8). On JD 173, the pond remained unstratified for the 24 hour monitoring period (Fig. 8). On JD 170, pond thermal stratification reached a maximum of 4.3 °C (Fig. 7). The level of solar irradiance and pond turbidity were nearly identical on both days. The main difference in weather was the wind speed during the pond heating period of each day (6:00 to 14:00 hrs). On JD 173, at 9:00 hrs, the wind speed increased to over 5.5 m/sec and remained constant until 18:00 hrs in the afternoon. On JD 170, during the heating period of the day, the wind speed was less than 3.5 m/sec. Although the wind speed eventually increased to 8.1 m/sec, the stratification process had already occurred. The comparison above shows that not only is the velocity of the wind important to the stratification and mixing processes in aquaculture ponds, but also the time at which the wind occurs.

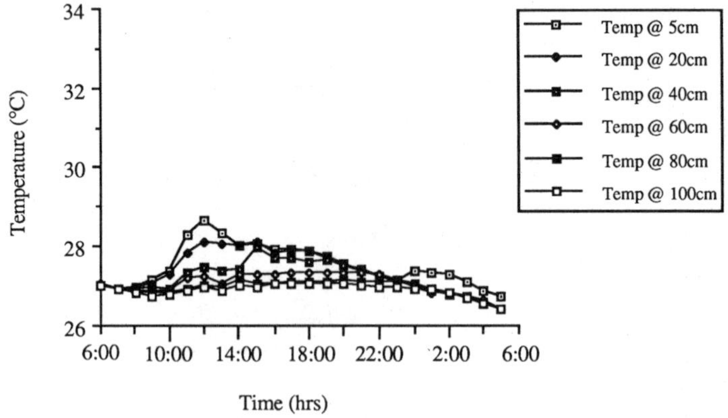

Figure 5. Diel temperature profile of a small 0.3 ha research production pond (A) in Hawaii resulting from low levels of solar irradiance, JD 231-232 1986.

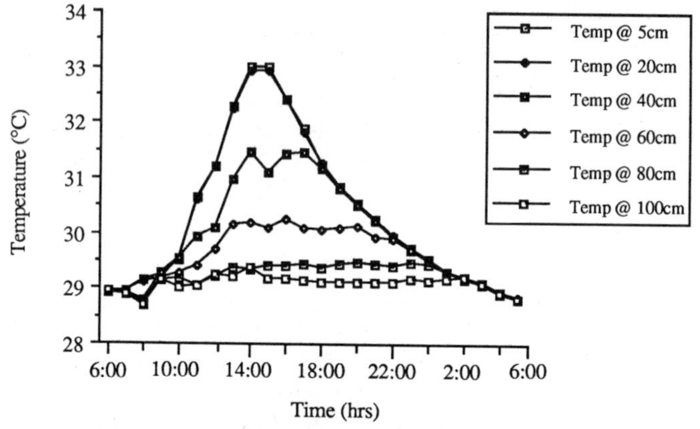

Figure 6. Diel temperature profile of a small 0.3 ha research production pond (A) in Hawaii resulting from high levels of solar irradiance, JD 239-240 1986.

The temperature and weather data from a smaller (0.82 ha) catfish production pond in northern California also provides a good example of the effect of wind speed on the stratification process. On JD 220 the pond became thermally stratified with a maximum surface to bottom (95 cm distance) temperature differential in excess of 5.4 C (Fig. 9). In contrast, the pond

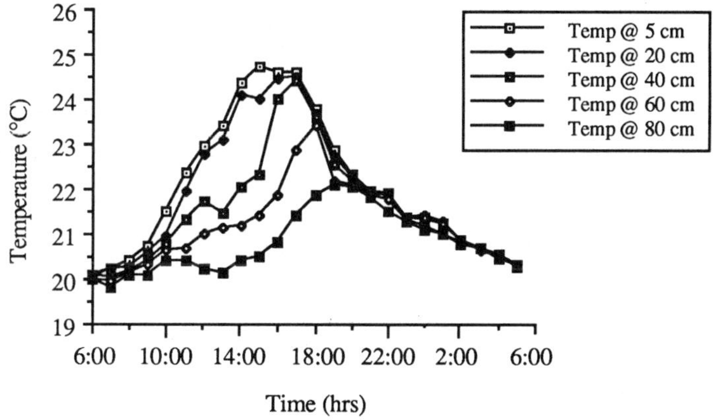

Figure 7. Diel temperature profile of a 4.0 ha catfish production pond in Northern California resulting from low wind conditions, JD 170-171 1987.

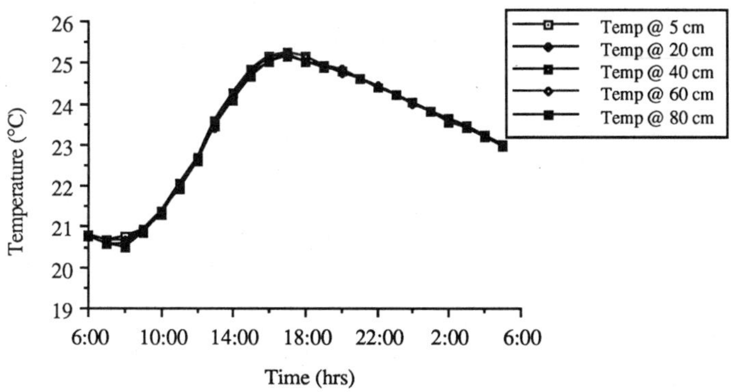

Figure 8. Diel temperature profile of a 4.0 ha catfish production pond in Northern California resulting from high wind conditions, JD 173-174 1987.

remained isothermal during the entire diel monitoring period of JD 225 - 226 (Fig. 10). A review of the solar irradiance measurements for both days indicates nearly identical conditions.

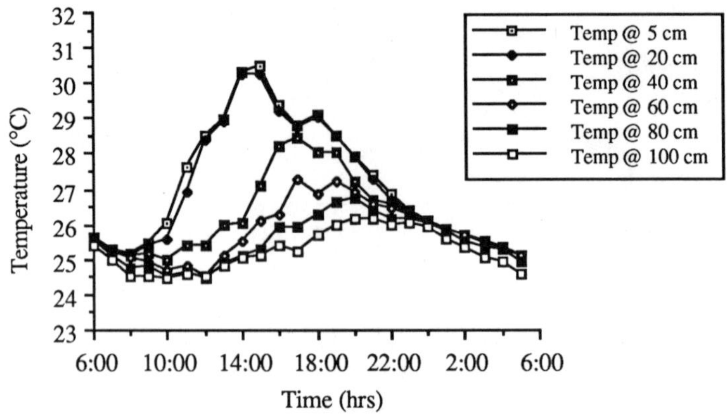

Figure 9. Diel temperature profile of a 0.8 ha catfish production pond in Northern California resulting from low wind conditions, JD 220-221 1987.

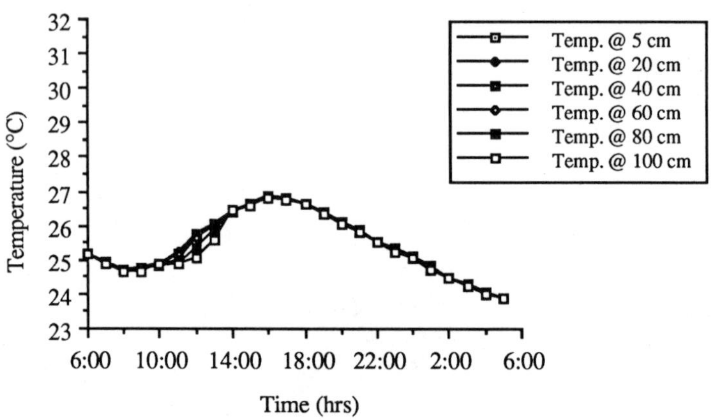

Figure 10. Diel temperature profile of a 0.8 ha catfish production pond in Northern California resulting from moderate wind conditions, JD 225-226 1987.

Again, as described above for the larger pond (Figs. 7 and 8), the cause of the differing temperature profiles on JD 220 and JD 225 was the amount of turbulent kinetic energy imparted to the water by the local wind conditions. On JD 220 (Fig. 9) the complete absence of wind in the morning allowed the stratification process to proceed. The slight increase in temperature of the water column at the 40, 60, and 80 cm depths (prior to 14:00 hrs) was due to direct solar irradiance (SDD ≈ 35 - 40 cm). The increase in temperature of the water column at the 40, 60, and 80 cm depths after 14:00 hrs was the result of an increase in the turbulent transport of heat in the water column. The increased turbulent mixing in the water column was due to higher wind speeds beginning at 14:20 hrs, reaching a maximum of 4.4 m/sec at 16:40 hrs.

On JD 225-226 (Fig. 10), the pond was well mixed by constant high winds in excess of 4 m/sec. From a theoretical viewpoint, the turbulent kinetic energy imparted to the water column by the wind at the surface (mixing the pond) far exceeded the buoyant potential energy of the water column "supplied" by the solar irradiance that causes stratification.

The Timing, Frequency and Magnitude of Thermal Stratification

During the 55 days of pond and weather monitoring, hourly temperature measurements at 6 depths in the water column were recorded. All of the ponds that were monitored became thermally stratified to some extent during the study. For reasons described above, the degree of stratification in each pond on a given day was affected by pond water quality (turbidity) and local weather conditions (wind speed and solar irradiance).

On over 65% of the days when conditions of thermal stratification were detected, the maximum pond thermal gradients were recorded at between 14:00 and 15:00 hrs in the afternoon. In the ponds that were monitored, thermal stratification was found to be a diel process. On 46 of the 48 days that thermal stratification occurred, the ponds became completely mixed by sunrise (6:00 hrs). The time of destratification was more variable than the time of maximum thermal stratification. On over 75% of the days that stratification occurred, the ponds had mixed completely by midnight (0:00 hrs).

The maximum daily level of pond thermal stratification (determined as the maximum difference between the surface (5 cm) and bottom (80 or 100 cm) temperature readings) were calculated from the pond temperature data base. On 24 of the 55 days of temperature measurements, the ponds became thermally stratified by over 4 °C. In contrast the ponds were stratified by less than 1 °C on only 9 of the 55 days and did not stratify at all (less than 0.5 °C) on only 4 days.

The maximum thermal gradient measured was 10.6 °C within a 0.75 m water depth. The largest temperature gradients generally developed in ponds with high turbidity (Secchi Disk depth readings of less than 0.15 m), during periods of intense solar irradiance and low wind speeds. Pond color (i.e., brown vs green) did not appear to influence the stratification process.

The Timing, Frequency and Magnitude of DO Stratification

Thermal stratification inhibits turbulent mixing in aquaculture ponds. The reduction of the vertical transport of dissolved oxygen, coupled with variable DO production and consumption in the water column, produces stratified DO conditions in ponds. For these reasons, the timing and magnitude of DO stratification followed that of the thermal stratification process.

Dissolved oxygen stratification was also found to be a diel process. Although the timing of the onset of stratified oxygen conditions coincided fairly closely with those of thermal stratification, the time of maximum DO stratification was not as well defined as that of thermal stratification. That is to say, that maximum stratified conditions did not occur with great frequency during any one specific time period. This was most likely because DO stratification is caused by biochemical reactions (photosynthetic oxygen production and respiration) as well as the physical processes that cause thermal stratification. The maximum level of DO stratification occurred between the hours of 10:00 and 0:00 hrs daily.

The time at which completely mixed conditions occurred also was not well defined. In general, stratified DO conditions persisted longer into the early morning hours than for thermal stratification (i.e., large DO stratification persisted with thermal stratification of less than 0.5 C). Only 60% of the ponds were homogeneous with respect to DO concentration by midnight (as compared with 75% for isothermal conditions). The ponds became completely mixed by 6:00 hrs the next morning on 37 of the 41 days during which DO stratification occurred.

The maximum daily level of pond DO stratification was determined as the maximum difference between any two sampling depths in the pond. Quite often, the maximum DO concentration was recorded at the 20 or 40 cm sampling depth. This phenomenon was most likely a result of dissolved oxygen being lost to the atmosphere at the surface and photoinhibition of photosynthetic oxygen production due to high light intensities at the 5 and 20 cm pond depths. On 23 of the 55 days of DO measurements, the ponds became DO stratified by over 10 mg/L (29 of 55 were stratified by 5 mg/L or greater). In contrast the ponds were stratified by less than 5 mg/L on 24 of the 55 days and did not stratify at all (less than 1 mg/L) on 12 days.

In relative terms, the magnitude of DO stratification was generally greater than that of thermal stratification (i.e., $\Delta DO/\max DO > \Delta T/\max T$). The magnitude of DO stratification in each pond was the result of the unique combination of the effects of the density gradient and the vertical variation of the photosynthetic and respiration processes in the water column at each site. Conditions of vertical DO stratification occurred only when thermally stratified conditions were present to some extent. The maximum level of DO stratification documented in this study was 21.5 mg/L within a 0.75 m vertical distance.

Horizontal vs Vertical Temperature Variations

In lakes and reservoirs that are well stratified, horizontal inhomogeneities are considered to be insignificant compared to the vertical temperature gradient. Newbold and Liggett (1974) indicated that in lakes, the horizontal diffusion rates may be as much as five orders of magnitude greater than vertical diffusion rates. The relative magnitude of mixing in the horizontal vs vertical direction is an important factor in determining the appropriate modeling approach for the simulation of water quality stratification in aquaculture ponds. If the horizontal variations in water temperature and water quality are small compared to vertical gradients, then a one dimensional modeling approach can be justified. A one dimensional approach is based on the assumption that the water column is made up of series of completely mixed, horizontal volumetric elements or slices (Ford and Thornton 1979).

To estimate the relative magnitude of mixing in the horizontal vs vertical direction in aquaculture ponds, the water temperature was monitored in 4 locations at 8 depths within the water column of 4 aquaculture ponds. The measurement system recorded (at hourly intervals) the vertical water temperature gradient in (horizontal) locations up to 80 m apart (Fig. 11). The system was deployed in 4 ponds for monitoring periods of 7 days each. One day of data was selected for analysis from each pond. The data sets were selected to estimate the magnitude of the horizontal temperature variation at the time of maximum vertical thermal stratification in each pond.

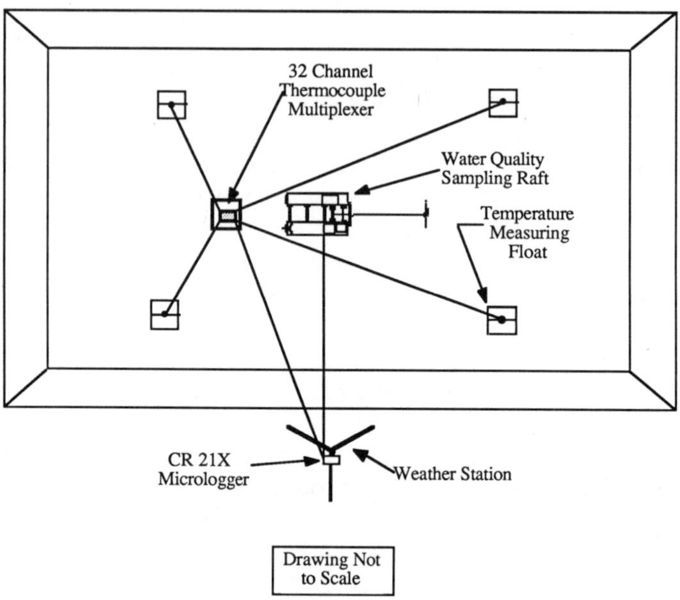

Figure 11. Automated water quality and temperature data aquisition system layout.

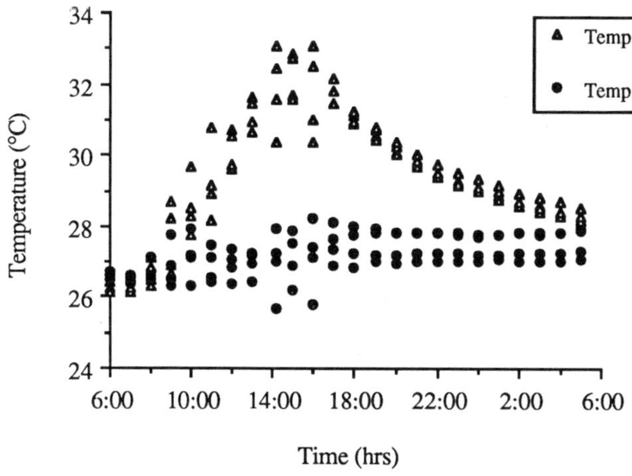

Figure 12. Diel temperature data at 4 locations and 2 depths in a small 0.3 ha research pond in Hawaii, JD 232-233 1986.

The data displayed in Figure 12 are water temperature measurements at the 5 cm and 80 cm depths measured in the 4 locations over a 24 hour monitoring period. Although the data displayed are from the small (0.3 ha) research pond in Hawaii, the pattern is typical of all of the ponds in which the temperature measuring array was deployed. In viewing the graphed data, the vertical variation in pond temperature appears to be far greater than the horizontal variation. A quantitative estimate of the difference in horizontal vs vertical temperature variation was determined by calculating the average difference between the temperatures at 5 cm and 80 cm for all of the floats, and the horizontal difference between the temperature reading at the same depths of the floats located diagonally across from each other at the time of maximum vertical pond stratification. Table 1 is a summary of the results of these calculations for 4 ponds.

The analysis of temperature data presented in Table 1 indicates that the vertical temperature variation during periods of intense thermal stratification is generally two orders of magnitude greater than that of the horizontal variation. These results are consistent with the findings presented by Cathcart (1987) for temperature variation in a small aquaculture pond in Maryland. These results suggest that in the aquaculture ponds studied, the magnitude of mixing in the vertical direction (during periods of intense thermal stratifica-

Table 1. Variation of temperature in the vertical vs horizontal direction at times of maximum pond thermal stratification.

Pond #	Pond Size (ha)	Time of Maximum Stratification	Vertical Gradient (°C/cm)	Horizontal Gradient (°C/cm)
1	0.3	15:00	6.5	0.06
2	4.0	14:00	12.9	0.02
3	0.8	13:00	7.5	0.01
4	0.8	17:00	6.3	0.01

tion) was greatly reduced and far less than in the horizontal direction. These results would justify the use of a one dimensional approach for modeling temperature and water quality stratification in small aquaculture ponds.

MODELING STRATIFICATION AND MIXING

The mathematical modeling of aquaculture ponds has developed slowly due to a lack of understanding of the basic principles governing the complex physical and biochemical interactions taking place within the system (Bolte et al. 1986). As a result, many early efforts in water quality modeling fall into the category of empirical or "black box" models. Only recently have "mechanistic" or "internally descriptive" models been developed for use in aquaculture. As noted previously, most aquaculture models to date have been based on the assumption of completely mixed conditions within the ponds. Dissolved oxygen simulation models have for the most part been approached from an empirical view point, while temperature prediction models have been more mechanistic in nature.

Although very little research on thermal and chemical stratification in aquaculture ponds has been reported, an extensive amount of limnological and engineering research has been focused on the nature and processes of lake and reservoir stratification. In addition, various temperature models have been developed for shallow, salt gradient stratified solar energy collection and storage ponds. The modeling techniques commonly used in the simulation of these water bodies provides a valuable point of reference for the development of models for stratified aquaculture ponds.

Many of the models developed for lakes, reservoirs and solar ponds have used a one dimensional approach. One dimensional models can be classified into two groups: diffusion models, and mixed-layer or integral energy models (Harleman 1982). These groups differ only in the manner in which the turbulent vertical transfer of energy and mass between the horizontal layers is simulated. The following is a brief description of the integral energy modeling approach, and a more detailed analysis of the diffusion modeling approach. The diffusion approach was used by the author (Losordo 1988) to model temperature and DO in shallow aquaculture ponds.

Integral Energy Models

Integral energy models have been used to simulate thermal and chemical stratification in deep lakes and reservoirs, and shallow salt gradient solar energy collection ponds. In this type of model the water body is assumed to be made up of many, completely mixed, horizontal layers usually referred to as volumetric elements. Each volumetric element except for the surface "mixed layer" is of equal and constant thickness (depth). In modeling deep lakes, the thickness (Δz) of the volumetric elements has commonly been from 0.25 to 2.0 meters (Stefan and Ford 1975). During the simulation process, the surface mixed layer varies in thickness depending upon the number of underlying volumetric elements that have been entrained into the surface layer by turbulent mixing.

In general, turbulent kinetic energy (TKE) supplied to the mixed layer by wind shear at the surface is calculated and compared to the buoyant potential energy (BPE) of the mixed layer during execution of the simulation. The BPE of the mixed layer can be thought of as the potential mixing work required to produce a change in the depth of the surface mixed layer by entraining denser water from below. The rate of change of the BPE of the water column is equal to the energy required to lift more dense water from the layer below to the center of mass of the mixed layer.

During each time step of the model execution, it is necessary to determine if there is enough TKE imparted at the surface (by wind shear) to overcome the potential ("buoyant") energy of the mixed layer to entrain the denser water from below. Stefan and Ford (1975) refer to this as the "integral stability criterion" (ISC), a ratio of TKE to BPE. If a critical value of the ISC is equaled or exceeded, then entrainment of the next volumetric element occurs. If the value of TKE to BPE is below the critical value then the TKE applied to the system is dissipated in mixing the mixed surface layer (Stefan and Ford 1975) and/or radiated into the quiescent layer below by internal waves (Harleman 1982). The critical ratio of TKE to BPE is usually taken as unity (i.e., TKE/BPE = 1) (Stefan and Ford 1975; Henderson-Sellers 1984)

The calculation of temperature profiles in most integral energy models is accomplished in two steps: 1) During each time interval (Δt), the temperature profile is calculated from estimating and summing heat energy inputs and losses at the surface of the water body. Some models account for the molecular diffusion of heat between volumetric elements while others omit this transfer process. 2) In the second step, the ratio of TKE to BPE for that time step is evaluated. If the ratio of TKE to BPE is less than one, the mixed layer depth (D) is left unchanged and the temperature profile calculated in the first step is utilized for initialization of the next time interval temperature calculations. However, if TKE/BPE $\geq$ 1, the thickness of the mixed layer is increased by Δz (to $D+\Delta z$), and the stability criterion is evaluated again. The surface heat balance, temperature, and density of the mixed layer and of the other horizontal layers must be recalculated as the mixed layer depth is changed. This process is continued within the time interval until the ratio of TKE to BPE is less than 1. At that point, the mixed layer depth is set, and the temperature profile is recalculated.

Diffusion Models

Models based on the concept of turbulent or eddy diffusion have been used extensively in the simulation of thermal and chemical stratification in deep lakes and reservoirs. The turbulent transport of energy and mass in lakes and reservoirs undoubtedly takes place on many scales depending upon the size of the eddies involved in the process. Turbulent eddy transport is commonly expressed with a Fickian diffusion equation in an effort to simplify calculations (Powell and Jassby 1974).

$$J_{sz} = -K_{sz}\, \partial s/\partial z \qquad (1)$$

where: J_{sz} = the flux of heat or mass in the z direction
K_{sz} = the coefficient of turbulent eddy diffusion of s at z
s = heat or mass concentration

In words, the rate of transport across a plane is the product of the gradient perpendicular to the plane and a turbulent eddy diffusion coefficient (K_{sz}). The coefficient is thus a measure of the rate of exchange or mixing intensity across the plane (Wetzel 1983).

In the absence of advective or convective vertical flows, a one dimensional (vertical) conservation equation for heat or mass can be used to calculate the change in temperature or mass concentration in a water body. A general form of the heat equation for a water body with a horizontal area independent of depth was given by Henderson-Sellers (1984) as

$$\partial T/\partial t = \partial/\partial z[(\varepsilon + K_z)\, \partial T_{wc} /\partial Z] + (\partial \phi/\partial z)/\rho_w\, c_{pw} \qquad (2)$$

where:			
	T_{wc}	= water temperature	(°C)
	t	= time	(sec)
	ε	= molecular diffuivity of heat	(m^2/sec)
	K_z	= coeeficient of the turbulent eddy diffusion of heat at depth z	(m^2/sec)
	ϕ	= direct radiative heat flux	(J/m^2/sec)
	ρ_w	= density of water	(kg/m^3)
	c_{pw}	= heat capacity of water	(J/kg/°C)

In many applications ε, ρ_w, and c_{pw} are considered constants within the range of temperatures encountered in lakes and reservoirs. As a result, it is common to combine the molecular diffusivity constant and the turbulent eddy diffusivity coefficient into a single term referred to as the effective diffusion coefficient (E_z) (Water Resources Engineers (WRE) 1968; Rossi et al. 1982).

Estimating the Value of E_z

The success of diffusion type models in simulating the turbulent transport of heat and mass is dependent upon the method and accuracy of

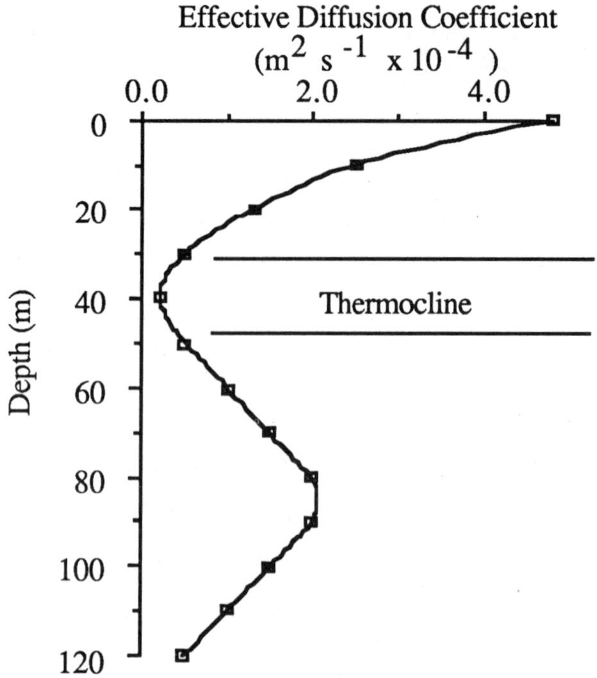

Figure 13. Typical relationship between the effective diffusion coefficient and depth in deep lakes and reservoirs.

estimating the value of E_z (Octavio et al. 1977; Henderson-Sellers 1984). Harleman (1982) states that models of shallow lakes with low flow-through exhibit a high degree of sensitivity to the value of E_z selected. Indeed, a significant amount of research has been devoted to the analytical estimation of the value of the effective diffusion coefficient in lakes and reservoirs (Hesslein and Quay 1973; Powell and Jassby 1974; Jassby and Powell 1975). Values for the effective diffusion coefficient have been calculated from the physical and chemical data of deep lakes (Jassby and Powell 1975; Powell and Jassby 1974; Rossi et al. 1982). However significant variation exists between the results of the calculations with the various methods. While the value of E_z is probably related to the unique hydrologic and thermal regime of each body of water studied, a similar pattern of the variation of the effective diffusion coefficient with depth is usually observed (WRE 1968) (Fig. 13).

In general, the effective diffusion coefficient is greatest at the surface and diminishes with depth to a minimum at the thermocline. Experimental results indicate that the value of E_z is a function of wind shear at the surface, vertical position, and the spatially-varying vertical stability of the water column due to temperature stratification and the density gradient (Harleman 1982; Henderson-Sellers 1976).

Many empirical and semi-empirical relationships have been proposed to estimate the value of E_z from characteristics of the water column, especially vertical stability (Thomann and Mueller 1987). Referring to results from laboratory studies of the atmospheric boundary layer, Jassby and Powell (1975) suggested that the effective diffusion coefficient in stratified fluids may be a function of a single dimensionless parameter, the Richardson number (Ri), defined as:

$$Ri = N^2/(u/ z)^2 \quad (3)$$

where:

$$N^2 = -(g/\rho)\,(u/ z) \quad (4)$$

and:

N^2	= Brunt-Vaisala frequency	($1/sec^2$)
u/ z	= horizontal velocity gradient with depth (z)	(1/sec)
g	= acceleration due to gravity	(m/sec^2)

It has been popular (Henderson-Sellers 1984; Jassby and Powell 1975; Harleman 1982; Sundaram and Rehm 1973) to relate the value of E_z to the value of the effective diffusion coefficient when the vertical stability of the water column is at a minimum ($E_{\theta,z}$) and the Richardson number such that:

$$E_z = E_{\theta,z}\, f(Ri) \quad (5)$$

where:

$E_{\theta,z} = E_z$ at neutral stability at depth z, (i.e., Ri = 0)

Sundaram and Rehm (1973) have suggested the following relationship for the effective diffusion coefficient.

$$E_z = E_{\theta,z} (1 + \sigma \, Ri)^{-m} \tag{6}$$

Where σ and m are empirical coefficients. Sundaram and Rehm (1973) considered equation (6) valid only above the thermocline. For mathematical convenience and simplicity, the value of E_z below the thermocline, was assumed to be constant and equal to the minimum value of E_z evaluated at the thermocline. The constants σ and m were estimated to be equal to 0.1 and 1, respectively (Sundaram and Rehm 1973). As an example of the variability that can be encountered in calculating the effective diffusion coefficient, Octavio et al. (1977) estimated the values σ and m to be 1.76×10^{-3} and 0.5, respectively.

Given that the function relating the Richardson number to the $E_{\theta,z}$ is accurately estimated, the value of $E_{\theta,z}$ must also be calculated. Sundaram and Rehm (1973) suggested that the value of $E_{\theta,z}$ at the surface can be estimated as:

$$E_{\theta,s} = c \, (W^*) \tag{7}$$

where:

$E_{\theta,s}$	= $E_{\theta,z}$ at the surface of the water body	(m^2/sec)
c	= a coefficient dependent on the water body	(m)
W*	= the frictional velocity due to wind stress	(m/sec)

The authors determined that the value of c was approximately 2.82×10^{-2}.

A major problem encountered in using the approach outlined above (equations 6 and 7) is one of accurately estimating the horizontal velocity gradient for the calculation of the Richardson number. The required data are only rarely available (Imboden 1978). Henderson-Sellers (1984) provides a complete review and analysis of many previous efforts to estimate the neutral value of the effective diffusion coefficient. Sundaram and Rehm (1973) suggest an alternative form for the Richardson number formulation to be:

$$R_i = a_v \, g \, z^2 / (w^*)^2 \, \partial T_{qwk} / \partial z \tag{8}$$

where:

a_v	= coefficient of expansion	(1/°K)

it is assumed that:

$$u/ z = W^*/z \quad (9)$$

and

$$1/\rho \;\; \rho/ z = -a_v \; T_{wk} / z$$

the coefficient of expansion is given by:

$$a_v = 1.5 \times 10^{-5} (T_{wk} - 277) - 2.0 \times 10^{-7} (T_{wk} - 277)^2 \quad (10)$$

where:

T_{wk} = temperature of the water (K)

This form of estimating Ri eliminates the need for determining the horizontal velocity gradient.

Noting the general shape of the effective diffusion curve in Figure 13, Orlob and Selna (1970) and WRE (1968) developed an algorithm for the estimation of E_z in the epilimnion that is based on the assumption that effective diffusion decreases exponentially with depth. In the epilimnion E_z was described by:

$$E_{z,z} = E_{\theta,s} \exp(-\delta z) \quad (11)$$

where:

$E_{z,z}$ = the effective diffusion coefficient at depth z (m^2/sec)

δ = a decay coefficient (1/m)

WRE (1968) indicated that the decay coefficient is uniquely determined by the ratio $E_{\theta,T}/E_{\theta,s}$ where $E_{\theta,T}$ is the neutral value of the effective diffusion coefficient at the thermocline. The authors noted that values of $E_{\theta,T}$ and $E_{\theta,s}$ are likely to be characteristics of a particular water body and whenever possible should be derived from previous studies of the specific water body. Below the thermocline, $E_{\theta,z}$ was set equal to $E_{\theta,T}$.

Taking a similar approach, Henderson-Sellers (1984) suggested a formulation for relating wind speed and pond depth to the effective diffusion coefficient in a neutrally buoyant water column (i.e., no density gradient). The value of the neutrally buoyant effective diffusion coefficient at any depth in the water column (Eq,z) can be calculated as (Henderson-Sellers 1984):

$$E_{\theta,z} = [(W_s^*)^2 / u_s k^*] \exp(-k^* z) \qquad (12)$$

where:

$$k^* = 6 W\text{-}1.84 \qquad (13)$$

$$u_s = 30 (W^*) \qquad (14)$$

$$W^* = (\tau_o / \rho w)^{1/2} \qquad (15)$$

$$\tau_0 = \rho_a C_z (W_{vz})^2 \qquad (16)$$

and:

k^*	= empirical decay coefficient	(1/m)
τ_0	= shear stress at the surface caused by wind	(N/m^2)
W_{vz}	= wind vector magnitude at a height of z meters above the water surface	(m/sec)

The empirical coefficient k* is used in equation (12) to approximate the change in the drift velocity (u_s) with depth in the pond. The decay of current speed with depth in lakes was investigated by Smith (1979) who determined the decay to be exponential and a function of wind speed for a given latitude (i.e., no change in the Coriolis effect). Cathcart (1987) notes that the influence (with regards to temperature distribution) of the Coriolis force on small water bodies such as aquaculture ponds should be minimal. Smith (1979) determined that the value of k* (latitude of 56° N) could be estimated using equation (13). Henderson-Sellers (1984) noted that the value of the surface drift velocity has been estimated to be approximately 30 times the frictional velocity (equation 14). The coefficient of aerodynamic resistance (C_z), with a reference height of 10 m (C_{10}) has been experimentally determined to be on the order of 1.0×10^{-3} for wind speeds of less than 5 m/sec (Octavio et al. 1977; Bloss and Harleman 1979; Fischer et al. 1979). Fischer and co-workers noted that for water bodies of less than 2.5 m in depth, the value of C_{10} remains close to 1.0×10^{-3} for all wind speeds.

Modeling Convective Mixing

Most authors using one-dimensional diffusion models have assumed that vertical advective and convective transport are minimal under stratified conditions. In periods of heating of the water column, this is true and vertical diffusivity is the dominant source of turbulent mixing (Octavio et al. 1977). However, during periods of net cooling of water bodies, convective transport (also called convective penetration) due to buoyancy instabilities in the water column becomes a dominant source of vertical mixing (Octavio et al. 1977). In diffusion type models, convective transport due to density instabilities is simulated in either one of two ways: (1) Sundaram and Rehm (1973) suggest setting E_z to a maximum value (E_{max}) in those regions where density instabilities exist. (2) A more common method of dealing with convective penetration is that of recognizing when buoyant instabilities exist in the water column and "numerically mixing" (averaging the temperatures) the unstable adjacent water layers. WRE (1968) noted that when the heat flow computation results in the "unnatural" situation of a warmer layer of water being below a cooler layer, then the temperatures of the two layers are averaged. The process of numerical mixing will also result in the averaging of mass concentrations in the water column.

Insuring Numerical Stability of Diffusion Models

Orlob (1981) suggested that to insure numerical stability in diffusion type models, time steps, Δt, should not exceed a calculated value of:

$$\Delta t \leq (\Delta z)^2 / 2\,E_z \qquad (17)$$

In using the concept of E_{max} to simulate convective mixing in an unstable water column, best results are produced (no residual temperature inversions in the water column) when the value of E_{max} is sufficiently large. Chapra and Richow (1983) listed calculated thermocline diffusion coefficients from

Lake Ontario ranging from 0.05 - 58.4 cm^2/sec (0.018 - 21.020 m^2/hr). WRE (1968) noted that the maximum effective diffusion coefficient at the surface of Fontana Reservoir was 0.81 cm^2/sec (0.292 m^2/hr). The selected value of E_{max} will, to a large extent, dictate the size of the time step used in the execution of the model. To reduce the time required for model execution, the smallest possible value of E_{max} that provides proper mixing of an unstable water column should be selected.

MODELING STRATIFIED AQUACULTURE PONDS

In modeling stratification in aquaculture ponds, it makes sense to learn from and build upon the broad body of knowledge developed over nearly half a century of studies on lakes, reservoirs, and ponds. The ponds that were studied and described earlier exhibited diurnal cycles of stratification and destratification similar to the seasonal cycles of lakes and reservoirs. The onset of stratification by the heating of the water column in aquaculture ponds is similar to the spring and early summer warming of natural water bodies. Similarly, the cooling and deepening of the mixed layer in the afternoon and evening resembles the cooling patterns experienced in lakes and reservoirs in the late summer and fall. However, pond water quality and temperature change much more rapidly in shallow aquaculture ponds than in deep lakes and reservoirs. If this is not kept in mind, and parameters are not updated accordingly (small values of Δt), errors can be propagated quickly. In this section, the modeling techniques described above are evaluated for application in aquaculture pond modeling.

Use of Integral Energy Models in Aquaculture

With further research and if properly employed, the integral energy approach could provide aquacultural modelers with a powerful tool in simulating energy and mass transfer due to turbulent mixing. However, the accurate estimation of the input of total kinetic energy at the surface by wind shear is essential for successful application of this method and is currently not possible. Implicit in the accurate calculation of wind shear on an aquaculture pond surface is the proper estimation of the coefficient of aerodynamic resistance or coefficient of drag. Fischer and co-workers (1979) noted that the coefficient of drag on a lake is a complex function of atmospheric and water surface conditions. Virtually no information is available on the estimation of proper values of the coefficient of drag for aquaculture ponds. Recent studies by the author (Losordo 1988) indicated that aquaculture pond mixing is extremely sensitive, not only to wind speed, but also wind direction.

Another problem with the integral energy method is the manner in which the turbulent energy transfer is treated. Turbulent energy transfer between the mixed surface layer and the lower volumetric elements is an all or none phenomenon . That is, energy is either completely shared with the next lower volume element or there is no transfer of energy at all. In the ponds that were studied, the transfer of heat and dissolved oxygen were found to be better described by a continuous function of the temperature or DO concentration gradient and a diffusion constant than by a step function. Additionally, proper selection of the volume element thickness (Δz) is essential to the success of the integral energy method because the simulated temperature profiles are sensitive to the pond volumetric element thickness and number (Octavio et al. 1977). Stefan and Ford (1975) suggested Δz values of between 0.25 and 2.0 m for lakes and reservoirs, water bodies that often exceeded 100 meters in depth. To realize the same spatial resolution in the modeling of a 1.0 m deep aquaculture pond, the value of Δz would have to be between 0.0025 and 0.02 m. Implementation of a pond model utilizing 50 to 400 volume elements would not be feasible based on the number of calculations required for the energy balance on each volumetric element and the time required for simulation execution.

Use of Diffusion Models in Aquaculture

The results of stratification characterization studies carried out by the author (Losordo 1988) indicated that turbulent diffusion contributes significantly to the transport of heat and oxygen to the lower strata of shallow aquaculture ponds. As noted previously, the transfer of heat and DO in aquaculture ponds would be better simulated by a continuous function of the temperature or oxygen concentration gradient and a diffusion constant than by a step function. With this in mind, a diffusion type model with sufficient volumetric elements to provide good spatial resolution of temperature and DO concentration (0.10 - 0.20 m) was developed to simulate thermal and DO stratification in shallow aquaculture ponds.

The model developed by Losordo (1988) was based upon a one dimensional "Effective" diffusion approach. The model was structured to simulate a 1.0 m water column using 6 completely mixed (horizontal) volumetric elements. The quantity of energy and mass transferred between volumetric elements by turbulent diffusion was estimated with an algorithm similar to that of equation (2). For each time step of the simulation, the effective diffusion coefficient used to calculate the energy or mass transfer between each volumetric element was estimated using an algorithm similar to that of equation (12). The calculated value of the effective diffusion coefficient for the neutrally buoyant water column ($E_{\theta,z}$) was adjusted for the density gradient between the volumetric elements using an algorithm similar to equation (6). The model was implemented using time steps of 0.025 hours (1.5 min). The short time step was used to assure numerical stability of the model and was found to be necessary to simulate the extremely dynamic nature of shallow aquaculture ponds.

Validation runs with the model indicated very accurate simulation of the timing of maximum stratification and completely mixed conditions. These results can be attributed to the mechanistic nature of the calculations which are used to estimate stratification based on the effects of pond environmental and atmospheric conditions (updated every 0.025 hours). Sensitivity analysis of the modeling results showed that the simulation of the thermal and DO stratification process was most sensitive to changes in wind speed and solar irradiance. These results are consistent with the findings of the characterization study about the importance of the dynamic interaction of wind speed and solar irradiance in determining the magnitude and timing of stratification in aquaculture ponds.

Validation of the model also reaffirmed the notion that the proper estimation of the turbulent diffusion of heat and mass in the water column be calculated as a function of surface wind speed, direction, pond depth, and the stability or density gradient of the water column. Estimation of the correct effective diffusion coefficient is far more important for aquaculture ponds than deeper or less stratified systems due to high vertical temperature and oxygen concentration gradients. Because of the magnitude of the gradients, improperly high or low values of the effective diffusion coefficient resulted in very large errors in energy and mass transfer estimations. During model execution, errors in temperature stratification simulations occurred most during periods of onset of intense thermal stratification. The results indicated that errors in estimating the effect of wind speed and direction at the surface of the pond on turbulent diffusion lead to significant error in the simulation. Similarly, the simulation runs indicated that error was due to the improper adjustment of the diffusion coefficient for density gradient changes.

Successful development of the model demonstrated that shallow aquaculture pond stratification can be simulated using a one dimensional effective diffusion approach. The model provides a "stratified" framework which will allow the incorporation of improved water quality simulation algorithms that have been developed for models using the assumption of completely mixed conditions.

The author shares Henderson-Sellers (1984) beliefs that although the effective diffusion approach is by no means a precise representation of the physics of stratification in water bodies, it has an established ability to give good and useful engineering answers.

FUTURE RESEARCH NEEDS

The information and studies described in this paper represent one of the first documentations of the characterization and modeling of stratification in shallow aquaculture ponds. As a result of these studies, numerous topics in need of additional research have been identified.

Further studies in the areas of wind / pond surface interactions are needed. A better understanding of the nature of wind speed profiles over aquaculture ponds and improved algorithms for calculating the wind speed at the pond surface, based upon wind speeds at higher elevations, are needed. Further studies are needed in estimating the magnitude and variation of the coefficient of drag for shallow aquaculture ponds.

The effect of wind fetch on the magnitude of turbulent mixing in aquaculture ponds is currently unknown. Experience with the stratification model described in this paper indicates that the adjustment of the effective diffusion coefficient for the direction of the wind is essential for the accurate simulation of turbulent mixing in shallow aquaculture ponds. Of equal importance is an improved algorithm to calculate the variation in turbulent diffusion due to density gradients in the water column.

An integral energy type of model for aquaculture ponds should be developed. Although the model may not be appropriate for the simulation of stratification in ponds, the integral energy modeling approach may allow the estimation of the magnitude of the turbulent kinetic energy (TKE) that is required to mix ponds under various stratified conditions. This type of model could simulate the density gradient in stratified ponds (depending on weather conditions), calculate the buoyant potential energy of the system, and estimate the TKE required to overcome the stratification. This information would be the first step in an engineering approach to determining the artificial mixing requirements of ponds and design criteria for new pond circulating equipment.

ACKNOWLEDGMENTS

The author acknowledges and thanks Dr. Raul H. Piedrahita for his efforts in the development and execution of the research described in this paper. Thanks are due to Ken Beer of The Fishery, Bill Jenkins and Tony Vaught of Shasta Fins of California, George and Fern Ray of Fish Breeders of California and the University of Hawaii for making the aquaculture ponds available for use in this study. The research described here was made possible by funds from the Aquaculture and Fisheries Program, the Department of Agricultural Engineering at the University of California at Davis, the Jastro-Shields Foundation and the Pauley Foundation. Support for manuscript preparation was provided by North Carolina State University.

LITERATURE CITED

Almazan, G. and C.E. Boyd. 1978. An evaluation of secchi disk visibility for estimating plankton density in fish ponds. Hydrobiologia 62:601-608.

Bloss, S. and R. F. Harleman. 1979. Effect of wind-mixing on the thermocline formation in lakes and reservoirs. Technical Report No. 249, Ralph M. Parsons Laboratory, Massachusetts Institute of Technology. Cambridge, Massachusetts, USA.

Bolte, J. P., R.O. Smitherman, and D. T. Hill. 1986. Applications of mathematical modeling to pond aquaculture: a state of the art review. Presented at the 1986 Summer Meeting American Society of Agricultural Engineers, San Luis Obispo, California, June 29 - July 2, 1986, ASAE Paper No. 86-5049.

Boyd, C. E., R.P. Romaire, and E. Johnston. 1978. Predicting early morning dissolved oxygen concentrations in channel catfish ponds. Transactions of the American Fish Society, 107 (3):484-492.

Boyd, C. E. 1979. Water quality in warmwater fish ponds. Auburn University Agricultural Experimentation Station, Auburn, Alabama, USA.

Busch, C.D., C.A. Flood, and R. Allison. 1978. Multiple paddlewheel's influence on fish pond temperature and aeration. Transactions of the American Society of Agricultural Engineers 21 (6):1222-1224.

Cathcart, T.P. 1987. Heat transfer and temperature prediction in small freshwater ponds. Ph. D. Dissertation. Department of Agricultural Engineering, University of Maryland, USA.

Chapra, S.C. and K.H. Reckhow. 1983. Engineering Approaches for Lake Management, Volume 2: Mechanistic Modeling. Butterworth Publishers, Boston, Massachusetts, USA.

Fast, A.W., D.K. Barclay, and G. Akiyama. 1983. Artificial circulation of Hawaiian prawn ponds. Sea Grant paper no. UNIHI-SEA-GRANT-CR-84-01, University of Hawaii Sea Grant Program. Honolulu, Hawaii, USA.

Fischer, H.B., E.J. List, R.C.Y. Koh, J. Imberger, and N.H. Brooks. 1979. Mixing in Inland and Coastal Waters. Academic Press, New York, New York, USA.

Ford, D.E. and K.W. Thornton. 1979. Time and length scales for the one-dimensional assumption and its relation to ecological models. Water Resources Research 15 (1):113-120.

Ganf, G.G. 1974. Incident solar irradiance and under water light penetration as factors controlling the chorophyll content of a shallow equatorial lake (Lake George, Uganda). Journal of Ecology 62:593-609.

Gebhart, G. E. and R.C. Summerfelt. 1976. Effects of destratification on depth of distribution of fish. Journal of the Environmental Engineering Division, American Society of Civil Engineers 102 (EE6):1212-1228.

Harleman, D. R. F. 1982. Hydrothermal analysis of lakes and reserviors. Journal of the Hydraulics Division, American Society of Civil Engineers 108 (HY3):302-325.

Hepher, B. and Y. Pruginin. 1981. Commercial Fish Farming. John Wiley and Sons, New York, New York, USA.

Henderson-Sellers, B. 1976. Role of eddy diffusivity in thermocline formation. Journal of the Environmental Engineering Division, American Society of Civil Engineers 102 (3):517-531.

Henderson-Sellers, B. 1984. Engineering Limnology. Pitman Advanced Publishing Program, Boston, Massachusetts, USA.

Hesslein, R. and P. Quay. 1973. Vertical eddy diffusion studies in the thermocline of a small stratified lake. Journal of the Fisheries Research Board Canada 30:1491-1500.

Imboden, D.M. 1978. Modelling of vertical temperature distribution and its implication on biological processes in lakes. Pages 545-560 In: S.E. Jorgensen editor. State-of-the-art in ecological modelling. 7, Pergamon Press, Oxford, UK.

Jassby, A. and T. Powell. 1975. Vertical patterns of eddy diffusion during stratification in Castle Lake, California. Limnology and Oceanography 20 (4):530-543.

Kersting, K. 1983. Bimodal diel dissolved oxygen curves and thermal stratification in polder ditches. Hydrobiologia 107:165-168.

Krom, M.D., C. Porter and H. Grodin. 1985. Cause of fish mortality in semi-intensively operated seawater ponds in Eilat, Israel. Aquaculture, 49:159-177.

Laska, M. 1981. Characteristics and modelling of physical limnology processes. Mitteilungen der Versuchsanstalt fur Wasserbau, Hydrologie und Glaziologie, Nr. 54, Zurich. (English).

Losordo, T.M. 1988. The characterization and modeling of thermal and oxygen stratification in aquaculture ponds. Ph.D. Dissertation, University of California, Davis, California, USA.

Losordo, T.M., R.H. Piedrahita and J.M. Ebeling. 1988. An automated water quality data acquisition system for use in aquaculture ponds. Aquacultural Engineering 7: 265-278.

Marjanovic, N. and G.T. Orlob. 1986. Modelling the hydromechanical and water quality responses of aquaculture ponds. Pages 207-230 in: Lannan, J. E., Smitherman, R. O. and G. Tchobanoglous, editors. Priciples and practices of Pond Aquaculture. Oregon State University Press, Corvallis, Oregon, USA .

Madenjian, C.P., G.L. Rogers, and A.W. Fast. 1987. Predicting night time dissolved oxygen loss in prawn ponds of Hawaii: Part II. A new method. Aquacultural Engineering 6 (3):209-225.

Markofsky, M. and D.R.F. Harleman. 1973. Prediction of water quality in stratified reservoirs. Journal of the Hydraulics Division, American Society of Civil Engineers 99 (HY5):729-745.

Newbold, D.J. and J.A. Liggett. 1974. Oxygen depletion model for cayuga lake. Journal of the Environmental Engineering Division, American Society of Civil Engineers 100 (EE1):41-59.

Octavio K.A., G.H. Jirka, and D.R.F. Harleman. 1977. Vertical transport mechanisms in lakes and reservoirs. Technical Report No. 227, Ralph M. Parsons Laboratory, Massachusetts Institute of Technology, Cambridge, Massachusetts, USA.

Orlob, G.T. 1981. Models for stratified impoundments. Pages 273 - 313 in: A.K. Biswas editor. Models for Water Quality Management. McGraw-Hill International Book Co., New York, New York, USA.

Orlob, G. T. and L.G. Selna. 1970. Temperature variations in deep reservoirs. Journal of the Hydraulics Division, American Society of Civil Engineers 96 (HY2):391-410.

Powell, T. and A. Jassby. 1974. The estimation of vertical eddy diffusivities below the thermocline in lakes. Water Resources Research 10 (2): 191-198.

Rossi, G., F. Beonio-Brocchieri, and J. Sangregorio. 1982. On the evaluation of the vertical eddy coefficient in lakes by means of measured temperature profiles. Paper presented at the Third International Conference on State-of-the-Art in Ecological Modelling, Colorado State University, May 24-28, 1982.

Smith, I.R. 1979. Hydraulic conditions in isothermal lakes. Freshwater Biology 9: 119-145.

Stahl, J.B. and D.S. May. 1967. Microstratification in waste treatment ponds. Water Pollution Control Federation Journal 39 (1):72-88.

Stefan, H. and D. E. Ford. 1975. Temperature dynamics in dimitics lakes. Journal of the Hydraulics Division, American Society of Civil Engineers 101 (HY1):97-114.

Sundaram, T.R. and R.G. Rehm. 1973. The seasonal thermal structure of deep temperature lakes, Tellus 25:157-167.

Thomann, R.V. and J.A. Mueller. 1987. Principles of Surface Water Quality Modeling and Control. Harper and Row, New York, New York, USA.

Veeningen, R. 1982. Temporal and spatial variations in dissolved oxygen concentrations in some Dutch polder ditches. Hydrobiologia 95:369-383.

Walters, R.A.. 1980. A time and depth-dependent model for physical, chemical and biological cycles in temperate lakes. Ecological Modelling, 8:79-96.

Water Resources Engineers, Inc.. 1968. Prediction of thermal energy distribution in streams and reservoirs. Report prepared for Department of Fish and Game, State of California, Sacramento, California, USA.

Welch, E.B. 1980. Ecological Effects of Waste Water. Cambridge Unversity Press, Cambridge, UK.

Wetzel, R.G.. 1983. Limnology. Saunders College Publishing, Philadelphia, PA. USA.

Young, J.O. 1975. Seasonal and diurnal changes in the water temperature of a temperate pond (England) and a tropical pond (Kenya). Hydrobiologia 47 (304):513-526.

LIST OF SYMBOLS

a_v	coefficient of expansion for water
BPE	buoyant potential energy
c	empirical coefficient dependent on the water body
c_{pw}	heat capacity of water
C_z	coefficient of aerodynamic resistance at height z above pond
d	average depth
D	depth of the mixed layer
DO	dissolved oxygen
E_z	effective diffusion coefficient
$E_{z,z}$	effective diffusion coefficient at depth z
$E_{\theta,z}$	E_z at neutral stability at depth z
$E_{\theta,s}$	$E_{\theta,z}$ at the surface of the water body
g	gravitational constant
ISC	integral stability criterion
JD	julian calendar day
J_{sz}	flux of heat or mass in the z direction
K_z	coefficient of the turbulent eddy diffusion of heat at depth z
K_{sz}	coefficient of turbulent eddy diffusion of s at z
k*	empirical decay coefficient
m	empirical coefficient
N^2	Brunt-Vaisala frequency
PAR	photosynthetically active radiation
Ri	Richardson number
s	heat or mass concentration
t	time
T_{wc}	water temperature, °C
T_{wk}	water temperarture, °K
TKE	total kinetic energy
u	horizontal velocity
u_s	drift velocity of the surface
W*	frictional velocity due to wind stress
W_{vz}	wind vector magnitude at a height of z meters above the water surface
z	vertical distance from surface

δ	an empirical decay coefficient for the effective diffusion coefficient
ε	molecular diffusivity of heat
ϕ	direct radiative heat flux
ρ_w	density of water
σ	empirical coeffiecient
τ_0	wind shear stress at the surface

Portions of this paper have been reproduced from T.M. Losordo and R.H. Piedrahita 1988. Modelling vertical water quality profiles in aquaculture ponds, review and evaluation. IChemE Symposium Series No. 111, pp 313-327, with permission from The Institute of Chemical Engineers, Rugby, England.

MECHANISTIC SIMULATION MODELING OF PHYTOPLANKTON-OXYGEN DYNAMICS IN AQUACULTURAL PONDS

Daniel Wilkins Smith

ABSTRACT

In theory, mechanistic simulation modeling offers a method to elucidate and predict the interaction between dissolved oxygen and phytoplankton dynamics in aquacultural ponds. In practice, however, these models do not make useful predictive tools, because the component processes are too poorly understood and because the models require data that cannot be obtained easily or precisely. Sensitivity analyses suggest that small errors in estimating gross primary production or plankton respiration can produce large errors in predicting dissolved oxygen concentrations. Mechanistic models of phytoplankton-oxygen dynamics serve primarily as heuristic and quasi-experimental devices. To demonstrate the quasi-experimental potential of these models, this work addresses the concept of reducing total pond respiration by shifting biomass from fast-respiring plankton to slow-respiring consumers (e.g., filter-feeding fish and shellfish). According to simulations, repackaging respiration in this way might improve dissolved oxygen concentrations in aquacultural ponds.

INTRODUCTION

Aquacultural ponds are complex systems with many components interacting in non-obvious ways. This complexity limits our understanding of pond processes, and our limited understanding, in turn, restricts our ability to optimize pond dynamics for aquacultural production (Piedrahita 1988). The interaction between phytoplankton and dissolved oxygen is but a small part of the pond ecosystem, but this interaction presents a good example of a process that is both difficult to understand and important to aquacultural production (Meyer and Brune 1982; Smith and Piedrahita 1988). Mechanistic simulation modeling offers a tool to enhance our understanding of complex interactions such as phytoplankton-oxygen dynamics and to help predict the outcomes of these interactions (Meyer and Brune 1982; Piedrahita 1988).

Mechanistic models — also called analytical, process, or theoretical models — divide the system into a series of compartments or reservoirs, called "state variables." Change in magnitude of state variables over time is described by linear or non-linear differential equations of the type

$$dS_1/dt = f(S_1, S_2 \ldots S_n, A_1 \ldots A_n)$$

where S_1 is the state variable in question, S_2 thru S_n are other state variables whose concentrations or levels affect S_1, and A_1 thru A_n are outside factors or forces that also affect the rate of change of S_1 over time. In the case of a model of phytoplankton-oxygen dynamics, the outside factors would include variables such as sunlight, temperature, and other pertinent environmental data.

In a mechanistic model, the form of the differential equation is guided by the modeler's perception of how the process works (i.e., by the "mechanisms" involved). In this sense, mechanistic models are quite different from empirical models (Lewis et al. 1986; Swartzman and Kaluzny 1987). The equations in empirical models are determined by statistical analyses, usually regression analyses, that derive the mathematical expression that best fits data from real systems. Hence, it is the ability to predict, in a statistical sense, that determines the forms of the equations of empirical models. In contrast to mechanistic model-building, the modeler's perception of how system components interact has little impact on the form of the empirical model.

These different processes of model-building influence the goals and uses of the two types of models. Since mechanistic models are preoccupied with mechanisms, they are often touted as good heuristic devices (i.e., helping us to understand the mechanisms of complex systems), but not as good predictors of real-world behavior (e.g., Bernard 1986). In contrast, empirical models are determined by the best fit to real data; consequently, these models are normally thought to be good at prediction but less useful at explanation.

Neither generalization is invariably true. The empirical models of Romaire and Boyd have been used in a heuristic manner (Boyd et al. 1978; Smith and Piedrahita 1988). Moreover, the greater understanding afforded by mechanistic models could produce better predictions than those obtained from black-box empirical models, and some have found that mechanistic models are better able to mimic real-world data (e.g., Losordo, this volume).

This chapter is not intended to be a detailed review of simulation modeling or a compilation of mechanistic models. (For recent reviews of simulation modeling in general and modeling of fish ponds in particular, the reader could consult the following: Straskraba and Gnauck 1985; Bernard 1986; Marjanovic and Orlob 1986; Swartzman and Kaluzny 1987; Piedrahita, this volume). Rather, the objective of this chapter is to assess the current state of mechanistic models of phytoplankton-oxygen dynamics in aquacultural ponds, and to determine the current uses and potential of these models in aquaculture and aquacultural science.

Within this context, two specific questions will be addressed in the following discussion:

1) Are current mechanistic models useful predictive devices for dissolved oxygen levels?

2) Are current mechanistic models useful heuristic devices?

METHODS

The work described below is based on a simple mass balance model for dissolved oxygen originally described by Meyer and Brune (1982), and modified by Smith and Piedrahita (1988). The model accounts for oxygen gains or losses resulting from phytoplankton production, mixing (reaeration) with the atmosphere, and respiration of the benthos, the fish, and the phytoplankton. Because the equation for primary production was obtained from the work of Baumert and Uhlmann (1983), this model was called the BAUMANN model. This terminology will be continued in this paper.

As formulated in Smith and Piedrahita, the model predicts oxygen dynamics of a pond stocked with 3200 kg/ha of channel catfish (*Ictalurus punctatus*) for a single day as a function of algal biomass, solar radiation, and water temperature. As in Smith and Piedrahita, simulation conditions were held at 400 Langleys per day solar radiation, 25 C, and wind like that found at Auburn in late summer. For the sake of brevity, the model will not be repeated here. The model was implemented in DYNAMO, a simulation language described more fully in Richardson and Pugh (1981) and Wolfe et al. (1986).

For a model to be a useful predictive device, three conditions must be satisfied:

1. The modeler must understand the system well enough to describe its behavior accurately.

2. The pond manager must be able to gather the necessary environmental data with sufficient accuracy, because a perfectly accurate model can produce good predictions only if it uses accurate environmental data.

3. The data required to run the model must not be too difficult for the pond manager to obtain.

Two types of sensitivity analyses were performed to address the first two conditions. A traditional sensitivity analysis assessed condition #2 by testing the response of the model to imprecise data inputs. The magnitudes of the model components (plankton production, plankton respiration, fish respiration, benthic respiration, and reaeration) were adjusted by plus or minus 10% to quantify the potential effect of small errors in data acquisition.

A variant of the traditional sensitivity analysis (hereafter called a literature sensitivity analysis) was performed to address the question underlying condition #1 — do we know enough about important system components to formulate an accurate model? In this second analysis, equations for the two most important components — plankton respiration and phytoplankton production — were replaced with alternative equations found in the literature.

Thus, Baumert and Uhlmann's equation for gross primary production (GPP) was replaced by comparable expressions found in the work of Smith

(1980) and the aquaculture models of Griffin et al. (1984) and Svireshev et al. (1984). Alternate values for plankton respiration were taken from Smith and Piedrahita (1988), Boyd et al. (1978), Grobellar and Soeder (1985), and Meyer and Brune (1982). The quantitative behavior of the BAUMANN model was then tested with the different equations. These substitutions assessed the degree of consensus and the state of our knowledge concerning the behavior of a component as well as the quantitative ramifications of our ignorance.

With both sensitivity analyses, the magnitude of response — i.e., the change in the model's prediction with a change of equation or a ± 10% perturbation — varied with algal biomass (e.g., see Fig. 1). Thus, an average response, called the mean range, was calculated for algal densities ranging from 25 to 625 mg/m^3, where the models generally predicted dawn oxygen to be above 0 ppm. The mean range for the traditional sensitivity analysis was calculated as the average difference between the predictions of the model when a component's magnitude was varied by ± 10% (Fig. 1). With the

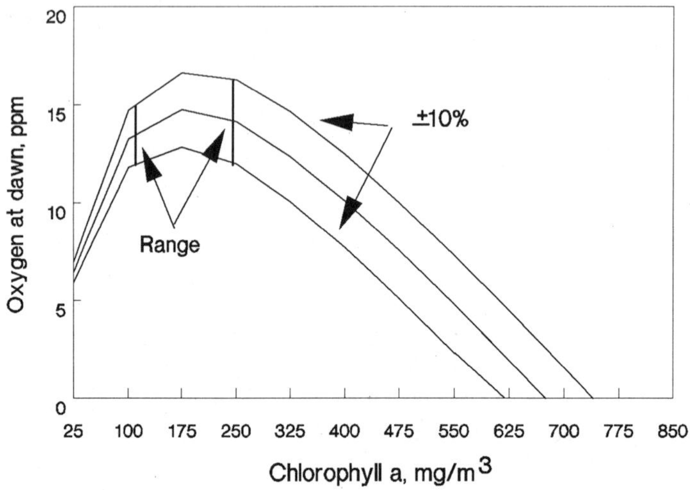

Figure 1. The response of the BAUMANN model to 10% overestimation, accurate calculation, and 10% underestimation of gross primary production (GPP). The range is the difference between predictions of dawn O_2 with GPP overestimated by 10% versus predictions with GPP underestimated by 10%.

literature sensitivity analysis, the mean range was the average difference between the two most different predictions of different versions of the BAUMANN model.

RESULTS AND DISCUSSION

Part 1. Are mechanistic models useful predictive devices?

As with previous work (Meyer and Brune 1982; Boyd et al. 1978; and Smith and Piedrahita 1988), the traditional sensitivity analysis suggests that dissolved oxygen at dawn responds most sensitively to — or depends most heavily on — phytoplankton production and phytoplankton respiration

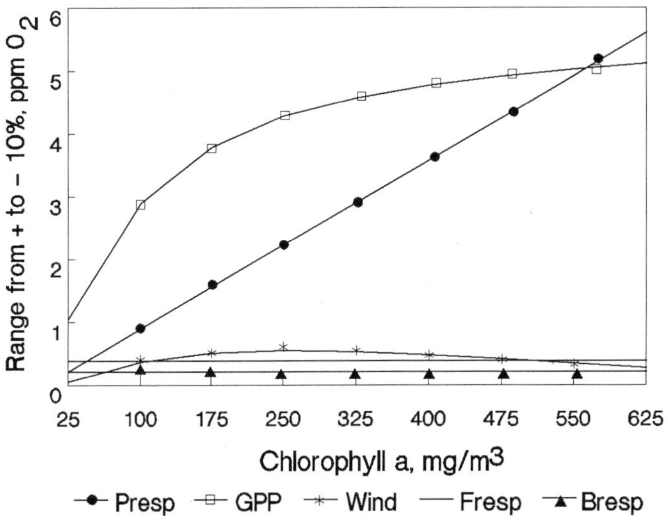

Figure 2. The range of the model's predictions (of dawn dissolved oxygen) with ± 10% errors in estimation of the components. The range is the difference between the model's predictions when a component was overestimated by 10% compared to predictions when the component was underestimated by 10%. See Figure 1.

(Table 1, Figs. 1 and 2). Fish respiration, reaeration, and benthic respiration had much less effect on the oxygen budget, especially at phytoplankton densities above 100 mg/m^3 of chlorophyll *a*.

Table 1. Results of sensitivity analyses and assessment ofdata requirements for model components. "Mean Range with ±10%" refers to the average difference beween the model's predictions of dissolved oxygen at dawn when the component was underestimated by 10% vs. predictions when the component was overestimated by 10%. "Mean Range with Literature" refers to the average difference between the most different predictions when the BAUMANN model was run with different equations for the component. The "Mean Range" was calculated for algal biomass between 25 and 625 mg/m^3. "Necessary data" refers to data that must be gathered to model the component.

COMPONENT	MEAN RANGE WITH ± 10%	MEAN RANGE WITH LITERATURE	NECESSARY DATA
GPP	3.9 g/m^3	12.0 g/m3	Sunlight* Temperature* Algal Biomass Non-algal Turbidity Algal Species? Algal Distribution?
ALGAL RESPIRATION	2.6 g/m^3	11.6 g/m3	Temperature* Algal Biomass Algal Species? GPP?
FISH RESPIRATION	0.4 g/m^3	————	Fish Species* Fish Size* Total Fish Mass* Temperature*
BENTHIC RESPIRATION	0.2 g/m^3	————	Temperature?* Dissolved Oxygen?* Algal Biomass? Benthos Density?
REAERATION	o.4 g/m^3	————	Dissolved Oxygen* Wind Speed*

* Accurate data on this parameter are easy to obtain.
? Data not required by current equations for the component, but may be necessary to accurately model the component.

Shifts of ± 10% in the magnitudes of either gross primary production (GPP) or plankton respiration caused large differences in the predictions of dawn oxygen (Figs. 1 and 2.) The average difference in model predictions for ± 10% GPP was about 4 ppm, while ± 10% errors in the estimation of plankton respiration caused an average 3.5 ppm difference in the model's prediction of dawn oxygen concentrations (Table 1). Thus, small inaccuracies in estimation of either GPP or plankton respiration could cause large errors in the prediction of the mechanistic model. In contrast, the inability to estimate accurately the other parameters (fish respiration, reaeration, or benthic respiration) had less impact on the accuracy of the model (Fig. 2, Table 1).

The literature sensitivity analysis presents an even more pessimistic picture for use of mechanistic models as predictive devices. The two components that most affect dissolved oxygen levels, GPP and plankton respiration, are apparently not well understood. For example, GPP is the most important source of oxygen to the pond, yet different expressions from the literature give very different predictions of the magnitude of primary production (Fig. 3). When different equations for GPP are integrated into the BAUMANN model, it gives a wide range of predictions of dissolved oxygen at dawn, varying by as much as 10 ppm. This range was well outside the range of predictions given by ± 10% perturbations (Fig. 4).

The discrepancy between GPP models may have been exaggerated by using the model from Baumert and Uhlmann (1983), since their model intends to predict the maximum GPP possible. In contrast, the other equations aim to predict real rates of GPP in the field. However, other models intended to predict real-world GPP can yield values like those predicted by Baumert and Uhlmann, e.g., Goldman's (1979) model for algal culture and Bannister's model (1979) as used in Laws and Malecha (1981). Even if the Baumert-Uhlmann equation were eliminated from the analysis, the choice of an equation for GPP significantly affects the prediction of the mechanistic model.

A modeler's choice of an equation for plankton respiration could also have a large effect on predictions of oxygen concentrations in aquacultural ponds. Different literature values for plankton respiration vary almost by an

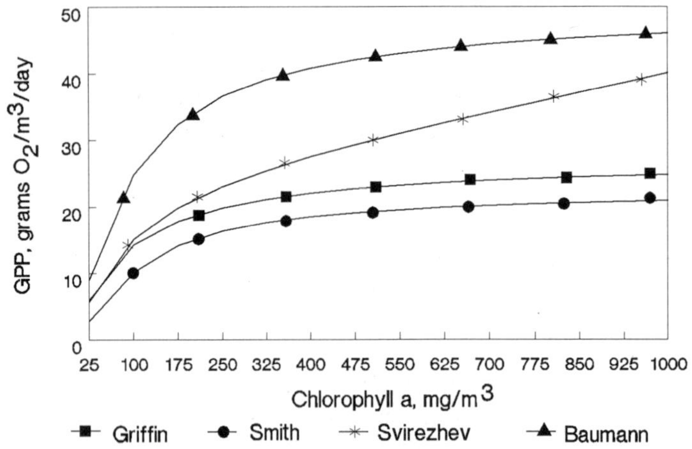

Figure 3. Values of gross primary production (GPP) as predicted by four equations found in the literature. Griffin comes from the equation for GPP found in Griffin et al. (1984), the Baumann equation is from Baumert and Uhlmann (1985), Smith is from Smith (1980), and Svirezhev GPP equation is taken from Svirezhev et al. (1984).

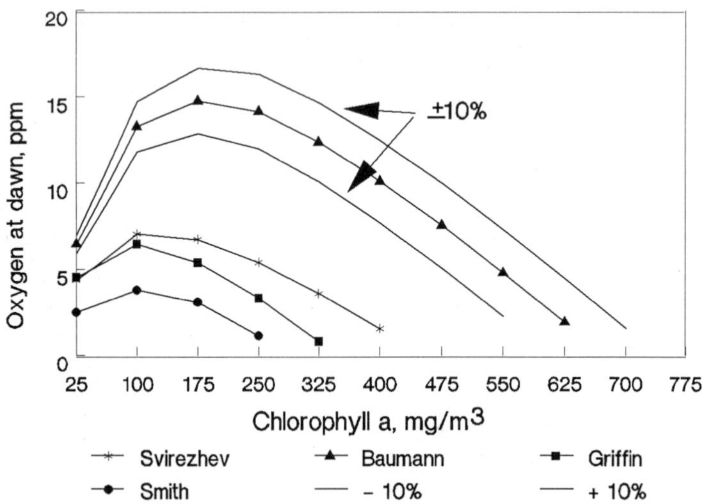

Figure 4. Dissolved oxygen at dawn, as predicted by the BAUMANN model with different equations for GPP. Equations are as in Figure 3, and the range of ± 10% is identical to that presented in Figure 1.

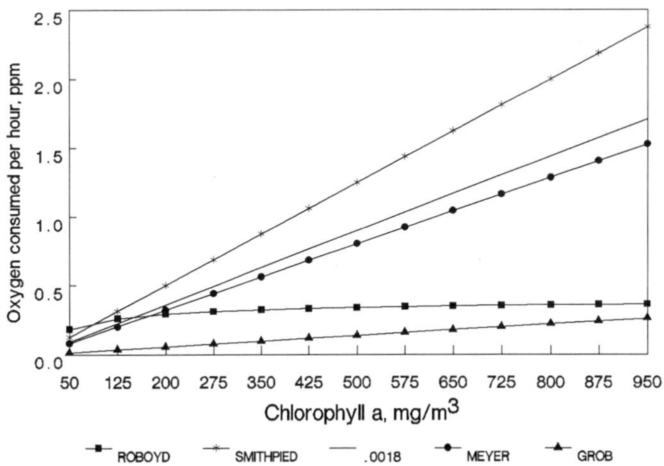

Figure 5. Values of plankton respiration as predicted by five equations or values found in the literature. Roboyd refers to the equation for plankton respiration found in Boyd et al. 1978, Smithpied is equal to 0.0025 mg O_2 per mg Chl *a*/hr (Smith and Piedrahita 1988), 0.0018 is a value calculated from Rhyther and Guillard (1965) and a Q_{10} of 2.6 from Jewson (1977), Meyer refers to a value taken from Meyer and Brune (1982), and Grob refers to the upper value from Grobbelaar and Soeder (1985).

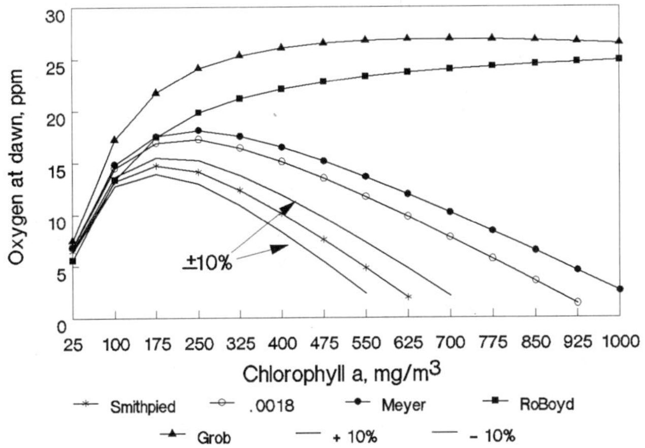

Figure 6. Dawn oxygen concentrations as predicted by the BAUMANN model with different equations for plankton respiration. Equations are as in Figure 5.

order of magnitude (Fig. 5). When integrated into the phytoplankton-oxygen model, the alternative expressions produce a wide range of predictions of dawn oxygen levels (Fig. 6). The average results for the two sensitivity analyses are summarized in Table 1.

To assess the amount of work involved in modeling a component, Table 1 also includes a listing of data required to model a component. A substantial amount of environmental data must be gathered to predict primary production or plankton respiration. For example, to accurately predict gross primary production, most models require precise quantification of sunlight intensity, algal biomass, non-algal color and turbidity, and the concentrations of essential plant nutrients. All but the first are relatively difficult to assess. The data needed to model plankton respiration are similarly extensive and difficult to gather.

These data requirements are clearly excessive in light of the reality of pond management — many ponds over many days — and the short-lived applicability of data to the rapidly changing plankton. The costs of gathering these data would probably outweigh any benefits of accurate prediction, even if the models were accurate predictive devices.

Since benthic respiration tends to have a minor effect on dissolved oxygen levels (Costa-Pierce et al. 1983), it was not subjected to a literature sensitivity analysis. However, this component also appears to be poorly understood, depends on many factors, and may require extensive data collection for accurate modeling (Edwards and Rolley 1965; Costa-Pierce et al. 1983; Bernard 1986; Colman, this volume; Table 1).

In contrast, the data needed to model reaeration and fish respiration are less extensive and easier to obtain. Thus, fish managers could probably model these components accurately and relatively easily. However, neither generally has major impacts on dissolved oxygen levels in fish ponds.

In conclusion, the sensitivity analyses illustrate several important points about mechanistic modeling of oxygen-phytoplankton dynamics in fish ponds. First, the two most important factors (GPP and plankton respiration) are too poorly described at this time to be modeled accurately. Second, small errors in the estimation of either GPP or plankton respiration can lead to large errors in the prediction of dissolved oxygen. Third, modeling these two components requires extensive and time-consuming collection of data.

These points imply that mechanistic models will not provide useful predictive tools for fish farmers now or in the near future. Calculation of primary productivity alone poses significant — probably insurmountable — procedural and theoretical problems for the use of mechanistic models as useful predictive devices. Accurate prediction of plankton respiration may also be too much work for too little gain.

In contrast, site-specific empirical models probably offer a more convenient and reliable method to predict dissolved oxygen in fish ponds. For example, the Romaire-Boyd model (Romaire and Boyd 1979; Boyd et al. 1978) has simple data requirements (temperature, Secchi disk depths, and solar radiation) that can be measured with ease and, except for Secchi disk (Almazan and Boyd 1978), great accuracy. However, because prediction errors as small as 1 to 0.5 ppm could lead to expensive fish kills, empirical models may also prove too unreliable for prediction of dissolved oxygen levels in aquacultural ponds.

Are mechanistic models useful heuristic devices?

In practice, mechanistic models of fish ponds have been used as clarifying or quasi-experimental devices (Meyer and Brune 1982; Wolfe et al. 1986; Piedrahita 1988; Smith and Piedrahita 1988). The next section will attempt to demonstrate how a mechanistic model can serve a heuristic, quasi-experimental purpose using a DYNAMO version of the BAUMANN model.

In a previous paper (Smith and Piedrahita 1988), the authors concentrated on methods to maximize oxygen inputs into a pond. Since oxygen levels are the difference between inputs and outputs, oxygen levels might also be improved by reducing the losses of dissolved oxygen. When considering which loss processes should be targeted for control, two criteria must be considered: can the loss process be controlled, and does the loss process have a significant impact on dissolved oxygen levels? Mechanistic models provide a framework to evaluate oxygen losses in light of these criteria.

According to sensitivity analyses and information presented in the first section, two loss processes can be eliminated from consideration. Reaeration can be eliminated, because it is usually a minor component of the oxygen budget. Furthermore, reaeration often represents a net gain of oxygen. For many ponds under many conditions, reaeration probably represents a net loss of oxygen to the air, especially during hot, sunny months. However, losses are most severe during sunny, windy days when oxygen depletions are less likely to occur, whereas reaeration probably represents a net gain — or insignificant loss — of oxygen on cloudy, calm days when oxygen depletions are more likely.

Controlling benthic respiration would also not seem profitable, since benthic respiration generally has little effect on the oxygen budgets of fish ponds. In addition, benthic respiration cannot be easily controlled. Benthic respiration probably depends on density of phytoplankton falling on the sediments, the concentration of oxygen overlying the water, and the density of infauna (Edwards and Rolley 1966; Costa-Pierce et al. 1983). None of these factors can be easily controlled, except possibly the density of benthic infauna, which might be controlled by stocking bottom feeders. Most aquacultural ponds already stock bottom-feeding fish, e.g., carp and catfish; thus, the potential for any further management is limited.

In contrast, selection of an aquacultural species is under the direct control of the aquaculturist, and growing certain species might improve dissolved oxygen if the species' rates of respiration were substantially lower than those of the species being cultivated. *Tilapia nilotica* and cichlids in general are reported to have metabolisms lower than most other fish (Verheyen et al. 1985). Silver carp (*Hypophthalmichthys molitrix*) reportedly also have low respiration rates (Mukhamedova 1977). Would the difference in metabolic rate have a significant effect on dissolved oxygen levels? The mechanistic model provides a useful framework to "test" such a hypothesis.

Figure 7 presents the result of such a test using BAUMANN to model dynamics of a pond stocked with identical biomasses of either *Tilapia nilotica*, silver carp, or channel catfish. The model's results suggest that dissolved oxygen levels in ponds with tilapia or silver carp would be almost 2 g/m^3 higher than levels in ponds stocked with catfish. Since 3200 kg of

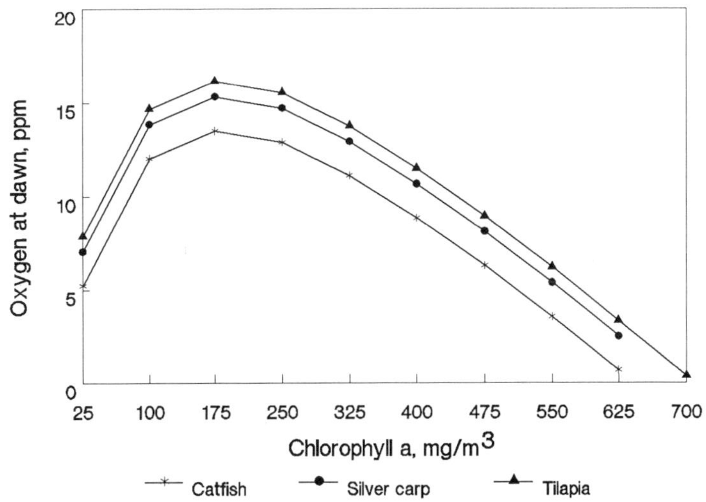

Figure 7. Dissolved oxygen at dawn for ponds stocked with 3200 kg/ha of either catfish, silver carp, tilapia, or mussels. The weight for mussels is wet weight, excluding the shell.

tilapia at 25 C consume less than 2 g/m^3 of oxygen daily, this extra two g/m^3 could represent a substantial increase in stocking rate or in the margin of safety. Stocking silver carp would also apparently improve dissolved oxygen levels.

Bivalves also tend to have very low respiration rates per unit biomass, probably because shellfish have reduced sensory and muscular demands (Jorgensen 1975; Hammen 1979; Mckee and Mackie 1983). Moreover, shellfish respiration tends to decrease with decreasing oxygen levels (Newell 1982), in contrast to most fish, which continue to respire at a high rate until some low-oxygen threshold (Warren 1971). Raising shellfish with respiration rates like the mussel, might also improve dissolved oxygen levels (Fig. 7).

Thus, the model suggests that the metabolic rate of the aquacultural species could have an economically meaningful impact on a pond's dissolved oxygen levels (Fig. 7), suggesting that metabolic rate should be considered when evaluating species for culture and polyculture. Before testing this supposition in the field, two important limitations must be redressed. First, the metabolic rates of tilapia were based on quiescent fish (Verheyen et al. 1985). Future models must account for the metabolic rates of actively feeding and growing fish. Second, shellfish, tilapia, and silver carp are all filter-feeders, and filter-feeders affect the plankton in myriad ways (Smith 1988), some of which could have negative impacts on dissolved oxygen levels. This second limitation could be investigated in complex models that include more realistic modeling of the interactions between the plankton and filter-feeding fish.

Plankton respiration might also be managed to improve dissolved oxygen in fish ponds. According to the models and experience (Romaire and Boyd 1979; Boyd 1982; Piedrahita 1989), plankton respiration represents the major loss of oxygen in most aquacultural ponds. Is there any way pond managers can control the respiration of the plankton?

In aquacultural operations where feeds are provided, the aquaculturist can avoid overfeeding and use feeds that fish can metabolize efficiently with minimal waste (Boyd 1982). As for the rest of the plankton, respiration appears to be most dependent on phytoplankton biomass and water temperature (Boyd et al. 1978). Since the pond manager has little control over water temperature, only plankton biomass remains as a manipulable element.

Flushing plankton from the pond effectively reduces the respiratory biomass in the pond. However, few operations have access to cheap, plentiful water, so few aquaculturists can economically flush ponds on a regular basis.

Herbicides or pathogens do not remove the plankton biomass from the pond. Rather, they transfer the respiratory biomass from live phytoplankton to non-living detritus and bacterioplankton. Because the latter have equal or greater rates of respiration than living phytoplankton, killing the phytoplankton generally worsens oxygen conditions (Tucker and Boyd 1979; Smith and Piedrahita 1988).

On the other hand, transferring biomass from plankton to fish or shellfish might reduce total respiration, because respiration per unit biomass has an inverse relationship to organism size. Total respiration might be reduced by transferring biomass (or carbon) from small, fast-respiring organisms to large, slowly-respiring organisms, via grazing and incorporation into grazer biomass. This process could be termed "respiratory repackaging."

The potential for repackaging respiration is constrained by two factors. First, transferring biomass from plants, with their minimal respiratory needs, to animals entails a metabolic penalty. A spherical alga 10 microns in diameter would have a metabolic rate only about 4 times that of a *Daphnia pulex,* almost 100,000 times the alga's mass[1]. Thus, grazers must be larger than zooplankton to significantly reduce respiration per unit biomass. Some very large planktivores — fish and shellfish — apparently have sufficiently low metabolic rates, as low as 0.5% to 3% of those reported for phytoplankton[2].

Secondly, for respiratory repackaging to work, the total respiration of the consumers must be less than the respiration of the phytoplankton that were consumed and repackaged. That is, the sum of the grazers' respiration plus the respiration of plankton not consumed must be less than the total plankton respiration expected in the absence of the grazer.

A stocking of 1000 kg/ha of 32 g silver carp would consume about 1 g/m^3 (1 ppm) of oxygen per day at 25 C. The same stocking rate could filter as much as 3/4 of the water volume per day at 25 C (Smith 1989). Assuming for the moment that silver carp could filter all plankton sizes efficiently or that the plankton community were composed only of relatively large particles (> 30 microns in diameter) that could be filtered efficiently, the BAUMANN model predicts that adding silver carp to a pond could improve oxygen levels by repackaging plankton biomass and respiration into silver carp biomass and respiration (Figs. 8 and 9). The degree of improvement in oxygen concentrations depends on various factors: the density of phytoplankton, the magnitude of plankton respiration vs. consumer respiration, and the rate at which the grazers consume plankton.

[1]. Plankton respiration was calculated, as in Smith and Piedrahita (1988), from the base value from Ryther and Guillard (1962) and Q_{10} from Jewson (1977). Zooplankton respiration was calculated from Richman (1958) assuming a Q_{10} of 2 and a dry weight to wet weight ratio of 0.25.

[2]. Algal respiration as above. Silver carp respiration predicted from Mukhamedova (1977) and shellfish respiration as in McKee and Mackie (1983).

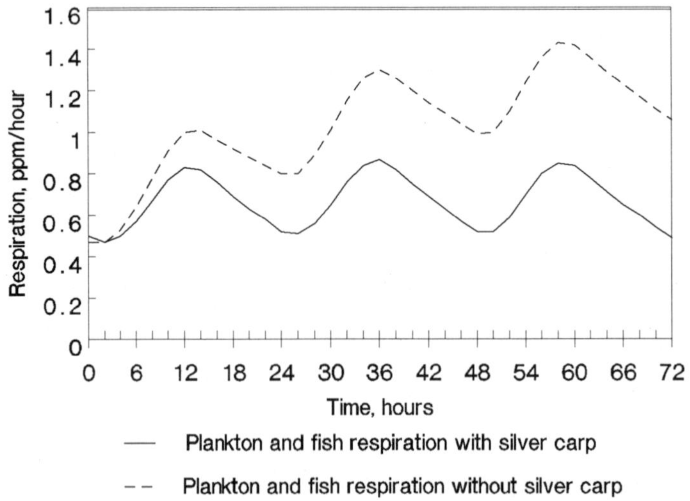

Figure 8. Total respiration due to phytoplankton and fish in ponds stocked with only catfish vs. ponds stocked with catfish and silver carp, as predicted by the BAUMANN model with phytoplankton biomass beginning at 200 mg/m³ and silver carp stocking of 1000 kg/ha.

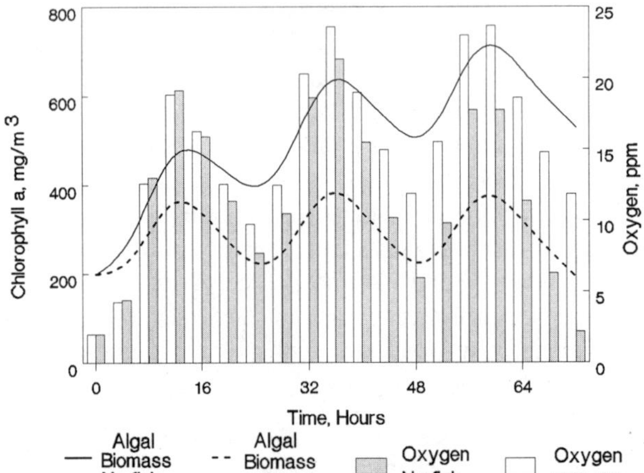

Figure 9. Dynamics of phytoplankton biomass and dissolved oxygen as predicted by the BAUMANN model, with conditions as described in Figure 8.

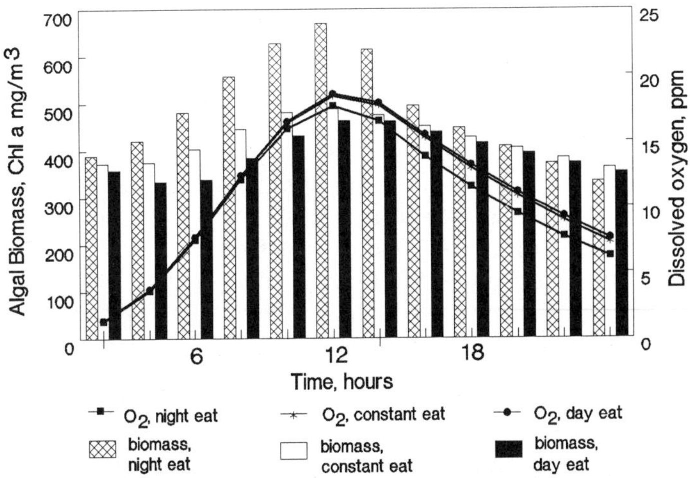

Figure 10. Dynamics of phytoplankton biomass and dissolved oxygen as predicted by the BAUMANN model for a pond stocked with 1000 kg/ha silver carp. The model was run with fish feeding for 12 hours during the light, for 12 hours during the night, or for the entire day at half the hourly rate as the other two conditions.

Shellfish might make even better agents of respiratory repackaging, because of their low rates of metabolism (Jorgensen 1975) per volume of water filtered. They also may take a wide range of particle sizes, which may allow them to harvest more of the phytoplankton than most filter- feeding fish (Smith 1988).

For the purpose of respiratory repackaging of plankton biomass, the best repackager would seem to be an organism that eats only during the night, because phytoplankton produce oxygen during the day. In the analyses above, grazers were assumed to filter at a constant rate throughout the day. Unfortunately, silver carp and many shellfish have been reported to feed only during the light period (Chiang 1971; Jorgensen 1975).

Mechanistic modeling can help assess the consequences of feeding periodicity (Fig. 10). For purposes of comparison with organisms that feed only during the light or during the dark period, a constantly-feeding grazer was assumed to filter the same amount in 24 hours as the day-grazers and night-grazers cleared in 12 hours. While feeding periodicity affects the hourly response of phytoplankton biomass and dissolved oxygen levels, the timing of grazing activity has little effect on either variable after twenty-four hours. Thus, the model suggests that grazers that feed during the day could function as respiratory repackagers.

Thus, the BAUMANN model suggests that some grazers may compensate for their own respiration by repackaging biomass from fast-respiring plankton to slow-respiring fish and shellfish biomass. Several caveats must be added to this conclusion. First, the model does not account for losses due to the oxidation of fish feces or metabolic products, which could offset any positive effects of respiratory repackaging. Equally important, most planktivores cannot filter all sizes of plankton (Smith 1989) and have generally proven unable to control algal biomass (Smith 1988). However, some untried grazers (e.g., shellfish) or grazer combinations (e.g., zooplankton and fish as in Smith 1985) may still prove capable of controlling algal biomass.

CONCLUSIONS

Mechanistic models of oxygen-dynamics in fish ponds will probably not serve as useful predictive devices in the management of fish ponds. The major components of oxygen dynamics are too poorly understood, the data needed to run mechanistic models are too extensive, and the accuracy of prediction needed by fish farmers is well beyond the capability of current mechanistic models. However, mechanistic models will continue to be important in shaping our ideas about how this complex interaction behaves in a qualitative sense, as well as enhancing our understanding of how the pond ecosystem functions.

These models can also be useful as quasi-experimental systems to pretest prospective pond-management techniques. As funding for science continues to decline, simulation modeling will offer a cost-effective alternative to preliminary field tests to determine priority of potential research projects.

ACKNOWLEDGMENTS

This work was funded in part by Pugh-Roberts Associates of Cambridge, MA, who supplied a copy of DYNAMO simulation language. I would also like to thank Ruth Anderson and Kip Cooper for their help in preparing this manuscript.

LITERATURE CITED

Almazan, G. and C. E. Boyd. 1978. An evaluation of Secchi disk visibility for estimating plankton density in fish ponds. Hydrobiologia 61: 205-208.

Bannister, T. T. 1979. Quantitative description of steady state, nutrient saturated algal growth, including adaptation. Limnology and Oceanography 24: 76-96.

Baumert, H. amd D. Uhlmann. 1983. Theory of the upper limit to phytoplankton production per unit area in natural waters. Internationale Revue gesamten der Hydrobiologie 6: 753-783.

Bernard, D. R. 1986. A survey of mathematical models pertinent to fish production and tropical pond aquaculture. Pages 231-242 J. E. Lannan, R. O. Smitherman, and G. Tchobanoglous, editors. Principals and practices of pond aquaculture. Oregon State University, Corvallis, Oregon, USA.

Boyd, C. E. 1982. Water Quality Management for Pond Fish Culture. Elsevier, Amsterdam, The Netherlands.

Boyd, C. E., R. P. Romaire, and E. Johnston. 1978. Predicting early morning dissolved oxygen concentrations in channel catfish ponds. Transactions of the American Fisheries Society 107: 484-492.

Chiang, W. 1971. Studies on the feeding and protein digestibility of silver carp . Joint Commission on the Reconstruction of Rural China, Fisheries Series 11: 95-114.

Colman, J. A. 1991. Modeling and manipulating processes at the sediment water interface. In D.E. Brune and J.R. Tomasso, Jr. (eds) Aquaculture and Water Quality (this volume).

Costa-Pierce, B. A., D. B. Craven, D. M. Karl, and E. A. Laws. 1983. Correlation of *in situ* respiration rate and microbial biomass in prawn (*Macrobrachium rosenbergii*) ponds. Aquaculture 37: 157-168.

Edwards, R. W. and H. L. J. Rolley. 1965. Oxygen consumption of river muds. Journal of Ecology 53: 1-19.

Goldman, J. C. 1979. Outdoor algal mass cultures II. Photosynthetic yield limitations. Water Research 43: 119-136.

Griffin, W. L., W. E. Grant., R. W. Brick, and J. S. Hanson. 1984. A bioeconomic model of shrimp maricultural systems in the U.S.A. Ecological Modelling 25: 47-68.

Grobbelaar, J. U. and C. J. Soeder. 1985. Respiration losses in planktonic green algae cultivated in raceway ponds. Journal of Plankton Research 4: 497-506.

Hammen, C. S. 1979. Respiration rates of marine bivalve molluscs determined by calorimetry. Comparative Biochemistry and Physiology 62A: 955-959.

Jewson, D. H. 1977. Light penetration in relation to phytoplankton content of the euphotic zone of Lough Neagh. Oikos 28: 74-83.

Jorgensen, C. B. 1975. Comparative physiology of suspension feeding. Annual Review of Physiology 37: 57-79.

Laws, E. A. and S. Malecha. 1981. Application of a nutrient saturated growth model to phytoplankton management in freshwater prawn ponds. Aquaculture 24: 91-101.

Lewis, W. M., and J. F. Sauders, D. W. Crumpacker, and C. Brendecke. 1986. Eutrophication and land use. Ecological Studies 46. Springer-Verlag, New York, New York, USA.

Losordo, T. M. 1991. Stratification and mixing in shallow aquaculture ponds. In D. E. Brune and J. Tomasso, Jr. (eds). Aquaculture and Water Quality (this volume).

Marjanovic, N. and G. T. Orlob. 1986. Modelling the hydromechanical and water quality response of aquaculture ponds. Pages 207-230. J. E. Lannan, R. O. Smitherman, and G. Tchobanoglous, editors. Principals and Practices of Pond Aquaculture. Oregon State University, Corvallis, Oregon, USA.

McKee, P. M. and G. L. Mackie. 1983. Respiratory adaptations of the finger nail clams, *Sphaerium occidenale* and *Musculium securis*, to ephemeral habitats. Canadian Journal of Fisheries and Aquatic Sciences 40: 763-776.

Meyer, D. I. and D. E. Brune. 1982. Computer modeling of the diurnal oxygen levels in a stillwater aquaculture pond. Aquacultural Engineering. 1: 245-261.

Mukhamedova, A. F. 1977. The level of standard metabolism of young silver carp, *Hypophthalmichthys molitrix*. Journal of Ichthyology 292-299.

Newell, I. E. 1982. Molluscan bioenegetics - a synopsis. Pages 252-271 G. D. Pruder, C. J. Langdon, and D. E. Conklin, editors. Proceedings of the second international conference on aquaculture nutrition: biochemical and physiological approaches to shellfish nutrition. Louisiana State University, Baton Rouge, Louisiana.

Piedrahita, R. H. 1988. Introduction to computer modelling of aquaculture pond ecosystem. Aquaculture and Fisheries Management 19: 1-12.

Piedrahita, R. H. 1991. Modeling water quality in aquaculture ecosystems. In D. E. Brune and J. Tomasso, Jr. (eds). Aquaculture and Water Quality (this volume).

Richardson, G. P, and A. L Pugh. 1981. Introduction to system dynamics modeling with DYNAMO. MIT Press, Cambridge, Massachusetts, USA.

Richman, S. R. 1958. The transformation of energy by *Daphnia pulex.* Ecological Monographs 28: 273-291.

Romaire, R. P. and C. E. Boyd. 1979. Effects of solar radiation on the dynamics of dissolved oxygen in channel catfish ponds. Transactions of the American Fisheries Society 108: 473-478.

Ryther, J. H. and R. R. L. Guillard. 1962. Studies of the marine planktonic diatoms. III. Some effects of temperature on five species. Canadian Journal of Microbiology 8: 447-453.

Smith, D. W. 1985. Biological control of excessive phytoplankton growth and enhancement of aquacultural productivity. Canadian Journal of Fisheries and Aquatic Sciences. 42: 1940-1945.

Smith, D. W. 1988. Phytoplankton and catfish culture: a review. Aquaculture 74: 167-189.

Smith, D. W. 1989. The feeding selectivity of silver carp *Hypophthalmichthys molitrix.* Journal of Fish Biology 34: 819-828.

Smith, D. W. and R. A. Piedrahita 1988. The relation between phytoplankton and oxygen in fish ponds. Aquaculture 68: 249-265.

Smith, R. A. 1980. The theoretical basis for estimating phytoplankton production and specific growth rate from chlorophyll, light and temperature data. Ecological Modelling 10: 243-264.

Straskraba, M. and A. Gnauck. 1985. Freshwater Ecosystems: Modeling and Simulation. Developments in Ecological Modeling, 8. Elsevier, Amsterdam, The Netherlands.

Svirezhev, Yu. M., V. P. Krysanova, and A. A. Voinov. 1984. Mathematical modeling of a fish pond ecosystem. Ecological Modelling 21: 315-337.

Swartzman, G. L. and S. P. Kaluzny. 1987. Ecological Simulation Primer. MacMillan, New York, New York, USA.

Tucker, C. S. and C. E. Boyd. 1978. Consequences of periodic application of copper sulfate and simazine for phytoplankton control in catfish ponds. Transactions of the American Fisheries Society 107: 316-320.

Verheyen, E., R. Blust, and C. Doumen. 1985. The oxygen uptake of *Sarotherodon niloticus* L. and the oxygen binding properties of its blood and hemolysate (Pisces: Cichlidae). Comparative Biochemistry and Physiology 81A: 423-426.

Warren, C. E. 1971. Biology and Water Pollution Control. W. B. Saunders, Philadelphia, Pennsylvania.

Wolfe, J. R., R. D. Zweig, and D. B. Engstrom. 1986. Computer simulation model of the solar-algae pond ecosystem. Ecological Modelling 34: 1-59.

REVIEW AND DEVELOPMENT OF AQUACULTURE MODELS FOR PREDICTING SOLUTE FLUX AT THE SEDIMENT-WATER INTERFACE

John A. Colman and Andrew R. Jacobson

ABSTRACT

A review of processes affecting solute flux at the sediment-water interface indicates that mixing likely would be dominated by direct wind stirring and by the swimming or burrowing of cultured organisms. Reactions in the sediment are primarily anaerobic (chemical and biological), typical of sediments in the natural environment that have high rates of organic-carbon input. The net effect of processing of settled material in the sediments contributes substantially to whole-pond budgets of elements such as oxygen, phosphorus, sulfur, and iron.

Existing aquaculture models make predictions at the sediment-water interface of oxygen flux primarily, although some models include phosphorus and nitrogen. The models commonly contain nonformal treatments of sediment-water interactions—for example, remineralization, and oxygen consumption of dead algae are modeled with first-order rate processes without distinguishing between the processes occurring on the sediment particles and from those occuring in the water column.

Selected nonaquaculture models that include sediment-water interactions, such as empirical lake eutrophication models, have been adapted for use in aquaculture but have uncertain predictive capabilities. Mechanistic models are available that could also be used in aquaculture, such as box models of lake sediment with mechanistic treatments of sediment chemistry, for nutrient release and sediment oxygen demand; and mechanistic models for sediment diagenesis, including terms for bioturbation, molecular diffusion, burial, and reaction kinetics, which were developed for marine and freshwater sedimentology.

An example mechanistic model development for calculating iron and sulfide cycling in *Clarias* catfish culture indicates how this type of model could be used for predictions that are location-specific, incorporate the effects of sediment chemistry, and vary with time and with changing conditions in the sediment. Although the example model indicated useful management strategies that could be applied, experimental data for verification of the predictions were lacking.

INTRODUCTION

Solutes released or absorbed by bed sediment during diagenesis (the sum of transformations of the sediments after deposition) can profoundly alter the quality of water in aquaculture ponds. For example, a large percentage of the nutrients required for plankton growth are released from organic matter during biochemical degradation at the sediment-water interface. Phytoplankton, in turn, are directly responsible for diurnal changes in dissolved-oxygen concentration, prediction of which is the objective of many aquaculture water-quality models. The sediments also directly influence water quality by release of toxic metabolites, chiefly hydrogen sulfide and ammonia, and by absorption of oxygen.

Despite the importance of sediment, few aquaculture models include the effects of sediment on water quality. Those that do are empirical and are applicable only to the particular culture conditions for which they were calibrated. They are less useful for modeling conditions that might result from experimental or altered management practices.

This paper reviews processes that control solute flux, and models that have been or could be used to predict fluxes of solutes moving between the sediment and water in aquaculture ponds. The paper includes an example application of a mechanistic, time-varying model to the prediction of iron release and interaction with sulfide in *Clarias* catfish culture to indicate how the predictive capability of existing aquaculture models might be improved.

THE SEDIMENT-WATER INTERFACE—DESCRIPTION AND PROCESSES

At the sediment-water interface, a more or less well-defined physical discontinuity is present between the deposited sediment and the overlying water. Few descriptions of the interface exist for fish ponds. Examination of undisturbed sediment cores collected from intensive, manure-fed aquaculture ponds in Israel (Shilo and Rimon 1982) indicated that a suspended-sediment layer divided the clearer water column above from a firmer sediment surface beneath. Shilo and Rimon (1982) found that the suspended layer was more active than the lower deposited layer in releasing ammonia to the water column. It is not known how common suspended layers are, or how their physical and chemical characteristics may vary over time and culture method.

The physical discontinuity brings about discontinuities in water flow and chemical gradients as well. In the absence of sediment mixing by organisms, currents, or waves, solute transport within the sediment occurs by molecular diffusion in the pore water (Berner 1980). Molecular diffusion is the result of the velocities associated with individual ions and water molecules, as opposed to whole water parcels that move about under turbulent conditions. These velocities, and hence molecular diffusion, are a function of temperature. Modifications of diffusion equations are required for modeling molecular diffusive transport within the sediments, the principal change being an incorporation of the effects of tortuosity (interruptions by sediment particles of a solute ion's straight-line path) into the diffusion coefficients (Berner 1980). Diffusion coefficients that have been corrected for tortuosity of the sediments are lower than the corresponding coefficients for water. Above the sediments, where turbulent flow is possible, mixing is several orders of magnitude faster than for molecular diffusion (Wimbush 1974), even for the most quiescent water bodies (Colman and Armstrong 1987).

The boundary between these widely differing mixing rates is at or near the sediment-water interface. Under relatively quiescent conditions, the molecular-diffusion mixing rate can extend a millimeter or two into the water column (Wimbush 1974; Jørgensen and Revsbech 1985). However, if the water column is vigorously stirred by wind or wave action, turbulent mixing can penetrate below the sediment surface by resuspension of deposited material (Heathershaw 1974; Vanderborght and Wollast 1976). Mixing rates increase with distance above the interface because larger-scale turbulence is possible.

Mixing at the sediment-water interface caused by organisms (bioturbation) is likely to be an important factor in aquaculture ponds. Although stirring or burrowing benthic organisms may be nearly eliminated from the sediment surface by foraging cultured species, the cultured species themselves may stir the sediments by excursions into the mud or simply from swimming near the interface. Solute exchange rates between the sediments and water may be limited by mixing rates, when mixing is at the molecular level, or by the production rate of solutes resulting from biological or chemical reactions in the sediments, when mixing is faster, such as in bioturbated sediments (Vanderborght et al. 1977a,b).

Bacterial degradation of sedimented organic matter is the principal driving force behind chemically and biologically mediated reactions of the bed sediments (Blackburn 1987). Oxygen is consumed during the degradation. Because of the abundance of organic carbon in the aquaculture pond sediment, oxygen would likely penetrate into the sediment 1 to 2 mm at most (Blackburn 1987). Thus anaerobic conditions and anaerobic microbial processes would dominate (Schroeder 1987).

There is a vertical stratification in the sediments of types of bacteriologically mediated carbon mineralization processes. The stratification depends on the available "electron acceptors," which are, in order of increasing depth, oxygen, nitrate, sulfate, and carbon dioxide (Berner 1980; Stumm and Morgan 1981; Blackburn 1987). In freshwater culture ponds that are well supplied with organic matter, putrefaction and ammonification of organic matter may be more important sources of hydrogen sulfide and ammonia production than reduction of electron acceptors as above (Fry 1987).

Other important reactions that occur in sediment are reduction and oxidation of iron and manganese (Robbins and Callender 1975; Davison and Heaney 1978), sorption and ion- exchange reactions between dissolved ions and solid-phase materials (refer to Li et al. 1972 for phosphate; Diab and

Shilo 1986 for ammonia), and precipitation and dissolution (Stumm and Morgan 1981).

There are too many sediment, chemical, and biological processes possible for all to be included in one model. Rather, a few key reactions which act to control pore-water solute concentrations must be identified. Types of information used to assess controlling processes are chemical phase diagrams, reaction rates, equilibrium constants, sorption isotherms, and the ecology of biologically mediated reactions in the sediments. Excellent reviews of sediment chemical processes (Berner 1980; Stumm and Morgan 1981; Davison 1985) and biological processes (Berner 1977; Anderson 1987; Blackburn 1987; Gächter et al. 1988) are available. Especially useful in these references are case-study examples that describe a sediment environment and indicate which reactions were considered important for the diagenesis of a given element or compound.

REVIEW OF MODELS

Aquaculture Models of the Sediment-Water Interface

No published aquaculture pond models include distinct treatments of sediment-water interactions. Most models, however, include terms that at least partially account for sediment-oxygen interactions occurring at the interface; some models include terms that partly account for nitrogen and phosphorus interactions.

Oxygen concentration, because of its importance for fish respiration and the close correspondence between observed fish distress and oxygen concentration for many species, is the water-quality parameter most commonly calculated in aquaculture models. Models of oxygen exchange at the sediment-water interface must predict the rate of oxygen transport from the water across the interface to the sediments, where it is used for respiration by benthic fauna and bacteria, or consumed by the chemical oxidation (generally biologically mediated) of sediment. The rate of oxygen transfer across the interface can be called the sediment oxygen demand (SOD).

The simplest model of oxygen consumption by sediments employs an empirically derived, constant rate for SOD (Boyd et al. 1978). The constant is the intercept of a linear regression of oxygen loss (less reaeration and fish respiration) against standing crop of phytoplankton. A constant (zero-order) rate process is not influenced, as through feedback loops (Wolfe et al. 1986), by changing conditions in the sediment or water. First used to predict nighttime water-column oxygen loss in channel-catfish culture ponds (Boyd et al., 1978), the constant SOD approach was later employed in a daytime model (Romaire and Boyd 1979) and in a model for predicting nighttime oxygen loss in tilapia ponds (Romaire et al. 1978). In the latter case, the constant SOD value could be considered amended for manure-fertilized systems in that oxygen consumption by manures (of order 30 mg O_2 m^{-2} h^{-1}) was treated as a separate term (Romaire et al. 1978). But to the extent that manures settle to the bottom, their oxygen use could be considered a sediment demand.

Actual pond SOD values are unlikely to be constant. For example, sediment oxygen demand would likely decrease as water-column oxygen decreases, especially at concentrations below 2 mg L^{-1} (Walker and Snodgrass 1986), and certainly becomes zero in the limit of the absence of oxygen. It is perhaps because the SOD is small, 60 mg O_2 m^{-2} h^{-1}, in comparison to the diel oxygen changes caused by phytoplankton, 360 mg O_2 m^{-2} h^{-1}, in channel-catfish ponds (Boyd et al. 1978), that oxygen models incorporating the constant-rate approximation have made accurate and useful predictions of whole-pond respiration. In addition, because plankton respiration rate is empirically derived, there may be a portion of oxygen use modeled as consumption by suspended algae that is in fact consumption by settled algae. In this way the "constant" SOD models would have, in effect, a component that varies depending on the standing crop of algae.

The constant SOD assumption has been used in general pond models by Smith and Piedrahita (1988), Chang et al. (1986), and in a *Clarias*-culture model by Diana et al. (1988). In other cases it has been found inadequate, particularly when a location-specific or time-varying demand is required (Meyer et al. 1983; Madenjian et al. 1987a). The effect of sediment SOD on the water column immediately adjacent to the sediments is increased if mixing with upper water is cut off by density stabilization of the bottom water through thermal stratification (Meyer et al. 1983). In these situations it may be more important to have a time-varying and location-specific estimation of SOD. Water conditions just above the interface would be particularly important for bottom-feeding organisms such as shrimp.

In Hawaiian prawn farms, SOD has been found to be variable and much greater than that estimated by Boyd et al. (1978), ranging from 0.136 to 0.748 mg O_2 L^{-1} h^{-1}, or assuming a 0.75 m pond depth, 102 to 561 mg O_2 m^{-2} h^{-1} (Madenjian et al. 1987a). It is unresolved whether the larger SOD in prawn ponds can be attributed to differences in measurement method (SOD was computed using the difference between whole-pond respiration and reaeration, and prawn and water column respiration) or to a feature of the culture system, such as increased mixing of the surface sediments by the prawns (bioturbation). Because SOD was a large part of whole-pond respiration in prawn culture (averaging 50%), Madenjian et al. (1987b) developed a method for calculating nighttime oxygen depletion based on surface reaeration, temperature, and measured evening oxygen depletion.

Nonconstant formulations of SOD also have been incorporated into aquaculture pond models. Modeling SOD as proportional to pond oxygen concentration (first-order kinetics) was suggested by Piedrahita (1988). The proportional formulation is supported by the results of Olah et al. (1978), who found pond respiration rate in polyculture carp ponds was related to oxygen concentration. Several models have assumed that pond oxygen demand is proportional to the amount of oxygen-demanding organic matter that has been generated in the pond (Meyer and Brune 1982; Svirezhev et al. 1984; Wolfe et al. 1986). Because these do not differentiate between use of oxygen by settled and by suspended organic matter, modeling the effects of localized depletion adjacent to the interface is not possible with this method.

The model by Wolfe et al. (1986), calibrated on a highly managed tank system rather than ponds, contains relatively sophisticated kinetics and is well verified for oxygen. Their model assumes the organic matter degradation rate is proportional to the concentration of organic particles and to the population of decomposing bacteria, and that a decomposition rate coefficient of 0.02 d^{-1} applies. In addition, a nonlinear oxygen factor is used which assumes that aerobic degradation drops off sharply when oxygen concentration falls below 2 mg L^{-1}. This parameterization of oxygen consumption in the water column could be useful for the sediment-water interface if combined with the mathematical structure for mixing and burial of the sediments as described in the mechanistic models section below.

Few water-column aquaculture models consider water-quality parameters in addition to oxygen. Nitrogen and phosphorus were included in those models above that kept track of the degradation of suspended organic matter. By use of carbon:nutrient ratios, the organic-matter degradation could simultaneously account for oxygen consumption and nutrient release. As was the case for oxygen, evaluation of these processes at the interface was not explicitly treated in these models.

Empirical Lake Models

The review of sediment-water interaction models can be extended beyond the few aquaculture examples available if limnological and oceanographic models are also considered. Empirical lake models of phosphorus and trophic state are well researched (Oglesby and Schaffner 1978; Schindler 1978; Stauffer 1985) and have been used in aquaculture to predict the effects of cage culture on phosphorus cycling in lakes and reservoirs (Beveridge 1984; Phillips 1985).

As reviewed by Stauffer (1985), the models have the general form—

$$TP_{lake} = a + b\frac{L}{X},$$

where TP_{lake} = the average total phosphorus concentration in the water column;

a = the fitted intercept coefficient;

b = the fitted slope coefficient;

L = the areal phosphorus loading; and

X = a function, depending on model, of q_s, the hydraulic flushing parameter (Reckhow and Simpson 1980); of τ_w, the hydraulic residence time, and q_s (Vollenweider 1976); or of h_{mix}, the mixed layer depth in midsummer (Oglesby and Schaffner 1978).

The models include implicit formulations for sediment-water interaction. The implication of the constant b is that part of the incoming phosphorus is permanently removed to the sediments and that the fraction is constant, on an areal basis, from lake to lake. The divisor term can account both for the export of phosphorus through the reservoir outflow and for the influence of lake morphology and hydrology on phosphorus sedimentation rate.

It is unlikely that empirical lake models could be applied to the relatively shallow ponds commonly used in aquaculture without recalibration. Phillips (1985) and Beveridge (1984), however, have described the application of these types of models in predicting increases in water-column phosphorus concentrations resulting from culturing trout in cages in lakes. Without a large data base to verify predictions, these authors advise caution in applying their results. One difficulty considered by Phillips (1985) is the high sedimentation rate of phosphorus-containing materials (uneaten food, fish feces, scales, and so forth) found localized under the cages. Self burial of the sediments under the cages, for instance, may slow the diffusion of remineralized phosphorus into the overlying water to an extent not accounted for in models calibrated for more uniform sediment deposition conditions.

Another general criticism of these models also concerns sediment-water interactions. Based solely on hydrographic and morphologic parameters, the models do not take into account the geochemistries of the sediments that would determine their ability to bind phosphorus. Stauffer (1985) argues that this omission accounts for the poor predictive capabilities of the models when applied to lake groups outside those on which they were calibrated. In particular, he refers to investigators (such as Williams et al. 1971a,b) who have found that phosphorus binding in sediments decreases with decreasing available iron content. The combination of iron in a solid sulfide phase can decrease the amount of available iron and, hence, phosphorus binding.

Dissolved phosphorus budgets of ponds depend largely on the degree to which phosphorus is remineralized from solids in the sediments, unlike ammonia, which is excreted in dissolved form from the gills of fish. Because phosphorus is probably limiting to primary production in culture ponds, accurate interface models would be required to model phosphorus and net productivity successfully on a long-term basis. Ponds dug in differing soil types may have widely different availabilities of iron. Furthermore, the ratio of iron to phosphorus at the interface would likely change during a culture grow-out cycle because of inputs of relatively high-phosphorus, low-iron feeds. It appears that accurate interface models would require terms that account for the influence of iron chemistry on sediment binding of phosphorus. Mechanistic models (described in the following two sections) provide a mathematical framework that could include accounting for sediment phosphorus chemistry.

Mechanistic Models of Diagenesis

Mechanistic models are detailed mathematical descriptions of processes that attempt to quantify all physical, chemical, and biological aspects of a modeling problem. Within the range of predominantly mechanistic to predominantly empirical models, the sediment-interaction components of the aquaculture and lake models discussed previously fall at the empirical extreme. However, several lake models of sediment-water interaction have been developed recently, that include mechanistic features. These will briefly be described, and followed by a description of a general mechanistic approach used to model sediment diagenesis and solute flux.

The lake mechanistic models are box models that divide the sediments into an oxygenated, faster-mixed, upper zone and an anaerobic, more slowly mixed lower zone. One model concerns phosphorus release from the organic matter in the lower, reduced-sediment zone and its upward diffusion and partial adsorption by iron oxides in the upper, oxic zone (Baccini 1985). The rates of the chemical reactions in the model were calibrated by measurements from laboratory-incubated sediment cores. Transport of dissolved species from the upper zone to the water column was assumed to be controlled by bioturbation and was evaluated empirically.

Several lake models with box-model treatments of the sediments are available for SOD. In one by Walker and Snodgrass (1986), the box-model formalism was used primarily to select those parameters of laboratory-incubated cores that could be most closely correlated to measured oxygen loss from the overlying water. Only the flux of oxygen was considered; movement of oxygen-consuming species was not. Consumption of oxygen was hypothesized to be the sum of aerobic degradation of organic matter in the surficial sediments, (assumed to be limited by a Monod-type reaction rate) and "chemical" oxygen use in a deeper anaerobic layer. The rate of chemical use was assumed to be limited by the diffusion of oxygen through the aerobic sediments to the lower layer. The two types of oxygen use were calibrated for every sediment that was modeled, by determination of the Monod rate constant, for the aerobic oxygen use, and diffusive transport coefficient, for the chemical use. In general, the resulting model, a function of temperature and oxygen concentration, agreed with in situ measurements made in Hamilton Harbour, on Lake Ontario, Canada, but underpredicted observations when oxygen concentrations exceeded 4 mg L^{-1}. The model remains largely empirical, although mechanistic considerations were used to chose the model parameters. The model can not respond to changes other than temperature and oxygen, such as the settling rate of organic matter.

A second model consists of the SOD part of a lake mechanistic model by DiToro and Connolly (1980), described in Snodgrass (1987). In this model, rates of formation of oxygen-consuming chemical species are computed in terms of reduced, soluble carbon equivalents, or in the case of nitrogen, of ammonia. It was assumed that oxygen demand is caused by the immediate oxidation of the reduced species as they diffuse across the interface. The model is more mechanistic than the model of Walker and Snodgrass (1986) in that rate constants were established for the generation of oxygen-consuming materials from several types of sediment constituents. The model also can respond to a variable sedimentation rate. Diffusion of the reduced materials from the upper box across the sediment-water interface was modeled by the product of an assumed diffusivity coefficient and a solute gradient, based on the average concentration in the box and the box depth.

The DiToro and Connolly (1980) sediment box model has some similarity to mechanistic models that have been developed to investigate sediment diagenesis (for example, Robbins and Callender 1975). Both account for sedimentation rate and model the generation or consumption of solutes using sediment chemical-reaction rates. The diagenetic models, however, attempt to model the vertical gradients of sediment solutes by analytical or numerical solution of depth- and time-dependent equations for solutes in the sediments.

In the case of numerical solutions, this amounts to a box model with hundreds of boxes rather than just two. Diagenetic mechanistic models are by definition concerned with processes in the sediments, but can be adapted for calculating rates of release or uptake of solutes across the sediment-water interface, since the rate of reaction in the sediments and/or transport within the sediments are the processes that control or limit transport across the interface.

The time scales of oceanographic sediment processes are such that processes in the surface sediments can be modeled using steady-state approximations. The approximate equations can often be solved analytically. In aquaculture, the approximation is less appropriate because of changing feeding rates and fish size during the grow-out period as well as sediment changes, such as desiccation, between culture cycles. An alternative for aquaculture would be time-varying simulations obtained by numerical computer solutions of the appropriate diagenetic equation.

The mathematical formulations for diagenesis and transport have been summarized by Berner (1980). Simplified forms of the diagenetic equations which Berner derived follow.

For models of the sediments, it is convenient to adopt a coordinate system with the origin at the sediment-water interface and distance, x, taken as positive downwards. With this convention, the origin moves up as the sediments accumulate on the bottom. The velocity of origin upward movement is incorporated into diagenetic equations to account for the relative advection of the sediment particles with respect to the coordinate system.

For a solid component, the equation describing change in the sediment with time is—

$$\frac{\partial C_s}{\partial t} = -\omega \frac{\partial C_s}{\partial x} + \sum R_s \qquad (1)$$

where C_s = the concentration of a solid component in terms of mass per unit mass of total solids,

$\sum R_s$ = the summed rates of change from all possible reactions of C_s,

ω = sedimentation rate as a velocity of upward movement of the sediment-water interface, and

t = time.

Assumptions for this equation are that the sediment density and porosity do not change with depth, and that mixing of the sediment particles is negligible.

For a dissolved component, the equation describing change in the sediment with time is—

$$(1+K)\frac{\partial C_d}{\partial t} = D_T\frac{\partial^2 C_d}{\partial x^2} - (1+K)\,\omega\frac{\partial C_d}{\partial x} + \Sigma R_d \quad (2)$$

where C_d = concentration of a dissolved material in units of mass per volume of pore water);

R_d = the summed rates of change of C_d resulting from chemical or biological processes;

K = a linear adsorption isotherm constant; and

D_T = the coefficient for the sum of diffusive transport resulting from molecular diffusion, bioturbation, and wave and current stirring.

Assumptions for Eq. 2 are as above, and that equilibrium adsorption of solutes to solid-phase material is present and follows a simple linear relation with proportionality K, and that D_T is constant with depth.

For solute transfer across the sediment-water interface the appropriate equation is—

$$J = -\phi D_T\left[\frac{\partial C_d}{\partial x}\right]_0 \quad (3)$$

where J = the flux from the sediment to the water column, and

ϕ = sediment porosity.

For Eq. 3, it is assumed that advective transport is much smaller than diffusive transport.

To illustrate the reaction terms for Eq. 2, the example of phosphorus diagenesis is considered. Possible reactions for R_d would be R_{biol}, the rate of generation of soluble phosphorus from organic sediment; and R_{precip}, the rate of formation of authigenic minerals, such as apatite ($Ca_5(PO_4)_3(OH,F,Cl)$) or vivianite ($Fe_3(PO_4)_2 \cdot 8H_2O$), which would be subtracted.

In Eq. 2, absorption reactions are separated out and accounted for by the term (1 + K). If it were assumed that the generation rate of soluble phosphorus depended only on the concentration of the particulate phase, Eq. 1 could be solved first to find the profile of particulate phosphorus in the sediments that would be substituted into Eq. 2 in the appropriate ΣR_d term.

An Example Mechanistic Model Development

A complete application of the mechanistic model in an aquacultural setting is not possible because of a lack of detailed information required on sedimentation rates, composition of settling material, and reaction rates in the sediment. To illustrate important features of the development and use of a mechanistic model for predicting and manipulating sediment-water interactions in aquaculture, a more limited problem, the potential for sulfide buildup, toxicity and interaction with iron in *Clarias* catfish culture in Thailand, is considered. In this case, ΣR_d is known from a study of flooded soils in rice culture. Mass-balance calculations indicate that significant amounts of sulfide are released from uneaten food and fish feces in *Clarias* ponds. Processes regulating sulfide buildup include precipitation with iron that is released from the sediments. The model accounts for sulfur input from feed and attempts to predict the rate of iron release from the sediments.

Although the model incorporates realistic parameter values and conditions, it has not been verified with field data. The purpose of the following discussion is to illustrate model development rather than to predict actual pond conditions.

Clarias Culture Description

The culture of *Clarias batrachus* catfish in Thailand is an aquacultural extreme as far as water quality is concerned. Fish stocking densities (50-200 individuals m^{-2}) and feeding rates (7 to 50 kg trash fish m^{-2} $crop^{-1}$) (Panayotou et al. 1982) are as high as can be found in static-water pond culture. Yields are correspondingly high, up to 9.7 kg m^{-2} y^{-1} (Bardach et al. 1972).

Tropical heat and lack of precipitation during the northeast monsoon season stress the pond environment and cause a relatively low concentration of oxygen, even at saturation (7.6 mg L^{-1} at 30 °C), and little available surface water for exchange. *Clarias batrachus* are equipped to survive in these conditions by having accessory respiratory organs so that they can breathe air at the pond surface (Munshi 1962). Observations of "zero" or undetected dissolved oxygen are common in ponds containing fish well advanced in their 6-month growing season (fig. 1).

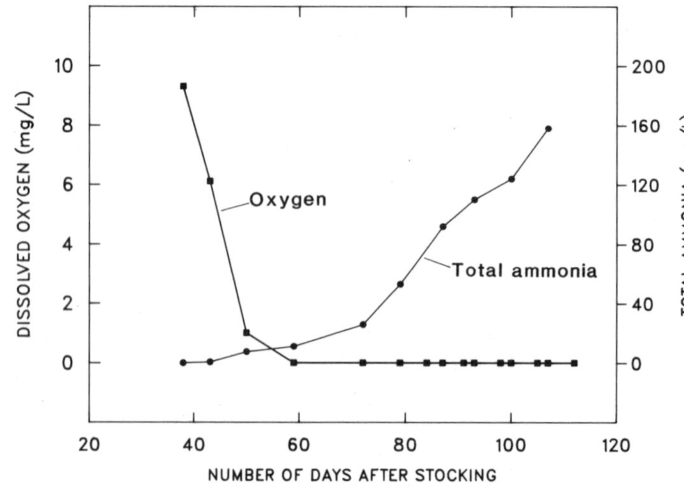

Figure 1. Example oxygen and ammonia concentrations for *Clarias* ponds as a function of time after stocking -- from culture ponds of the Asian Institute of Technology, Bangkok, Thailand.

Oxygen concentration in the water column is not by itself a good predicter of mortality or yield in *Clarias* culture (Diana et al. 1988). It is intriguing to consider consequences of further water quality degradation. Ammonia may build up without oxygen present for conversion to nitrate (fig. 1), but relatively low pH in the culture system, approximately 7.4 for the pond of Figure 1, would allow only 2% of the total, at 30 C, to remain in the toxic un-ionized form.

Also usual in the absence of oxygen is the buildup of sulfide. For sulfide, which forms a negative ion, a decrease in pH would increase the fraction of the un-ionized form (at pH 7.4, approximately 27%). Un-ionized sulfide is highly toxic to fish. Sulfide build-up in organic-rich aquatic systems after oxygen depletion depends on the supply of sulfur to the system as compared to the supply of reduced iron (Berner 1980). Reduced iron and sulfide readily form an insoluble precipitate (FeS) in hypoxic freshwater systems, which removes whichever ion is in shorter supply to a level in accordance with the solubility product of the precipitate (Davison and Heaney 1978). The model evaluates this "titration" between iron and sulfide by accounting for sulfur added in fish feed and computing the iron released from the sediments.

Model Development

The principal effort in developing a mechanistic model of sediment-water interaction is in establishing the proper values of the mixing- and reaction-rate parameters for Eq. 2. Subsequently, Eq. 2 may be solved to yield C_d (here the concentration of soluble iron) as a function of time and depth in the sediment, and the gradient $(\partial C_d/\partial x)_o$, which is used in Eq. 3 to calculate the flux (of iron) across the sediment-water interface.

As reviewed above, mixing rates and reaction rates are uncertain and variable in the bioturbated zone of the interface. Because of high stocking density in *Clarias* ponds, it is likely that fish movement will disturb the sediments at the interface. Without experimental observation as to the penetration of disturbance into the sediments or rate of mixing induced, it has been assumed that the mixing coefficient D_T increases by two orders of magnitude in the top 3.5 cm of sediments as compared to the strictly molecular mixing rate which exists below this region. These values, assumed by Vanderborght et al. (1977a) in an investigation of marine sediments stirred by wind and wave mixing, serve in this aquaculture application only until measured values are available to replace them.

Sulfur input was assumed to be dominated by the trash-fish feed commonly used in *Clarias* culture. The rate of feed input was determined from observations of farm practices by Srisuwantach et al. (1981). A sulfur input of 0.16% of feed input was estimated from the assumptions that 20% of the feed would be incorporated into fish (Srisuwantach et al. 1981) and that the feed was 0.2% sulfur (Ockerman 1978).

Availability of dissolved iron (ΣR_d from Eq. 2) depends on release of iron from the pond sediments and, in turn, on the soil type in which a pond is dug. In this case study, the acid-sulfate soils in the Chao Phraya River delta north of Bangkok, Thailand are considered. Data on dissolved ferrous iron production and diffusion in these soils have been published by Harmsen and Van Breemen (1975) and Van Breemen and Harmsen (1975), which are the sources of the following information.

The surficial soils are typically noncalcareous, dark-gray, clayey sediments containing 1 to 2.5% sulfur as pyrite (FeS_2) and 8% organic carbon. Because of alternate flooding and drying of these soils with the yearly monsoon seasons, aerobic and anaerobic conditions vary in the upper 160 cm of soil. Oxidation of pyrite during the dry, aerated period results in iron oxide and the characteristic yellow jarosite ($KFe_3(SO_4)_2(OH)_6$), which can become available as soluble ferrous iron, after reduction during subsequent flooding. The rate of ferrous iron production depends on depth in the soil and on time after flooding.

In this example, it is assumed that conditions in the pond sediment after pond filling correspond to soil conditions after monsoon flooding, with the pond sediment-water interface corresponding to the soil profile beginning at 120 cm below the soil surface. (Refer to iron concentration in sediment at zero elapsed time, Fig. 2, and the time and depth functional dependence of soluble iron generation as in Table 1, ΣR_d.)

Results obtained for the applications described above follow numerical solutions of Eq. 2. The method of solution was by second-order, forward-time, centered-space, finite-difference approximation (Lapidus and Pinder 1982, pp. 155-156). Solutions were obtained using a popular spread-sheet program on a microcomputer. Model conditions, cell size of 0.1 cm depth, time step of 50 seconds, sedimentation rate of 2 cm y^{-1}, and $D_T = 10^{-6}$-10^{-4} cm^2 s^{-1} (refer to table 1), were tested by comparing analytical and numerical solutions of a steady-state case. Average error in the concentration profile of the numerical solution for a 2-month simulation was 0.24%.

The simulated profiles of dissolved-iron concentration in sediment at successive days after pond filling are shown in Figure 2. The large

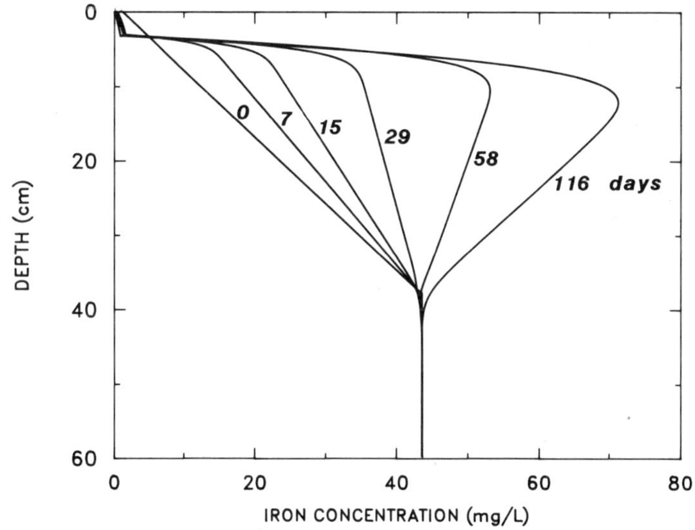

Figure 2. Predicted ferrous iron concentrations in the pore waters of pond sediments at selected days (indicated in figure) after pond filling.

production of soluble iron in the 5- to 20- cm depth interval reflects the iron-generating potential of the soil at these depths (Harmsen and Van Breemen 1975) and the steep gradient of iron from 5 to 3.5 cm depth. Above 3.5 cm iron was "swept" out into the water column because of the higher mixing coefficient used near the interface.

The predicted fluxes of iron and sulfur into the water column are shown in Figure 3. The iron flux came primarily from the zone above a depth of 3.5 cm in the sediment and decreased with time as this zone was depleted of iron. The sulfide input rate changed as a step function according to approximately weekly adjustments in feed rates made by the pond manager. The cumulative loading of iron and sulfide, assuming quantitative precipitation of FeS in a pond, is shown in Figure 4. For the first 38 days of the culture period, iron is predicted to be in excess. Subsequently, sufficient sulfur entered to remove the incoming iron, and sulfur is predicted to accumulate.

The model results are highly speculative because they are not verified by field measurement. Mixing coefficients and reaction rates actually occur-

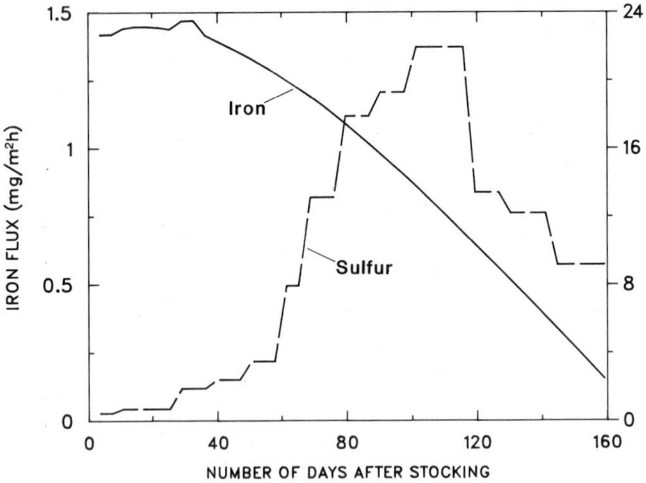

Figure 3. Computed rates of net sulfur input from feed and predicted rate of iron input from sediments to a *Clarias* pond water-column as a function of time after stocking.

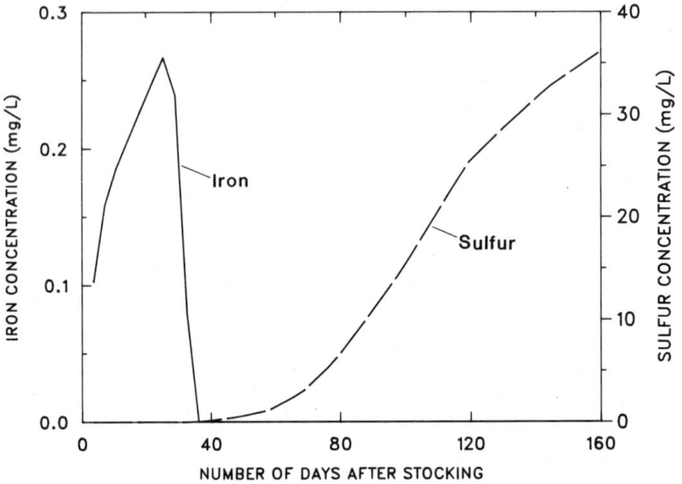

Figure 4. Cumulative loadings of iron and sulfur to the pond water-column as a function of time after stocking assuming quantitative precipitation of FeS.

ring in *Clarias* ponds have not been measured. Furthermore, the only mechanism considered for removal of sulfide was by precipitation with iron; other removal mechanisms, such as escape of hydrogen sulflide to the air or precipitation with manganese, are also possible.

The model, nevertheless, does describe the basic structure of the sulfide-iron interaction and produces results that may be useful for pond management. It is shown, for example, that only the iron released in the bioturbated zone is likely to be available for precipitation during the diffusion time available in one grow-out cycle. The model indicates that the sediment source could be supplemented by iron application at the beginning of the

Table 1. Conditions for simulation of *Clarias* pond

Zone	Model algorithm	Reference
Sediment		
	$\omega = 2$ cm yr^{-1}	(personal observation)
	D_T	(Vanderborght et al. 1977a)
	for x = 0-3.5 cm, $D_T = 10^{-4}$ cm^{2}s^{-1}	
	for x > 3.5 cm, $D_T = 10^{-6}$ cm^{2}s^{-1}	
	R	(Harmsen and Van Breemen 1975)
	for t = 0-150 d,	
	$R = (k_0 - k_1x)(1-Bt)$ mg Fe L^{-1} d^{-1}	
	for t = >150 d, R = 0	
	$k_0 = 1.21$	
	$k_1 = 3.02 \times 10^{-2}$	
	$B = 6.7\ 10^{-3}$	
	initial conditions	(Harmsen and Van Breemen 1975)
	for x = 0-40 cm,	
	$C_0 = 1.11 + 1.12x$ (mg Fe L^{-1})	
	for x > 40-60 cm; $C_0 = 43.6$ (mg Fe L^{-1})	
	boundary conditions	
	x = 60 cm; $C_d = 43.6$ mg Fe L^{-1}	
Interface		
	$\phi = 0.7$	(representative value for near-interface sediments)
Water column		
	sulfur input:	(see text, p. 477)
	S = 0.0016 (feed input)	
	boundary conditions:	
	for x = -100 cm (at the water surface),	
	$D(dC/dx)_{-100} = 0$	

grow-out period; the added iron would likely remain at the sediment surface so that one application should be sufficient for the full grow-out period.

The importance of a time-varying model is emphasized by the results, which indicate that a pond may change from iron sufficiency to insufficiency during one growth cycle. Changes during the growth cycle mean that initial conditions for a second growth cycle would be substantially different from those used in this simulation, for a newly dug pond. If sufficient aeration time were allowed for the iron sulfide precipitate to oxidize between cycles, the iron would be available at the surface in the next grow-out cycle to precipitate additional sulfur. In this case, the sulfuric acid formed from the iron sulfide oxidation could be removed by pond flushing before filling for the next growth cycle.

The iron-sulfide example has shown how a time-varying, mechanistic model of the interface might be constructed. The model structure can be updated as needed using more exact information pertaining to a particular culture situation. Features of the mechanistic model structure, such as the consideration of reaction and mixing variation throughout the sediment column, as well as in time, make possible the accurate predictions of sediment-water interactions that are required in aquaculture. In addition, because the model is based on accurate descriptions of processes occurring at the interface, it may be used after field calibration as a research tool to predict the outcome of experimental changes in management practices.

SUMMARY

The review of processes affecting the sediment-water interface indicate that mixing would likely be dominated by direct wind stirring and by the swimming or burrowing of cultured organisms. As such, it must be measured under field conditions rather than extrapolated from investigations of natural, deeper systems (lakes, oceans). Reactions in the sediment are primarily anaerobic (chemical and biological), typical of sediments in the natural environment that have high rates of organic carbon input. The net effect of processing of settled material in the sediments would contribute substantially to whole-pond budgets of elements such as oxygen, phosphorus, sulfur, and iron.

Existing aquaculture models make predictions of solute flux at the sediment-water interface for oxygen, primarily, although some include phosphorus and nitrogen. The models commonly contain nonformal treatments of sediment-water interactions— for example, remineralization and oxygen consumption of dead algae are modeled with first-order rate processes without distinguishing between the processes on the sediments and in the water column. Thus, they cannot predict location-specific effects of sediment (such as oxygen removal), which may be greater for water adjacent to the interface than water farther away. The models, which are empirical, can accurately predict average concentrations for management practices and culture systems on which they were calibrated, but may fail in predicting responses to changes in management practices, and cannot be transferred to different culture systems. Some of the most-used models use zero-order kinetics for sediment processes, which are time invariant. Because the models do not include the effects of sediment composition, their prediction of nutrient flux, especially phosphorus, which is influenced by iron chemistry, are unlikely to be accurate.

Selected non-aquaculture models that include sediment-water interactions such as empirical lake eutrophication models have been adapted for use in aquaculture, but have uncertain predictive capabilities. These could be supplemented by mechanistic models, also available but not yet used in aquaculture. The latter include box models of lake sediment with mechanistic treatments of sediment chemistry, for nutrient release and sediment oxygen demand; and mechanistic models for sediment diagenesis, including terms for bioturbation, molecular diffusion, burial, and reaction kinetics, which were developed for marine and freshwater sedimentology.

An example mechanistic model development for calculating iron and sulfide cycling in *Clarias* catfish culture indicated how this type of model could be used to predict and exploit changes at the sediment-water interface resulting from management techniques in aquaculture. Predictions were location-specific, incorporated the effects of sediment chemistry, and varied with time and with changing conditions in the sediment. Although the example model indicates useful management manipulations that could be applied, experimental data for verification of the predictions are lacking.

LITERATURE CITED

Anderson, J. M. 1987. Production and decomposition in aquatic ecosystems and implications for aquaculture. Pages 123-147 in D. J. W. Moriarty and R. S. V. Pullin (eds.), Detritus and microbial ecology in aquaculture. International Center for Living Aquatic Resources Management. Manila, Philippines.

Baccini, P. 1985. Phosphate interaction at the sediment-water interface. Pages 189-205 in W. Stumm (ed.), Chemical processes in lakes. Wiley-Interscience, New York, New York, USA.

Bardach, J. E., J. H. Ryther, and W. O. McLarney. 1972. Aquaculture, The Farming and Husbandry of Fresh Water and Marine Organisms. Wiley-Interscience, New York, New York, USA.

Berner, R. A. 1977. Stoichiometric models for nutrient regeneration in anoxic sediments. Limnology and Oceanography 22:781-786.

Berner, R. A. 1980. Early Diagenesis: A Theoretical Approach. Princeton University, Princeton, New Jersey, USA.

Beveridge, M. C. M. 1984. Cage and pen fish farming carrying capacity models and environmental impact. Food and Agriculture Organization, Fisheries Technical Paper 255. Rome, Italy.

Blackburn, T. H. 1987. Role and impact of anaerobic microbial processes in aquatic systems. Pages 32-53 in D. J. W. Moriarty and R. S. V. Pullin (eds.), Detritus and microbial ecology in aquaculature. International Center for Living Aquatic Resources Management. Manila, Philippines.

Boyd, C. E., R. P. Romaire, and E. Johnston. 1978. Predicting early morning dissolved oxygen concentrations in channel catfish ponds. Transactions of the American Fisheries Society. 107:484-492.

Colman, J. A., and D. E. Armstrong. 1987. Vertical eddy diffusivity determined with ^{222}Rn in the benthic boundary layer of ice-covered lakes. Limnology and Oceanography. 32:577-590.

Chang, W. Y. B., J. S. Diana, and W. Chuapoehuk. 1986. Modeling production systems for commonly cultured fishes of Southeast Asia. Pages 269-290 in R. Day, and T. L. Richards (eds) Proceedings of the second international conference on warm water aquaculture—finfish. Brigham Young University Hawaii Campus Office of Continuing Education, Laie, Hawaii, USA.

Davison, W., and S. I. Heaney. 1978. Ferrous iron-sulfide interactions in anoxic hypolimnetic waters. Limnology and Oceanography 23:1194-1200.

Davison, W. 1985. Conceptual models for transport at a redox boundary. Pages 31-53 in W. Stumm (ed.) Chemical processes in lakes. Wiley. New York, New York, USA.

Diab, S., and M. Shilo. 1986. Transformation of nitrogen in sediments of fish ponds in Israel. Bamidgeh 38:67-88.

Diana, J. S., S. L. Kohler, and D. R. Ottey. 1988. A yield model for walking catfish production in aquaculture systems. Aquaculture. 71:23-35.

DiToro, D. M. and J. P. Connolly. 1980. Mathematical models of water quality in large lakes, Part 2—Lake Erie. Environmental Engineering and Science Program Report, Manhattan College, Bronx, New York, USA.

Fry, J. C. 1987. Functional roles of the major groups of bacteria associated with detritus. Pages 83-122 in D. J. W. Moriarty and R. S. V. Pullin (eds.), Detritus and microbial ecology in aquaculature. International Center for Living Aquatic Resources Management. Manila, Philippines.

Gächter, R., J. S. Meyer, and A. Mares. 1988. Contribution of bacteria to release and fixation of phosphorus in lake sediments. Limnology and Oceanography 33:1542-1558.

Harmsen, K., and N. Van Breemen. 1975. Translocation of iron in acid sulfate soils: II. Production and diffusion of dissolved ferrous iron. Soil Science Society of America Proceedings 39:1148-1153.

Heathershaw, A. D. 1974. "Bursting" phenomena in the sea. Nature. 242:394-395.

Jørgensen, B. B., and N. P. Revsbech. 1985. Diffusive boundary layers and the oxygen uptake of sediments and detritus. Limnology and Oceanography 30:111-122.

Lapidus, L., and G. F. Pinder. 1982. Numerical Solution of Partial Differential Equations in Science and Engineering. Wiley , New York, New York, USA.

Li, W. C., D. E. Armstrong, J. D. Williams, R. F. Harris, and J. K. Syers. 1972. Rate and extent of inorganic phosphate exchange in lake sediments. Soil Science Society of America Proceedings 36:279-285.

Madenjian,C. P., G. L. Rogers and A. W. Fast. 1987a. Predicting night time dissolved oxygen loss in prawn ponds of Hawaii: Part I. Evaluation of traditional methods. Aquacultural Engineering 6: 191-208.

Madenjian,C. P., G. L. Rogers and A. W. Fast. 1987b. Predicting night time dissolved oxygen loss in prawn ponds of Hawaii: Part II. A new method. Aquacultural Engineering. 6: 209-225.

Meyer, D. I., and D. E. Brune. 1982. Computer modeling of the diurnal oxygen levels in a stillwater aquaculture pond. Aquacultural Engineering. 1:245-261.

Meyer, D., D. E. Brune, and B. Haight. 1983. Modelling the diel oxygen response of a small fish pond. Journal of the World Mariculture Society 14:336-343.

Munshi, J. S. D. 1962. On the accessory organs of *Clarias batrachus*. Journal of Morphology. 109:115-140

Ockerman, H. W. 1978. Source Book for Food Scientists, vol. 1. AVI, Westport, Connecticut, USA.

Olah, J., A. Zsigri, and A. V. Kintzly. 1978. Primary production estimations in fishponds by the mathematical evaluation of daily O_2 curves. Aquacultura Hungarica (Szarvas). 1:3-14.

Oglesby, R.T., and W. R. Schaffner. 1978. Phosphorus loadings to lakes and some of their responses. Part 2. Regression models of summer phytoplankton standing crops, winter total P, and transparency of New York lakes with known phosphorus loadings. Limnology and Oceanography 23:135-145.

Panayotou, T., S. Wattanutchariya, S. Isvilanonda, and R. Tokrisna. 1982. The economics of catfish farming in central Thailand. International Center for Living Aquatic Resources Management technical reports. Kasetsart University Research and Development Institute. Manila, Philippines.

Piedrahita, R. H. 1988. Introduction to computer modelling of aquaculture pond ecosystems. Aquacultural and Fisheries Management 19:1-12.

Phillips, M. J. 1985. The environmental impact of cage culture on Scottish freshwater lochs. Pamphlet. University of Stirling, Stirling, Scotland.

Reckhow, K. H., and J. T. Simpson. 1980. A procedure using modeling and error analysis for the prediction of lake phosphorus concentrations from land use information. Canadian Journal of Fisheries and Aquatic Science 37:1439-1448.

Robbins, J. A., and E. C. Callender. 1975. Diagenesis of manganese in Lake Michigan sediments. American Journal of Science 275:512-533.

Romaire, R. P., and C. E. Boyd. 1979. Effects of solar radiation on the dynamics of dissolved oxygen in channel catfish ponds. Transactions of the American Fisheries Society 108:473-478.

Romaire, R. P., C. E. Boyd, and W. J. Collis. 1978. Predicting nighttime dissolved oxygen decline in ponds used for tilapia culture. Transactions of the American Fisheries Society 107:804-808.

Schindler, D. W. 1978. Factors regulating phytoplankton production and standing crops in the world's freshwaters. Limnology and Oceanography 23:478-486.

Schroeder G. L. 1987. Carbon pathways in aquatic detrital systems. Pages 217-236 in D. J. W. Moriarty and R. S. V. Pullin (eds.), Detritus and microbial ecology in aquaculture. ICLARM. 420 pp.

Shilo, M., and A. Rimon. 1982. Factors which affect the intensification of fish breeding in Israel, 2. Ammonia transformation in intensive fish ponds. Bamidgeh 34:101-114.

Smith, D. W. and R. H. Piedrahita. 1988. The relationship between phytoplankton and dissolved oxygen in fish ponds. Aquaculture. 68:249-265.

Snodgrass, W. J. 1987. Analysis of models and measurements for sediment oxygen demand in Lake Erie. Journal of Great Lakes Research 13:738-756.

Srisuwantach, V., R. Soungchomphan, and P. Sae-Eng. 1981. Water quality conditions as disease related stressors in Clarias pond culture. Thai Department of Fisheries—United Nations Development Program/ Food and Agriculture Organization/THA/75/012, Bangkok, Thailand.

Stauffer, R. E. 1985. Relationships between phosphorus loading and trophic state in calcareous lakes of southeast Wisconsin. Limnology and Oceanography 30:123-145.

Stumm W., and J. J. Morgan. 1981. Aquatic Chemistry, 2nd ed. Wiley, New York, New York, USA.

Svirezhev, Yu. M., V.P Krysanova, and A. A. Voinov. 1984. Mathematical modelling of a fish pond ecosystem. Ecological Modelling 21:315-337.

Vanderborght, J. P., and R. Wollast. 1976. Mass transfer properties in sediments near the benthic boundary layer. Pages 207-219 in J. C. J. Nihoul (ed.) Bottom Turbulence. Elsevier, New York, New York, USA.

Vanderborght, J. P., R. Wollast, and G. Billen. 1977a. Kinetic models of diagenesis in disturbed sediments. Part 1. Mass transfer properties and silica diagenesis. Limnology and Oceanography 22:787-793.

Vanderborght, J. P., R. Wollast, and G. Billen. 1977b. Kinetic models of diagenesis in disturbed sediments. Part 2. Nitrogen diagenesis. Limnology and Oceanography 22:794- 803.

Van Breemen, N., and K. Harmsen. 1975. Translocation of iron in acid sulfate soils: I. Soil morphology, and the chemistry and mineralogy of iron in a chronosequence of acid sulfate soils. Soil Science Society of America Proceedings 39:1140-1148.

Vollenweider, R. A. 1976. Advances in defining critical loading levels for phosphorus in lake eutrophication. Memorie Istituto Italiano di Idrobiologia 33:53-83.

Walker, R. R. and W. J. Snodgrass. 1986. Model for sediment oxygen demand in lakes. Journal of the Environmental Engineering Division Proceedings ASCE 112:25-43.

Williams, J. D., J. K. Syers, D. E. Armstrong, and R. F. Harris. 1971a. Characterization of inorganic phosphate in noncalcareous lake sediments. Soil Science Society of America Proceedings 35:556-561.

Williams, J. D., J. K. Syers, R. F. Harris, and D. E. Armstrong. 1971b. Fractionation of inorganic phosphate in calcareous lake sediments. Soil Science Society of America Proceedings 35:250-255.

Wimbush, M. 1974. The physics of the benthic boundary layer, Pages 3-10 in I. N. McCave (ed.). The benthic boundary layer. Plenum. New York, New York, USA.

Wolfe, J. R., R. D. Zweig, and D. G. Engstrom. 1986. A computer simulation model of the solar-algae pond system. Ecological Modelling 34:1-59.

NUTRIENT FLOW IN POND AQUACULTURE SYSTEMS

Gerald L. Schroeder, Anat Alkon, and Malik Laher

ABSTRACT

Two fundamentally different systems of pond aquaculture are currently used: the standing water pond and the pond with circulated or stirred water. The ecologies of these two systems are quite different.

The standing water pond must process or store all organic matter that enters. This includes, on a square meter basis, daily additions of 2 to 5 g carbon (C) as manure and/or similar amounts of organic matter in feed, and 3 to 5 g C as net photosynthesis. Of this carbon, 5% to 15% is recovered in fish growth. Another 10% to 20% is oxidized by metabolic activities of the fish. Microbial activity processes the remainder. Reserves of carbonate and mineralized nitrogen in the water column are sufficient for less than 3 days of photosynthesis. Rapid mineralization and resuspension of residual organic matter is essential.

The high fish density used in ponds with circulated water requires daily feed inputs often exceeding 100 g/m^2. The circulated water suspends light residues and concentrates the larger particles at the bottom vortex from where they are drained daily. Suspension of the particulates keeps them in the oxic region of the pond (sediments of both standing and circulated water ponds become anoxic by 1 mm depth). This encourages growth of a pelagic heterotrophic microbial food web without the accumulation of toxic, reduced end-products such as ammonia and sulfides which inhibit fish growth. Approximately 20% of the added feed is recovered in fish growth and another 10% in drained residues. The remainder is mineralized to carbon dioxide by the metabolic processes of the fish (20% to 40%) and the microbial community.

INTRODUCTION

As with most, and perhaps all, forms of animal husbandry, the costs of producing fish for market depend to a large extent on the cost of the supplied feeds. However, unlike most forms of animal husbandry, the feed, as supplied by the farmer to the target fish, is not always the food upon which the fish grows. If not consumed soon after addition to the pond, processing of the supplied feed within the pond may alter its characteristics prior to consumption, with soluble parts of the feed entering the ecosystem of the pond. Such processing may be biological as in the case of microbes populating the uneaten feed or physical as in dissolution of portions of the feed. In either case, the contribution of the supplied feed to the growth of the target fish can be quite different than that which was intended by the pond manager.

Differences might result from the microbes actually upgrading the quality of the feed or from the intensification of the natural food web of the pond by the fertilizing effects of the leached feed stuffs. When confronted with evidence that, in pellet-fed ponds on the coastal plain of Israel, half or more of the common carp growth and 70% or more of the hybrid tilapia growth originated, not with the supplied feed, but rather with the natural foods of the ponds, the then head of the Israel Fish Growers Association, Dan Mires, responded that "it appears that our supplied feeds have been a very expensive form of fertilizer!"

Fertilization of ponds is obviously an integral part of fish farming. As Boyd (1982) points out, it is especially important in areas with infertile watersheds. But to be most cost-effective, the fertilization should be accomplished according to a plan and not be an inadvertent benefit of a less-than-effective feeding program. It is possible that the added fertility of the water might cause problems of effluent pollution that outweigh the benefits toward fish production.

The extent to which a given food or fertilizer is effective in the fertilization of a pond depends upon the existing pond conditions. There are regions with such fertile soils and mineral-rich waters, that fertilization is superfluous. Unfortunately, for most of us active in aquaculture, this is not the case. The added feeds and fertilizers are essential to attain a marketable yield of fish.

To evaluate the potential effect of a program of pond fertilization, it is useful to have a general idea of the sources and sinks of nutrients within the pond. It is the general scheme in the flow of nutients that I discuss herein.

The simplest aquacultue system to consider is the flow-through pond, the raceway. It may also be the most costly for raising fish since the raceway itself makes little or no contribution toward the feeding of the fish. It serves much as a cage or feedlot would serve a terrestrial animal. Aside from the microbial slimes that grow on the walls and bottom, the organic matter that is added to the raceway is rapidly washed out along with excreta and other catabolites.

In the cases of rainbow trout and channel catfish, where research has provided what is believed to be a complete feed, the feedlot approach to aquaculture may be adequate. But not every fish can command the market price that would justify either the high cost of developing a complete feed or the operating expenses involved in supplying a pellet containing greater than 30% protein and all the needed lipids and vitamins. Even after years of effort to develop a balanced feed for the Penaeids, it has been this author's observation that, in raceway culture, between feedings the shrimp actively scour the raceway walls for the small amounts of slime that may be present. Furthermore, shrimp growth rate and survival increase when the raceway is allowed to receive direct sunlight. This latter effect may be due to photosynthetic growth on the walls. Even with the highly truncated ecological system of a raceway, it appears that a flow of nutrients beyond the direct one of pellet-to-fish has an effect on the yield of the cultured animal.

THE ECOSYSTEM OF THE STANDING WATER POND

At the end of the aquaculture spectrum, opposite to the raceway, is the standing water pond. Whereas the raceway acted largely as a well-ventilated cage, the standing water pond is a cage, a pasture and a garbage disposal unit integrated into one complex environment.

The surface of the pond is a two-way interface with volatiles entering or leaving depending upon their relative concentrations. An obvious and extreme loss from the pond across the water/air interface is that of water. With water temperatures of 25 C to 30 C, water loss in clear weather is approximately 0.5 cm per day. That amounts to 50 m^3 of water/ha/day^{-1}, an amount able to irrigate many cash crops.

The abundance of oxygen in the atmosphere might imply a rapid influx of oxygen to water which is depleted of dissolved oxygen. This, unfortunately, is not the case. Flow of gases across the water/air interface is quite slow. This is because the surface layer of the pond, as is typical of most interface layers, is laminar. It does not convectively mix with the water just below it unless vigorously stirred, as for example with a paddle wheel aerator.

Even though the laminar surface layer may be only 100 microns thick (Kanwisher 1963), as this laminar layer becomes saturated with oxygen from the atmosphere, no further influx of oxygen is possible until the very slow process of diffusion through this layer moves the oxygen from the surface to the deeper region where convective water eddies rapidly transport the oxygen away from the surface. The result of this laminar layer is to limit the rate at which oxygen can enter the water to a maximum of about 2 g oxygen/m^2/day. This amounts to an increase of 2 ppm dissolved oxygen in a one meter deep pond.

When the pond is supersaturated, the flow of oxygen is out of the pond to the atmosphere. The rate of flow across the interface is proportional to the differences in concentrations in the air and in the water. Since undersaturation can not exceed 100%, that is zero dissolved oxygen in water, but pond water supersaturation reaching in excess of 300% is not uncommon in fertile waters at the end of a sunny day, there is a greater potential for oxygen loss than for oxygen gain.

What is true for the transfer of oxygen is true for all gases. Ammonia in the un-ionized form diffuses as a gas. At a pH of 8, approximately 7% of the total ammonia in warm water ponds will be un-ionized (Boyd 1982). At this pH in a pond having 0.5 ppm total ammonia in the water, 50 mg ammonia/m^2/day will diffuse from the pond water into the atmosphere. This represents, in a one meter deep pond, a daily loss of 10% of the ammonia in the water column.

Considering the role of photosynthesis in production of natural foods and dissolved oxygen, perhaps the most important natural input across the water/air interface is that of sunlight. Solar radiation reaching a pond surface is the same to within 20% for latitudes within 30 degrees of the equator (von Arx 1962). This implies that the maximum photosynthesis is also similar in these regions, at least for the warm months. Goldman (1979) reported continuous yields of 15 to 20 g algal carbon/m^2/day in high rate outdoor algal cultures. This is similar to the calculated theoretical maximum of approximately 20 g carbon net photosynthesis/m^2/day (Schroeder 1987) and is about 4 times greater than values for net photosynthesis observed in intensely fertilized tropical and sub-tropical fish ponds. The shortfall in the primary productivity actually attained relative to the theoretical maximum value may be due to lack of stirring and hence the autoshading of the lower layers of pond water by the overlying algae (Hepher 1962).

Flux of nutrients across the sediment/water interface is difficult to quantify. The actual diffusive processes follow theoretical laws. Thus the flux of oxygen into undisturbed, submerged sediments at fixed temperatures is quite constant (Schroeder 1975) and varies in a consistent manner with changes in temperature (Hargrave 1969). Maximum uptake of oxygen by sediments from the water, in fresh and salt water at 30 C containing more than a few ppm dissolved oxygen, is 3 g oxygen/m^2/day. The temperature dependency of the uptake rate is such that lowering the temperature to 20 C reduces the dissolved oxygen uptake to about 1.5 g oxygen/m^2/day. This 50% decrease in microbial activity for a 10 C decrease in this temperature range is not uncommon. It is similar to the response in the microbial processing of manure (Schroeder 1974).

The influence of bio-turbulence on nutrient transfer across the sediment/water interface is not constant. Bottom feeders such as common carp, mullet, and hybrids of tilapia feeding in the sediments resuspend considerable amounts of bottom matter. At stocking densities of one to 3 fish/m^2, in one meter deep earthen ponds at the Dor Research Station, pans placed on the bottoms of the ponds collected 200 to 500 g dry matter/m^2/day. Organic matter in this collected material accounted for 3% to 4% of the total dry weight. In these ponds, net primary productivity added 5 g algal carbon or about 10 g algal dry weight/m^2/day. Manuring added organic matter at about that rate. A simple material balance shows that most of the matter collected in the pans was the precipitation of re-suspended bottom matter. There were only gentle breezes during these experiments. Therefore, fish-induced bioturbulence and not wind-induced turbulence was the most likely cause of the high rates of collection in these pans.

The transition from oxic to anoxic conditions at the bottom interface (anoxia is reached within a few millimeters of sediment depth at most and often at less than one mm) causes reduction of the insoluble ferric compounds to the reduced soluble ferrous forms. This liberates nutrients initially carried into the sediments as insoluble ferric compounds. The resuspension of these nutrients, and the subsequent re-enrichment of the water column, results in the increase in yield of pelagic filter feeding silver carp grown in polyculture with bottom feeding fish relative to the silver carp yields when grown in monoculture (Yashouv and Halevi 1972).

Concentrations of mineral forms of nitrogen and phosphorous are usually 10 to 1000 times greater in the first few centimeters of sediment interstitial water than in the overlying water column. This high concentration of chemicals makes possible the intense microbial activity just within the sediment surface relative to the water even close to the bottom. Kirchman (1987) reported that there are roughly a million bacteria/ml of pond water compared to a billion bacteria/gram of sediment in the aquaculture ponds with which he works in Thailand. We find a similar relationship of bacteria concentrations in fish ponds in Israel. Microbial digestion of cellulose at the sediment surface is 2 to 5 times more rapid than in the overlying water. This is true for fresh and salt water.

The water column provides the immediate source of nutrients for algal production. With secchi depths of 30 to 40 cm common, the depth of compensation (where seston respiration equals algal photosynthetic production) is 60 to 70 cm. There is no autotrophic production at the pond bottom in a fertile meter deep pond. This means that essentially all the autotrophic natural foods of the pond are produced within the water column. This production alone can support fish yields in excess of 25 kg/ha/day, averaged over the entire growout season, with algae being harvested live in the water column or after precipitation to, and microbial processing on, the pond bottom.

Most of the organic carbon in the water column is not in the form of particulate matter. In fertilized fish ponds, with total organic carbon concentrations in the order of 100 mg/l, 60% to 70% of this organic carbon passes through a 0.45 micron filter. In lake water, Wetzel (1983) reported that 90% of the organic carbon was in the dissolved form and that most of this organic carbon was in the refractory forms of humic and fulvic acids, not in forms easily available to microbial digestion. The lake water is considerably less rich than that of fertilized fish ponds. The labile forms of organic matter exuded from the abundant algal concentrations of fish ponds, such as amino acids and fatty acids have turnover times measured in hours or less and therefore are present in very low concentrations. An estimated 5% to 10% of the total carbon fixed in primary productivity is released as exudate dissolved organic matter (Carlson and Carlson 1984; Mague et al. 1980).

PONDS WITH CIRCULATED WATER

The ecology of a pond with circulated water (also referred to as stirred water) is dominated by the regular, circular motion of the water. This motion prolongs the suspension of particulate organic matter within the water column and concentrates the precipitated organic matter at the vortex. From here it is drained daily. To justify the cost of construction of a pond with firm walls and bottom and circulated water, the density of fish stocking must be 10 to 30 times greater than the 1 to 3 fish/m^2 stocking density typical of the standing water pond. The increased number of fish implies a proportional increase in added feed and released catabolites.

The gentle stirring of the water, while sufficient to cause an accumulation of precipitated organic matter at the center of the pond, is not sufficient to break the laminar layer at the pond surface. Therefore, the slow process of diffusion (slow relative to the more radid process of convection) limits the rate of transfer across the water/air interface. Amounts of oxygen able to diffuse into the pond during periods when dissolved oxygen concentrations are less than 100% are no greater in the stirred pond than in the standing water pond (about 2 g oxygen/m^2/day). The demands of the fish plus seston respiration exceed the net supply of dissolved oxygen. Aeration is almost always essential in the stirred pond.

Unlike the standing water pond, here there is no long term accumulation of organic matter at the bottom of the pond. This does not, however, preclude anoxia at the pond bottom. The biological oxygen demand (BOD) of the abundant feed residues and excreta is sufficient to reduce the dissolved oxygen in slurries of particles to zero prior to their being drained (Avnimelech et al. 1986). The release of anoxic metabolites occurs, but unlike the standing water pond where these metabolites are in part fixed in the sediments, the firm bottom of the stirred pond does not allow transfer across its interface. Thus, any turbulence at the bottom releases these often toxic substances (ammonia, sufides) directly into the water column. This is one of the reasons for frequent draining of the bottom.

Hepher (1985) reported that the conditions of these pond types, with proper management, encouraged a growth of tilapia hybrid (*Tilapia nilotica* x *T. aurea*) in excess of 2 g/day. This equals or exceeds growth rates in standing water ponds although the density of the fish in standing water ponds is less than 10% that of the stirred pond. It is possible that the organic matter suspended by the circular motion of the water may serve as a rich source of natural foods. Avnimelech et al. (1986) reported that in stirred ponds approximately 70% of the added organic carbon was metabolized to carbon dioxide. This implies a high rate of bacterial activity in the water column and hence an abundant supply of vitamin-rich natural foods. Since in standing water ponds, 80% or more of the growth of this hybrid of tilapia is attributable to the natural food web, the enhancement of the growth of these fish by suspended natural foods is not an unreasonable assumption.

ADDED NUTRIENTS: FEEDS, MANURES, FERTILIZERS

Feeds used in aquaculture range from unprocessed grains to pellets containing 40% protein and a mix of vitamins and lipids. As the density of fish stocking increases, there is the trend to rely on refined feeds to ensure adequate fish growth. The production of a kilogram of live weight fish usually is accompanied by the addition of 2 kg dry feed. It is not correct to say that the fish production "required" this amount of feed since for many fish cultures, half or more of the fish growth results from the natural foods of the pond and not from the feed. Even if the entire 2 kg feed were consumed, the balance is approximately 8 kg dry feed added for each one kg dry weight fish growth. The 7 kg difference between added feed and fish growth enters the pond ecosystem in various forms. With a typical 5,000 kg/ha annual (250 day season) live weight fish yield, the excess feed matter handled by the pond exceeds 8,000 kg/ha.

The added feeds contain 2% to 7% nitrogen (as protein) and in the order of 1% to 2% phosphorous. With this rate of input, the ponds rapidly come to represent highly eutrophic water bodies.

Added manures have lower nutritive value, not only in relation to target fish, but also in relation to microbes. The feed pellets used in Israeli aquaculture have a BOD of approximately 140 g oxygen/kg/24 hours at 30 C. This is the typical July temperature of the pond water in Israel. Poultry and cow manures have BOD values of 20 to 60 g oxygen/kg dry weight/24 hours at 30 C. The lower BOD of the manure is a measure of their lower digestibility by bacteria. Of the dry weight of the maure, 15% to 30% is fiber. This fiber contains 60% to 70% cellulose. Nitrogen content ranges from 1% to 4% of the dry weight. Phosphorous contents are similar. Nitrogen is usually in the reduced form of ammonia compounds. When added to the pond, either as a slurry or field dried, 75% of the organic matter settles to the pond bottom. Since the bottoms of the ponds become anoxic within a millimeter of depth, much of the processing of the manure organic matter releases metabolites in their reduced form. Daily manuring, with no addition of feed or mineral fertilizers, can produce fish yields reaching 35 kg/ha/day averaged over the entire grow-out season. Studies of the distribution of stable carbon isotopes (delta C) indicates that 80% of this yield can be attributed to the enhancement of the natural food webs by the mineral fraction of the manure. The organic matter of the manure contributes only marginally except in the case of very soft (low carbonate) water where the carbon dioxide released in the microbial processing of the manure can contribute to the needs of primary production.

The three most common fetilizers used in aquaculture are lime, ammonium and phosphate rich minerals. Liming is used to adjust the pH of the soil. It releases carbonates into the water and aids in primary production. Ammonia and phosphate enrich both the autotrophic and the heterotrophic food webs. Rates of fertilization depend upon the fertility of the inlet water and the pond soil. With 0.4 g of mineral nitrogen and phosphorous/m^2 added daily, average fish yields of 25 kg/ha/day can be sustained for the entire grow-out season with no added feeds (Ling 1986; Schroeder and Buck 1987).

CARBON, NITROGEN AND PHOSPHOROUS BUDGETS IN STANDING WATER PONDS

Far more carbon, nitrogen and phosphorous are added to the pond in the form of feed and/or fertilizers than are recovered in the fish.

In catfish ponds, if contributions from natural foods are not considered, then recovery of carbon, nitrogen and phosphorous are approximately 25%, 27% and 30%, respectively (Boyd 1985). This high fractional recovery is due in part to the use of floating feed pellets, all of which are consumed by the fish. If the organic matter of the net photosynthesis is considered, the recovery of organic carbon in the form of fish growth is 16%. On Israel's coastal plain, with a polyculture of common carp, silver carp and tilapia hybrid (*T. nilotica* x *T. aurea*) stocked at a total of 20,000 fish/ha, recovery of net primary productivity organic carbon in chemically fertilized ponds (no feed pellets or organic manures added, fish yield 27 kg/ha/day) is 7%. Net primary productivity in these ponds is 5 times greater than that reported by Boyd (1985) for catfish ponds in Alabama.

In pellet-fed, fertilized ponds, the recovery of carbon in fish growth (when net primary productivity is included as a form of added organic matter available to the fish) is 5%. If primary productivity is not included, carbon recovery (based on the carbon of the added feed only) approaches 25%. This is, of course, fiction since most of the target fish growth in these fertilized ponds comes not from the feed pellets but from the natural food web. Nitrogen recovery in these ponds is 10% to 20% of the total nitrogen added as feed, manures and fertilizer. Phosphorous recovery is 10% to 15% of the total amount added in pellet-fed plus fertilized ponds and 15% to 20% in ponds receiving chemical fertilization alone. The higher recovery of nutrients in the catfish ponds relative to the polyculture ponds reflects the better fit of the floating pellets to the demands of the fish.

If all the added nitrogen had accumulated in the water column, the nitrogen concentration at the end of the season would have approached 80 g/

m^2 or in a one meter deep pond, 80 mg/l. Phosphorous concentrations would approach 50 mg/l and organic carbon 800 mg/l. None of these entities approach such high concentrations. In spite of their continual addition in the forms of feed and fertilizers, it appears that an equilibrium is rapidly reached in which the rates at which the pond processes the added organic matter equals the rates at which it is added.

Figure 1 shows representative sources and sinks of organic carbon, nitrogen, phosphorous and dissolved oxygen in a standing water pond receiving feed pellets or daily manuring and chemical fertilization once each two weeks.

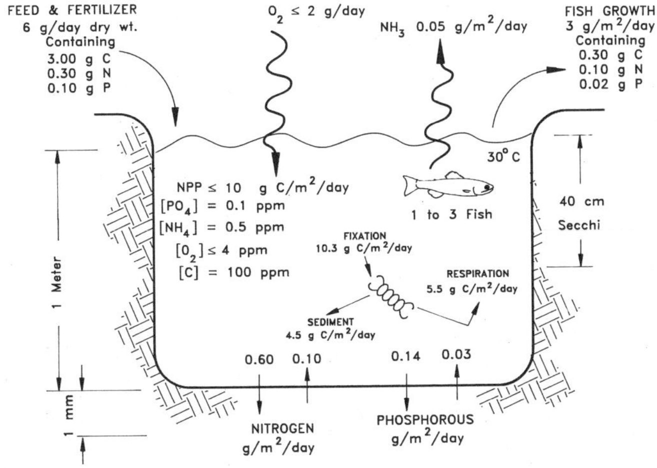

Figure 1. Representative sources and sinks for organic carbon, nitrogen, phosphorous and dissolved oxygen in the water column of a standing water fish pond receiving feed daily and fertilization once per 2 weeks. Fish yield is 30 kg/ha/day or 3 g/sq m/day. Blue green algae when present add not more than 0.05 g nitrogen/sq m/day. Note the importance of bioturbulence in maintaining the nitrogen and phosphorous balance in the water column.

Precipitation of organic matter from the water column to the pond bottom is a key item in the flow of nutrients in a standing water pond. This removes uneaten and unassimilated feeds and feces and algae. Because the bottom becomes anoxic in a millimeter or less, the microbial processing of this precipitated matter releases organic molecules that may be used by sediment microbes or by algae and microbes when stirred back into the water column. The anoxic nature of the bottom keeps iron in the soluble ferrous form and so allows water seepage (about 0.5 cm/day) to carry phosphate as well as ammonia deeper into the sediments.

Only when ammonia concentrations in the water column reach 1 mg/l does loss of ammonia by diffusion to the atmosphere become significant. The dynamic uptake of ammonia by sestonic algae and bacteria usually keeps ammonia concentrations at less than 0.5 mg/l. Only in the event of an ecological upset such as an algal crash or in the fall when water temperatures drop to 20 C and so microbial metabolic rates decrease, do ammonia concentrations exceed 1 mg/l.

The lack of build-up of nitrogen and phosphorous in the water column, relative to the rate at which they are added to the pond is the result of their rapid transfer onto and into the sediments. The lack of build-up of carbon in the water can be attributed to its precipitation to the pond bottom and to its respiration to carbon dioxide both in the water column and on the bottom. Of the total average of 3 to 4 g organic carbon/m^2 added in daily manuring or feeding plus the gross primary productivity (10 g carbon/m^2/day), some 10 g carbon/m^2/day are respired or fermented to carbon dioxide. Approximately one gram of carbon/m^2 is fixed daily in fish growth. The remainder is either fermented within the upper few centimeters of sediments or accumulated in the sediment. If each day for 250 days, one gram of organic matter is accumulated and mixed in sediments over a 10 cm depth, this will only add a total of 0.08% organic matter to this sediment layer. The heterogeneous nature of the sediment will probably make this small accumulation unnoticable except over periods of many years and in that several year period, much of the organic matter would likely be leached or decayed.

In commercial fish ponds receiving feeds and fertilizers, approximately 0.3 g phosphorous/m^2 are added daily. Algae contain in the order of 1% to 2% phosphorous (Eren et al. 1977). A gross daily primary productivity of 10 g carbon or 20 g algal dry weight/m^2 fixes aproximately 0.2 g phosphorous/ m^2/day within algal cells. Half of this precipitates to the sediments. Ferric compounds form insoluble flocculent precipitates. Phosphorous concentrations within the sediments of Israeli fish ponds are in the order of 0.1% to

0.3%. In catfish ponds, values are similar to this. The lack of increase in sediment phosphorous with time during the growing season is attributable to the dilution within and the leaching from the sediments as ferric compounds are reduced to the ferrous form.

The concentrations of adsorbed nitrogen compounds in sediments at the start of a season are not very different from the concentrations at the end of the season. One might have thought that the several month fallow period of the pond prior to the start of the season would have allowed oxidative release of the nitrogen. This is not the case. Drying of the sediment, even in the extreme of a drying oven, only marginally reduces the amount of adsorbed nitrogen. The benefical effect of pond drying on fish growth in the season following the drying appears to be related to compounds other than ammonia.

Although the dynamics of the pond imply rapid processing of nutrients in the water column, we were surprised to observe in 3 experiments (work in progress) that reducing the insolation to the pond by as much as 60% (by artificially shading the pond) for 6 to 8 hours in mid-day did not cause consistent or significant changes in dissolved organic matter or phosphate or ammonia within the water column, even though the shading during these tests reduced the primary productivity by as much as 50%.

CARBON, NITROGEN AND PHOSPHOROUS BUDGETS IN PONDS WITH WATER CIRCULATION

The sources of organic matter and minerals in ponds with stirred water are largely confined to the added feeds. Stocking densities reach 50 fish/m^2. If harvest weights are 600 g/fish, the average fish biomass during the season is 15 kg/m^2. Daily feeding rates are about 2% of the fish biomass and can exceed a rate of 300 g feed/m^2/day. At once it is clear that primary productivity (net approximately 10 g/m^2/day) can contribute only marginal amounts of macronutrients to fish growth. Micronutrients such as vitamins originating with the natural foods may still be significant. Aeration and food must come from without. As such, the stirred pond is a cage, not a pasture.

In hybrid tilapia growth in stirred ponds and in raceways containing penaeids, comparison of the stable carbon isotopes (delta C) of target animal and feed confirm that essentially all growth comes from the supplied feed, little or none from the natural foods. This is in sharp contrast with feeding dynamics in standing water ponds where greater than 80% of the hybrid tilapia growth originates with the natural foods of the pond.

The feeding adds in excess of 130 g carbon, 12 g nitrogen and 3 g phosphorous/m^2/day. For each kilogram of fish (dry weight) produced, 8 kg of feed (also dry weight) are added. Fish growth fixes 20 g carbon, 5 g nitrogen and 0.7 g phosphorous/m^2 daily. Of the added feed, 80% of the carbon, 60% of the nitrogen and 80% of the phosphorous enters the pond water. This loading of excreta and catabolites exceeds the ability of the pond to microbially process these wastes. Water from such ponds is typically exchanged several times daily with water held in a standing water pond having 5 times or more the volume of the stirred pond. Both the stirred pond and the holding pond have central outlets from which accumulated organic wastes are drained daily.

Studies by Avnimelech et al. (1986), in small (3.4 m diameter) stirred ponds showed that as much as 70% of the added organic carbon can be metabolized to carbon dioxide by the fish and the microbial community. This high conversion to carbon dioxide is in accord with the suspension of particulate matter and the subsequent oxic microbial processing within the water column.

TRYING TO IMPROVE THE FISH YIELDS

In aquaculture systems which base fish production on added feeds, increasing water circulation and water change, as in the stirred ponds, results in yields per unit area of pond that are 10 to 100 times greater than yields in standing water ponds. Unfortunately, the water consumption in these stirred ponds increases in accord with the fish yield. The gain is in yield per unit area of pond, not necessarily in yield per unit volume of water used in the production. The loss is that natural foods no longer contribute to the growth and so a diet complete in nutrients must be supplied.

In standing water ponds with no added feeds, intense fertilization plus intense manuring and increased stocking densities have not resulted in fish yields above the 30 to 35 kg/ha/day reported by workers in Asia, Europe and the USA. There appears to be a maximum rate at which the pond can produce usable natural foods. Perhaps an artificial means of resuspending the precipitated organic matter might increase availability of the natural foods, but this increase would be accompanied by the added costs to induce resuspension.

ACKNOWLEDGEMENTS

This work was supported by the GIARRA Fund.

LITERATURE CITED

Avnimelech, Y., B. Weber, B. Hepher, A. Milstein and M. Zorn. 1986. Studies in circulated fish ponds: organic matter recycling and nitrogen transformation. Aquaculture and Fisheries Management 17:231-242.

Boyd, C. 1982. Water quality management for pond fish culture. Elsevier, Amsterdam, The Netherlands.

Boyd, C. 1985. Chemical budgets for channel catfish ponds. Transactions of the American Fisheries Society 114:291-298.

Carlson, D. and M. Carlson. 1984. Reassessment of exudation of fucoid macroalgae. Limnology Oceanography 29:1077-1078.

Eren, Y., T. Tsur and Y. Avnimelech. 1977. Phosphorous fertilization of fishponds in the upper Galilee. Bamidgeh 29:87- 93.

Goldman, J. 1979. Outdoor algal cultures. II. Photosynthetic yield limitations. Water Research 13:119-136.

Hargrave, B. 1969. Similarity of oxygen uptake by benthic communities. Limnology Oceanography 14:801-805.

Hepher, B. 1962. Primary production in fishponds and its application to fertilization exeriments. Limnology Oceanography 7:131-136.

Hepher, B. 1985. Aquaculture intensification under land and water limitations. Geojournal 10.3:253-259.

Kanwisher, J. 1963. On the exchange of gases between the atmosphere and the sea. Deep Sea Research 10:195-207.

Kirchman, D. 1987. Quoted on page 231 in D. J. W. Moriarty and R. S. V. Pullin, editors. Detritus and microbial ecology in aquaculture. ICLARM, Manila, Philippines.

Ling. C. 1986. Preliminary study on culturing fish with the application of chemical fertilizers. Journal of Aquaculture in the Tropics 1:43-48.

Mague, T., E. Friberg, D. Hughes and I. Morris. 1980. Extracellular release of carbon by marine phytoplankyton: a physiological approach. Limnology Oceanography 25:262-279.

Schroeder, G. 1974. The use of fluid cowshed manure in fish ponds. Bamidgeh 26:84-96.

Schroeder, G. 1975. Night-time material balance for oxygen in fish ponds receiving organic wastes. Bamidgeh 27:65-74.

Schroeder, G. 1987. Carbon and nitrogen budgets in manured fish ponds on Israel's coastal plain. Aquaculture 62:259-279.

Schroeder G. and H. Buck. 1987. Estimates of the relative contributions of organic and mineral contents of manure to fish growth. Journal of Aquaculture in the Tropics 2:133-138.

von Arx, W. 1962. Introduction to Physical Oceanography. Addison-Wesley, Reading, Massachusetts, USA.

Wetzel, R. 1983. Limology; 2nd Edition; Saunders College Publishing Company. Philadelphia, Pennsylvania, USA.

Yashouv, A. and A. Halevi. 1972. Experimental studies of polyculture in 1971. Bamidgeh 24:31-39.

A QUANTITATIVE AND QUALITATIVE ASSESSMENT OF WASTES FROM AQUATIC ANIMAL PRODUCTION

M. C. M. Beveridge, M. J. Phillips, and R. M. Clarke

ABSTRACT

This paper describes the origin of the wastes produced in aquatic animal production. It examines a range of culture methods and species and attempts to quantify wastes, particular attention being paid to those which are of greatest environmental significance.

In general, there is good correspondence between laboratory-generated physiological data and field measurements of excretion, and it is possible to formulate mass balance equations for various waste parameters. It remains impossible, however, to predict effluent characteristics with any accuracy. Enormous variations in effluent quality have been recorded and much more work is required in order to explain the roles of system and management. Standardization of sampling and analytical methods and the way in which data are expressed is necessary.

Effluent quality from land-based systems can be improved by treatment, although the cost effectiveness of different treatment options remains to be investigated.

INTRODUCTION

A brief search of dictionaries for a definition of the term "wastes" which might be used in an aquaculture context generated a number of closely-related meanings, describing that which is left over after use or thrown away because it is not water (i.e. the materials used in aquatic animal production which are not removed during harvesting and which may eventually find their way into the environment at large). There have been a number of studies of waste production in aquaculture, although most have been carried out only in the last ten years or so. Part of the reason for this is that the industry until recently was small and employed fairly extensive methods. It has grown and intensified dramatically in recent years, and with this has come waste management problems. Among the first studies of aquaculture wastes were those of Liao and colleagues in the early 1970's (eg. Liao and Mayo 1974). Many of the subsequent studies have been initiated through international organizations such as the International Council for the Exploration of the Seas (ICES) (eg. Rosenthal et al. 1987) and the European Inland Fisheries Advisory Council (EIFAC) of FAO (eg. Alabaster 1982).

This paper sets out to describe and quantify wastes associated with aquaculture, mainly of the intensive and semi-intensive kinds, by reviewing available material. It does not, however, attempt to deal with wastes from the processing or other associated industries (eg. net cleaning). It briefly examines how wastes may be reduced and attempts to draw attention to the importance of developing a sound methodology for evaluating wastes.

QUALITATIVE AND QUANTITATIVE ASSESSMENT OF WASTES

There are three principal sources of aquaculture wastes: uneaten food, excreta, and chemicals and therapeutics. Mortalities and escapes may also be included.

Uneaten foods

During both semi-intensive and intensive aquaculture, food is provided; in the former, as a supplement to available natural food and in the latter, as a more or less complete source of nutrition. In semi-intensive aquaculture, which predominates in the tropics, various combinations of organic and inorganic fertilizers and agricultural by-products such as rice bran are used. Although trash fish is still used in some industries, most intensive aquaculture now relies on formulated dry feeds largely compounded from fishmeal, soya and fish oil with various vitamins and minerals added.

A proportion of the feed given is not eaten, either because too much is given or because the system and the way in which it is managed does not optimize ingestion. There have been few attempts to estimate directly the proportion of uneaten food, partly because it is difficult to distinguish between the food and fecal components of solids collected. Most work to date has been on rainbow trout (*Salmo gairdneri*) and on cage or earth pond systems and no data are available for crustaceans. Results show that the proportion of uneaten food can be as little as 1% or as much as 30% and confirm that system, type of feed and management are important determinants of waste (Table 1). The proportion of uneaten food in intensively cultured trout ponds appears to be greatest when trash fish is used and lowest when dry pelleted feeds are used, a result of differences in stability of the feeds in water and of problems in food handling by the fish. If the dry matter content of feeds is taken into account, however, the significance of these differences is negligible.

Table 1. Estimates of unconsumed fish feed from intensive salmonid farming.

System	Species[a]	Feed	Feeding Method	% Lost	Reference
Earth pond	RT	Trash fish	Hand	10-30	Warren-Hansen 1982
Earth pond	RT	Moist Pellet	Auto	5-10	Warren-Hansen 1982
Earth pond	RT	Dry pellet	Auto	1-5	Warren-Hansen 1982
Freshwater cage	RT	Moist & dry pellet	-	30	Penczak et al. 1982
Marine cage	AS	Moist & dry pellet	-	20	Braaten et al. 1983
Marine cage	AS	Dry pellet	-	15-20	Gowen & Bradbury 1987

[a]RT = rainbow trout; AS = Atlantic salmon

The data suggest that feed losses from cages are considerably greater than those from ponds. This is confirmed when Food Conversion Ratios (FCR) values from cages are compared with those from other systems (see Beveridge 1984). There are also many incidental reports of high feed losses from cages (Loyacano and Smith 1976; Muller and Varadi 1980; Phillips et al. 1985).

Manufacturers of salmon and trout pelleted feeds claim that the dust content is no more than 1-2%, although measurements have shown that dust can account for as much as 3.7% (Clark et al. 1985). This suggests that most of the food that remains uneaten is a result of other factors related to management and system.

It is at present impossible to be any more specific about feed losses and much more work is required to identify where and why they occur. The results from Warrer-Hansen's (1982) and Makinen's (1988) studies, however, show much variability within treatments, which is probably largely attributable to differences in management practices and sampling methodology.

Excreta

The undigested fraction of food, together with mucus, sloughed intestinal cells, and bacteria, is voided as feces, while the digested fraction is absorbed and metabolized. Nutrients absorbed in excess of requirements may be excreted together with end-products (ammonia and urea) derived from the catabolic breakdown of protein for energy purposes.

Feces production can be estimated from digestibility studies of the principal constituents of diets. In Table 2 data are presented for salmonids which suggests that around 74% of a typical commercial diet is digestible on a dry weight basis, i.e. around 25-30 g of dry matter, typically comprising 17% protein, 3% lipid, 62% carbohydrate and 17% ash, is voided as feces per 100 g ingested food. This is similar to values obtained by direct measurement (Butz and Vens Cappell 1981; Clarke et al. 1985). Calculations are also shown for channel catfish and prawns.

This approach can be applied to any of the dietary constituents. However, it ignores the effects of temperature and other environmental variables, body size, health, feeding rate and the synergistic/antagonistic effects of one dietary component on the digestibility of another. It also does not consider the effects of the quality of nutritional components and the influence of processing and non-nutritional constituents such as binders on digestibility (Hilton et al. 1981; Storebakken 1984).

Table 2. Estimates of diet digestibilities (%) and fecal production (g dry wt 100 g^{-1} ingested) calculated using data for dietary components and proximate analyses (% dry weight) of typical diets. Data from various sources.

dietary component	salmonid		catfish		carp		shrimp	
	diet	digestibility	diet	digestibility	diet	digestibility	diet	digestibility
protein	45	90	35	80	35	85	31	82
lipid	19	95	7	97	15	70	8	52
carbohydrate	27	40	48	50	40	70	51	77
ash	9	50	10	50	10	50	10	50
whole diet digestibility		73%		59%		73%		74%
fecal production		27		41		27		26

Data on feces and pseudofeces (rejected food particles) for mussels and oysters are difficult to interpret because of differences in culture methods, analytical techniques, and in the way in which data has been expressed. Kautsky and Evans (1987) showed that 1.76 g dry weight of feces are voided per g dry weight of mussel (*Mytilus edulis*) each year and that 130 mg of carbon, 1.7 mg of nitrogen and 0.26 mg of phosphorus are produced. Most data, however, has been collected on carbon production and, expressed on an areal basis, it shows a fifty-fold variation (0.5-27 gC m^{-2} d^{-1})(Cabanas et al. 1979; Tenore et al. 1982 in Kautsky and Evans 1987).

Data on ammonia and urea production for various farmed species are summarized in Table 3. Most of the nitrogen excreted is in the form of ammonia. There is a great deal of variability not only between but also within species, since excretion rate is principally dependent upon nitrogen consumption and is to some extent affected by temperature (Paulson 1980).

Little work has been done to evaluate the importance of excretion of gut bacteria from farmed organisms on the resultant bacterial loading from aquaculture operations. All aquatic animals possess a rich gut flora of microorganisms and the guts of fish can act as proliferation sites for

Table 3. Nitrogen, ammonia and urea excretion rates for various farmed aquatic species (figures in brackets for molluscs are -1g shell-free dry wt).

Species	Component	Production
(a) Molluscs		
Mytilus edulis[a]	total-N	0.07-3.8 mg N kg^{-1} h^{-1} (1.4-75.5)
(b) Crustacea		
Penaeus esculentus[b]	NH_3	17-27 ug N g^{-1} day^{-1} (6-40 ug NH_3 g^{-1} day^{-1})
(c) fish		
Salmo gairdneri[cd]	NH_3	20-78.5 g N kg $food^{-1}$ d^{-1}
	NH_3	433-895 mg NH_3 kg^{-1} d^{-1}
	urea	73-474 mg urea kg^{-1} d^{-1}
	total	511-958 mg N kg^{-1} d^{-1}
Salvelinus fontinalis[e]	NH_3	3.2-3.6 mg NH_3-N $kg^{-1}h^{-1}$
Onchorhyncus nerka[f]	NH_3	8.2-35 mg N $kg^{-1}h^{-1}$
Anguilla anguilla[g]	NH_3	5-50 mg N $kg^{-1}h^{-1}$
	urea	0-25 mg N $kg^{-1}h^{-1}$
Anguilla rostrata[h]	NH_3	7-17 mg N $kg^{-1}h^{-1}$
Gadus morhua[i]	NH_3	12.3-41.2 mg N $kg^{-1}h^{-1}$
Cyprinus carpio[d]	NH_3	110-581 mg N kg^{-1} d^{-1}
Oreochromis mossambicus[j]	NH_3	1.72 mg N $kg^{-1}h^{-1}$
Oreochromis niloticus[k]	NH_3	1.7-9.4 mg N $kg^{-1}h^{-1}$
Sparus aurata[l]	NH_3	25.2-70 mg N $kg^{-1}h^{-1}$

[a]Kautsky and Wallentinus (1980); [b]Dall and Smith (1986); [c]Meade (1985); [d]Kaushik (1980); [e]Paulson (1980); [f]Brett and Zala (1975); [g]Knights (1985); [h]Gallagher and Mathews (1987); [i]Ramnarine et al. (1987); [j]Musisi (1984); [k]McKinney (pers. comm.); [l]Porter et al. (1987).

Enterobacteriaceae, *Aeromonas*, and fecal streptococci which enter the culture systems through the influent water or feed (Trust and Money 1972; Niemi and Taipilinen 1982; Austin and Allen-Austin 1985).

Chemicals

A wide range of chemicals is used in aquaculture, including therapeutics, vaccines, hormones, flesh pigments, anesthetics, disinfectants and water treatment compounds. Chemicals present in materials used in the fabrication of aquaculture systems can also find their way into the water (Table 4).

The use of chemicals varies greatly with species, intensity of culture and location. Few chemicals are used in mollusc culture, for example, in contrast to the salmon and trout industries. Only three antibiotics are used in the USA (Schnick et al. 1986) compared with ten in West Germany (Rasmussen 1988). There is some evidence from Norway that the 7-fold increase (on a per-unit-weight basis) in the use of chemicals that has occurred in mariculture since 1974 is in part due to intensification of production (Solbe 1987).

There are few published data on the quantities of chemicals used in aquaculture. Solbe's (1982) survey of 148 farms in the UK found 23 chemicals in use, the most common being malachite green (used on 60% of the farms) and formalin (used on 38% of farms). The most commonly used antibiotic was oxytetracycline, used on 11% of farms. Data for Norway show that chemicals are heavily used in their salmon industry (Table 5). It has been predicted that by 1990 the use of antibiotics in aquaculture in Norway would far exceed the totals used in general practice and agriculture (Solbe 1987)(Progress in the development of successful vaccines for a number of the more important salmon diseases, however, may mean that these predictions are not fulfilled). Penaeid prawn hatcheries in many parts of the tropics are heavily reliant on the prophylactic use of antibiotics.

Therapeutics enter the environment in several ways, depending upon how they are administered and how they are subsequently eliminated form the body. Treatment is by bath, injection or inclusion of chemicals in food. Many compounds used to treat skin and gill problems, such as malachite

Table 4. Chemicals used in aquaculture.

Type	Comments
THERAPEUTICS	
Parasiticides	
(e.g. neguvon, formalin, malachite green)	Widely used,
Antimicrobials	although many are
(e.g. exytetracycline, furazolidone)	controlled by legis- lation
VACCINES	Several commer- cially available for salmonid diseases
HORMONES	
(e.g. methyltestosterone, LHRH, HCG)	Widely used in carp and tilapia hatcher- ies
PIGMENTS (e.g. synthetic carotenoids)	Use in Europe under revision
ANESTHETICS (e.g. MS222, benzocaine, CO2)	Widely used in small quantities
DISINFECTANTS	
(e.g. iodophore compounds, hypochlorite)	Widely used
WATER TREATMENT CHEMICALS	
Oxidizers (potassium permanganate)	Uncommon
pH regulators (lime, acid)	Acid used occasion- ally in inflows
Herbicides and Algicides (e.g. copper sulphate)	Widely used
Pesticides (e.g. dipterex)	Used in tropical ponds
Predator control (tea seed cake, rotenone)	Control predators in shrimp ponds
ANTIFOULANTS (e.g. tributyltin, copper)	Used in cage and pen nets
PLASTIC ADDITIVES	
(e.g. stabilizers, pigments, UV absorbents)	Present in plastics and styrofoam

Table 5. Chemicals used in Norwegian fish farms in 1986 and predicted for 1990.

	kg/year	
	1986	1990
Sulphamerizine	10	10
Oxytetracycline	13050	45100
Furazolidone	6100	21100
Tribrissen	4250	17700
Neguvon	38450	132900
Formalin	78000	271000

source: Solbe 1987

green, formalin and the organophosphorus compounds Neguvon and Nuvan, are simply given in both form, the fish being transferred to a treatment container for the appropriate period of time. As far as can be determined, these compounds are usually allowed to soak away through the soil once used. In cage and pen culture, however, treatment is at present usually carried out in situ (Beveridge 1987; Egidius and Moster 1987), with the resultant release of all of the chemical into the environment.

Losses of therapeutics administered in feed occur both as a result of leaching and as a result of feed losses. Leaching of oxytetracycline from food has been shown to be dependent upon temperature, pH and pellet surface area : volume ratio (Fribourgh et al. 1969), with up to 20% being leached in 15 mins.

The uptake, distribution and elimination of therapeutics are determined by species, the formulation and physico-chemical characteristics of the therapeutic, feed composition, and environmental factors (Rasmussen 1988). Recent work by Cravedi et al. (1987) among others has shown that digestibility of chloramphenicol in diets fed to rainbow trout was 99%, while the digestibility of oxytetracycline was only around 7-9%. Digestibility is also greatly modified by inclusion rate in some antimicrobials (Cravedi et al. 1987). Elimination can take anything from a few days to a few months (Rasmussen 1988). Studies of hormones commonly used in aquaculture, such as methyltestosterone, suggest that they too are completely eliminated from the animals' body, although it takes several weeks (Johnstone et al. 1983).

Pesticides, particularly organophosphate compounds, are used in some areas of the world not only to control ectoparasitic infestations but also pests (e.g. fish in shrimp ponds)(Table 4). There are a few data on quantities used. Solbe's survey (Solbe 1982) of UK farms suggests that only around 1% use pesticides. Numbers in the tropics, however, are very much higher.

These antifoulants which worked by poisoning fouling organisms and which were widely used to treat cage and pen nets throughout the 1970s and early 1980s (Beveridge 1987) are now much less commonly used as fears about the accumulation of organotin- and copper-based compounds in farmed fish flesh and the effects of these chemicals on the environment have grown (Waldock and Thain 1985; Short and Thrower 1986; Davies et al. 1987). A ban on the use of TBT-based compounds is in operation in a number of countries.

MASS BALANCE ESTIMATES OF WASTE PRODUCTION

It is possible to estimate the amounts of uneaten food, feces and excreta produced from data on feed qualities and quantities, FCR values, digestibilities and fecal composition and to produce mass balance equations for various waste parameters, such as nitrogen, phosphorus or carbon, solids or biological oxygen demand.

Consider nitrogen balance in rainbow trout as an example. According to Penczak et al. (1982), the nitrogen content of rainbow trout is 3% and that of trout food is around 8%. Assuming an FCR value of 1.6:1, for every ton of trout produced approximately 98 kg of nitrogen is lost to the environment.

By incorporating other data and a number of assumptions, a more detailed picture emerges. If 20% of food is uneaten, then 102.4 kg of nitrogen is consumed but only 30 kg is retained per ton of trout produced. From Table 2, approximately 260 g of feces with nitrogen content of 4% (Penczak et al.

1982) are produced per kg food consumed. Thus, 13.4 kg of nitrogen are voided as feces, while the balance, 59 kg, is excreted as ammonia and urea.

Many such mass balance equations have been produced for fish (e.g. Braaten et al. 1983; Beveridge 1984; Gowen and Bradbury 1987; Heinsbroek 1988), although none are available for prawns. They can be fairly accurate. For example, the figure estimated above for nitrogen excretion approximates those estimated by Warrer-Hansen (1982) and Solbe (1982)(45 kg and 55.5 kg respectively). No examples, however, directly account for food losses, and this must be a major source of error. Moreover, this approach cannot predict effluent composition in terms of suspended or settleable solids, nitrogen or phosphorus species, since both system and management play a major role in determining how much and in what forms the wastes might appear. The form in which wastes are discharged and temporal variability in effluent quality are important determinants of environmental impact.

Most models also give no indication of diurnal or seasonal variability, although a recent document by Hakanson et al. (1988) on Atlantic salmon shows how waste output might vary over the course of the growing season (Fig. 1).

EFFLUENT FROM AQUACULTURE OPERATIONS

The parameters that are typically measured in relation to aquaculture effluents include suspended solids (SS), biochemical oxygen demand (BOD), chemical oxygen demand (COD), ammonia (NH3), phosphorus (P), and dissolved oxygen (DO). Levels of bacteria, chemicals or therapeutics in effluents have not been routinely measured.

Most aquaculture uses a great deal of water and consequently aquaculture effluents are characterized by their relatively large volume. They also tend to have relatively low concentrations of wastes compared with effluents from other industries (Alabaster 1982; Muir and Beveridge 1987). From studies of European fish farms, effluents have around 9 mg l^{-1} suspended solids, 4 mg l^{-1} 5-day BOD, 18 mg l^{-1} COD, < 1.0 mg l^{-1} ammonia, 1.0 mg l^{-1} oxidized nitrogen and 0.1 mg l^{-1} phosphate phosphorus (Alabaster 1982).

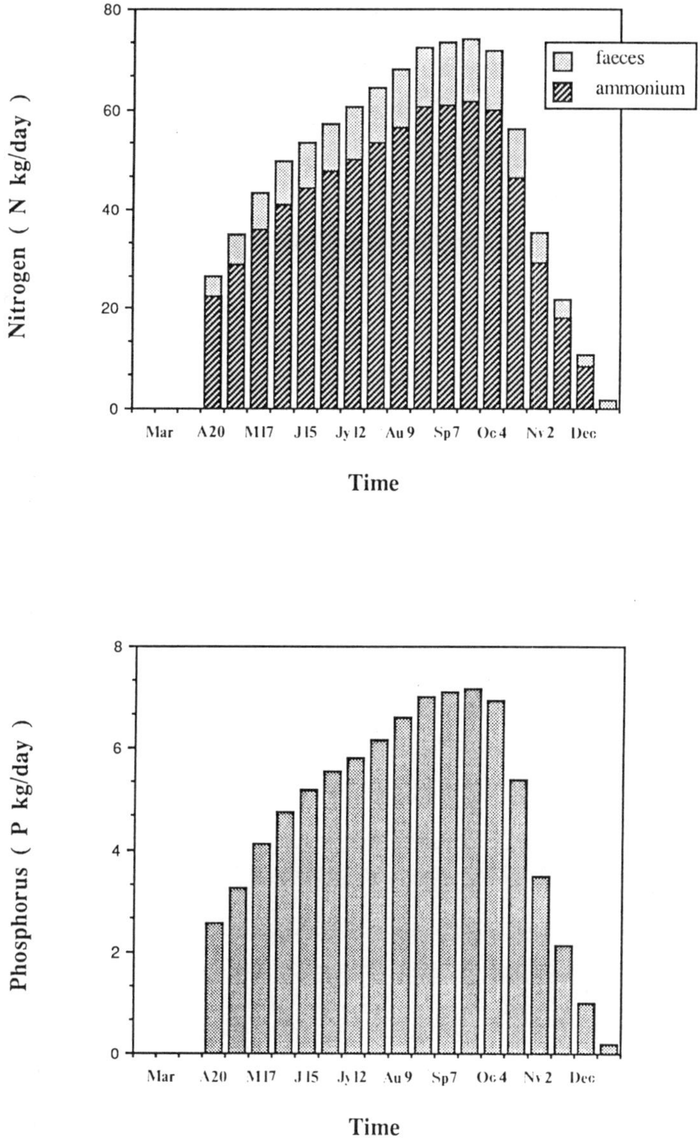

FIGURE 1. Predicted total nitrogen and total phosphorus loadings from a Scandanavian rainbow trout farm (redrawn from Hakanson et al. 1988).

From an analysis of these and other surveys, however, it is apparent that effluent quality is highly variable. Alabaster (1982) noted that effluent concentrations of suspended solids, BOD and DO in particular varied enormously. When effluent waste concentrations are standardized on a per unit production or biomass basis, the variability persists, although it is greatest (as much as several orders of magnitude) among N and P data (Table 6).

TABLE 6. Loadings of suspended solids (ss), nutrients and Biological Oxygen demand from fresh water salmonid culture.

SS	BOD	NH_4-N	NO_2-N	NO_3-N	Total N	PO_4	Total P	Ref.
(a) g kg fish^{-1} day^{-1}								
0.0-7.1	1.6-2.7	-	-	-	0.1-3.8	0.02-0.27	0.01-0.43	a
-	1.4-8.1	-	-	-	-	-	-	c
0.8-0.9	-	0.3-0.8	-	0.13-0.21	-	0.07-0.17	-	b
-	-	0.03	-	0.05	0.12	0.033	0.10	d
(b) kg ton fish^{-1} annum^{-1}								
474-4015	510-990	37-180	-	0-548	-	-	22-110	e
1350	285	55.5	1.81	10.2	-	-	15.7	f
	350	45	-	-	83	-	11	g
(c) g kg feed^{-1}								
-	80-300	-	-	-	-	-	-	c
183	165	25	0.27	0	-	26	4.0	h
80-280	100-370	-	-	-	37-48	-	4.7-10.8	i

[a]Bergheim et al. (1982); [b]Clark et al. (1985); [c]Butz and Vens Cappell (1982); [d]Korzeniewski et al. (1982); [e]Alabaster (1982); [f]Solbe (1982); [g]Warrer-Hansen (1982); [h]Butz (1988); [i]Makinen (1988).

There are many variables governing effluent quality, including species, size, method and intensity of culture, management, temperature, mode of discharge and the extent of treatment or dilution prior to discharge.

Strong correlations have been found between a number of parameters, such as BOD and SS, and between NH_3 and COD. The former occurs because much of the BOD is present in the solids fraction (Alabaster 1982). In most cases, however, there remains an insufficient understanding of the interrelationships between parameters.

Species

Most studies to date on aquaculture effluents have been on salmonids and only limited work has been done on characterization of wastes from other fish species (Ellis et al. 1978; Bohl 1988) or crustacean culture (Tchobanoglous and Shlesher 1974 in Rosenthal et al. 1987). Theoretical calculations indicate that wastes from one species may be several times as great as those from another (Heinsbroek 1988). From direct measurements, however, the range of values for each waste parameter in each species is so wide that it is impossible at this stage to evaluate the role of species in determining effluent quality.

Body Size

The production of metabolic wastes is proportional to metabolic rate which decreases with increasing body size and increases with temperature. Ammonia and phosphate waste production have been shown to fall with increasing body size in rainbow trout, although no change was apparent in suspended solids production (Clark et al. 1985), possibly because of the role of the system in determining suspended solids characteristics (see below).

Feed

Many studies and surveys have demonstrated strong relationships between the quantities and qualities of feed given and food conversion ratio with effluent characteristics such as suspended solids and BOD (e.g. Knosche 1972 in Rosenthal et al. 1987; Butz and Vens Cappell 1982; Pursiainen 1988). These studies demonstrate that total BOD increased with the amounts of food fed, and that increases in digestibility did not appear to reduce organic load, although fecal dry matter production, a major constituent of waste solids, did fall.

System and management

The most important factors, however, in determining effluent characteristics are probably the system and its management. Water exchange rate influences the probability with which a food item is encountered and thus the proportion of food which is uneaten (see above), while flow characteristics will in part determine whether uneaten food and fecal particles remain intact and the proportion that settle within the system.

Unfortunately, there has been little work on system-specific characterization of wastes. Nevertheless, it is possible to make some generalizations. Ponds differ from most other systems in that water exchange is typically low, even in fairly intensive systems. Effluent discharges are small and effluent waste loadings are low, except if ponds are drained to facilitate harvesting. Extremely high levels of solids, BOD and nutrients in fish pond effluents during harvest have been documented (Boyd 1978; Ellis et al. 1978; Solbe 1987; Bohl 1988; Butz 1988). Correlations between biomass harvested and effluent solids loadings are weak as a result of pond soil erosion. According to Tucker (1985), the current practice in the US is to harvest catfish without draining ponds.

There is a more or less constant exchange of water through tank and raceway systems, determined by biomass and environmental conditions. Surprisingly, perhaps, suspended solids and BOD effluent loadings in concrete raceways, expressed in terms of % food fed, have been observed to fall with increased water exchange (Butz 1988).

Effluent characteristics can change dramatically as a result of routine operations. Studies carried out at the University of Stirling and elsewhere show that during cleaning effluent loadings increase dramatically (Alabaster 1982; Butz 1988; Kadri 1988). Alabaster (1982) observed that 70% of the daily BOD, 75% of total-P and 10% of total-N could be discharged during a 30-minute cleaning period.

Mortalities and Escapes

Both mortalities and escapes may be considered as net waste loadings since they may constitute materials not removed during harvesting. The amounts may be considerable, particularly from cage farms. Penczak et al. (1982) estimated that 10% of the phosphorus loadings from a cage rainbow trout farm were accounted for by dead fish. Escaped animals also pose problems associated with the introduction of a novel species or strain to an environment (see Weston 1986).

Bacteria and Other Microorganisms

Results from studies of bacteria in effluents from freshwater trout farms indicate that these are generally similar to those in the inflows. In studies carried out in the U.K. by Austin and co-workers, on only 8 out of 29 occasions were numbers substantially higher (Allen et al. 1985; Austin 1985; Austin and Allen-Austin 1985), and during periods of chemotherapy numbers in the effluent were less (Austin 1985). Preliminary studies of freshwater cage sites in Scotland agree with these findings (Smith, unpublished data).

By contrast, a study of a pump-ashore turbot (*Scophthalmus maximus*) farm showed that bacterial numbers in the effluent were 3-50 times greater than in the inflows (Austin and Allen-Austin 1985), although this is insufficient evidence to generalize on bacterial concentrations in marine fish culture as a whole.

Qualitative changes in the bacterial microflora of trout, channel catfish, carp and turbot farm effluents have been observed (Tackett 1974; Niemi and Taipalinen 1982; Austin 1985; Austin and Allen-Austin 1985; Sugita et al. 1985). Diversity increased with effluent nutrient content, although decreased markedly during chemotherapy (Austin 1985). Studies on fecal coliforms are contradictory. Recorded in low numbers in farm inflows, they were significantly higher in some trout farm effluents (Niemi and Taipilinen 1982), but lower in channel catfish farm effluents (Tackett 1974).

Viruses have also been detected in fish farm effluents. Leong and Turner (1979) measured effluent concentrations of infectious hematopoietic necrosis virus (IHNV) as high as 400 pfu ml^{-1} during a disease outbreak at a salmonid hatchery.

Chemicals and Therapeutics

The few data available on the amounts of chemicals present in aquaculture effluents suggest that they are low. Average effluent concentrations of malachite green, formalin and hyamine at UK freshwater fish farms were 0.61, 15.2 and 0.55 mg l^{-1} respectively (Solbe 1982), although the standard deviations were greater than the mean values, indicating high variability and that most are used on an irregular basis.

It is possible to estimate the concentrations of antibiotics which might occur under worst possible conditions during chemotherapy, based upon dosage, stocking density and flow rate data. The figures calculated in Table 7 assume minimal water flow (500 m^3 ton^{-1}) and that all of the therapeutic passes out in the effluent over a 24 hour period. They also do not account for degradation or biometabolism prior or during the retention period. Nevertheless, the estimated effluent concentrations are very small and in reality are likely to be much smaller, since retention times are in days rather than hours (Rasmussen 1988). There has been only one published account of measuring effluent antibiotic concentrations, partly because techniques have only recently been developed Crathorne et al. (1986). Concentrations of oxolinic acid were found to be extremely small (1200 ng l^{-1}), even during chemotherapy, a factor of ten lower than those estimated by the above method.

TABLE 7. Dose rates and estimated effluent concentrations of antimicrobial compounds following chemotherapy at land-based salmonid farms.

Antimicrobial compound	Dosage (mg kg body wt^{-1} day^{-1})	Duration (days)	Effluent concentration (ug l^{-1})
Erythromycin phosphate	75-125	7-21	250
Oxolinic acid	10	10	20
Oxytetracycline	75	10	150
Potentiated sulphonamide	30	5-7	60

modified from Austin (1985)

There are no data available on hormone or vaccine concentrations in aquaculture effluents. Vaccines, however, are generally applied either by injection or as a bath treatment (Ellis 1988). According to Solbe (1987), spent vaccine is disposed of to soil and hence is unlikely to be present in effluents.

DISCUSSION AND CONCLUSIONS

The vast bulk of data on aquaculture wastes and effluents comes from freshwater rainbow trout culture. The reasons for this are in part historical: it is a large and comparatively long-established industry which relies on intensive methods and it was here that environmental problems were first observed.

Although it is possible to formulate mass balance equations for various waste parameters, such as in Table 2, it remains impossible to predict effluent characteristics with any accuracy. The roles of species, system and management, which are recognized as being of paramount importance (Alabaster 1982), remain to be quantified. Thus, while it is possible to characterize effluents from systems in a general sense - pond effluents, for example, tending to be small with very low concentrations of wastes, in contrast to the comparatively large volumes of dilute wastes discharged from raceways - the degree of variability within a system type makes it impossible to quantify these differences (see, for example, Butz 1988).

Part of the problem is lack of data. Little work has been done on a number of important aquaculture species (particularly shrimps), on feed losses, or on the physical characterization of solid wastes (particle size, density, stability in water etc.). However, much of the problem lies in the way in which samples have been collected and analyzed and in the way data are expressed.

Much of the variability apparent in survey data in particular is due to differences in timing of sampling. A number of studies, including desk studies, have shown how effluent characteristics change over the growing season as body size, biomass, temperature and feeding rate change (e.g., Bergheim et al. 1982; Hakanson et al. 1988). Short-term fluctuations can also be dramatic. While effluent suspended solids and BOD levels tend to peak at or shortly after feeding, ammonia excretion, which can increase as much as eight-fold after a meal, typically peaks some 2-4 h later (e.g. Brett and Zala 1975). Few studies relate sampling times to feeding. Other routine tasks affecting solids, such as cleaning, further compound sampling difficulties. Thus, considerably more work is required to quantify seasonal and diurnal fluctuations.

Collection of samples is not without problems; if a sample is taken near the surface in an outflow channel, for example, then solids which have already settled lower in the water column will be missed, leading to an under-estimate of solids discharged (Solbe 1987).

Carlsson (1988) has recently made a number of recommendations on analytical methods pertinent to aquaculture effluent monitoring. While these are appropriate for routine studies, more attention must be paid to other components, such as pigments, vitamins, etc., which although present in low concentrations, may have important effects on receiving waters. Some standardization of how effluent characteristics are expressed is also necessary. The least acceptable is probably in terms of biomass, since it is difficult to estimate standing crop with any accuracy, and the best is probably in terms of food fed. Moreover, it is not always clear whether effluent discharges are being reported as net increases (i.e. effluent concentration minus inflow concentration) or not.

Effluent waste concentrations can be reduced and there are two principal approaches; prevention and treatment. In both, solids removal, in a manner which is both gentle and rapid, is a key objective, since not only suspended solids, but also much of the BOD and nutrients are in the solids component of the wastes (Alabaster 1982; Butz and Vens and Cappel 1982; Kadri 1988). In theory, prevention can be achieved by reductions at each step in the production cycle; better formulated, more digestible and water-stable feeds, better feeding methods, improved system design, etc. There is considerable economic pressure to minimize feed losses and this has in part been responsible for the more than 50% drop in effluent solids loadings, on a per unit production basis, that occurred from UK land-based farms between 1980 and 1987 (Solbe 1987). It is also in the farmer's interests to install self-cleaning designs of tanks, since resuspension of settled solids within a system is undesirable.

In the absence of any legislative controls, there is little economic pressure to treat wastes and there are comparatively few farms that have any form of waste treatment. The situation is changing, however, and in many parts of the world controls on effluent waste concentrations are being introduced, although there is little consistency in the levels that are being demanded or in the types of legislation that are being applied (see Pursiainen 1988, for example). In part, this is because many authorities attempt to relate effluent discharges to the nature of the receiving water bodies. In northern and highland Europe, for example, concerns are more with phosphorus loadings, while in southern Europe, concern is principally with solids (Solbe 1987; Pursiainen 1988).

Conventional methods of waste treatment from land-based systems have involved using settlement and or pressurized sand filters. There have been a number of recent innovations, however, such as swirl concentrators and triangle filters. Claims as to their effectiveness vary: according to recent reports, however, new designs of swirl concentrators can remove between 40 and 80% of solids and 20 to 50% of total phosphorus (Lefferstra and Kryvi 1988; Makinen 1988). Triangle filters have been found to reduce solids by up to 95% and total phosphorus by up to 80% (Kadri 1988; Lefferstra and Kryvi 1988).

Prospects for effluent treatment from cages or mussel rafts seem poor. While up to 45% of solids and 15-20% of nitrogen and phosphorus have been recovered from fish cages in Poland using funnel-shaped collectors (Tucholski and Wotno 1980), efforts elsewhere, particularly at exposed sites, have proven much less effective and costly (Beverdige 1984; 1987; Solbe 1987; Makinen 1988). The removal of or dispersal of sedimented wastes from under cages has been attempted in Norway (Braaten et al. 1983), but remains to be thoroughly evaluated. One of the most important ways of minimizing the impact of cage and mussel culture remains the siting of cages in fast-flushing waters which rapidly dilute and disperse wastes.

In a study carried out some years ago, Muir (1982) found that the cost-effectiveness of different effluent treatment methods varied considerably and there is a great need for the economics of waste treatment to be re-evaluated in the light of recent technical developments. Aquaculture wastes are rich in nutrients and these can be utilized by integration with other aquaculture or agriculture production systems (Willett and Jacobsen 1986; Little and Muir 1987). Beveridge (1984) estimated that recovery of escaped fish by anglers in a Scottish loch reduced total phosphorus loadings by around 1.5%. Much more work is needed on recovery and utilization of wastes.

ACKNOWLEDGEMENTS

We thank a number of our colleagues who kindly helped in the preparation of this paper: Ed Branson, Matt Briggs, Kim Jauncy, Richard McKinney, Mark Northcott, Erkki Schwanck and Pete Southgate. M. C. M. Beveridge is funded by the Overseas Development Administration of the British Government.

LITERATURE CITED

Alabaster, J. S. 1982. Report of the EIFAC workshop on fish-farm effluents. EIFAC Technical Paper 41. FAO, Rome Italy.

Allen, D. A., B. Austin, and R.R. Colwell 1985. Numerical taxonomy of bacterial isolates associated with a freshwater fishery. Journal of General Microbiology 129: 2043-2962.

Austin, B. 1985. Antibiotic pollution from fish farms: effects on aquatic microflora. Microbiological Sciences 2: 113-117.

Austin, B. and D. Allen-Austin 1985. Microbial quality of water in intensive fish rearing. Journal of the Applied Bacteriology Symposium Supplement 1985: 207-226.

Bergheim, A., A. Silversten, and A.R. Selmer-Olsen 1982. Estimated pollution loadings from Norwegian fish farms. 1. Investigations 1978-79. Aquaculture 28: 347-361.

Beveridge, M. C. M. 1984. Cage and pen fish farming. Carrying capacity models and environmental impact. FAO Fisheries Technical Paper 255: 131 pp.

Beveridge, M. C. M. 1987. Cage Aquaculture. Fishing News Books Ltd., Farnham, Surrey, U. K.

Bohl, M. 1988. National report for the Federal Republic of Germany on suspended solids from land-based fish farms. Pages 29-36 in M. Pursiainen, editor. National contributions on suspended solids from land-based fish farms. Monistettuja Julkaisuja, Helsinki, Finland.

Boyd, C. 1978. Effluent from catfish ponds during fish harvest. Journal of Environmental Quality 7: 59-62.

Braaten, B., A. Ervik, and E. Boje 1983. Pollution problems on Norwegian fish farms. Aquaculture Ireland 14: 6-7.

Brett, J. R. and C.A. Zala 1975. Daily pattern of nitrogen excretion and oxygen consumption of sockeye salmon (*Oncorhynchus nerka*) under controlled conditions. Journal of the Fisheries Research Board of Canada 32: 2479-2486.

Butz, I. 1988. Situation of fish farm effluents in Austria. Pages 4-12 in M. Purainen, editor. National contributions on suspended solids from land-based fish farms. Monistettuja Julkaisuja, Helsinki, Finland.

Butz, I. and B. Vens-Cappell 1982. Organic load from the metabolic products of rainbow trout fed with dry food. Pages 73-82. In J. S. Alabaster, editor. Report of the EIFAC workshop on fish-farm effluents. Silkeborg, Denmark, 26-28 May 1981. EIFAC Technical Paper 41, FAO, Rome, Italy.

Cabanas, J. M., J.J. Gonzalez, J. Marino, A. Perez, and G. Roman 1979. Estudio de mejillon y de su epifauna en los cultivos flotantes de la Ria de Arosa. Boletin del Instituto Espanol de Oceanographica 5: 45-50.

Carlsson, S. A. 1988. Standardization of methodology in fish-farm effluent research and monitoring. Pages 90-93. In M. Purainen, editor. National contributions on suspended solids from land-based fish farms. Monistettuja Julkaisuja, Helsinki, Finland.

Clark, E. R., J.P. Harman, and J.R.M. Forster 1985. Production of metabolic and waste products by intensively farmed rainbow trout, *Salmo gairdneri* Richardson. Journal of Fish Biology 27: 381-393.

Crathorne, B., M. Fielding, C.P. James, M. Righton, J.A. Stratford, and C.D. Watts 1986. Fish-farm chemicals: Development and application of methods for the analysis of anti-microbial compounds. Water Research Centre, Medmenham, U.K.

Cravedi, J.-P., G. Choubert, and G. Delous 1987. Digestibility of chloramphenicol, oxolinic acid and oxytetracycline in rainbow trout and influence of these antibiotics on lipid digestibility. Aquaculture 60: 133-141.

Dall, W. and D.M. Smith 1986. Oxygen consumption and ammonia-N excretion in fed and starved tiger prawns *Penaeus esculentus* Haswell. Aquaculture 55: 23-33.

Davies, I. M., S.K. Bailey, and D.C. Moore 1987. Tributyltin in Scottish sea lochs, as indicated by degree of imposex in the dogwhelk, *Nucella lapillus* (L.). Marine Pollution Bulletin 18: 400-404.

Egidius, E. and B. Moster 1987. Effect of Nevugon and Nuvan treatment on crabs (*Cancer pagurus*, *C. maenus*), lobster (H*omarus gammarus*) and blue mussel (*Mytilus edulis*). Aquaculture 60: 165-168.

Ellis, A. E. 1988. Fish vaccination. Academic Press, London, U.K.

Ellis, J. E., D.L. Tackett, and R.R. Carter 1978. Discharge of solids from fish ponds. Progressive Fish-Culturist 40: 165-166.

Fribourgh, J. H., F.P. Meyer, and J.A. Robinson 1969. Oxytetracycline leaching from medicated fish feeds. U.S. Bureau of Sport Fisheries and Wildlife Technical Paper 40: 7 pp.

Gallagher, M. L. and A.M. Matthews 1987. Oxygen consumption and ammonia excretion of the American eel *Anguilla rostrata* fed diets with varying protein energy ratios and protein levels. Journal of the World Mariculture Society 18: 107-112.

Gowen, R. J. and N.B. Bradbury 1987. The ecological impact of salmonid farming in coastal waters: a review. Oceanography and Marine Biology Annual Review 25: 563-575.

Hakanson, L., A. Ervik, T. Makinen, and B. Moller 1988. Basic concepts concerning assessments of environmental effects of marine fish farms. Nordic Council of Ministers, Copenhagen, Denmark.

Heinsbroek, L. T. N. 1988. Waste production and discharge of intensive culture of African catfish, *Clarias gariepinus*, European eel, *Anguilla anguilla*, and rainbow trout *Salmo gairdneri*, in recirculation and flow through units in the Netherlands. Pages 55-61 in M. Pursiainen, editor. National contributions on suspended solids from land-based fish farms. Monistettuja Julkaisuja, Helsinki, Finland.

Hilton, J. W., C.Y. Cho,. and S.J. Slinger 1981. Effect of extrusion processing and steam pelleting diets on pellet durability, pellet water absorption and the physiological response of rainbow trout (*Salmo gairdneri* R.). Aquaculture 25: 185-194.

Johnstone, R., D.J. Macintosh, and R.S. Wright. 1983. Elimination of orally administered 17a-methyltestosterone by *Oreochromis mossambicus* (tilapia) and *Salmo gairdneri* (rainbow trout) juveniles. Aquaculture 35: 249-257.

Kadri, S. 1988. A study of a screen filter's performance in treating salmonid hatchery effluent. Master's thesis, University of Stirling, Scotland, U.K.

Kaushik, S. J. 1980. Influence of nutritional status on the daily patterns of nitrogen excretion in the carp (*Cyprinus carpio* L.) and the rainbow trout (*Salmo gairdneri* R.). Reproduction and Nutriton Developments 20: 1751-1765.

Kautsky, N. and S. Evans 1987. Role of biodeposition by *Mytilus edulis* in the circulation of matter and nutrients in a Baltic coastal ecosystem. Marine Ecology Progress Series 38: 201-212.

Kautsky, N. and I. Waalentinus 1980. Nutrient release from a Baltic *Mytilus*-red algal community and its role in benthic and pelagic pollution. Ophelia 1: 17-30.

Knights, B. 1985. Energetics and fish farming. Pages 309-340 in P. Tytler and P. Calow, editors. Fish energetics. Croom Helm, London.

Korzeniewski, K., Z. Banat, and A. Moczulska 1982. Changes in water of the Uniesc and Skotawa rivers, caused by intensive trout culture. Polish Archives of Hydrobiology 29: 683-691.

Lefferstra, H. and H. Kryvi 1988. Report on suspended solids from fish farms in Norway. Pages 62-73 in M. Purainen, editor. National contributions on suspended solids from land-based fish farms. Monistettuja Julkaisuja, Helsinki, Finland.

Leong, J. and S. Turner 1979. Isolation of waterborne infectious hematopoietic necrosis virus. Fish Health News 8: 6-8.

Liao, P. B. andR.D. Mayo 1974. Intensified fish culture combining water reconditioning with water pollution abatement. Aquaculture 3: 61-85.

Little, D. and J.F. Muir 1987. Integrated aquaculture. Institute of Aquaculture, University of Stirling, Stirling, U.K.

Loyacano, H. A. and D.C. Smith 1976. Attraction of native fishes to catfish culture cages in reservoirs. Proceedings of the 29th Annual Conference of the South Eastern Association of the Game and Fisheries Commission. Columbia, South Carolina, U.S.A. 63-73.

Makinen, T. 1988. Report on suspended solids from fish farms in Finland. Pages 17-28 in M. Purainen, editor. National contributions on suspended solids from land-based fish farms. Monistettuja Julkaisuja, Helsinki, Finland.

Meade, J. W. 1985. Allowable ammonia for fish culture. Progressive Fish Culturist 47: 135-144.

Muir, J. F. 1982. Economic aspects of waste treatment in fish culture. Pages 123-136 in J. S. Alabaster, editor. Report of the EIFAC workshop on fish-farm effluents. Silkeborg, Denmark, 26-28 May 1981. EIFAC Technical Paper 41, FAO, Rome, Italy.

Muir, J. F. and M.C.M. Beveridge 1987. Water resources and aquaculture development. Archiv fur Hydrobiologia 28: 321-324.

Muller, F. and L. Varadi. 1980. The results of cage fish culture in Hungary. Aquacultura Hungarica 2: 154-167.

Musisi, L. 1984. The nutrition, growth and energetics of tilapia, *Sarotherodon mossambicus*. Ph.D. thesis, University of London, UK.

Niemi, M. and I. Taipalinen 1982. Fecal indicator bacteria at fish farms. Hydrobiologia 86: 171-175.

Paulson, L. J. 1980. Models of ammonia excretion for brook trout (*Salvelinus fontinalis*) and rainbow trout (*Salmo gairdneri*). Canadian Journal of Fisheries and Aquatic Science 37: 1421-1425.

Penczak, T., W. Galicka, M. Molinski, E. Kusto, and M. Zalewski 1982. The enrichment of a mesotrophic lake by carbon, phosphorus and nitrogen from the cage aquaculture of rainbow trout *Salmo gairdneri*. Journal of Applied Ecology 19: 371-393.

Phillips, M. J., M.C.M. Beveridge, and L.G. Ross 1985. The environmental impact of salmonid cage culture on inland fisheries: present status and future trends. Journal of Fish Biology 27 (Supplement A): 123-137.

Porter, C. B., M.D. Krom, M.G. Robbins, L. Brickell, and A. Davidson 1987. Ammonia excretion and total N budget for gilthead seabream *Sparus aurata* and its effect on water conditions. Aquaculture 66: 287-297.

Pursiainen, M. 1988 (ed.). National contributions on suspended solids from land-based fish farms. Monistettuja Julkaisuja, Helsinki, Finland.

Ramnarine, I. W., J.M. Pirie, A.D.F. Johnstone, and G.W. Smith 1987. The influence of ration size and feeding frequency on ammonia excretion by juvenile Atlantic cod, *Gadus morhua* L. Journal of Fish Biology 31: 545-559.

Rasmussen, F. 1988. Therapeutics used in fish production: pharmacokinetics, residues and withdrawal periods. EIFAC/XV/88/Inf. 13. Food and Agriculture Organization, Rome, Italy. 25 pp.

Rosenthal, H., D. Weston, R. Gowen, and E. Black (eds.). 1987. Report of the ad hoc Study Group on "Environmental impacts of Mariculture." Cooperative Research Report 154, International Council for the Exploration of the Sea, Copenhagen, Denmark. 83 pp.

Schnick, R. A., F.P. Meyer, and D.F. Walsh 1986. Status of fishery chemicals in 1985. Progressive Fish-Culturist 48: 1-17.

Short, J. W. and F.P. Thrower 1986. Accumulation of butyltins in muscle tissue of chinook salmon reared in sea pens treated with tri-n-butyltin. Proceedings of Oceans 86, Washington, D.C., 23-25 September 1986: 1177-1181.

Solbe, J. F. de L. G. 1982. Fish farm effluents: A United Kingdom survey. Pages 29-56 in J. S. Alabaster, editor. Report of the EIFAC workshop on fish-farm effluents. Silkeborg, Denmark, 26-28 May 1981. EIFAC Technical Paper 41, FAO, Rome, Italy.

Solbe, J. F. de L. G. 1987. Report of the European Inland Fisheries Advisory Commission Working Party on fish farm effluents. Water Research Centre, Medmenham, U. K. 20 pp.

Storebakken, T. 1984. Binders in the feed of rainbow trout. Norsk Fiskeoppdrett. 9(3): 30-33.

Sugita, H., S. Ushoioka, D. Kihara, and Y. Deguchi 1985. Changes in the bacterial composition of water in a carp-rearing tank. Aquaculture 44: 243-247.

Tackett, D. L. 1974. Yield of channel catfish and composition of effluents from shallow-water raceways. Progressive Fish-Culturist 36: 46-48.

Trust, T. J. and V.G. Money 1972. Bacterial population of diets for aquarium fishes. Journal of the Fisheries Research Board of Canada 29: 429-433.

Tucholski, S. and T. Wojno 1980. Studies on removal of wastes produced during cage-rearing of rainbow trout (*Salmo gairdneri* Richardson) in lakes. 111. Budgets of materials and some nutrient elements. Zesz. Nauk. Akad. Roln. Szcz. 82: 31-50.

Tucker, C. S. (ed.) 1985. Channel Catfish Culture. Elsevier, Amsterdam, Netherlands.

Waldock, M. J. and J.F. Thain 1985. The comparative leach rates and toxicity of two fish net antifouling preparations. International Council for the Exploration of the Seas Statutory Meeting, CM 1985/E:29.

Warrer-Hansen, I. 1982. Evaluation of matter discharged from trout farming in Denmark. Pages 57-64 in J. S. Alabaster, editor. Report of the EIFAC workshop on fish-farm effluents. Silkeborg, Denmark, 26-28 May 1981. EIFAC Technical Paper 41, FAO, Rome, Italy.

Weston, D. P. 1986. The environmental effects of floating mariculture in Puget Sound. Washington Department of Fisheries and Washington Department of Ecology, Washington, U.S.A.

Willett, I. R. and P. Jacobsen 1986. Fertilizing properties of trout farm waste. Agricultural Wastes 17: 7-13.

THE EFFECTS OF AQUACULTURE ON INDIGENOUS BIOTA

Donald P. Weston

ABSTRACT

With the growth of the aquaculture industry has come the recognition that aquaculture can substantially alter natural communities. The potential effects of aquaculture considered here are: (1) organic enrichment of the substratum and alteration of the macrofaunal community; (2) nutrient enrichment and eutrophication; (3) effects on indigenous fish communities; (4) chemical usage by the industry; and (5) introduction of exotic species and interbreeding between cultured and wild populations. The deposition of feed and/or feces causes changes in the composition and abundance of the infaunal invertebrate populations including a reduction in species richness, increased densities of opportunistic species, and shifts in the vertical profile of biomass within the sediment column. Nutrient enrichment can stimulate phytoplankton productivity with adverse consequences to both the farm and the natural ecosystem. Such problems are more likely in freshwater culture than in mariculture, and the potential for eutrophication is greatest when feed is provided to the cultured animal. Density and species richness of fishes increases in the immediate vicinity of the culture site because of the increased availability of food resources and behavioural attraction to solid objects. Chemicals used in aquaculture can have effects on native fauna, but existing data are meager. There are several examples of mortalities in non-target native fauna due to toxic effects of aquaculture chemicals, and the stimulation of antibiotic resistance in the microbial community by use of chemotherapeutants. "Biological pollution" from aquaculture, including the introduction of exotic species and the genetic consequences of interbreeding between cultured and wild populations, can potentially have dramatic regional impacts, yet our predictive capabilities in these areas are poor. With expansion of the aquaculture industry and technological advances that permit an increased production per unit area, it is becoming increasingly imperative that consideration be given to alteration of the natural biota, both for reasons of environmental protection and viability of the industry.

INTRODUCTION

Aquaculture is commonly perceived as a "clean" industry. Aquaculturists, however, farm marine and aquatic organisms at tremendous densities, and attain levels of productivity that are greater per unit of area than the productivity of any other form of animal protein on earth (Ryther 1969). The resultant waste production often causes substantial physical and chemical changes in the area surrounding the farm site. These changes can result in structural and functional changes in the biota indigenous to the area, and, in some cases, may have adverse consequences for the growth and survival of the cultured organisms.

The potential environmental effects of aquaculture encompass a broad range on both spatial and ecological scales. At one end of the spectrum is simple displacement; the presence of the aquaculture facility makes that area unavailable to native fauna. This effect is unavoidable and is usually considered inconsequential except in areas of special concern such as biological preserves. At the next level of significance are localized impacts such as those resulting from depletion of dissolved oxygen or nutrient enrichment. These effects are observed on a scale ranging from within a few meters of the farm site up to a few kilometers distant. Effects on this scale are comparatively well documented, and we have acquired some degree of success in estimating the magnitude of impact before the farm operation is in place. "Biological pollution," such as the introduction of exotic species or alteration of the wild gene pool by interbreeding of cultured and wild stock, are potentially the most serious effects of aquaculture. These impacts can have ecosystem-wide repercussions and be observed far beyond the immediate vicinity of the farm. Unfortunately, it is in these areas that existing knowledge and predictive capabilities are seriously lacking.

In this review I will consider the biological implications of waste production by aquaculture on the native flora and fauna. Emphasis will be placed upon: (1) organic enrichment of the substratum and alteration of the macrofaunal community; (2) nutrient enrichment and eutrophication; (3) effects on indigenous fish communities; and (4) chemical usage by the industry and its potential biological effects. Information will be drawn from many forms of aquaculture, but much of the discussion is devoted to marine net-cage culture of salmonids. This bias is in large part a consequence of the fact that much of the research on the environmental effects of aquaculture has been directed specifically towards this segment of the industry. Such emphasis, however, is not misdirected for the recent explosive growth of the net-cage culture industry makes an understanding of environmental effects increasingly essential to sound environmental management.

BENTHIC ENRICHMENT

Potential effects

Mariculture structures can influence sedimentation simply by modifying current velocity and direction. As with any other structure placed in a moving fluid, mariculture structures alter the flow regime and may create zones of both erosion and deposition. In areas of intensive culture, bathymetry or even coastline topography can be changed significantly. Hanging and stick shellfish culture in the Philippines has increased sedimentation to the point that progradation of coastlines has occurred in some areas (Davis 1956). Mollusc culture in France has caused sedimentation due to the reduced current velocities among the oyster racks and bouchot poles (Ottman and Sornin 1985; Sornin 1979; Sornin 1981). About 30% of the oyster and mussel farms of France face problems of sedimentation, forcing occasional relocation and abandonment of the beds.

In addition to sedimentation attributable to changes in water flow, all forms of animal aquaculture produce organic-rich fecal wastes, and pseudofeces in the case of bivalves. A typical oyster (*Crassostrea gigas*) raft in Hiroshima Bay, for example, holds 420,000 oysters. Over a nine month culture period such a raft can produce 16 metric tons of feces and pseudofeces (Arakawa et al. 1971). With about 1,000 rafts in operation in Hiroshima Bay (Arakawa 1973) culture may have a major impact on sediment deposition in the bay. Fish culture will not only generate fecal waste, but if feed is provided, uningested feed also contributes to the total particulate load. The amount of unutilized feed in salmon culture is poorly documented and will vary between farms, but has been estimated to range from 1-40% of the feed provided (VKI 1976; Penczak et al. 1982; Braaten et al. 1983; Gowen et al. 1985; Phillips and Beveridge 1986). Estimates of solid waste production from salmonid net-cage culture range from 0.3 to 0.7 kg dry weight of waste feed and feces per kg of fish produced (Phillips and Beveridge 1986; Weston 1986; Gowen and Bradbury 1987).

The accumulation of feed and feces in the vicinity of the culture site results in physical and chemical changes in the substrate. These effects include increased organic carbon, nitrogen and phosphorus content; increased rate of sediment oxygen consumption; decreased sediment redox potentials; the generation of hydrogen sulfide and methane; and changes in the rates of nutrient cycling (Pamatmat et al. 1973, Dahlbäck and Gunnarson 1981; Cho et al. 1982; Trojanowski et al. 1982; Enell and Löf 1982; Jana and Patel 1984; Kaspar et al. 1985, 1988; Hall and Holby 1986; Brown et al. 1987; Blackburn et al. 1988). These changes are similar in both shellfish and fish culture and in both fresh and saltwater culture.

The effects of marine fish culture and suspended mussel culture on benthic communities have been examined in Japan (Kitamori 1977; Tsutsumi and Kikuchi 1983), Europe (Tenore et al. 1982; Mattsson and Linden 1983; Doyle et al. 1984; Olsgard 1984; Stewart 1984; Ervik et al. 1985; López-Jamar 1985; Brown et al. 1987), North America (Pease 1977; Weston 1990), and New Zealand (Kaspar et al. 1985). The effects of mariculture on benthic infaunal communities are consistent with the generalized trends described for other sources of organic enrichment (Pearson and Rosenberg 1978). Macrofauna may be lacking entirely in the area directly under the culture site. At the perimeter of the farm and for a short distance beyond, there is a zone characterized by a low number of species, high infaunal density, the presence of species generally characterized as opportunistic (most notably members of the *Capitella capitata* species complex), and a concentration of the infaunal biomass at or near the sediment-water interface. A white mat on the sediment surface, formed by the sulfur-oxidizing bacteria *Beggiatoa* is often reported in a halo around the farms, although the mat is often poorly developed directly below the site (personal observation, Earll et al. 1984). The cause of the reduced density of *Beggiatoa* directly under the culture operations has not been investigated although there may be several contributing factors such as a rapid sedimentation rate, hypoxic water above the sediment-water interface, or antibiotics in the waste feed and feces. The same type of changes including reduced species richness, predominance of enrichment-tolerant faunal groups (e.g. tubificid oligochaetes), and the appearance of bacterial growth (*Sphaerotilus* and *Beggiatoa*) are reported in pond, lacustrine and riverine aquaculture (Hinshaw 1973; Mantle 1982; Markmann 1982; Skogheim and Bremnes 1984; White 1986).

The areal extent of benthic impacts from culture will depend upon farm production, culture practices and bathymetric and hydrographic conditions. Most investigations of finfish and shellfish mariculture sites have shown effects extending 20-45 m from the culture site (Mattsson and Linden 1983; Doyle et al. 1984; Brown et al. 1987), although at one atypically large salmon net-cage site in the northwestern United States dramatic effects were evident at least 45 m from the cages and more subtle effects were measurable at least 150 m distant (Weston 1990).

Case studies

The effects of salmon net-cage culture on local macroinvertebrate communities are well-illustrated by examples from two farm sites: one in Scotland containing 35 metric tons of Atlantic salmon (*Salmo salar*); and a second in the northwestern United States holding 200-400 metric tons of coho salmon (*Onchorhyncus kisutch*). At both sites the number of macrofaunal species within a few meters of the farm perimeter was reduced to less than 20% of the number of species characteristic of undisturbed conditions (Fig. 1). Total abundance was elevated in close proximity to the farm sites. At the Scottish farm this peak in abundance was due to high densities of the polychaetes *Mediomastus fragilis* and *Capitella* cf. *capitata*. At the North American farm site nematodes and, to a much lesser extent, *Capitella* cf. *capitata* were responsible for the observed increase in density. An increase in total community biomass often accompanies an increase in abundance in organically-enriched areas (Pearson and Rosenberg 1978). Increased biomass was observed at the Scottish farm, but biomass was depressed near the North American site because of the small size of the dominant nematodes.

The effects of organic enrichment attributable to mariculture are reflected in the vertical distribution of infaunal biomass within the sediment column as well as in species composition and density. As organic enrichment increases, the depth of the aerobic surficial sediment stratum

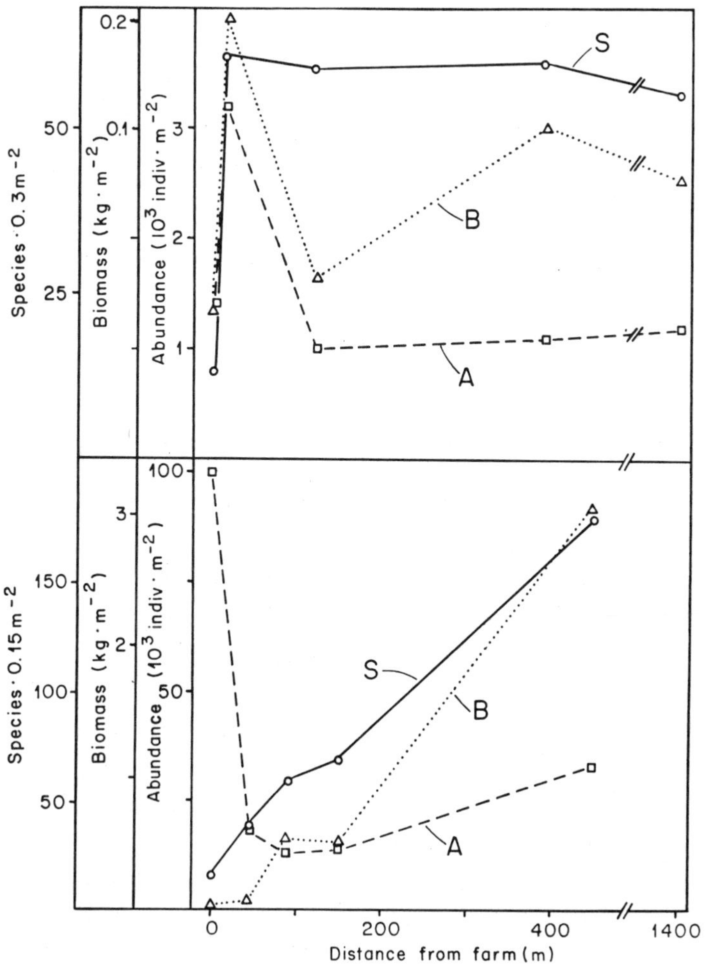

FIGURE 1. Trends in areal species richness (S), biomass (B) and abundance (A) of macrofauna with distance from two salmon net-cage mariculture sites. Data in upper figure from Loch Spelve, Scotland, from Brown et al. (1987). Data in lower figure from Puget Sound, northwestern United States, from Weston (1990).

becomes progressively shallower, and the sediment column becomes either too depleted in oxygen or enriched in sulfides for many macrofaunal species. The most obvious impact of these physical and chemical changes is a

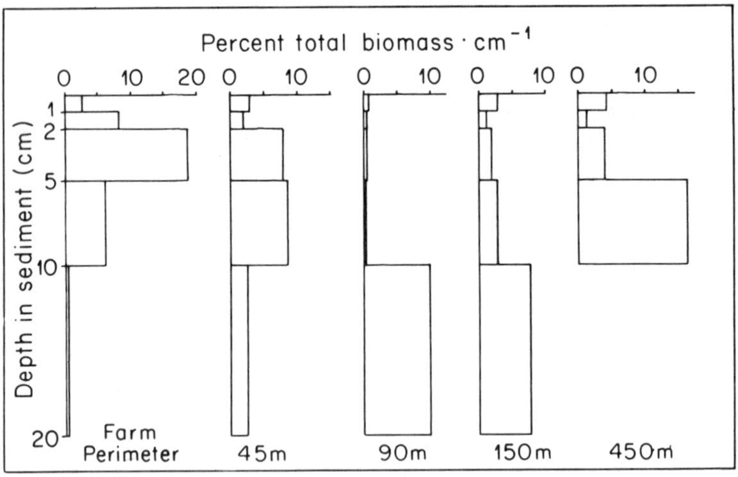

FIGURE 2. Trends in vertical distribution of infaunal biomass in the sediment column with distance from a salmon net-cage site. Figures portray the percentage of total biomass at the station found within each 1 cm stratum (i.e. at the farm perimeter 54% of the biomass was found in the 2-5 cm sampling interval so it was assumed that each of the 1 cm strata within the interval contained 18% of the biomass). The corer did not penetrate below 10 cm at the 450 m site. From Weston (1990).

disappearance of the large, deep-burrowing infauna, causing the vertical biomass profile to shift upwards within the sediment column. This trend was evident (Fig. 2) at the same North American farm site mentioned above. At the farm perimeter the majority of the infaunal biomass was present between depths of 2 and 5 cm in the sediment column. Over 90% of the total biomass was present at depths shallower than 10 cm. With increasing distance from the farm site, progressively more of the biomass was found in the deeper sediment strata. In marked contrast to conditions at the farm perimeter, at distances of 90 to 150 m from the farm only 3-22% of the total biomass was present above a depth of 10 cm and most of the biomass, consisting primarily of large bivalves, was found in the 10-20 cm stratum. Qualitative observations suggest the same pattern persisted 450 m from the farm.

The rates at which benthic conditions change after installation of a culture operation and the rate at which normal conditions return after removal of the farm will vary among sites, but it appears that disturbance occurs over a period of months while recovery requires several years. After installation of mussel (*Mytilus edulis*) longlines at a Swedish site, the original fauna were replaced with an infaunal assemblage characteristic of enriched conditions over a period of 6 to 15 months (Mattsson and Linden 1983). After installation of salmon net-cages, *Capitella* cf. *capitata*, which had previously been absent, appeared in less than 3 to 6 months at sites in both Scotland (Gowen et al. 1988) and the United States (Weston and Gowen 1988). Following cessation of organic input after harvest or removal of the culture structure, some period of time will be required for the benthic population to recover from the disturbance. Studies to date have not continued for a sufficiently long period of time to fully document the recovery process — recovery was incomplete 18 months after removal of mussel longlines in Sweden (Mattson and Linden 1983) and 20 months after removal of salmon net-cages in Scotland (Gowen et al. 1988). Recovery of benthos following closure of a pulp mill has been found to require three to eight years (Rosenberg 1976), and a similar period of time would probably be required for recovery from the effects of organic enrichment attributable to mariculture.

Mitigation strategies

Culture in open water, either net-cage culture of fish or suspended culture of shellfish, is not amenable to many of the solid waste retention techniques available to culturists operating in more confined systems (e.g., the use of settling basins in raceway culture). Several strategies have been employed to minimize sediment accumulation beneath marine net-cage sites, although generally the motivation behind these efforts has been to protect the cultured organisms from release of toxic substances such as hydrogen sulfide, rather than for reasons of environmental protection. Large funnel-like devices from which solid wastes can be removed have been suspended beneath net-cages, but the technique has so far proven inefficient or impractical (Enell et al. 1984; Leminen et al. 1986; Kehitysaluerahasto 1987). The operator is also faced with difficulty in disposing of this material, because the salt content prohibits its use for agricultural fertilizer or similar upland applications. Dredging or submersible fan-like devices are used at some sites to remove or disperse accumulated sediments, but the only benefits of such tactics are to the culture operation rather than to the natural infaunal community. Site rotation has been used as a management technique to exploit marginal environments, but again this practice only benefits the culturist. Since the period required for disappearance of the natural community after initiation of culture is on the order of months, and the frequency of site rotation is on the order of years, the practice only results in the disturbance of a greater area of the bottom.

The only alternatives presently available to net-cage culturists to minimize benthic impacts are proper site selection and husbandry practices. Historically, most net-cage sites have been placed in semi-enclosed embayments with water depths of 20 m or less. There is an industry-wide trend towards location in more exposed areas with deeper water and faster currents. Such conditions would promote the dispersal of solid wastes and reduce sediment enrichment in the vicinity of the farm site. Husbandry practices which reduce the amount of wasted feed also reduce benthic impacts. These practices include the use of more efficient feed delivery systems or the use of low density feeds which would sink at slower speeds, maximizing both the opportunity for ingestion and dispersal by currents.

EUTROPHICATION

Potential effects

Fecal and excretory wastes, together with waste feed, will alter water quality by nutrient enrichment and depletion of dissolved oxygen. The biological manifestation of nutrient additions can include local changes in the standing stock, species composition, or productivity of phytoplankton and macroalgae. Laboratory studies have demonstrated waste products from aquaculture can stimulate the growth of certain phytoplankton species (Nishimura 1982; Takahashi and Fukazawa 1982). This algal growth may threaten the farmed crop as well as wild fauna if they can not escape the affected area. Certain species of phytoplankton can kill fish and shellfish by physical damage to sensitive tissues, clogging of gills, or production of toxins. Algal respiration or decomposition can reduce dissolved oxygen concentrations to critical levels (Smayda and Packard 1979). The farm can also be affected by macroalgal growth which reduces water flow through the culture site. Marketability can be affected either by tainting of the flesh with odorous metabolites of certain algae or by the presence of algal toxins harmful to the consumer such as paralytic and diarrhetic shellfish poisoning.

Intensive culture systems

One of the principal determinants of the type and magnitude of impact on algal populations is whether the culture system requires the supply of food to the cultured species (intensive culture) or whether the organisms subsist entirely on natural food sources (extensive culture). Eutrophication is most likely in intensive culture systems where the input of feed results in a net increase in the amount of nutrients in the surrounding environment. This type of culture, represented by many forms of fish culture, can affect phytoplankton and macroalgal communities by supplying a nutrient that is limiting in the natural environment. Nutrients, most notably phosphorus and nitrogenous compounds, are often limiting in natural systems and have received the greatest amount of study, but micronutrients like vitamins may be important as well (Gowen and Bradbury 1987; Rosenthal et al. 1988).

There are numerous examples of measurable water quality changes in the vicinity of freshwater farm sites including decreased dissolved oxygen concentrations and increased ammonia, organic and total nitrogen, phosphate, and BOD (Eley et al. 1972; Hinshaw 1973; Bergheim and Selmer-Olsen 1978; Korzeniewski et al. 1982; Mantle 1982; Beveridge 1984). Phytoplankton growth stimulated by nutrient additions in freshwater systems is well-documented, and sometimes used to the advantage of the farm operator. Application of inorganic or organic fertilizers is used in some extensive pond culture systems to increase primary productivity and increase the food supply to the cultured species (Boyd 1982). Nutrient enrichment from intensive culture, on the other hand, can accelerate eutrophication with adverse consequences to both the culture operation and the ecosystem as a whole. There is concern that intensive aquaculture in some lakes will alter the trophic state of the water bodies and result in the disappearance of native fish species (Korycka and Zdanowski 1981; Penczak et al. 1982).

While nutrient loadings per unit production from marine culture facilities are no less than from freshwater farms, measurable changes in water quality from marine culture are not as commonly observed nor as great in magnitude or spatial extent. Effects have been noted in some cases such as increased ammonia concentration around salmon net-cages in Norway (Ervik et al. 1985), or decreased dissolved oxygen concentrations of 0.2-2.5 $mg\,l^{-1}$ associated with yellowtail and sea bream net-cages in Japan (Kadowaki et al. 1978; Kadowaki and Hirata 1984). In other marine situations, however, the effects of culture on water quality have been unmeasurable or detectable only under special circumstances (Hagstrom and Larsson 1982; Kaspar et al. 1985; Larsson 1985; Müller-Haeckel 1986). The principal reasons for these differences between freshwater and marine systems are the lesser degree of water column mixing (and therefore, dilution of wastes) in freshwater, and the fact that in freshwater systems a large proportion of the water body is often routed through the culture operation (e.g. 70% of river flow at a Norwegian site; Bergheim and Selmer-Olsen 1978).

Even at those mariculture sites where nutrient enrichment is observed, there is usually no measurable effect on phytoplankton communities (Fig. 3). No effects of mariculture on phytoplankton abundance, biomass or productivity were found near salmon farms in the northwestern United States (Pease 1977 as measured by chlorophyll a), Sweden (Müller-Haeckel 1986 as measured by cell density), or Scotland (Gowen et al. 1988 as measured by chlorophyll a and productivity). Measurable effects of intensive marine finfish culture on algal communities have been observed only for attached macroalgae (Ruokolahti 1988) or planktonic microalgae when water mixing is unusually restricted such as in marine pond culture (Krom et al. 1985) or in coastal areas lacking appreciable tidal currents (Halonen 1985).

Ammonium produced by mariculture is readily assimilated by microalgae, and, if growth is nitrogen-limited, nutrient additions should readily be converted to algal biomass. Individual phytoplankton cells, however, require several hours to divide even at rapid reproduction rates, and whole populations generally require a day or more to double in number (Parsons et al. 1977). In most marine areas, the nutrient-enriched water would be rapidly diluted, and any effect on phytoplankton biomass would be undetectable. A localized and measurable effect on marine microalgal communities would be expected only in cases of unusually dense culture activity in poorly flushed coastal embayments.

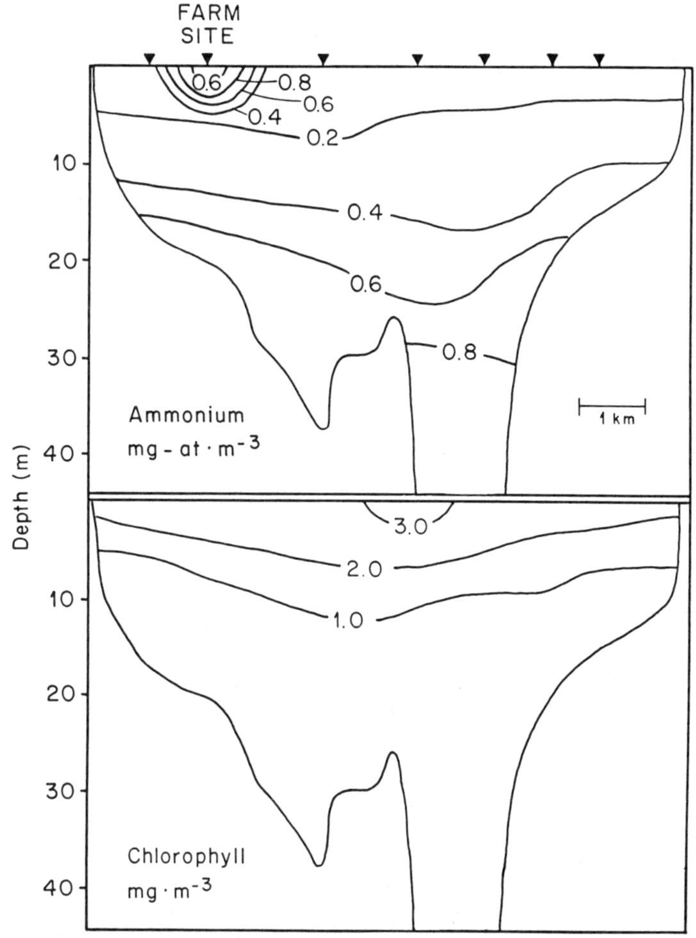

FIGURE 3. Cross-sectional profile of Loch Spelve, Scotland illustrating the distribution of ammonium and chlorophyll in August 1985. Location of profiling stations indicated on upper figure. From Gowen et al. (1988).

Extensive culture systems

Harvest of organisms results in a net loss of nutrients from extensive culture systems when neither feed nor fertilizer is added. For example, in mussel culture the nutrient needs of the cultured organisms are derived from consumption of phytoplankton and suspended detritus. About 30% of the nutrients extracted by the mussels are removed with the harvest, 30% fall to the bottom as feces and pseudofeces, and 40% are released to the water column as soluble metabolites (Ackefors and Grip 1985, Ackefors and Södergren 1985).

Although extensive culture systems such as bivalve culture do not cause eutrophication attributable to allochthonous nutrient additions, there are other effects of culture activities on the natural microalgal communities. The potential effects include either or both a decrease in phytoplankton biomass and an increase in primary productivity.

Shellfish culture can reduce phytoplankton biomass through the filter-feeding activity of the bivalves. The effectiveness with which oysters remove particulate material, including phytoplankton, is evident in the work of Imai (1971). Seston concentration was reduced 23% from initial concentration after passage through a single oyster raft, and eventually reduced 94% after passage through 11 rafts. The consumption of phytoplankton by cultured bivalves could potentially reduce the food supply available to zooplankton and result in major ecological consequences of bivalve culture. Work at Swedish (Incze and Lutz 1980; Rosenberg and Loo 1983) and Irish (Rodhouse and Roden 1987) mussel culture sites has suggested that shellfish production could be greatly increased before the mussels would dramatically reduce the concentration of phytoplankton available to other herbivores. In Spain's Ria de Arosa, however, culture has dramatically changed the composition of the natural planktonic community. The culture of 100,000 metric tons of mussels per year has caused replacement of zooplankton by cultured mussels as the dominant herbivore, and replacement of copepods with the larvae of epifaunal organisms as the dominant zooplankters (Tenore and Gonzàlez 1975; Tenore et al. 1985; Hanson et al. 1986).

Shellfish culture may enhance primary productivity through an increased rate of nutrient cycling (Tenore and González 1975; Campos and Mariño 1982; Kaspar et al. 1985). Bivalve culture will consume phytoplankton biomass and release nutrients capable of supporting additional primary production. These nutrients will be made available either directly, as excretion of ammonia and other soluble metabolites, or indirectly through remineralization of fecal and pseudofecal wastes.

Mitigation strategies

Mechanical methods to reduce the nutrient loading from aquaculture are largely limited to devices for the collection of solid wastes, such as sediment retention ponds (upland culture) or funnel-shaped waste catchment devices (net-cage culture). In freshwater environments, phosphorus is generally the nutrient limiting phyhtoplankton growth and such waste retention systems may be appropriate since most of the phosphorus released from the culture site will be in a particulate form. In marine environments, however, nitrogen typically limits phytoplankton growth and nitrogen inputs from aquaculture are primarily in the form of soluble metabolites such as ammonia and urea. Removal of these soluble metabolites is not economically feasible.

Environmental loading of nutrients from intensive culture systems can be reduced by formulating feeds to reduce the amount of unassimilable phosphorus. Considerable progress has been made in reducing the phosphate content of fish feed without impairing growth. It has been estimated that while nitrogen loading to Finnish waters from aquaculture has increased three-fold over the past decade, paralleling industry growth over the same period, the phosphorus loading has only slightly increased because of a concomitant reduction in the phosphorus content of fish feed (Table 1).

TABLE 1. Estimated historical trends in phosphorus and nitrogen loadings from inland and brackish water aquaculture in Finland (from ICES 1988).

Year	Production (tons/yr^{-1})	Feed composition (%) Phosphorus	Nitrogen	Loading (tons/yr^{-1}) Phosphorus	Nitrogen
1978	3,205	2.0	6.5	115	337
1979	3,275	1.9	6.5	105	323
1980	4,670	1.7	6.5	132	460
1981	5,383	1.5	6.5	124	495
1982	6,325	1.3	6.5	123	582
1983	7,511	1.1	6.5	119	691
1984	9,493	1.1	6.5	140	812
1985	10,913	1.0	6.5	131	861
1986	10,913	1.0	6.5	142	933

Management of nutrient loading from aquaculture and the potential effects of eutrophication would be facilitated by the development of holding capacity models that predict the degree of ecosystem alteration at a given nutrient loading. Unfortunately, precise and reliable quantification of holding capacity is not currently possible, and we are limited to gross approximations. The approach of Vollenweider (1968) has been used to relate primary productivity to some permissible aquaculture-related increase in phosphorus loading in freshwater lakes (Beveridge and Muir 1982; Beveridge 1984). There are, however, no comparable models for marine environments, so the determination of holding capacity in estuaries and coastal embayments has been limited to a trial and error approach of little predictive utility.

NATIVE FISH POPULATIONS

In general, aquaculture results in an increased density and species richness of wild fish fauna in the vicinity of culture operations. Exceptions to this may include areas where excessive culture has caused eutrophication (Penczak et al. 1982) or areas where culture activities occupy much of the water body, displacing the native fauna.

Several authors have reported higher densities of wild fish in close proximity to fish net-cages than in reference areas. Cage culture of channel catfish (*Ictalurus punctatus*) and rainbow trout (*Onchorhyncus mykiss*) in freshwater lakes in the United States has been found to increase the density of wild fish such as bluegills (*Lepomis macrochirus*) and redear sunfish (*L. microlophus*) in the vicinity of the cages (Collins 1970; Loyacano and Smith 1976; Hays 1980). Rainbow trout culture in Scottish lochs has increased the growth rate of native roach (*Rutilus rutilus*), brown trout (*Salmo trutta*) and stocked rainbow trout (Forbes 1981; Phillips et al. 1985). The density of roach and white bream (*Blicca bjoerkna*) increased around a Finnish fish farm (Koivisto 1987).

Such effects have been reported in the vicinity of shellfish culture sites as well. Kaspar et al. (1985) reported that the bottom beneath mussel longlines attracted predators such as asteroids and fish. In the Ria de Arosa, abundance of some gobiid species increased in areas near raft culture, although there were no other consistent differences between the fish populations of raft and nonraft areas (Chesney and Iglesias 1979; Iglesias 1979). In species common to both areas, however, there was a shift in diet from infaunal to raft epifaunal prey (López-Jamar et al. 1984).

There are several reasons to expect an increase in fish abundance, biomass, or species richness in the vicinity of aquaculture facilities. The culture site is likely to represent an area of enhanced food availability; feed pellets not ingested by cultured fish are utilized by the wild fish community (Collins 1970; Phillips et al. 1985). Epifaunal and epiphytic growth on the culture structure could be utilized as a food resource (López-Jamar et al. 1984). The dense populations of opportunistic macrofaunal species characteristic of organically-enriched sediments comprise a major portion of the diet of many demersal fishes (Becker and Chew 1987). In addition there is a behavioural tendency for fish to congregate around structures (Gooding and Magnuson 1967), and the culture structure may act in ways comparable to an artificial reef. Finally, fertilization of lakes has been shown to increase fish productivity (Barraclough and Robinson 1972; Mills 1985) and nutrient inputs from aquaculture could have a comparable effect.

CHEMICAL USAGE

Many of the chemicals used in the aquaculture industry present environmental risks if misused or if their dangers are unrecognized. Aquaculturists may become aware of toxic effects as a result of mortalities of the cultured organisms, but impacts to wild fauna are usually overlooked.

Construction materials may contain many potentially toxic chemicals, such as fungicides that have been implicated in mortalities of juvenile salmonids (Zitko et al. 1985). Tributyltin (TBT) antifoulants are the best-publicized example of a substance incorporated in mariculture construction materials that has subsequently been found to be toxic to both the cultured animals (Short and Thrower 1987) and wild fauna (Alzieu et al. 1980; 1981). Although the impacts of the TBT on shellfish and other fauna was a consequence of its use as an antifoulant on boat hulls, the use of the compound as a net antifoulant in mariculture has been prohibited in many countries.

Many of the chemicals of potential concern are deliberately used for their biocidal properties, yet could have adverse effects on non-target organisms. Examples would include the antibiotics (e.g. oxytetracycline, sulphamerizine, trimethoprim), parasiticides and fungicides (e.g. dichlorvos, formalin, malachite green), disinfectants (e.g. calcium hypochlorite), algicides and herbicides (e.g. copper sulfate), and chemicals used for predator or pest control (e.g. carbaryl).

The scale of potential environmental problems can not be estimated since the quantities of chemicals used in the aquaculture industry has not been determined. Statistics on chemical usage are available for only a very few countries and for only a very few industry segments. The Norwegian salmon mariculture industry produced 22,000 metric tons of fish in 1984, and in the same period used 18 metric tons of Neguvon for the treatment of salmon lice, and 20 tons of antibiotics including 6.2 tons oxytetracycline, 7.8 tons Tribrissen, 5.5 tons nitrofurazolidone, and 0.009 tons sulphamerizine (weight given are those of the active substances alone - Midtlyng 1985). The Swedish rainbow trout culture industry, which produced about 4000 tons of trout in 1987, used 0.325 tons of oxytetracycline in the same year (ICES 1988). Antibiotic use per ton of fish produced may be no different in these countries than elsewhere, but Sweden and Norway are among the few countries for which statistics are available.

Antibiotic usage is an area of special concern because of the potential for stimulation of antibiotic resistance in the microbial community. Use of antibiotics in finfish culture has resulted in the appearance of resistant bacterial strains in the vicinity of culture sites (Bullock et al. 1974; Aoki et al. 1980; 1985; Austin 1985). Of 170 bacterial strains isolated from cultured rainbow trout, 87.6% showed resistance to at least one chemotherapeutant, most frequently ampicillin, sulfadiazine, nitrofurantoin and chloramphenicol (Toranzo et al. 1984). Drug resistance may be transferred to other bacteria through plasmid or transposon exchange, and even in the absence of positive selection, resistance may persist for long periods, being lost ultimately only by dilution with non-resistant strains (Brazil et al. 1986; Levy 1986). The development of antibiotic resistant strains is of obvious concern to the culturist, but it is not clear if resistance in aquatic and marine bacteria is of major significance to the efficacy of antibiotic therapy in human medicine.

An important factor determining the degree of antibiotic resistance found in bacteria is the magnitude of selective pressure for that resistance, i.e. the frequency of antibiotic usage. Unlike many forms of animal husbandry in which antibiotics are used on a long-term chronic basis, aquaculturists employ antibiotics only for short term therapeutic purposes. This distinction breaks down, however, if antibiotic residues persist in the environment for long periods following treatment. Recent laboratory work has shown that the half-life of oxytetracycline in marine sediments is 10 weeks, and antimicrobial activity is measurable for up to 12 weeks following treatment (Jacobsen and Berglind 1988).

The principal environmental concern for chemicals other than antibiotics is toxicity to non-target organism. For example, use of Neguvon for treatment of salmon lice (*Lepeophtheirus salmonis*) at a Norwegian fish farm is suspected to have resulted in the death of captive lobsters in the vicinity, and subsequent laboratory studies demonstrated it was highly toxic to lobsters and slightly less toxic to crabs and mussels (Egidius and Møster 1987). Death of animals beyond the confines of the fish cages was discovered only because the lobsters were held in crates, and it is likely that adverse effects to wild fauna in this or similar instances would have gone undetected.

Chemical usage in aquaculture and potential impacts to indigenous fauna is not limited to the fish industry. In the northwestern United States the oyster industry uses the carbamate pesticide Sevin to kill upogebiid and callianassid shrimp whose burrowing causes burial of young and adult oysters. The beds are treated by hand or by helicopter as needed, typically about every six years. Treatment inevitably results in mortalities to wild fauna other than the target species, particularly cancrid crabs (WDF/WDOE 1985; 1989).

The environmental impact of chemical use in aquaculture needs much further research. In review and approval of new therapeutants, much attention is given to potential risks to human consumers, but information on environmental effects is often minimal. Information on drug retention is needed to determine the proportion that is released to the environment. For example, the digestibility of chloramphenicol is 99%, but only 7-9% for oxytetracycline so much of the latter compound is released to the environment in a biologically-active form (Cravedi et al. 1987). Information on environmental persistence is needed under varying conditions of salinity, temperature, and oxidation. Finally data on environmental fate (e.g. solubility, breakdown products) and toxicity to aquatic and marine biota are essential to an evaluation of the environmental risks of chemical usage in aquaculture.

OTHER ENVIRONMENTAL ISSUES

It is generally assumed that genetic adaptations of native populations confer a selective advantage in the habitat in which the populations naturally occur. Populations introduced for aquaculture are presumably less fit, either because their genome is best adapted to their original habitat rather than the locality to which they are introduced, or because of selective breeding for attributes such as enhanced growth efficiency or high fecundity that are desirable in the culture environment. It has been demonstrated that hatchery trout have a lower survival rate after release than wild populations (Reisenbichler and McIntyre 1977; Chilcote et al. 1985). Should cultured animals escape and interbreed with native populations, there may be a temporary reduction in the reproductive capacity of the population as a whole, since reproductive effort may be wasted in producing genotypes that would be subject to significant negative selective pressures.

The genetic consequences of interbreeding between cultured and wild populations is of concern, but there are very little data with which to assess the magnitude of potential impact. Obviously, the extent of impact to indigenous populations will depend principally on the probability of interbreeding with cultured animals (extent of escapes, ecological or behavioural barriers to interbreeding) and the relative size of the breeding population of escaped fish in comparison to the wild breeding population. There has been some movement in the aquaculture industry towards the development of sexually sterile stocks for cultivation. Sterility of the animals has several advantages from a husbandry perspective, and it also would be advantageous in eliminating the possibility of genetic alteration of native populations.

The use of exotic species in aquaculture or the reliance on distant sources for eggs or juveniles involve risks of ecological change resulting from introduction of a species into an environment where it had not previously occurred. There are many examples of accidental introductions attributable to either aquaculture or fisheries enhancement. At least 48 species of freshwater fish parasites have been transferred among continents because of the importation of live or frozen fish (Hoffman 1970). Transfer of trout and infected eggs has spread the IHN virus throughout the northwestern and north central United States and into Japan (Amend et al. 1973; Egidius 1984). Furunculosis may have been introduced into Europe by the importation of rainbow trout from North America (Rosenthal 1980). Transplantation of shellfish has resulted in widespread dispersal of associated pests, predators and pathogens (Rosenfield and Kern 1979). A recent example is the appearance of the parasite *Gyrodactylus salaris* in many Norwegian rivers. The parasite is believed to have been introduced by stocking of infected fish (Johnsen and Jensen 1988).

The environmental consequences of the introduction of exotic species for purposes of aquaculture have been reviewed by Rosenthal (1980) and potential impacts are well documented. Regulatory mechanisms are already in place to assess and minimize the potential environmental effects of introductions, although the extent of environmental review and import restrictions vary widely among countries. Of all possible effects of aquaculture on indigenous fauna, deliberate or accidental introduction of exotic species has the greatest ramifications, because the effects may be continental in scale and in most cases are irreversible.

CONCLUSIONS

Aquaculture, like other industries, generates wastes that can have a significant effect on the surrounding physical, chemical and biological environment. With expansion of the industry and technological advances which permit an increased production per unit area, it is becoming imperative that consideration be given to the extent to which culture alters the structure and functioning of natural communities. Fortunately, many of the potential adverse effects of culture can be mitigated or eliminated by care in site selection and operation.

Aquaculture can cause environmental change without any impact to the viability of the culture facility itself. Such changes include the introduction of exotic species, toxicity from aquaculture chemicals, and genetic changes in wild populations. In many other respects, however, minimization of environmental change is in the interest of the industry. The farm itself can be adversely affected by many of the same changes that affect the indigenous communities including organic enrichment of sediments, eutrophication, stimulation of antibiotic resistance and introduction of pathogens. Consequently it is not surprising that many industry trends such as movement to more open, exposed sites with better flushing, reduction of phosphorus content of feed, increased use of vaccination rather than antibiotic treatment, and culture of sexually sterile organisms are advantageous both from the perspectives of husbandry and environmental protection.

ACKNOWLEDGMENTS

This work was supported by Cooperative Agreement No. CX-813705-01 between the U.S. Environmental Protection Agency and the University of Washington, and by a contract from the Washington Department of Fisheries to the University. The suggestions of D. Penry and two anonymous reviewers helped to improve the manuscript. Contribution No. 1808 from the School of Oceanography, University of Washington.

LITERATURE CITED

Ackefors, H. and K. Grip. 1985. Musslor och Ostron. Swedish Council for Planning and Coordination of Research (FRN). 84 pp.

Ackefors, H. and A. Södergren. 1985. Swedish experiences of the impact of aquaculture on the environment. International Council for the Exploration of the Seas C.M. 1985/E:40. 7 pp.

Alzieu, C., M. Heral, Y. Thibaud, M.-J. Dardignac and Y. Feuillet. 1981. Influence des peintures antisalissures a base d'organostanniques sur la calcification de la coquille de l'huitre *Crassostrea gigas*. Revue des Travaux de l'Institut des Peches Maritimes 44(4): 301-348.

Amend, D. F., W. T. Yasutake, J. L. Fryer, K. S. Pilcher and W. H. Wingfield. 1973. Infectious hematopoietic necrosis (IHN). Pages 80-87 in W. A. Dill, editor. Symposium on the Major Communicable Fish Diseases in Europe and their Control. EIFAC Technical Paper 41.

Aoki, T., Y. Jo and S. Egusa. 1980. Frequent occurrence of drug resistant bacteria in Ayu (*Plecoglossus altivelis*) culture. Fish Pathology 15(1): 1-6.

Aoki, T., T. Kanazawi and T. Kitao. 1985. Epidemiological surveillance of drug resistant *Vibrio anguillarum* strains. Fish Pathology 20: 199-205.

Arakawa, K. Y. 1973. [Aspects of eutrophication in Hiroshima Bay viewed from transition of cultured oyster production and succession of marine biotic communities]. Niho Kaiyo Gakkai-Shi 11(2): 43-48.

Arakawa, K. Y., Y. Kusuki and M. Kamigaki. 1971. [Studies on biodeposition in oyster beds: I. Economic density for oyster culture]. Venus 30(3): 113-128.

Austin, B. 1985. Antibiotic pollution from fish farms: effects on aquatic microflora. Microbiological Sciences 2(4): 113-117.

Barraclough, W. E. and D. Robinson. 1972. The fertilization of Great Central Lake. II. Effect on juvenile sockeye salmon. Fishery Bulletin 70: 37-48.

Becker, D. S. and K. K. Chew. 1987. Predation on *Capitella* spp. by small-mouthed pleuronectids in Puget Sound, Washington. Fishery Bulletin 85(3): 471-479.

Bergheim, A. and A. R. Selmer-Olsen. 1978. River pollution from a large trout farm in Norway. Aquaculture 14: 267-270.

Beveridge, M. 1984. Cage and pen fish farming: carrying capacity models and environmental impact. FAO Fisheries Technical Paper 255. 131 pp.

Beveridge, M. and J. F. Muir. 1982. An evaluation of proposed cage fish culture on Loch Lomond, an important reservoir in central Scotland. Canadian Water Resources Journal 7(2): 181-196.

Blackburn, T. H., B. Aa. Lund and M. D. Krom. 1988. C- and N-mineralization in the sediments of earthen marine fishponds. Marine Ecology - Progress Series 44: 221-227.

Boyd, C. E. 1982. Water Quality Management for Pond Fish Culture. Developments in Aquaculture and Fisheries Science, Volume 9. Elsevier, Amsterdam, The Netherlands.

Braaten, B., J. Aure, A. Ervik and E. Boge. 1983. Pollution problems in Norwegian fish farming. International Council for the Exploration of the Sea C.M. 1983/F:26.

Brazil, G., D. Curley, F. Gannon and P. Smith. 1986. Persistence and acquisition of antibiotic resistance plasmids in *Aeromonas salmonicida*. Pages 107-113 in S. B. Levy and R. P. Novick, editors. Banbury Report 24: Antibiotic Resistance Genes: Ecology, Transfer, and Expression. Cold Spring Harbor Laboratory, New York, New York, USA.

Brown, J. R., R. J. Gowen and D. S. McLusky. 1987. The effect of salmon farming on the benthos of a Scottish sea loch. Journal of Experimental Marine Biology and Ecology 109: 39-51.

Bullock, G. L., H. M. Stuckey, D. Collias, R. L. Herman and G. Maestrone. 1974. In Vitro and in vivo efficacy of a potentiated sulphonamide in control of furunculosis in salmonids. Journal of the Fisheries Research Board of Canada 31: 75-82.

Campos, M. J. and J. Mariño. 1982. Comparison of phytoplankton species composition and abundance in the Rias Arosa and Muros (NW Spain). Marine Ecology 3(1): 1-12.

Chilcote, M. W., S. A. Leider and J. J. Loch. 1985. Hatchery summer-run steelhead natural reproductive success in the Kalama River. Washington Department of Game, Report No. 85-24, Olympia, Washington, USA.

Cho, C.-H., H.-S. Yang, K.-Y. Park and M.-K. Youm. 1982. Study on bottom mud of shellfish farms in Jinhae Bay. Bulletin of the Korean Fisheries Society 15(1): 35-41.

Chesney, E. J., Jr. and J. Iglesias. 1979. Seasonal distribution, abundance and diversity of demersal fishes in the inner Ria de Arosa, North West Spain. Estuarine and Coastal Marine Science 8: 227-239.

Collins, R. A. 1970. Cage culture of catfish in reservoir lakes. Proceedings Annual Conference Southeastern Association of Game and Fish Commissioners 24: 489-496.

Cravedi, J.-P., G. Choubert and G. Delous. 1987. Digestibility of chloramphenicol, oxolininc acid and oxytetracycline in rainbow trout and influence of these antibiotics on lipid digestibility. Aquaculture 60: 133-141.

Dahlback, B. and L. A. H. Gunnarson. 1981. Sedimentation and sulfate reduction under a mussel culture. Marine Biology 63: 269-275.

Davis, J. H. 1956. Influence of man upon coast lines. Pages 504-521 in W. L. Thomas, editor. Man's Role in Changing the Face of the Earth. University of Chicago Press, Chicago, Illinois, USA.

Doyle, J., M. Parker, T. Dunne, D. Baird and J. McArdle. 1984. The impact of blooms on mariculture in Ireland. International Council for the Exploration of the Sea, Special Meeting on the Causes, Dynamics and Effects of Exceptional Marine Blooms and Related Events, Copenhagen, 4-5 October 1984. ICES, Copenhagen, Denmark.

Earll, R. C., G. James, C. Lumb and R. Pagett. 1984. A report on the effects of fish farming on the marine environment of the Western Isles. Marine Biological Consultants, Kempley, Gloucestershire, Scotland.

Egidius, E. 1984. Diseases of salmonids in aquaculture. Helgolander Meeresuntersuchungen 37: 547-569.

Egidius, E. and B. Møster. 1987. Effect of Neguvon and Nuvan treatment on crabs (*Cancer pagurus, C. maenas*), lobster (*Homarus gammarus*) and blue mussel (*Mytilus edulis*). Aquaculture 60: 165-168.

Eley, R. L., J. H. Carroll and D. DeWoody. 1972. Effects of caged catfish culture on water quality and community metabolism of a lake. Proceedings Oklahoma Academy of Science 52: 10-15.

Enell, M. and J. Löf. 1983. Miljoeffekter av vattenbruk - sedimentation ach narsaltbelastning fran fiskkasseodlingar. Vatten 39: 364-375.

Enell, M., J. Löf and T.-L. Bjorklund. 1984. Fisk-kasseodling med Rening, Teknisk Beskrivning och Reningseffekt. Institute of Limnology, Lund, Sweden.

Ervik, A., P. Johannessen and J. Aure. 1985. Environmental effects of marine Norwegian fish farms. International Council for the Exploration of the Sea. C. M. 1985/F:37. 13 pp.

Forbes, I. J. 1981. A preliminary investigation into the effects of cage culture on the wild fish of Loch Fad, Bute. B.Sc. thesis. University of Stirling, Stirling, Scotland.

Gooding, R. M. and J. J. Magnuson. 1967. Ecological significance of a drifting object to pelagic fishes. Pacific Science 21: 486-497.

Gowen, R. J. and N. B. Bradbury. 1987. The ecological impact of salmonid farming in coastal waters: a review. Oceanography and Marine Biology Annual Review 25: 563-575.

Gowen, R. J., N. B. Bradbury and J. R. Brown. 1985. The ecological impact of salmon farming in Scottish coastal waters: a preliminary appraisal. International Council for the Exploration of the Sea C. M. 1985/F:35. 13 pp.

Gowen, R. J., J. R. Brown, N. B. Bradbury and D. S. McLusky. 1988. Investigations into benthic enrichment, hypernutrification and eutrophication associated with mariculture in Scottish coastal waters (1984-1988). Department of Biological Sciences, University of Stirling, Stirling, Scotland. 289 pp.

Hagstrom, A. and A-M. Larsson. 1982. Pelagiska bakterier och musselodling. Meddelande Havsfiskelaboratoriet Lysekil, No. 286. 21 pp.

Hall, P. and O. Holby. 1986. Environmental impact of a marine fish cage culture. International Council for the Exploration of the Sea C. M. 1986/F:46. 12 pp.

Halonen, L. 1985. Kalankasvatuksen vaikutukset rannikkovesissä, - Vesihallituksen monistesrja nro 346.

Hanson, R. B., M. T. Alvarez-Ossorio, R. Cal, M. J. Campos, M. Roman, G. Santiago, M. Valera and J. A. Yoder. 1986. Plankton response following a spring upwelling event in the Ria de Arosa, Spain. Marine Ecology - Progress Series 32: 101-113.

Hays, T. 1980. Impact of net pen culture on water quality and fish population on Bull Shoals Reservoir. Completion Report Arkansas Game and Fisheries Commission.

Hinshaw, R. N. 1973. Pollution as a result of fish cultural activities. EPA-R3-73-009. United States Environmental Protection Agency, Washington, D.C., USA.

Hoffman, G. L. 1970. Intercontinental and transcontinental dissemination and transfaunation of fish parasites with emphasis on whirling disease (*Myxosoma cerebralis*). Pages 69-81 in S. F. Snieszko, editor. A Symposium on Diseases of Fishes and Shellfishes. American Fisheries Society, Special Publication No. 5.

ICES. 1988. Report of the working group on environmental impacts of mariculture: Hamburg, Federal Republic of Germany, 19-21 April 1988. International Council for the Exploration of the Sea, Copenhagen, Denmark.

Iglesias, J. 1979. Spatial and temporal changes in the demersal fish community of the Ria de Arosa (NW Spain). Marine Biology 65: 199-208.

Imai, T. 1971. Aquaculture in Shallow Seas: Progress in Shallow Sea Culture. Koseisha Koseiku, Tokyo, Japan. Translated by the National Marine Fisheries Service and republished in 1977 by Amerind Publishing Company, New Dehli, India.

Incze, L. S. and R. A. Lutz. 1980. Mussel culture: an east coast perspective. Pages 99-140 in R. A. Lutz, editor. Mussel Culture and Harvest: a North American Perspective. Developments in Aquaculture and Fisheries Science, 7. Elsevier, Amsterdam, The Netherlands.

Jacobsen, P. and L. Berglind. 1988. Persistance of oxytetracycline in sediments from fish farms. Aquaculture 70: 365-370.

Jana, B. B. and G. N. Patel. 1984. Measurements of some groups of microbial populations with special reference to denitrification in water and sediments of fish farming ponds under monoculture and polyculture systems. Internationale Revue der Gesamten Hydrobiologie 69: 231-240.

Kadowaki, S. and H. Hirata. 1984. Oxygen distribution in the coastal fish farm. I. Effects of feeding on the distribution patterns. Suisan Zoshoku 32(3): 142-147.

Kadowaki, S., T. Kasedo and T. Nakazono. 1978. Continuous records of DO contents by cruising in the coastal culture farm. I. Relation between DO content and fish density in cages. Memoirs of the Faculty of Fisheries Kagoshima University 27(1): 273-280.

Kaspar, H. F., P. A. Gillespie, I. C. Boyer and A. L. MacKenzie. 1985. Effects of mussel aquaculture on the nitrogen cycle and benthic communities in Kenepuru Sound, Marlborough Sounds, New Zealand. Marine Biology 85: 127-136.

Kaspar, H. F., G. E. Hall and A. J. Holland. 1988. Effects of sea cage salmon farming on sediment nitrification and dissimilatory nitrate reductions. Aquaculture 70: 333-344.

Kehitysaluerahasto, OY. 1987. Kalanvilfelyn verkkoallaskasvatuksen vesistökuormituksen vähaentäminen. Kehitysalurahasto OY, Tutkimus A 7: 1987. 73 pp.

Kitamori, R. 1977. [Changes in the species composition (principally benthic organisms]. Pages 67-76 in Senkai Yoshoku to Jika Osen, Suisangaku Shirizu 21. Koseisha Koseikaku, Tokyo, Japan.

Korycka, A. and B. Zdanowski. 1981. Some aspects of the effect of cage fish culture on lakes with special reference to heated lakes. Pages 131-137 in K. Tiews, editor. Aquaculture in Heated Effluents and Recirculation Systems. Heenemann Verlagsgesellschaft, Berlin, Federal Republic of Germany.

Koivisto, V. 1987. Sammanfattande arbetsrapport, Levande skärgärd - levande vatten, fiskodlings konsekvenser i kust - och skärgärdsomrädet i SW Finland. 25 s., moniste.

Korzeniewski, K., Z. Banat and A. Moczulaka. 1982. Changes in water of the Uniesc and Skotawa Rivers, caused by intensive trout culture. Poliskie Archiwum Hydrobiologii 29: 683-691.

Krom, M. D., C. Porter and H. Gordin. 1985. Description of the water quality conditions in a semi-intensively cultured marine fish pond in Eilat, Israel. Aquaculture 49: 141-157.

Larsson, A-M. 1985. Blue mussel sea farming - effects on water quality. Vatten 41: 218-224.

Leminen, E., T. Mäkinen and J. Junna. 1986. Kalanviljelyn vesistökuormituksen vähentäminen varkkokassilaitoksella - kenttätutkimus meriolosuhteissa. Vesihallituksen monistesarja nro 6., 1986. 32 pp.

Levy, S. B. 1986. Ecology of antibiotic resistance determinants. Pages 17-29 in S. B. Levy and R. P. Novick, editors. Banbury Report 24: Antibiotic Resistance Genes: Ecology, Transfer, and Expression. Cold Spring Harbor Laboratory, New York, New York, USA.

López-Jamar, E., J. Iglesias, and J. J. Otero. 1984. Contribution of infauna and mussel-raft epifauna to demersal fish diets. Marine Ecology - Progress Series 15: 13-18.

Loyacano, H. A., Jr. and D. C. Smith. 1976. Attraction of native fish to catfish culture cages in reservoirs. Proceedings Annual Conference of Southeastern Association of Game and Fish Commissioners 29: 63-73.

Mantle, G. J. 1982. Biological and chemical changes associated with the discharge of fish-farm effluent. Pages 103-112 in J. S. Alabaster, editor. Report of the EIFAC Workshop on Fish-farm Effluents, Silkeborg, Denmark, 26-28 May 1981. Food and Agricultural Organization of the United Nations, EIFAC Technical Paper No. 41.

Markmann, P. N. 1982. Biological effects of effluents from Danish fish farms. Pages 99-102 in J. S. Alabaster, editor. Report of the EIFAC Workshop on Fish-farm Effluents, Silkeborg, Denmark, 26-28 May 1981. Food and Agricultural Organization of the United Nations, EIFAC Technical Paper No. 41.

Mattson, J. and O. Linden. 1983. Benthic macrofauna succession under mussels, *Mytilus edulis* L. (Bivalvia), cultured on hanging long lines. Sarsia 68: 97-102.

Midtlyng, P. V. 1985. Bruk av medikamenter of desinfeksjonsmidler i norske fiskoppdrettsanlegg. Vann 3: 177-180.

Mills, K. H. 1985. Responses of lake whitefish (*Coregonus clupeaformis*) to fertilization of Lake 226, the experimental lakes area. Canadian Journal of Fisheries and Aquatic Science 42: 129-138.

Muller-Haeckel, A. 1986. Control of water quality around a cage fish farm in the Norrby Archipelago (Northern Bothnian Sea). Vatten 42: 205-209.

Nishimura, A. 1982. Effects of organic matters produced in fish farms on the growth of redtide algae (*Gymnodinium* type-'65 and *Chattonella antiqua*. Bulletin Plankton Society of Japan 29(1): 1-7.

Olsgard, F. 1984. [Pollution effects on soft bottom fauna from a nearby marine fish culture site]. Norsk Fiskeoppdrett. 9(10): 32-38.

Ottman, F. and J.-M Sornin. 1985. Observations on sediment accumulation as a result of mollusk culture systems in France. Pages 329-337 in N. L. Chao and W. Kirby-Smith, editors. Proceedings of the International Symposium on Utilization of Coastal Ecosystems: Planning, Pollution and Productivity, 21-27 November 1982, Rio Grande, Brazil. Editora da Furg, Rio Grande, Brazil.

Pamatmat, M. M., R. S. Jones, H. Sanborn and A. Bhagwat. 1973. Oxidation of organic matter in sediments. EPA-660/3-73-005. United States Environmental Protection Agency, Washington, D.C., USA.

Parsons, T. R., M. Takahashi and B. Hargrave. 1977. Biological Oceanographic Processes. Pergamon Press, New York, New York, USA. 332 pp.

Pease, B. 1977. The effect of organic enrichment from a salmon mariculture facility on the water quality and benthic community of Henderson Inlet, Washington. Ph.D. thesis, University of Washington, Seattle, Washington, USA.

Pearson, T. H. and R. Rosenberg. 1978. Macrobenthic succession in relation to organic enrichment and pollution on the marine environment. Oceanography and Marine Biology Annual Review 16: 229-311.

Penczak, T., W. Galicka, M. Molinski, E. Kusto and M. Zalewski. 1982. The enrichment of a mesotrophic lake by carbon, phosphorus and nitrogen from the cage aquaculture of rainbow trout, *Salmo gairdneri*. Journal of Applied Ecology 19: 371-393.

Phillips, M. and M. Beveridge. 1986. Cages and the effect on water condition. Fish Farmer 9(3): 17-19.

Phillips, M. J., M. C. M. Beveridge and L. G. Ross. 1985. The environmental impact of salmonid cage culture on inland fisheries: present status and future trends. Journal of Fish Biology 27 (Supplement A): 123-127.

Reisenbichler, R. R. and J. D. McIntyre. 1977. Genetic differences in growth and survival of juvenile hatchery and wild steelhead trout, *Salmo gairdneri*. Journal of the Fisheries Research Board of Canada 34: 123-128.

Rodhouse, P. G. and C. M. Roden. 1987. Carbon budget for a coastal inlet in relation to intensive cultivation of suspension-feeding bivalve molluscs. Marine Ecology - Progress Series 36: 225-236.

Rosenberg, R. 1976. Benthic faunal dynamics during succession following pollution abatement in a Swedish estuary. Oikos 27: 414-427.

Rosenberg, R. and L-O. Loo. 1983. Energy-flow in a *Mytilus edulis* culture in western Sweden. Aquaculture 35: 151-161.

Rosenfield, A. and F. G. Kern. 1979. Molluscan imports and the potential for introduction of disease organisms. Pages 165-189 in R. Mann, editor. Exotic Species in Mariculture. MIT Press, Cambridge, Massachusetts, USA.

Rosenthal, H. 1980. Implications of transplantations to aquaculture and ecosystems. Marine Fisheries Review 42: 1-14.

Rosenthal, II., D. Weston, R. Gowen and E. Black. 1988. Report of the ad hoc study group on the environmental impact of mariculture. International Council for the Exploration of the Sea, Cooperative Research Report No. 154.

Ruokolahti, C. 1988. Effects of fish farming on growth and chlorophyll a content of *Cladophora*. Marine Pollution Bulletin 19: 166-169.

Ryther, J. H. 1969. The potential of the estuary for shellfish production. Proceedings National Shellfisheries Association 59: 18-22.

Short, J. W. and F. P. Thrower. 1987. Toxicity of tri-n-butyl-tin to chinook salmon, *Oncorhynchus tshawytscha*, adapted to seawater. Aquaculture 61: 193-200.

Skogheim, O. K. and T. Bremnes. 1984. [Effects on the environment from fish farms in lakes]. Norsk Fiskeoppdrett. 9(2): 37-38.

Smayda, T. and T. Packard. 1979. The phenomenon of anoxia as related to dinoflagellate blooms. Pages 456-458 in D. L. Taylor and H. H. Seliger, editors. Toxic Dinoflagellate Blooms. Proceedings of the Second International Conference on Toxic Dinoflagellate Blooms. Elsevier, Amsterdam, The Netherlands.

Sornin, J. M. 1979. Enquete sur la sedimentation et l'exhausment des fonds dans les zones conchylicoles des cotes de France. Rapport DEA, Université de Nates, France.

Sornin, J. M. 1981. Sedimentary process and biodepositon linked to different shellfish culture methods. Laboratorie de Geologie Marine et Appliquee 1981. 188 pp.

Stewart, K. I. 1984. A study on the environmental impact of fish cage culture on an enclosed sea loch. M.Sc. thesis, University of Stirling, Stirling, Scotland.

Takahashi, M. and N. Fukazawa. 1982. A mechanism of "redtide" formation. II. Effect of selective nutrient stimulation on the growth of different phytoplankton species in natural water. Marine Biology 70: 267-273.

Tenore, K. R., L. F. Boyer, R. M. Cal, J. Corral, C. Garcia-Fernandez, N. González, E. González-Gurriarán, R. B. Hanson, J. Iglesias, M. Krom, E. López-Jamar, J. McClain, M. M. Pamatmat, A. Perez, D. C. Rhoads, G. deSantiago, J. Tietjen, J. Westrich and H. L. Windom. 1982. Coastal upwelling in the Rias Bajas, NW Spain: contrasting the benthic regimes of the Rias de Arosa and de Muros. Journal of Marine Research 43: 701-772.

Tenore, K. R., J. Corral, N. González and E. López-Jamar. 1985. Effects of intense mussel culture on food chain patterns and production in coastal Galacia, NW Spain. Pages 321-328 in N. L. Chao and W. Kirby-Smith, editors. Proceedings of the International Symposium on Utilization of Coastal Ecosystems: Planning, Pollution and Productivity, 21-27 November 1982, Rio Grande, Brazil. Editora da Furg, Rio Grande, Brazil.

Tenore, K. R. and N. González. 1975. Food chain patterns in the Ria de Arosa, Spain: an area of intense mussel aquaculture. Tenth European Symposium on Marine Biology, Ostend, Belgium, 17-23 September, 1975, Vol. 2: 601-619.

Toranzo, A. E., P. Combarro, M. L. Lemos and J. L. Baria. 1984. Plasmid coding for transferrable drug resistance in bacteria isolated from cultured rainbow trout. Applied and Environmental Microbiology 48: 872-877.

Trojanowski, J., C. Trojanowska and H. Ratajczyk. 1982. Effect of intensive trout culture in Lake Letowo on its bottom sediments. Polskie Archiwum Hydrobiologii 29: 659-670.

Tsutsumi, H. and T. Kikuchi. 1983. Benthic ecology of a small cove with seasonal oxygen depletion caused by organic pollution. Publications of the Amakusa Marine Biology Laboratory 7(1): 17-40.

VKI. 1976. Vandkvalitetsinst. & Jydsk Teknologisk Inst.: Forskellige driftsparameters indflytelse pa forureningen fra dambrug. Rapport til Teknologiradet. Horsholm, Denmark.

Vollenweider, R. A. 1968. Scientific fundamentals of the eutrophication of lakes and flowing waters, with particular reference to nitrogen and phosphorus as factors of eutrophication. Paris, OECD, Technical Report DA5/SU/68.27. 250 pp.

WDF/WDOE. 1985. Use of the insecticide Sevin to control ghost and mud shrimp in oyster beds of Willapa Bay and Grays Harbor. Final Environmental Impact Statement. Washington Departments of Fisheries and Ecology, Olympia, Washington, USA.

Weston, D. P. 1986. The environmental effects of floating mariculture in Puget Sound. School of Oceanography, University of Washington, Seattle, Washington, USA.

Weston, D.P. 1990. Quantitative examination of macrobenthic commumity changes along an organic enrichment gradient. Marine Ecology - Progress Series 61:233-244.

Weston, D. P. and R. J. Gowen. 1988. Assessment and prediction of the effects of salmon net-pen culture on the net-pen culture on the benthic environment. Technical Report 414, School of Oceanography, University of Washington, Seattle, Washington, USA.

White, D. 1986. Biological principles of pond culture: sediment and benthos. Pages 15-19 in J. E. Lannan, R. O. Smitherman and G. Tchobanoglous, editors. Principles and Practices of Pond Aquaculture. Oregon State University Press, Corvallis, Oregon, USA.

Zitko, V., L. E. Burridge, M. Woodside and V. Jerome. 1985. Mortalities of juvenile Atlantic salmon caused by the fungicide OBPA. Canadian Technical Report, Fisheries and Aquatic Sciences 1358. 29 pp.

IMPACT OF AQUACULTURE ON WATER RESOURCES

M. J. Phillips, M. C. M. Beveridge and R. M. Clarke

ABSTRACT

Water is essential for all forms of aquaculture. However, growing demand for water from an expanding aquaculture industry is resulting in increasing competition with other water users for limited water resources. The aim of the present paper is to examine the use of water resources for aquaculture, both in terms of availability and the water demand for different aquaculture systems and the potential for developing more resource efficient means of using water for aquaculture.

INTRODUCTION

Water serves several purposes within an aquaculture system. Cultured aquatic animals and plants require oxygen and all aquaculture systems must be designed to convey oxygen to the cultured stock (in the widest sense of the word, aquaculture also includes the culture of microorganisms such as bacteria and yeasts which may not require oxygen; these are not considered in any detail in this review). The water supply is the only source of this oxygen for most cultured species, although there are exceptions. Culture of air breathing fish, such as *Ophicephalus* spp. and *Clarias* spp. may be carried out successfully in anaerobic ponds in Southeast Asia, with fish switching to air breathing in the absence of dissolved oxygen (Muir 1981; Bevan 1986).

The availability of dissolved oxygen to cultured organisms depends on the amount of dissolved oxygen entering the aquaculture system and/or processes within the aquaculture system itself. The relative importance of these two sources is highly dependent on system design.

Intensive aquaculture systems with short water residence times rely heavily on inflowing water to supply the dissolved oxygen requirements of the cultured stock. In this situation, the carrying capacity (amount of stock biomass that can be held) is determined largely by the respiratory requirements of the stock and their physiological tolerance to low concentrations of dissolved oxygen in relation to the supply of dissolved oxygen in inflowing water (Haskell 1955; Shepherd and Bromage 1987; Fivelstad 1988).

The water supply contributes less dissolved oxygen in systems with long water residence times, such as extensive or semi-intensive ponds, where internal processes, such as phytoplankton photosynthesis and turbulent mixing (Boyd 1982) are more important sources of dissolved oxygen.

A water supply can be an important source of harmful toxins but it is also an important medium for receiving and dispersing excretory products and other aquaculture waste materials, which may prove harmful if allowed to accumulate (Haskell 1955). Water borne organisms, such as phytoplankton and bacteria, are also important in control of water quality in slow flushing systems. The external medium also plays a direct role in detoxifying waste materials. The pH of most fresh and sea waters allows fish to excrete ammonia, which is rapidly converted to non-toxic ammonium ions. Terrestrial animals have been forced to adopt more energy intensive mechanisms for removal of ammonia. Recent evidence has also shown that an African cichlid (*Oreochromis alcalicus grahami*) has been forced to adopt urea as opposed to ammonia excretions to survive ammonia toxicity in highly alkaline soda lakes (Randall et al. 1989). The carrying capacity of many aquaculture systems may be limited by the rate of dispersal of metabolic wastes, especially when supplementary oxygenation is supplied (Fivelstad 1988).

Apart from certain aquatic farmed organisms that may forage or spend part of their time in terrestrial habitats (e.g., frogs, crocodiles), the physical and chemical characteristics of the water also play an important role in aquaculture nutrition and growth. Intensive aquaculture relies on supplementary feed being added to the water supply, and the physical and chemical characteristics of the water supply (water velocity, density, turbidity) may significantly affect ingestion of feed and conversion efficiency. Aquaculture systems defined as semi-intensive or extensive, such as many low technology fish, shellfish and seaweed operations also rely on water as a source of nutrition (either nutrients as in seaweed culture or food organisms).

CHARACTERISTICS OF WATER SUPPLIES

Water resource availability is a key factor in determining where aquaculture may develop. The world's water resources can be roughly divided into freshwater and marine resources (Table 1). A simple comparison between the production statistics from freshwater and marine aquaculture in relation to resource volume demonstrates the low utilization of marine resources in relation to resource availability. In theory, freshwater aquaculture may utilize inland resources at most stages of the hydrological cycle. Intensive aquaculture of high value species gives intensive production systems more scope to utilize a wider range of water resources than low cost extensive or semi-intensive systems; however, in practice, most freshwater aquaculture depends on static or running surface waters which form less than 0.3% of inland freshwater resources. Shallow aquifers and springs are utilized to a lesser extent, even though groundwater forms a very significant reservoir of the Earth's freshwater resources. Concentrating on readily available resources clearly increases the potential for conflicts with other users.

Table 1. World water resources, in relation to annual aquaculture production (source: Wilson 1984 and FAO)

	Volume	Aquaculture Production (1986) (mt)			
	($10^{12}m^3$)	Finfish	Molluscs	Crustacea	Seaweed
Freshwater lakes	125				
Rivers & streams	1.25	4,820,880	9,682	117,200	0
Groundwater	8,250				
Saline lakes & inland seas	105	-	-	-	-
Soil moisture	65	-	-	-	-
Atmosphere	13	-	-	-	-
Polar ice caps, glaciers, snow	29,200	-	-	-	-
Seas & oceans	1,320,000	375,267	3,199,452	266,193	2,626,572

A survey of resource utilization in the UK in 1982 showed that the majority of aquaculture was based on river water supplies (Table 2), although farms commonly made use of spring and well supplies, particularly for hatching and growing juvenile fish (Solbe 1982). Similar statistics from other countries show variable relative utilization although surface waters usually dominate (Alabaster 1982). Recent trends in Scotland show increasing utilization of freshwater lochs (lakes) for salmonid cage culture. Cage culture of freshwater and marine fish has expanded throughout many regions of the world in recent years (Beveridge 1987). Groundwater is less important on a global scale, although it may be a significant source of water for aquaculture in several localities, for example, channel catfish farms in the southern USA (Tucker et al 1984; Parker 1988) and carp hatcheries in Bangladesh.

Table 2. Freshwater supplies of United Kingdom fish farms, based on a 1982 survey. Table shows the number of farms using only one type of water supply, and the number using combinations of two supplies (e.g., 84 using solely river water and 6 using a combination of river and well water). (Source: Alabaster 1982).

	Spring	Well	River	Other
Spring	12	3	5	2
Well		9	6	0
River			84	2
Other				10

Quality as well as accessibility also plays a role in the suitability of a water supply for aquaculture development. In theory, all water supplies are available for aquaculture given appropriate physical and chemical pre-treatment. In practice, water treatment is limited by operational costs and most aquaculture supplies are utilized without pre-treatment. Characteristics of different water resources are discussed below.

Surface freshwaters are particularly vulnerable to pollution, although this vulnerability is closely related to catchment area size and the degree of anthropogenic influence. Pollution of remote catchments by acid rain and radiation has also demonstrated that freshwater pollution by man may occur at some considerable distance from the pollutant source. Water quality of streams and rivers will be determined largely by the topography and solubility of the catchment area, climate and the nature and extent of any pollution sources (Wheaton 1985). Pollutant and natural solute concentrations tend to increase from source. Water flow also tends to increase and the amplitudes of short and long-term flow play an important role in the suitability of a water supply for aquaculture; for example, development may be constrained by minimum flows, rather than mean flows.

Static freshwaters such as ponds, lakes and man-made reservoirs all tend to have slightly more stable physical and chemical characteristics than flowing waters. This stability, which is closely related to flushing time, allows the development of highly fertile pond ecosystems which have been successfully used for production of fish in many tropical and subtropical regions. This same stability and lack of flushing can also result in an accumulation of wastes from intensive aquaculture, for example, cage culture in lakes and reservoirs (Beveridge 1984; Phillips et al 1985; Beveridge et al, and Weston this volume).

Marine water supplies also tend to have fairly stable characteristics, although coastal water, particularly near large freshwater sources, tend to be more variable, with a risk from pollution and salinity and temperature fluctuations. Additional problems associated with the use of marine water includes the presence of fouling organisms and blooms of toxic algae.

Groundwater supplies (artesian or other wells, springs and groundwater fed ponds) usually have more stable chemical characteristics than surface waters, although they may contain undesirable levels of dissolved salts and metal ions, low levels of dissolved oxygen and high levels of toxic gases (carbon dioxide, hydrogen sulfide). There is also increasing concern in many regions about contamination of groundwater supplies by persistent pollutants which may also prove detrimental to aquaculture (there is also concern about aquaculture contributing to this pollution; e.g., nitrogen from aquaculture operations in Bavaria; Solbe 1987).

Aquaculture has also been successful in utilizing water from industrial sources (e.g., warm cooling water from nuclear power stations) (Coche

1981) and municipal tap water supplies (Wheaton 1985). Heated effluents are frequently supersaturated with nitrogen and other gases (Colt 1986) and other industrial and municipal waters may be contaminated with various inorganic and organic compounds which may require pre-treatment or culture of appropriately tolerant organisms.

WATER VOLUME REQUIREMENTS

The amount of water required by a given aquaculture system depends on several factors: the physiological tolerances of cultured stock, the nature of the aquaculture system and the quality of the water supply.

Several surveys of the water demand of aquaculture have been carried out which show the extreme variability in requirements. A survey of seven European countries reported by Alabaster (1982) illustrates the very large variation in the amount of water used by European farms for producing one ton of fish (Fig. 1). Unfortunately, it is difficult to assess the reasons for this

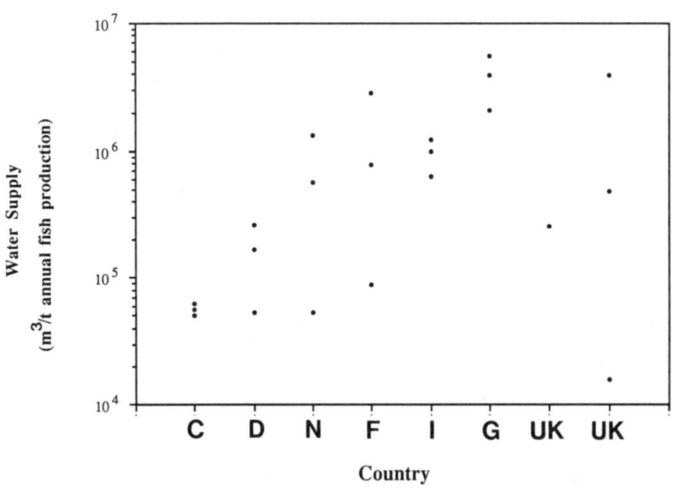

Figure 1. Water demand of European aquaculture, based on a survey of Europe reported in Alabaster (1982). Graphs show the mean and range of results. Countries and numbers of farms sampled in each country are given below.
C, Cyprus (2); D, Denmark (11); N, Norway (10); F, Finland (5); I, Italy (4); G, FDR Germany (3); UK, United Kingdom (85); UK, United Kingdom (22).

variability without more detailed information on fish production and management techniques.

The results of a more detailed analysis of water requirements for a range of aquaculture systems and fish species are shown in Table 3. This analysis demonstrates the very wide range in water requirements for different culture systems.

Table 3. Water requirements per mt of aquaculture production.

Species and System	Prod./Yr (mt/ha)	Water Req. (m^3/mt)	Source
Clarias batrachus Intensive static pond system (Thailand)	100-200	50-200	Muir (1981)
Tilapia (*Oreochromis niloticus*) rainfed extensive ponds (static)	0.05-0.3	3,000-5,000	Muir and Beveridge (1987)
Tilapia (*O. niloticus*) sewage fed, minimal water exchange (Thailand)	6.8	1,500-2,000	Edwards et al. (1987)
Tilapia (*O. niloticus*) intensive, mechanically stirred ponds (Taiwan)	17.4	21,000	Hepher (1985)
Common carp/tilapia/ mullet/silver carp/ conventional pond culture (Israel)	3	12,000	Sarig (1988a)
Common carp/tilapia/ mullet/silver carp semi-intensive pond (Israel)	9	5,000	Sarig (1988a)
Common carp/tilapia intensive pond (Israel)	20	2,250	Sarig (1988a)

Common carp intensive raceways (Japan)	1443	740,000	Kawamoto (1957) in Hepher 1985
Catfish (*Ictalurus punctatus*) intensive pond culture (USA)	3	6,470	Boyd (1982)
Catfish (*I. punctatus*) intensive raceway culture (USA)	-	29,000-14,500	Beleau (1985)
Various European farms (European survey, 1982)	-	15,768-5,544,029	Alabaster (1982)
Rainbow trout (*Salmo gairdneri*) raceways (USA)	150	210,000	Bardach et al. (1972)
Salmonids pond and tank culture (UK)	-	252,000	Solbe (1982)
Salmonids cage culture (Scotland)	40-200	2,260,000	Present study
Penaeid shrimp semi-intensive pond culture (Taiwan)	4.2-11	11,000-21,430	Wickins (1986) Chien et al. (1988)
Penaeid shrimp intensive pond culture (Taiwan)	12.6-27.4	29,000-43,000	Wickins (1986) Chien et al. (1988)
Penaeid shrimp intensive raceway culture (Mexico)	11.8	55,125	Salser et al. (1978)

The lowest water requirements per unit production are for farming air breathing walking catfish *Clarias batrachus* in Thailand. This very low water requirement can only be sustained because the fish are physiologically able to tolerate the poor water quality in the anaerobic culture conditions

(Muir 1981). Above this very low minimum, various tropical and subtropical pond systems have a comparatively low unit water demand. The highest water demands are for culture of salmonids (and carp) in intensive flow-through systems, associated with the water being used to supply dissolved oxygen and remove metabolites from the culture system.

It is apparent from this analysis that intensification of aquaculture (defined as an increase in production per unit area) has resulted in different approaches in different systems. Intensification in farming of most species (salmonids, channel catfish and shrimps) has been accompanied by an increase in the water requirement (Table 3). In contrast, intensification of common carp and tilapia culture in Israel has resulted in a decrease in water requirements (Hepher 1985; Sarig 1988a), primarily because of the high cost of water in Israeli aquaculture (17.5-21.3% of production costs; Sarig 1988b) and priority being given to water conservation.

The very high water requirement of salmonid culture can in part be explained by the sensitivity of salmonids to poor water quality and the need to maintain high dissolved oxygen and low concentrations of metabolites within the aquaculture system. However, the very large differences between these "flow-through" systems and other pond systems cannot be explained solely on the grounds of physiological tolerance and suggest that water quality control within pond systems (through production of oxygen and assimilation of metabolites) is a major factor in improving the water use efficiency of these type of culture systems.

Seasonal variations in water requirements may also be important. The demand of fish for dissolved oxygen increases with water temperature (Table 4) and is also inversely related to fish size (Liao 1971). Consequently water requirements change throughout the production cycle, particularly in aquaculture systems in temperate climates. In addition, the production capacity of farms may even be limited by seasonal low flows. The importance of this seasonal variation can be demonstrated with Atlantic salmon smolt farming in Scotland, where maximum biomass and oxygen demand occur at a time of high temperatures and low flows (Fig. 2). To overcome such difficulties, some farms may manipulate stocking densities to maximize fish production at times of high flows, which will enable fish production to exceed theoretical

Table 4. Theoretical water requirements for 1 mt of 200 g rainbow trout, in relation to temperature (Source: Shepherd and Bromage 1988).

Water Temperature (°C)	Oxygen Consumption (kg/day)	Water Requirements* (l/s)	(m^3/day)
6	2.6	4.3	371.5
8	3.4	6.2	535.7
10	4.3	8.6	743.0
12	5.1	11.2	967.7
14	6.0	14.3	1235.5
16	6.8	17.7	1529.3
18	7.7	20.9	1805.8

* Water requirements are based on the assumption of 100% saturated air entering the system and on an effluent concentration of 5.5 mg/l.

maximum carrying capacity. Shepherd and Bromage (1988) suggest that for northern European and US rainbow trout farms, carrying capacity to production ratio is normally around 1:1.6.

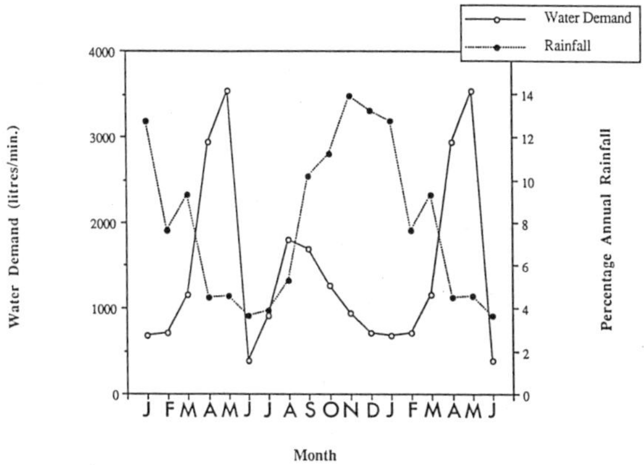

Figure 2. Water demand for an Atlantic salmon farm producing 100,000 smolts per annum (4 metric tons) under typical Scottish conditions, in relation to rainfall patterns.

Table 5. Typical water requirements of industry and agriculture in comparison with selected aquaculture systems. Figures for water requirement of pork and beef production refer to the total water requirement (i.e., feed plus drinking water). (Modified from Schwab et al. 1971 and Muir and Beveridge 1987).

Product	Water Use (m^3/mt)	Nominal Value ($)	Water Value ($/$m^3$)
Alcohol	125-170/m^3	2000/m^3	12-16
Cotton	90-450	1000/mt	2.2-11
Paper	9-450	300/mt	0.7-33
Steel	8-250	200/mt	0.8-25
Beef	42	2000/mt	48
Pork	54	2000/mt	37
Petroleum	21.6-810/m^3	500/m^3	0.6-23
Aquaculture			
Shrimp ponds	11,000-55,000	6000-12000/mt	0.1-1.1
Salmonids	252,000	1650-4000/mt	0.006-0.018
Channel catfish ponds	6470	1650/mt	0.25
Clarias ponds	50-200	1000/mt	5-20

Table 6. Production loading from aquaculture and other industrial and agricultural sources. (Source: Muir and Beveridge 1987).

Product	Kg BOD/mt	BOD/$ Value
Textiles	120-180	0.12-0.18
Tannery	60-240	0.06-0.24
Brewery	18-24	0.02
Slaughter	8-24	0.002-0.006
Pulp	15-21	0.15-0.21
Aquaculture	200-1000	
Shrimp		0.02-0.06
Salmonids		0.05-0.60
Clarias		0.2-1.0

COMPARISON OF WATER REQUIREMENTS WITH OTHER INDUSTRIES

Compared with other industries, aquaculture has a very high water demand, except for the atypical *Clarias* systems in Thailand (Table 5). In terms of value, aquaculture also adds fairly little value to water resources (Table 5), showing that it may be justified to give aquaculture a low priority when in competition with other industrial or agricultural resource users with similar water quality requirements.

Costs associated with aquaculture pollution may also be high in comparison with other industries (Table 6), further justifying careful consideration of the use of aquatic resources for aquaculture (Muir and Beveridge 1987). These analyses argue strongly for integration of aquaculture with other industries or agriculture but ignore some of the positive effects aquaculture may have on water resources. Indeed, it is possible for aquaculture to add value to a water resource. For example, fish production in sewage fed ponds may improve water quality and generate income from a waste material (Edwards et al. 1987; Little and Muir 1987). Researchers in Scandinavia are also attempting to improve water quality in acidified freshwater lakes through selective "pollution" by cage culture (Solbe 1987).

REGULATIONS AND THE USE OF WATER FOR AQUACULTURE

Licenses for water abstraction are required in many countries before either surface or groundwater resources can be utilized for aquaculture. The main feature of such regulation is their extreme variability. No regulations exist in some countries, whereas tight control over water resources exists in others, with significant incountry variations. In some instances abstraction may also be controlled indirectly by controlling the quality of effluent discharged, rather than the actual volume of water abstraction (as in the case of the Scottish Purification Boards). Use of seawater appears to be less restricted than freshwater resources. Examples of regulations governing water abstraction for aquaculture are given in Table 7.

Table 7. Examples of legislation governing use of water for freshwater aquaculture.

Region	Regulation	Comments	Source
British Columbia (Canada)	Water License	Required by all commercial users of surface water in BC (groundwaters do not require license; cost = $10 Can/yr)	Tillapaugh and Edwards (1980)
Idaho (USA)	Water Rights Permit (Idaho State Dept. of Water Resour.)	Required to abstract water in the state of Idaho	Anonymous (1982)
Norway	Watercourse Act	Require license to regulate or change water level in lakes and to abstract river water	Leffertstra and Kryvi (1988)
Denmark	Danish Environmental Act (1974)	Controls the volume of water (surface and groundwater) abstracted by fish farming	Jorgensen (1982)
United Kingdom	Control of Pollution Act (2) (1974)	Abstraction not controlled directly, but the volume of discharge is controlled to reduce abstraction impacts (maximum of 50% of Q (95) flows) (Scotland) Abstraction controlled in some areas by abstraction licenses. Table fish production exempt in some areas. Variable charges for abstraction.	Present Study

Costs associated with water use are also very variable, from low or nonexistent in many countries (e.g., $10 Can per annum for an abstraction license in British Columbia, Canada; no volume restrictions) to high costs in other areas (e.g., Israel; S$0.04/m^3; Sarig 1988a). The cost associated with water may play a significant role in the development of water conservation. For example, high costs of water in Israel have stimulated the development of pond systems with high productivity but very low water requirement (Table 3). Low costs associated with an abundance of water in most other areas has undoubtedly allowed systems with very high water demands to develop.

Local regulations governing the quality of effluent discharge may also influence the volume of water required for aquaculture. In the United Kingdom, the Control of Pollution Act 2 (1974) is used to control the discharge of agricultural and industrial effluent and stringent conditions may be applied to the quality of aquaculture effluent in order to minimize ecological impact in receiving waters. Consent to discharge is based on different parameters depending on the resource being exploited for farming. In freshwater lochs (lakes) in Scotland, consent is usually given on the basis of phosphorus loadings, because of concern over possible eutrophication from fish farming. In rivers, consent may be based primarily on suspended solids, BOD and ammonia loadings. These differences result in a very high apparent water requirement for lake based cage systems in comparison with river based fish farms (Table 3).

It is extremely difficult to predict future trends in legislation concerning water use for aquaculture. New laws are being formulated specifically for aquaculture in several countries (e.g., Canada; Wildsmith 1985). There is also increasing demand in Europe for consistent regulations which treat aquaculture on a fairer basis, although this basis has yet to be defined. In at least one country (Denmark), new regulations are being formulated which will force farmers to pump their discharge water back to the point of abstraction. It seems inevitable that aquaculture will have to face increasingly tight restrictions over development in many areas of the world, as concern over environmental impact and competition for resources grow.

IMPACTS OF AQUACULTURE WATER USE

A major impact on utilization of water for aquaculture is the impact on water quality. This aspect is discussed elsewhere in this symposium (Beveridge et al. and Weston) and will not be considered further here. Abstraction of water <u>per se</u> may also have an adverse impact on the environment, with water being lost or temporarily diverted from the catchment area.

Loss of water resources may occur through transfer of water between different stages of the hydrological cycle. Transfer of groundwater to surfacewater can result in loss of groundwater resources, with possible serious consequences for other groundwater uses and land resources. For example, utilization of groundwater for penaeid shrimp culture has resulted in salinization of groundwaters and subsidence of coastal land by up to 3 m in some parts of Taiwan (Chien et al. 1988).

Evaporation (and to a lesser extent transpiration) may also be a significant loss of surface water from aquaculture facilities. Evaporation occurs at all water-air interfaces and is dependent on water and air temperatures, relative humidity of the air and wind velocity. Although evaporation will occur in all facilities, it is only likely to be a significant part of the water budget in ponds with long retention times. Evaporation losses from open water varies from 0 - 0.8 cm/day in Europe to 2.5 cm/day or more in tropical regions (Huet 1970).

Seepage, which involves transfer of water from surface water to groundwater, will be proportional to soil porosity and may also be a significant loss in earthen ponds. Reported rates for fish ponds vary from 0.01 - 0.66 cm/day (Hepher and Pruginin 1981; Boyd 1985). Combined losses from seepage and evaporation form a very significant part of water losses from ponds with long retention times (up to 100% in seasonal rainfed ponds) and may have a significant impact on resources where pond culture is extensive.

In most fast flowing intensive systems (e.g., salmonid farms), low evaporation and seepage losses are anticipated but severe impacts may arise as a result of abstraction and temporary diversion of a surface water course. This localized depletion may result in areas of reduced flow in irrigation canals, rivers or streams, with associated negative impacts on indigenous fisheries (particularly migratory fish), potable or other industrial or agricultural water supplies and recreational water users (Alabaster 1982). Abstraction may also result in conflicts between aquaculturists, as has been noted in irrigation canals used as water supplies for penaeid shrimp culture (ASEAN 1978). Such problems may become particularly acute in areas where aquaculture water demands are greatest at times of minimum water availability (e.g., Atlantic salmon smolt farming in Scotland; Fig. 2).

OPTIMIZING AQUACULTURE WATER USAGE

Water conservation is practiced in many aquaculture systems and can be implemented in various ways. Several simple pond management techniques have been used to save water. Deepening of ponds to increase storage of rainwater has been recommended to conserve seasonal water resources (Hepher and Pruginin 1981); Boyd 1982; Hepher 1985). Loss of water during fish harvesting can also be reduced by avoiding pond draining, although this technique may result in long-term deterioration of water quality (Parker 1988).

More direct integration with agriculture also offers scope for more efficient use of water resources. In Israel, integration has been forced on fish producers because of the high value associated with irrigation water, with the result that more than 20% of fish ponds have now been turned over to integrated fish/irrigation schemes (Sarig 1988a). This change has involved a shift to more intensively managed pond systems or more recently, cage and pen culture in larger irrigation reservoirs. Integration of aquaculture and agriculture offers some scope for increasing the fertility of irrigation water to improve yields or maintain yields with lower levels of fertilization. Utilization of fertile fish pond water has been reported as a fertilizer for maize (Edwards 1980) and for rice and vegetables in the Philippines (Eusobio et al. 1978), and fish pond sediments and water have been widely used as fertilizer for many years in China (FAO 1983).

Fish culture in rice fields in tropical regions also illustrates the potential for using aquaculture to increase the value of agricultural crops. The costs associated with the water required for fish culture in rice fields are very low because water is already available. However, economic benefits can be significant (up to a 74 times increase in profitability over rice-only systems in Thailand; Kloke and Potaras 1975 in Muir and Beveridge 1987), through production of fish and increased rice yields (Koesoemadinata and Costa-Pierce 1988).

Fish culture also has potential for improving the availability of water resources for aquaculture. For example, stocking of irrigation canals with macrophyte feeders such as grass carp (*Ctenopharyngodon idella*), java carp

(*Puntius gonionotus*) and *Tilapia zilli* has also been used in several regions for removing machrophytes and improving water flow to traditional agricultural crops (Little and Muir 1987).

Cage and pen culture offers particularly good scope for improved use of water resources, because these systems do not require water abstraction. Cages have been used successfully in irrigation reservoirs in several countries (Beveridge 1987; Beveridge and Phillips 1988) and in a wide range of lakes, reservoirs and running waters (Beveridge 1987). Freshwater reservoir resources will increase significantly in many parts of the world over the next decade (e.g., six-fold in Southeast Asia; Fernando 1980) suggesting niches for cage aquaculture to utilize a freshwater resource that is actually expanding. Problems associated with environmental impact appear to be a major concern (Weston this volume), although greater understanding of interactions between cage culture and the environment and development of models for predicting stocking levels (Beveridge 1984) should assist optimal use of resources. In warm waters, integration of extensive with intensive cage culture offers scope for minimizing eutrophication from intensive cage farming. For example, cages of filter feeding bighead carp have been used to remove phytoplankton from potable water reservoirs in Singapore (Beveridge 1984).

Pond farming could also be advocated as a water efficient culture technique in comparison with intensive flow-through systems, although large land resources may be required for large scale pond systems, because of relatively low areal production. There may also be significant management problems with large volumes of water and low densities of fish stock. These problems may be partly overcome by culture of fish in cages in large ponds, as is developing now in several regions (e.g., Hungary and Israel). Water can also be conserved by culturing fish species with improved tolerance to low dissolved oxygen and high levels of waste metabolites (e.g., tilapias).

Water treatment may be used to improve stocking densities and reduce water requirements. Natural water treatment already exists in many low-cost extensive and semi-intensive pond systems which are managed to promote phytoplankton growth as fish feed organisms or as a carbon source at the base of the food chain. Phytoplankton have the added benefit of producing oxygen by photosynthesis and assimilating waste metabolites which may otherwise accumulate and prove harmful to fish stock. However, as pond stocking densities increase, more precise water quality control is needed, such as aeration and efficient mixing, to ensure adequate supplies of dissolved oxygen and efficient dilution and assimilation of metabolites and aerobic metabolism of organic wastes (Hepher 1985). The tendency towards increasing water flows and decreasing water retention time in more intensive pond systems (e.g., for shrimp culture; Wickins 1986) results in a loss of many of the internal water quality control systems of pond culture which must inevitably lead to even greater water demand.

Stocking densities of intensive tank systems can be increased by varying degrees of water treatment, from simple aeration through to recycling of water accompanied by biological filtration, solids removal and oxygenation (Muir 1982). These systems have the added benefit of improved water economy (Table 8), although they involve additional cost burdens (Muir 1982). Flow regimes within intensive systems can also be improved to maximize stocking densities per unit flow by appropriate design (Burley and Klapis 1988).

Table 8. Water requirements for intensive tank culture of salmonids, with varying degrees of recycling (source, Muir and Beveridge).

% Recycle	Water Demand (m^3/mt)
0	200,000
80	40,000
90	20,000
95	10,000
99	2,000

ACKNOWLEDGMENTS

We would like to acknowledge our funding agencies, the Highland and Islands Development Board (MJP), the Overseas Development Administration (MCMB) and the University of Stirling (MJP and RMC) for supporting this work. Thanks are also due to Dr. Mark Northcott for drawing the figures and to Miss Mary Hennessey for collating some of the data presented.

LITERATURE CITED

Alabaster, J. S. 1982. Report of the EIFAC Workshop on fish-farm effluents. Silkeborg, Denmark, 26-28 May 1981. EIFAC Technical Paper, (41). FAO, Rome, Italy.

Anonymous. 1982. Catfish Farming in Idaho. United States Soil Conservation Service, Boise, Idaho.

ASEAN. 1978. Manual on pond culture of penaeid shrimp. ASEAN National Coordinating Agency of the Philippines, Ministry of Foreign Affairs, Manilla, Philippines.

Bardach, J. E. J. H. Ryther, and W. O. McLarney. 1972. Aquaculture: The Farming and Husbandry of Marine and Freshwater Organisms. Wiley Interscience, New York.

Beleau, M. H. 1985. High density culture systems. Pages 85-105 in C. S. Tucker, editor. Channel catfish culture. Developments in aquaculture and fisheries science, volume 15. Elsevier, Amsterdam, The Netherlands.

Bevan, D. J. 1986. The effect of dissolved oxygen and water depth on the respiratory behaviour and growth of *Clarias macrocephalus* (Pisces, Claridae). Unpublished MSc thesis, McGill University, Canada. 63 pp.

Beveridge, M. C. M. 1984. Cage and pen fish farming. Carrying capacity models and environmental impact. FAO Fisheries Technical Paper 255.

Beveridge, M. C. M. 1987. Cage Aquaculture. Fishing News Books, Surrey, England.

Beveridge, M. C. M. and M. J. Phillips. 1988. Aquaculture in reservoirs. Pp. 234-243 in S. S. de Silva (ed.). Proceedings of the IDRC Workshop on Tropical Reservoir Management. Kathmandu, Nepal, 21-28 November 1987. IDRC, Ottawa, Canada.

Beveridge, M. C. M., M. J. Phillips and R. M. Clarke. 1989. A quantitative and qualitative assessment of wastes from aquatic animal production. (this volume).

Boyd, C. E. 1982. Water Quality Management for Pond Fish Culture. Developments in Aquaculture and Fisheries Science, Volume 9. Elsevier, Amersterdam, The Netherlands.

Burley, R. and A. Klapis. 1988. Making the most of your flow (in fish rearing tanks). Pp. 211-239 in K. Murray (ed.), Aquaculture Engineering Technologies for the Future. Institute of Chemical Engineers Symposium Series 111. Hemisphere Publishing Corporation, New York, USA.

Chien, Y. H., I. C. Liao and C. M. Yang. 1988. The evolution of prawn grow-out systems and their management in Taiwan. Pp. 143-168 in K. Murray (ed.), Aquaculture Engineering Technologies for the Future. Institute of Chemical Engineers Symposium Series. No. 111. Hemisphere Publishing Corporation, New York.

Coche, A. G. 1981. Report of the symposium on new developments in the utilization of heated effluents and of recirculation systems for intensive aquaculture. Stavanger, 29-30 May 1980. EIFAC Technical Paper, 39. FAO, Rome, Italy.

Colt, J. 1986. Gas supersaturation - impact on the design and operation of aquatic systems. Aquaculture Engineering 5: 49-85.

Edwards, P. 1980. Food potential of aquatic macrophytes. ICLARM Studies Review 5. ICLARM, Manila, Philippines.

Edwards, P., C. Polprasert and L. W. Kok. 1987. Resource recovery and health aspects of sanitation. Asian Institute of Technology Research Report, number 205.

Eusobio, J. A., B. I. Rabino and E. C. Eusobio. 1978. Recycling system in integrated plant and animal farming. University of the Philippines Technical Bulletin 1.

FAO. 1983. Freshwater aquaculture development in China. Report of the FAO/UNDP study tour organized for French-speaking African countries. 22 April-20 May 1980. FAO Fisheries Technical Paper 215.

Fernando, C. H. 1980. The fishery potential of man-made lakes in Southeast Asia, and some strategies for optimization. BIOTROP Aniversary Publication: 25-38.

Fivelstad. S. 1988. Water flow requirements for salmonids in single pass and semi-closed land-based seawater and freshwater systems. Aquacultural Engineering 7:183-200.

Haskell, D. C. 1955. Weight of fish per cubic foot of water in hatchery troughs and ponds. Progressive Fish Culturist: 117-118.

Hepher, B. 1985. Aquaculture intensification under land and water limitations. Geojournal, 10: 253-259.

Hepher, B. and Y. Pruginin. 1981. Commercial fish farming. John Wiley and Sons, New York.

Huet, M. 1970. Textbook of Fish Culture, Breeding and Cultivation of Fish. Fishing News Books, Surrey, England.

Jorgensen. O. H. 1982. Legislative measures concerning trout farming in Denmark. Pp. 137-140 in J. S. Alabaster (ed.). Report of the EIFAC Workshop on Fish-Farm Effluents. EIFAC Technical Paper, 41. FAO, Rome, Italy.

Koesoemadinata, S. and B. A. Costa-Pierce. 1988. Development of rice-fish farming in Indonesia: past, present and future. Paper presented at the ICLARM-IRRI-CLSU "First International Workshop on Rice-Fish Culture", Ubon, Thailand, 21-25 March 1988.

Leffertstra, H. and H. Kryvi. 1988. Report on suspended solids from fish farms in Norway. Pp. 62-73 in M. Pursiainen (ed.). National Contributions on Suspended Solids from Land-Based Fish Farms. Papers presented at the first session on the EIFAC Working Party on Fish Farm Effluents. The Hague, Netherlands, 22-30 May and 1 June 1987. 93 pp.

Liao, P. B. 1971. Water requirements of salmonids. Progressive Fish Culturist 33: 210-215.

Little, D. and J. Muir. 1987. A guide to integrated warm water aquaculture. Institute of Aquaculture, Scotland.

Jorgensen, O. H. 1982. Legislative measures concerning trout farming in Denmark. Pp. 137-140 in Report of the EIFAC Workshop on Fish-Farm Effluents. Silkeborg, Denmark, 26-28 May 1981. EIFAC Technical Paper 41, 166 pp.

Muir, J. F. 1981. Management in *Clarias* culture, Thailand. Programme for the Development of Pond Management Techniques and Disease Control (DoF - UNDP/FAO THA/75/012) Thailand. National Inland Fisheries Institute, Bangkok, Thailand. 30 pp.

Muir, J. F. 1982. Recirculated water systems in aquaculture. Pp. in J. F. Muir and R. J. Roberts (eds.). Recent Advances in Aquaculture 1. Croon Helm, Kent, England.

Muir, J. F. and M. C. M. Beveridge. 1987. Water resources and aquaculture development. Arch. Hydrobiol. Beih. 28:321-324.

Parker, N. 1988. Fish Culture in the United States. Pp. 268-301 in J. Shepherd and N. Bromage (ed.). Intensive Fish Farming. BSP Professional Books, Oxford, England.

Phillips, M. J. M. C. M. Beveridge and L. G. Ross. 1985. The environmental impact of salmonid cage culture on inland fisheries: present status and future trends. Journal of Fish Biology 27 (Supplement A): 123-137.

Randall, D. J., C. M. Wood, S. F. Perry, H. Bergman, G. M. O. Maloiy, T. P. Mommsen and P. A. Wright. 1989. Urea excretion as a strategy for survival in a fish living in a very alkaline environment. Nature, 337: 165-166.

Salser, B. R., L. Mahler, D. Lightner, J. Danald, C. Brand, N. Stamp, D. Moore and B. Colvin. 1978. Controlled environment aquaculture of penaeids. In P. N. Kaul and C. J. Sinderman (eds.). Drugs and Food from the Sea. University of Oklahoma, Norman, Oklahoma, USA.

Sarig, S. 1988a. The development of polyculture in Israel: a model of intensification. Pp. 302-331 in C. J. Shepherd and N. R. Bromage (eds.). Intensive Fish Farming. BSP Professional Books, Oxford, England.

Sarig, S. 1988b. Is an increase in yields in intensive commercial fish farming always accompanied by an increase in profitability? The Israeli Journal of Aquaculture 40: 122-124.

Schwab, G. O., R. K. Frevert, K. K. Barnes and T. W. Edminster. 1971. Elementary Soil and Water Engineering. John Wiley and Sons, New York.

Shepherd, J. and N. Bromage. 1988. Intensive Fish Farming. BSP Professional Books, Oxford, England.

Solbe, J. F. de L. G. 1982. Fish Farm Effluents: A United Kingdom Survey. Pp. 29-55 in J. S. Alabaster (ed.). Report of the EIFAC Workshop on Fish-Farm Effluents. Silkeborg, Denmark, 26-28. May 1981. EIFAC Technical Paper (41).

Solbe, J. F. de L. G. 1987. Report of the European Inland Fisheries Advisory Commission Working Party on Fish Farm Effluents. Water Research Centre, Medmenmham, U. K. 20 pp.

Tillapaugh, D. L. and J. C. Edwards. 1980. A permit and license guide for the prospective aquaculturist. Information Report of the Marine Resources Branch, No. 1. Ministry of Environment, British Columbia, Canada.

Tucker, C. S., T. E. Schwedler and T. Taylor. 1984. Water quality in channel catfish (*Ictalurus punctatus*) hatcheries. Mississippi Agricultural and Forestry Experimental Station, Mississippi State, Mississippi, Research Report 8, Number 16.

Weston, D. P. 1990. The effects of aquaculture on indigenous biota. pp. 534-567. *In*: D. E. Brune and J.R. Tomasso (eds.), Aquaculture and Water Quality. World Aquaculture Society.

Wheaton, F. W. 1985. Aquaculture Engineering. Robert E. Krieger Publishing Company, Florida, USA.

Wickins, J. F. 1986. Prawn farming today: opportunities, techniques and developments. Outlook on Agriculture. 15: 52-60.

Wildsmith, B. H. 1985. Toward an appropriate federal aquaculture role and legislative base. Canadian Technical Report of Fisheries and Aquatic Sciences, Number 1419.

Wilson, E. M. 1984. Engineering Hydrology. 3rd Edition. Macmillan Publishers Ltd., Basingstoke, United Kingdom.

INDEX